Food and Sustainability

Edited by

Paul Behrens

Leiden University College The Hague, Leiden University and
Institute of Environmental Sciences, Leiden University

Thijs Bosker

Leiden University College The Hague, Leiden University and
Institute of Environmental Sciences, Leiden University

David Ehrhardt

Leiden University College The Hague, Leiden University

OXFORD
UNIVERSITY PRESS

OXFORD

UNIVERSITY PRESS

Great Clarendon Street, Oxford, OX2 6DP,
United Kingdom

Oxford University Press is a department of the University of Oxford.
It furthers the University's objective of excellence in research, scholarship,
and education by publishing worldwide. Oxford is a registered trade mark of
Oxford University Press in the UK and in certain other countries

Published in the United States of America by Oxford University Press
198 Madison Avenue, New York, NY 10016, United States of America

British Library Cataloguing in Publication Data
Data available

Library of Congress Control Number: 2019940876

ISBN 978–0–19–881437–5

Printed in Great Britain by
Bell & Bain Ltd., Glasgow

PREFACE

How can we provide sufficient and sustainably produced food to all? And how might we do this in the context of economic growth and population increases around the world?

Our current production systems are often not sustainable. They have significant impacts on the natural environment —biodiversity is being lost at an alarming rate, agro-chemical pollution has a variety of short- and long-term impacts, water resources are being depleted, fertile topsoils are being lost rapidly, and food systems are affecting, and being affected by, climate change.

Nor are our current consumption patterns sustainable. Many countries have high rates of malnutrition, including undernutrition, obesity, or both. How can we ensure everyone has access to sufficient, and healthy, food? And can we promote diets that are both healthy and environmentally sustainable? If so, what governance arrangements are most effective at promoting sustainable production and consumption of food?

As with many other complex global challenges, the transition towards sustainable food and food systems defies easy solutions. The aim of this book is to present students, policymakers, researchers, and other interested readers with state-of-the-art knowledge of the main challenges and opportunities in the transition into food sustainability. Originally, this book grew out of an interdisciplinary undergraduate course on sustainability at Leiden University College in the Netherlands. Thanks to a committed team of contributing experts, the text has become a broader exploration of the challenges and opportunities for sustainable food.

Do we provide 'the' solution to transform unsustainable food systems? Yes and no. Small-scale solutions and best practices appear in every chapter of the book. Perhaps combining all these insights can help us produce sufficient good-quality food for our large and growing population, far into the future. But every chapter also highlights large and, as yet, unresolved obstacles. We hope that by introducing both the existing solutions and unresolved challenges, the book inspires readers to consider creative and transdisciplinary innovations, which are needed in the transition of our current food systems.

LEARNING FROM THE BOOK

The book includes several features that facilitate learning and critical thinking. In each chapter, readers will encounter:

- **Chapter overviews** summarizing the key points of the chapter
- **Boxes** that provide further details about concepts in the main text
- **Case studies** which illustrate how insights from the main text apply to real-life examples
- **Food controversies** that highlight the complexity of 'hot topics' in the field of the chapter
- Three types of question that encourage the reader to actively engage with the concepts and issues being discussed:
  - **Pause and think** questions prompt the reader to consider the wider implications of a concept
  - **Connect the dots** helps the reader to think about the relevance of the chapter in the context of the rest of the book, so that they can better think about food and sustainability as a 'web' of interconnected ideas, issues, and challenges
  - **In your experience** questions challenge the reader to think about their own interactions with food systems.

- Each chapter finishes with **Questions** and suggestions for **Further reading**, the latter including books, reports, and journal articles.

Food and Sustainability does not end with the textbook itself—a range of resources to augment the text and to further support teaching and learning are available online.

The resources include:

For students:
- Online quizzes for each chapter to support independent study
- Guidance for answering the end-of-chapter questions.

For instructors:
- Figures from the book ready to download, in JPG and PowerPoint formats
- Suggested exam questions for each chapter
- A teaching guide to offer guidance and suggestions to those teaching this subject.

Go to http://www.oup.com/uk/behrens to find out more.

ACKNOWLEDGEMENTS

We would like to thank the peer reviewers for their insightful comments and thoughts on the drafts of the textbook.

Thank you also to Caoilinn Hughes for her input on the introduction and conclusion. Jon Crowe and the team at Oxford University Press deserve huge thanks for their hard work in preparing the book and for always being so responsive and helpful.

The Faculty of Global Governance and Affairs offered financial support to complete this textbook. Thank you to Dr Eveline Stilma for her contribution of the textbox in *Chapter 2: Biodiversity*. We received funding and, more importantly, support from Leiden University to complete this book. Finally, a very special thank you to the students of Leiden University College.

ABOUT THE EDITORS

Dr Paul Behrens is an Assistant Professor in Energy and Environmental Change at Leiden University, the Netherlands. A physicist by training, Paul received his MSc in Physics and Astronomy from the University of Sheffield (UK) after conducting research at the Isaac Newton Group of Telescopes (Spain). His PhD is from the University of Auckland (New Zealand) where he developed remote sensing techniques for the harnessing of wind energy. Before joining Leiden University, Paul worked at the Royal Society of New Zealand, providing expert scientific advice to the New Zealand Government on environmental sustainability and other topics. He has worked in industry, NGOs, and academia. His current work focuses on the environmental impact of human consumption, including energy, food, and water. Paul is passionate about science outreach and his recent research, published in the *Proceedings of the National Academy of Sciences of the United States* and *Nature Energy*, has appeared in *The New York Times*, the *BBC*, and *Scientific American*.

Dr Thijs Bosker is an Associate Professor in Environmental Science at Leiden University, the Netherlands. He obtained his BSc and MSc in Plant Science at Wageningen University (the Netherlands), specializing in sustainable agriculture. Next, Thijs worked for an NGO where he focused on reducing pesticide emissions to the environment. In 2005, he moved to Canada to complete a PhD in aquatic toxicology at the University of New Brunswick. He continued as a postdoctoral fellow at the Canadian Rivers Institute and became an Assistant Professor at the University of Connecticut (USA), before moving to Leiden University. In his current research, he uses laboratory and field techniques to study impacts of contaminants on ecosystem health. Thijs has worked on a variety of environmental issues, with a special focus on the combined impacts of contaminants and environmental stressors on the reproduction and development of organisms. At Leiden University, he teaches courses in Ecotoxicology, Environmental Science, and Sustainability.

Dr David Ehrhardt is an Assistant Professor in International Development at Leiden University, the Netherlands. With a DPhil in Development Studies from the University of Oxford (UK), David subsequently worked as a postdoctoral Research Officer in Oxford's Department of International Development. In his current research, he uses qualitative and quantitative research methods to understand the efficacy and development of 'hybrid' governance in Nigeria, as well as comparatively. In the past, his research has explored ethnic and religious conflict and cooperation, the development of political authority beyond the state, and the ways in which inequality and governance interact in the formation of group identities and violent conflict. At Leiden University, David coordinates a BSc programme in Governance, Economics, and Development.

ABOUT THE CONTRIBUTING AUTHORS

Dr Caroline Archambault is an Assistant Professor in Cultural Anthropology at Leiden University College, the Netherlands. She received her PhD in Anthropology from Brown University, USA, with an investigation into the changing role of formal education for the pastoral Maasai of Southern Kenya. She went on to continue as a postdoc at McGill University, focusing on how changes in land governance impacted Maasai livelihood. Concurrently, she took a teaching appointment at the University College Utrecht and developed and directed an Africa field studies and internship programme. She then joined International Development Studies of Utrecht University, the Netherlands to continue her research on gender and land governance. At Leiden University, she pursues her passion in teaching, research, experiential learning, and interdisciplinary collaborations in East Africa and the Netherlands.

Dr Gerard Breeman is an Assistant Professor at the Institute of Public Administration at Leiden University, the Netherlands. In 2006, he received his PhD from Leiden University on the research topic '*How to trust public policies?*' Between 2006 and 2015, he was Assistant Professor at Wageningen University (also in the Netherlands), and since 2015, he has been back at the Institute of Public Administration of Leiden University. His research interests are comparative policy analysis, public trust, the politics of attention, agenda-setting, food governance, and the European Union. Gerard is co-founder of the research team on comparative agenda-setting.

Anke Brons , MSc, is a PhD candidate at Aeres University of Applied Sciences in Almere, the Netherlands, in collaboration with the Environmental Policy Group at Wageningen University & Research, also in the Netherlands. Her PhD project focuses on inclusiveness in access to healthy and sustainable food in an urban context. She is interested in food systems, social equity questions, and social practice theories. During her MSc International Development Studies at Wageningen UR, Anke conducted a systematic literature review on food system governance as her first master thesis, which sparked her interest in the many complexities and opportunities surrounding the study of food. Since then, food has fascinated her, as she continues to study people's diverse consumption practices and their creativity in the context of everyday life.

Dr Molly E. Brown is a Research Professor at the Department of Geography of the University of Maryland College Park, USA. Molly has 18 years of experience in interdisciplinary research using *satellite* remote-sensing data and models with socio-economic and demographic information, to better understand food security drivers, publishing over 100 journal articles in a variety of disciplines and two books. In 2015, she was the lead author of a US Climate Assessment report published by the US Department of Agriculture entitled 'Climate Change, Global Food Security and the U.S. Food System'. Previously, Molly worked for 13 years at the NASA Goddard Space Flight Center in the Biospheric Sciences Branch. In addition to her work with the University of Maryland, she is the Chief Science Officer of 6th Grain Corporation.

Dr Ellen Cieraad is an Assistant Professor at the Institute of Environmental Science at Leiden University. She studied Conservation Biology at the University of Amsterdam (the Netherlands) and worked as a research technician at Landcare Research in New Zealand. Her PhD research (Durham University, UK) concerned the effects of temperature on the physiology and distributional limits of trees. In her later role as a researcher at Landcare Research, her investigations focused on the interactions between plants and their environment, focusing on the effects of a changing environment on ecosystem processes, biodiversity, and species' distributions. At Leiden University, Ellen coordinates the interdisciplinary Minor programme Sustainable Development, and her research investigates the human impacts on ecosystem processes, particularly in agricultural and urban settings.

Dr Peter Houben is an Assistant Professor in Environmental Earth Sciences and Sustainability at Leiden University College, the Netherlands. He is a broadly trained physical geographer whose main interests are within the interaction between soil and river systems, and human activities. Peter's interests touch on topics of soil and water resources management, landscape functioning, and ecosystem services in the Anthropocene. His teaching draws on his own field-oriented data recorded using information from a range of related fields such as geology, climatology, archaeology, and agriculture. At Leiden University College, Peter teaches courses in Earth System Science, Climate Change, Field Methods, and Soils.

Dr Paul F. Hudson is an American geographer and Associate Professor at Leiden University College, the Netherlands. Growing up in Florida, Paul developed a passion for water and riverine environments that continues to motivate his research. He has graduate degrees from the University of Florida (MSc) and Louisiana State University (PhD) where he examined riverine dynamics of the Lower Mississippi. From 1998 to 2010, Paul was Assistant and Associate (tenured) Professor at the University of Texas at Austin in Geography and Environment. In 2010, Paul relocated to the Netherlands. His intertwined research and teaching utilize field-based approaches, augmented with GIS analysis, to examine environmental issues of flooding, land use change, sediment transport, and lowland river dynamics in Louisiana (USA), Mexico, and the Netherlands.

Dr Jessica Kiefte-de Jong is an Assistant Professor in Global Public Health at Leiden University, the Netherlands. She received her MSc in Public Health Research and Nutrition from the VU University (the Netherlands), and her PhD in Early Life Nutrition from the Sophia Children's Hospital/Erasmus Medical Centre (the Netherlands). She was a visiting researcher at Harvard School of Public Health (USA) and a postdoctoral researcher in Nutrition and Health at the Department of Epidemiology, Erasmus Medical Centre. Her research focuses on the role of nutrition in chronic diseases along the lifespan and psychosocial aspects of dietary behaviour, including the impact of food insecurity. At Leiden University, Jessica teaches courses in Epidemiology, Nutrition, and Prevention within the Global Public Health Program.

Dr Meredith T. Niles is an Assistant Professor in the Food Systems program and the Department of Nutrition and Food Sciences at the University of Vermont, USA. She studies farmer perceptions of climate change, farmer adoption of adaptation and mitigation practices, and food security and climate change. Meredith previously worked for the United States Department of State and for several non-profit organizations. Meredith is passionate about making academic research more publicly available and is a Board Member at the Public Library of Science (PLOS), a non-profit open-access publisher. Meredith has a BA in Politics from The Catholic University of America and a PhD in Ecology from the University of California at Davis, and was a postdoctoral fellow in Sustainability Science at Harvard University.

Dr Peter Oosterveer is a Professor in the Environmental Policy Group at Wageningen University, the Netherlands. His research interests are in global public and private governance arrangements on sustainable food production and consumption such as labelling and certification practices in global supply chains. Furthermore, he is studying food consumption practices from a sociological perspective and is particularly interested in how consumers access sufficient, sustainable, and healthy food, including the role of retail. His publications cover a range of subjects on globalization, sustainability, and food, with particular attention to seafood and palm oil. He is currently involved in several research projects around the world, including in South East Asia and Africa.

Dr Krijn Trimbos is an Assistant Professor in Conservation Biology at the Institute of Environmental Sciences of Leiden University, the Netherlands. Krijn obtained both his MSc and PhD degrees at Leiden University. During his PhD, he focused on conservation genetics of the iconic black-tailed godwit, looking at how population dynamics of this threatened species had been affected by land use change and nature management in agricultural landscapes, through the use of genetics. After his PhD, he worked at an NGO where he focused on meadowbird protection. Krijn returned to Leiden University in 2014, and his research now focuses on the use of DNA as a measuring tool for biodiversity to ultimately improve conservation success.

Dr Daniela Vicherat-Mattar is an Assistant Professor of Sociology at Leiden University College, the

Netherlands. Trained as a sociologist in Chile, she completed her PhD at the European University Institute (Florence, Italy), with a thesis focused on the role of public spaces as social underpinnings of democracy. In 2008, she took up a Marie-Curie postdoctoral fellowship at the University of Edinburgh to carry out research on European cities. At Leiden University College, Daniela studies and teaches the uses of theory in everyday life and the impact of large socio-political processes, such as democratization or migration, affecting public spaces in cities in Europe and Latin America. In particular, she is interested in processes of border-making and the politics of belonging.

Dr Martina G. Vijver is a Professor at the Institute of Environmental Science at Leiden University, the Netherlands. She obtained her BSc in Environmental Sciences at Hogeschool IJsselland and completed her PhD at Vrije Universiteit in the Netherlands. Her expertise is in ecotoxicology, and she has focused on the bioaccumulation kinetics of metals in soil invertebrates, along with the mechanisms underlying metal-induced ecological effects. In 2015, she received a VIDI-NWO grant to study the fate and effects of nanomaterials. Martina is also a project leader of several projects, including on pesticides in the environment, in which she advises policymakers at (inter)national level.

Dr Bríd Walsh is an Assistant Professor in Environmental Geography at Leiden University College, the Netherlands. She has an interdisciplinary background with a BA in Geography and an LLB in Law from the National University of Ireland, Galway, and a Master in Geography from the University of California, Los Angeles. Her PhD work explored the social sustainability of wind energy in Ireland, drawing on lessons from Denmark and the Netherlands. More broadly, she is interested in questions related to energy and society interactions (for example, the role of the public in the energy transition) and rural community development (for example, can the energy transition facilitate rural economic revitalization?).

To Ken, Liz, and Libby for all their support.
Paul Behrens

To Anina, Thura, and Immar for always being up for an adventure—and to Richard and Hanneke, for making me curious about the world around me.
Thijs Bosker

To Johanneke, Elin, and Nina for being the source of so much joy—and to all my parents for all their never-ending support and enthusiasm.

David Ehrhardt

CONTENTS

ABBREVIATIONS

AES	agri-environmental schemes
AFOLU	agriculture, forestry, and other land use
AIDS	acquired immune deficiency syndrome
AR5	Fifth Assessment Report (of the IPCC)
BECCS	bioenergy carbon capture and sequestration
BMI	body mass index
CAP	Common Agricultural Policy
CBO	community-based organization
CCS	carbon capture and sequestration
CEC	cation exchange capacity
CF_4	carbon tetrafluoride
CFP	Common Fisheries Policies (EU)
CFW	cash-for-work
CH_4	methane
CO_2	carbon dioxide
CO_2e	carbon dioxide equivalent
COLI	cost of living index
CRISPR	Clustered Regularly Interspaced Short Palindromic Repeats
CSPO	Certified Sustainable Palm Oil
CWA	Clean Water Act
DDT	dichlorodiphenyltrichloroethane
DNA	deoxyribonucleic acid
EASAC	European Academies Science Advisory Council
EBT	Electronic Benefit Transfer
EEZ	exclusive economic zone
EU	European Union
FAO	Food and Agricultural Organization of the United Nations
FBO	faith-based organization
FCPI	food consumer price index
g	gram
GDP	gross domestic product
GHG	greenhouse gas
GJ	giga joule
GM	genetically modified
GMO	genetically modified organism
GPP	gross primary production
GWP	global warming potential
H^+	hydrogen (ion)
ha	hectare

HANPP	human appropriated net primary production
HIV	human immunodeficiency virus
H_2O	water
ICDP	Integrated Conservation and Development Project
IEA	International Energy Agency
IPCC	United Nations Intergovernmental Panel on Climate Change
IUCN	International Union for Conservation of Nature
IWM	integrated watershed management
IWN	integrated watershed framework
J	joule
K	potassium
KACE	Kenya Agricultural Commodity Exchange
kcal	kilocalorie
kg	kilogram
km	kilometre
L	litre
LCA	life cycle analysis/assessment
m	metre
mg	milligram
Mha	million hectare
MJ	million joules
mm	millimetre
MPA	marine protection area
MSC	Marine Stewardship Council
N	newton
N_2	nitrogen (gas)
Na^+	sodium
NGO	non-governmental organization
N_2O	nitrous oxide
NPP	net primary production
NRD	national recommended diet
O_2	oxygen
ODA	official development assistance
OECD	Organisation for Economic Cooperation and Development
OLU	other land use
P	phosphorus
PCB	polychlorinated biphenyl
PNAE	*Programa Nacional de Alimentação Escolar*

POP	persistent organic pollutant
PSNP	Productive Safety Net Program
RBC	red blood cell
RSPO	Roundtable on Sustainable Palm Oil
s	second
SDG	Sustainable Development Goal
SNAP	Supplemental Nutrition Assistance Program
SOM	soil organic matter
TB	tuberculosis
UK	United Kingdom
UN	United Nations
UNCCD	United Nations Convention to Combat Desertification
UNDP	United Nations Development Programme
UNEP	United Nations Environment Programme
USA	United States
USAID	United States Agency for International Development
USD	US dollars
UV	ultraviolet
VAD	vitamin A deficiency
W	watt
WEF	World Economic Forum
WFD	Water Framework Directive
WFP	World Food Programme
WHO	World Health Organization
WIC	US Women, Infants, and Children (programme)
WWF	World Wildlife Fund

Introduction
Can we feed the world sustainably?

Thijs Bosker, Paul Behrens,
and David Ehrhardt

Chapter Overview

- The human population grew dramatically after two major revolutions: the Agricultural Revolution and the Industrial Revolution. This population growth has required resources to be diverted from natural systems to human societies, including food, water, and energy.

- This growth was enabled by a period of relative stability in the Earth's climate called the Holocene. Due to the global and pervasive impacts of human activities on the environment, many scientists believe that we have moved into a new period—the Anthropocene.

- The impact of humans on the environment can be quantified, for example by using the concept of an ecological footprint. Stocks and flows are another useful concept to describe the difference between sustainable and unsustainable resource use.

- Malthus' early predictions (1798) of unsustainable population growth have been outpaced by the technological innovation and the Green Revolution; yet under quantitative and qualitative measures of sustainability, we are not using resources in a way that safeguards future natural or human systems.

- Scientific research investigates complex systems as a whole (holism) or by distilling their essential components (reductionism). Research on these systems ranges from mono-disciplinary to transdisciplinary.

Introduction

Humans are transforming planet Earth at an unprecedented scale—altering landscapes, creating artificial habitats, outcompeting all other direct competitors, and even changing the world's climate. This has happened over a relatively short period of time. The human species, *Homo sapiens*, evolved around 1,000,000 to 200,000 years ago. For a very long time, humans were a species in the margins—neither at risk of extinction nor dominating the global landscape.

Then something remarkable happened—humans began to settle down and become farmers, cultivating plants and producing sufficient amounts of food to allow for population growth. This would later be referred to as the Agricultural Revolution. The results were staggering; in 10,000 BC, 2 to 3 million individuals roamed Earth, but by the year 1,000 BC, this number had grown to around 100 million individuals (see Figure 1.1). The human population doubled every 400 years and spread all over the globe, into almost every ecosystem (Boivin et al., 2016).

While this population growth was impressive, it was nothing compared to what occurred from the nineteenth century onwards. In the early 1800s, the Industrial Revolution began, with a simultaneous burst of technological, political, economic, and social innovation. Advances in productive capacity, transportation, public sanitation, and medical sciences allowed the size of the human population to double not every 400 years, but every 33. In 2018, the global population reached 7.6 billion. While growth appears to be slowing down, projections suggest we will grow for another few decades to more than 10 billion people (UN, 2015).

One important condition for the rapid growth and development of the human population is the relative stability in the Earth's climate over the past 10,000 years. Previously, during the **Pleistocene** (2.6 million–10,000 years ago), Earth's climate experienced large fluctuations in temperature—ice ages were followed by short, warm periods (see Figure 1.2). Around 10,000 years ago, this unstable Pleistocene climate made way for a period of relative stability called the **Holocene**. The Holocene is an interglacial period with a very stable climate and temperature (see Figure 1.2). The Holocene provided a stable environment in which life could flourish, allowing for the development of agriculture and the rise of civilizations on many continents.

Industrialization made it possible for humans to exploit Earth's resources at a much greater scale than before, and divert an ever increasing share of food, water, and energy to their societies. The environmental impacts of this development have included large shifts in biodiversity, biochemical cycles, and land use (see Figure 1.3). Many academics now think that this new period of change marks a departure from the Holocene to a period called the **Anthropocene** (derived from the Greek *anthropo-*, human, and *-cene*, new), when global climate and environmental processes are being driven by human activities.

Pause and think

According to most scientists, we have entered a new geological epoch—the Anthropocene. An epoch is a geological timescale separated by significant changes in the global geological record, for example changes in rock layers. Imagine a geologist in the distant future. She is looking for significant changes that might indicate the start of the Anthropocene. For example, she could identify the first appearance of plastics (see Figure 1.3c) and, based on this, date the start of the Anthropocene to the end of the twentieth century. What other indicators can you think of that she could use?

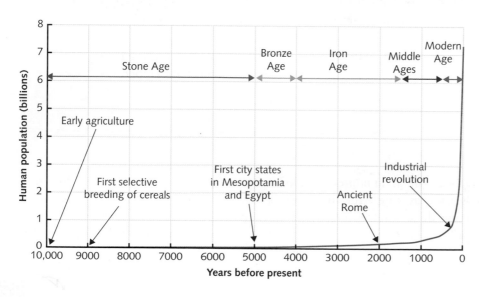

Figure 1.1 Human population growth over the past 10,000 years.

Source: Authors' own work.

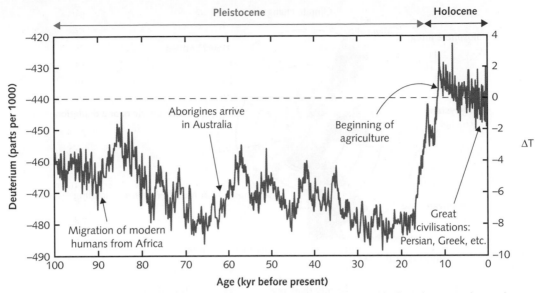

Figure 1.2 The last glacial cycle of Deuterium concentration (an indicator of temperature) and selected events in human history. The Holocene is characterized as the last 10,000 years of relatively stable temperature.

Source: Adapted from Young, O. R., and Steffen, W. (2009). The Earth system: sustaining planetary life-support systems. Pages 295–315 in F. S. Chapin, III, G. P. Kofinas, and C. Folke, editors. *Principles of ecosystem stewardship: resilience-based natural resource management in a changing world*. By permission of Springer Nature, New York, New York, USA.

Figure 1.3 Several examples of the ways human activities can impact the environment: (a) an industrial landscape; (b) the destruction of a primary forest for agriculture production; (c) plastic pollution along a lake shoreline; and (d) open pit mining for natural resources.

Source: (a) CCO License; https://www.pexels.com/photo/air-air-pollution-climate-change-dawn-221012/; (b) NASA LBA-ECO Project; (c) Wilfredor/ Wikimedia Commons/CC BY-SA 3.0; (d) Roşia Montană/CC BY 2.0.

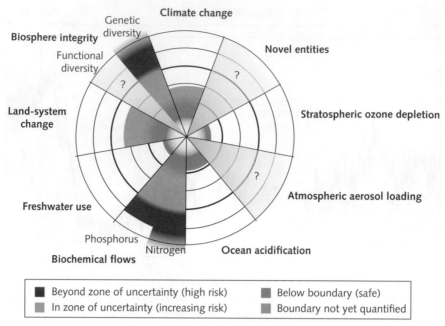

Figure 1.4 The concept of planetary boundaries provides a safe operating space for key environmental processes. The green zone in the figure indicates the safe operating space; the yellow zone represents the zone of uncertainty (increasing risk), and the red zone is a high-risk zone. The planetary boundary itself lies at the intersection of the green and yellow zones. Note that, for several processes, the boundaries cannot yet be quantified (represented by grey wedges).

Source: Adapted from Rockström, J., W. Steffen, K. Noone, Å. Persson, F. S. Chapin, III, E. Lambin, T. M. Lenton, M. Scheffer, C. Folke, H. Schellnhuber, B. Nykvist, C. A. De Wit, T. Hughes, S. van der Leeuw, H. Rodhe, S. Sörlin, P. K. Snyder, R. Costanza, U. Svedin, M. Falkenmark, L. Karlberg, R. W. Corell, V. J. Fabry, J. Hansen, B. Walker, D. Liverman, K. Richardson, P. Crutzen, and J. Foley. 2009. Planetary boundaries: exploring the safe operating space for humanity. *Ecology and Society* 14(2): 32. Copyright © 2009 by the authors. Image: Azote Images/Stockholm Resilience Centre.

Human activity may result in a shift away from the stable climate of the Holocene, which has allowed humanity to thrive and flourish. To quantify this risk, the concept of **planetary boundaries** has been developed (Rockstrom et al., 2009) (see Figure 1.4). These boundaries are a set of indicators and threshold levels for nine processes, including climate change, ocean acidification, land use change, freshwater use, and biochemical flows (Rockstrom et al., 2009; Steffen et al., 2015). Staying within these boundaries would allow us to persist in a Holocene-like state and is therefore a requirement for the sustainability of our environment and the societies that we have built. We are, however, currently exceeding several of these boundaries, for example through the disruption of biochemical flows, changes in global vegetation, and the loss of genetic biodiversity, as well as severe impacts on our climate.

Food production is one of the main contributors to the exceedance of planetary boundaries, which we explore in this book. In this introductory chapter, we introduce several key concepts, before moving into detail in the chapters. First, we define the concept of sustainability (Section 1.1) and describe important quantitative tools researchers have at their disposal to measure and analyse sustainability (Section 1.2). Subsequently, we introduce the key food sustainability challenges this book discusses: the challenges related to food and the environment (Section 1.3), food consumption (Section 1.4), and food governance (Section 1.5). Finally, we explain how these challenges are integrated in this book (Section 1.6).

1.1 What is sustainability, and how does it relate to food production?

Several key drivers push us towards exceeding the planetary boundaries. One such driver is the use of fossil fuels in industry and transportation, which has a direct link to ocean acidification and climate

Figure 1.5 The 17 Sustainable Development Goals, as formulated by the UN, focus on environmental, economic, and social well-being and constitute a transdisciplinary approach to sustainability.

Source: Sustainable Development Goals © United Nations 2018.

change. Another is the increasing demand for quantity, variety, and quality of food, due to an increase in overall population size and its increasing affluence. The availability of sufficiently high-quality food has been central to the development and growth of human societies. To this end, we have learned how to divert a larger and larger share of the available, finite natural resources to producing our food. But the consequences for our environment have been devastating. Therefore, there is now widespread agreement that we need to find ways to make our food system more sustainable. But what would a sustainable food system look like?

In its most general form, **sustainability** refers to a situation in which both human and natural systems are able to survive and flourish in the very long-term future. This conception of sustainability explicitly incorporates both human and natural systems, acknowledging that while the sustainability of Earth's natural system is crucial to human survival, having human societies that are stable and flourishing in the long run is equally important. Moreover, it reflects the fact that to attain sustainability, it is crucial to take into account the ways in which natural and human systems interact.

This definition is also consistent with the **UN's Sustainable Development Goals**—the 17 interrelated goals for organizing international efforts around the three pillars of sustainable development: environmental, economic, and social well-being (see Figure 1.5). For all these reasons, this book will focus on challenges related to both natural and human systems when it comes to food production. In doing so, it will bring together a wide range of disciplines from the natural and social sciences around the challenge of how to make food sustainable. This transdisciplinary approach has advantages over a mono-disciplinary stance, given the nature of the global sustainability challenge (see Box 1.1).

In your experience

Think of the degree programme you are pursuing at the moment. Link your programme to the concepts of reductionism and holism, as well as to the concepts of discipline, as shown in Figure A in Box 1.1. Can you think of ways in which you could complement your studies (for example, during a study abroad or by additional courses) in light of Box 1.1?

BOX 1.1

Holism, reductionism, and disciplinarity

There are many different approaches to investigate the world around us. One important distinction is between reductionist and holistic approaches. **Reductionism** explains systems in terms of their individual constituent parts and their interactions. In contrast, **holism** focuses on studying systems and their properties as a whole, and assumes that the whole is greater than the sum of its parts. In practice, you often need both—you need to have a solid understanding of the different components of complex systems, before you can understand the added complexity of these systems.

The subject of sustainable food, as well as this book, is full of examples of the usefulness of both reductionism and holism. Understanding pollution requires both a reductionist focus on the processes by which pollutants are taken up by organisms and holistic understanding of the wider ecological system within which they are produced and disseminated. Getting a grip on climate change means we need both a reductionist explanation of how greenhouse gases trap sunlight and a holistic model to understand the far-reaching social, economic, and environmental impacts of climate change and ways to adapt to these changes.

A second distinction between different ways of conducting research relates to the academic disciplines. The dominant way to conduct research is what we call disciplinary research (see Figure A). Take the example of research on tropical deforestation. If an ecologist conducts research on how this would impact species in that area, we call this disciplinary research. In addition to the ecologist, scientists from other disciplines could also investigate the impacts—a soil scientist may look at the way the soil profiles are impacted; a hydrologist may research the new water-holding capacity; a social scientist may focus on why local communities feel the need to deforest a specific area; an anthropologist may determine the impact on native communities; and a political scientist could investigate which rules and regulations are in place regarding logging.

In this scenario, there are different ways these groups of researchers could collaborate, coordinate, and communicate. When there is limited communication and coordination between these disciplines, we talk about multidisciplinary research (see Figure A). Although important information can be obtained in this way, the lack of communication and coordination can prevent a true understanding of the issue.

A more holistic approach is possible by combining disciplines in a more connected way, through interdisciplinary research (see Figure A). In this case, there is an overarching concept that drives the researchers within their discipline. In the deforestation case, the soil scientist, ecologist, and hydrologist collaborate to reach an understanding of the combined environmental impacts of deforestation. Finally, if we also involve social scientists, historians, and policy researchers to understand not only the impact on the environment, but also the socio-political drivers which cause deforestation, we talk about transdisciplinary research (see Figure A).

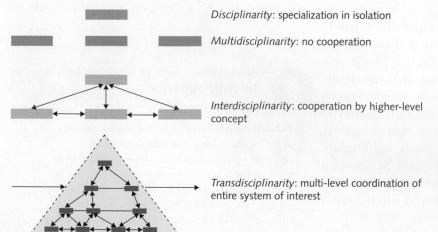

Disciplinarity: specialization in isolation

Multidisciplinarity: no cooperation

Interdisciplinarity: cooperation by higher-level concept

Transdisciplinarity: multi-level coordination of entire system of interest

Figure A A schematic outlining the levels of disciplinarity from specialization in isolation to multi-level coordination of an entire system of interest.

Source: Based on Jantsch (1972) and Odum and Barrett (2005).

1.2 How many people can we fit on Earth? Introducing the carrying capacity

7

1.2 How many people can we fit on Earth? Introducing the carrying capacity

One of the earliest, rigorous attempts to examine the interaction between humans and the environment came from Thomas Malthus (1766–1834) in his classic work *An Essay on the Principle of Population* (1798). Malthus had realized that the human population grows in a different way to food production. He observed that the production of food grew in a straight line (linear), whereas population growth appeared to be exponential (non-linear), as demonstrated in Figure 1.6. He foresaw a future where the population would exceed the amount of food available and would then collapse in famine and war. In this **Malthusian catastrophe**, he suggested that the human population would be in a perpetual state of famine and that it was preferable to act preventatively by restricting the number of children born.

Today, we have applied this theory to both plant and animal (including human) populations. One important term is **population density**, which is defined as the total number of individuals (human or otherwise) in a population living in a defined unit of space (e.g. Earth, a lake, or a nature reserve). The factors that drive population density vary and can be divided into two main types: density-dependent and density-independent factors. Density-dependent factors change when the population density changes, and include the amount of food, water, or shelter available to individuals. For example, if the population of deer increases in a forest, the available food, water, and shelter per individual decreases. In contrast, density-independent factors are those that do not change as the population density increases or decreases, and include the weather and natural disasters. For example, a prolonged drought will kill all members of a population, regardless of that population's size or density.

Based on density-dependent and -independent factors, a habitat can support a maximum number of organisms sustainably over the long term. This number of organisms is referred to as the **carrying capacity**. When populations exceed the carrying capacity within an ecosystem, they reach an unsustainable state, with insufficient resources to support the population. Often, there are natural mechanisms in place that prevent this from having long-term adverse impacts on populations. However, in some extreme cases, without these mechanisms, we can see severe impacts on populations.

Pause and think

Recall that the carrying capacity is the maximum number of organisms a habitat can support over a long period of time. You have been asked to determine the carrying capacity for humans in your habitat. How would you define your habitat (a region, a country, a continent, Earth)? And what key resources would you use to determine the carrying capacity (food resources only or also other resources such as fossil fuels)?

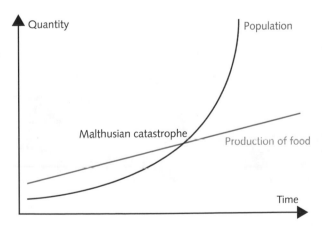

Figure 1.6 The Malthusian theory visualized: the population grows exponentially, while food production grows linearly.
Source: Authors' own work.

A good example is the population explosion and concurrent crash of a herd of reindeer on Saint Paul Island in Alaska [United States (US)]. In 1911, 25 individual reindeer were introduced on an island that had no natural predators and plenty of food. By 1938, the reindeer population had grown to over 2000 individuals. Unfortunately, this number of reindeer exceeded the carrying capacity of the system, and, as a result, there was an overuse of the available resources, leading to a shortage of food. The reindeer population experienced a dramatic population crash, wiping out nearly the entire population of reindeer (see Figure 1.7). This demonstrates that overshooting the carrying capacity of a system can result in a collapse of the population.

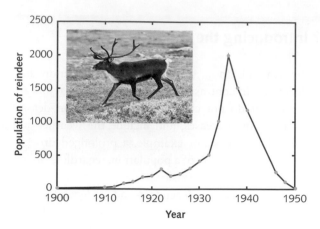

Figure 1.7 Reindeer population over time on Saint Paul Island, Alaska, USA.

Source: Reindeer: Alexandre Buisse (Nattfodd)/CC BY-SA 3.0. Graph adapted from: Scheffer, Victor B. The Rise and Fall of a Reindeer Herd. *The Scientific Monthly*, vol. 73, no. 6, 1951, pp. 356–362.

The Malthusian theory also hypothesizes that there is a maximum number of humans Earth can support, that is a human carrying capacity. Malthus believed that when the human population exceeded the carrying capacity, we would experience a population collapse and the aforementioned Malthusian catastrophe. Even though his book was published over 200 years ago, the catastrophe has not occurred, and, in fact, the global human population has continued to grow exponentially (at even faster rates than Malthus predicted).

How is this possible? Partly, this is because we are using more land for producing our food. In the period between 1961 and 2007, arable land use has increased by 1.3 million km², and the total area of pastures increased by 2.9 million km² (FAO, 2009). If the combined increase of land for agriculture and pastures (4.2 million km²) would be a country, it would be the seventh largest country on the planet.

Next to increased land use for food production, scientific innovations have led to a **Green Revolution** within our food system. Unlike other animals, humans have been able to innovate, inventing new technologies to avoid the Malthusian trap, and expand their carrying capacity. Between the 1930s and 1960s, there were an unprecedented number of technological advances in agriculture, including the mass production of chemical pesticides, growth in high-yield crops, and the development of powerful machinery such as tractors (see Figure 1.8). All these innovations have been fuelled by large quantities of cheap energy in the form of fossil fuels. As a result, crop yields exploded, dramatically increasing the overall food production per hectare. In the United Kingdom (UK), for example, crop yields for wheat grew from 2 tonnes per hectare in 1930 to 5 tonnes in 1960, and to 8 tonnes per hectare in 1996. This pattern has occurred all over the world (see Figure 1.9).

To put the impact of technological advances in perspective, it has been estimated that if humans were to rely solely on hunting and gathering, without traditional or modern agricultural systems, Earth could only support 10 million individuals (Burger and Fristoe, 2018)—approximately the population of Sweden. In 2018, Earth plays host to 7.6 billion human individuals.

a)

b)

Figure 1.8 Technological advances—such as the development of chemical pesticides (a) and powerful machineries (b)—have been important innovations during the Green Revolution and have contributed to the increase in crop yields.

Source: (b) Charles Knowles/CC BY 2.0.

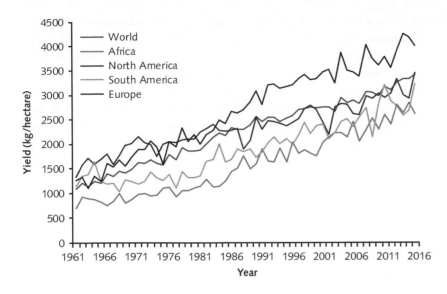

Figure 1.9 Average wheat yield has increased significantly after the onset of the Green Revolution in 1961.

Source: Based on data from the Food and Agriculture Organization (FAO) of the United Nations.

1.2.1 How can we quantify the impact of individuals on the environment?

Since we have been able to avoid the Malthusian trap, does this mean that humans do not have a carrying capacity? Is there a maximum number of individuals that can be sustained in a given area—say in India, Tanzania, or the US? Or does technological innovation allow such a population to grow indefinitely? The answer to this question is uncertain, because we cannot predict what future technological innovation will bring

us. However, what is clear is that carrying capacity does not work in the same way for humans as it does for other animals, mainly because we can use technology and trade goods to bypass the overuse of resources.

Importantly, though, not all impacts of human population growth have been mitigated by technological innovation. This can be illustrated with the concept of an **ecological footprint**. This is the amount of biologically productive land and sea used by an individual (see Box 1.2 for an example). Tools

BOX 1.2

How to calculate a footprint?

To calculate an ecological footprint, data are collected on how many resources are used within a country, which is expressed in the number of hectares needed to produce those resources. Data are collected from major databases, including the Food and Agricultural Organization (FAO), the United Nations (UN), and the International Energy Agency (IEA). These numbers are regularly updated when new data are available, resulting in changes in footprint.

For example, based on data from the US Department of Agriculture from 2016, there were 26 million tonnes of wheat consumed in the US. The average production was 3.5 tonnes of wheat per hectare (10,000 m^2). This means a total of 7.4 million hectares of wheat are needed to produce sufficient wheat to meet the US demand. As

there were 324 million US citizens in 2016, this means that, on average, each US citizen needed 0.02 hectares for producing wheat.

In Tanzania, by contrast, a total of 10.1 million tonnes of wheat were consumed in 2016. According to the World Bank, the average wheat yield in Tanzania in 2016 was 1.54 tonnes per hectare. As a result, a total of 6.5 million hectares were needed to meet the Tanzanian demand. The population of Tanzania was 55.6 million in 2016, which means that, on average, each Tanzanian needed 0.12 hectares for producing wheat.

We can do this for all resources a US or Tanzanian citizen uses, and adding up these hectares will give you an indication of the pressure these citizens place on the environment.

are available for people to calculate their own footprint, including the footprint of specific goods they might consume (for example, the amounts of land and sea used in the production of wheat, milk, or mobile phones). But they can also be used to calculate the ecological footprints for cohorts of people, for example for the average British, Tanzanian, or Indian citizen.

In your experience

Calculate your ecological footprint using an online resource (for example, http://www.footprintcalculator.org). How big is your impact? And what part of this footprint is related to your food consumption? What are ways you could reduce your food footprint?

Pause and think

Based on Box 1.2, the average US citizen requires 0.02 hectares of wheat for their diet, while the average Tanzanian needs 0.12 hectares. This is a sixfold difference. Think of several reasons that could explain this difference. Include reasons related to differences in production, processing, and consumption of wheat.

Carrying capacity and ecological footprint serve a similar analytical goal, but the difference is the unit of expression—carrying capacity is expressed in individuals per hectare, which provides a limit to the maximum number of individuals per hectare. The ecological footprint is expressed in hectares per individual and is not limited to the land area of a specific country. Using the ecological footprint, we can calculate how much a certain subpopulation of humans (a city, a province/state, a country, a continent) needs (ecological footprint) and how many resources they have available in the specific area (**biocapacity**). If the ecological footprint exceeds the biocapacity, a country has a deficit. If the footprint is less than the biocapacity, there is a surplus or reserve.

Furthermore, we can also determine the total amount of productive land available per person. The total amount of productive land on Earth in 2010 was approximately 11.9 billion hectares, which was available for a total global population of approximately 6.7 billion. Therefore, the average amount of productive land available per person was around 1.8 hectares per person. The average global usage in 2010 was 2.7 hectare per person, clearly exceeding the amount of productive land available. This means we would need approximately 1.5 planets (2.7/1.8) to support our current lifestyles in the long run (see Figure 1.10).

The ecological footprint approach, as well as other approaches (see Box 1.3 for another important tool), are useful to understanding the pressure we put on the environment. How do we account for the fact that we are using more than one planet's worth of resources, according to the ecological footprint calculations? Here, we need to highlight the difference between resource stocks and resource flows. On the one hand, resources that require time to accumulate are called **resource stocks**. Fossil fuels are a good example, requiring millions of years at high temperatures and pressures to accumulate (see Figure 1.11a). Like saving money in a bank account that can be withdrawn in the future when we need it, fossil fuels are the savings from natural processes. As these resources take a long time to renew themselves (millions of years in the example of fossil fuels), we also call these **non-renewable resources**.

Resource flows, on the other hand, are **renewable resources** and, when used sustainably, do not run out. However, flows cannot be guaranteed to be always available. Sunlight is a good example of a flow; it does not run out, and trees can make energy from it in their leaves, or we can use solar panels to generate electricity (see Figure 1.11b). However, as we all know, sunlight is not available all the time.

Returning to the concept of the ecological footprint and sustainability, we are heavily exploiting the stocks of resources available to us. By some measurements, we are already exceeding the total amount of resources available in a year, resulting in the use of more than one planet, as calculated by the ecological footprint approach.

Given that we might need up to 1.5 planets to ensure that our food production and consumption

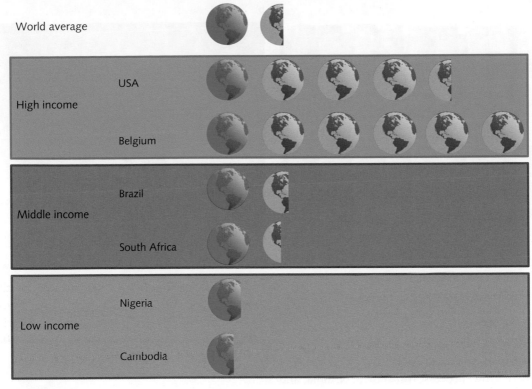

Figure 1.10 The difference in ecological footprint across high-, middle-, and low-income nations in 2010, expressed as the number of Earths required if the entire world would have a similar lifestyle as the residents of the countries listed in the figure.

Source: Data from: Global Footprint Network, http://www.footprintnetwork.org/

BOX 1.3

Other ways to quantify the impact of human activity on the environment

The ecological footprint is one approach to assess the impacts of individuals, cities, and even countries on the environment, but it is important to realize that there are many other perspectives we can take, using other methods. A popular approach is called a decomposition analysis where we take observed data, such as the amount of an environmental pollutant (that is, carbon dioxide concentrations) and (attempt to) explain the trends of the pollutant levels in terms of the important drivers.

Changing population size, affluence, and technological advances have all had a huge influence on humanity's impact on the environment. In 1970, a basic equation was developed (Ehrlich and Holdren, 1971)—the 'IPAT' equation—to aid our understanding of how these factors influence the human impacts on the environment:

$$I = P \times A \times T$$

This formula states that environmental impacts by humans (I) are the product of the human population size (P), the affluence of individuals (A), and the (harmful and beneficial) impacts of technology (T). Both an increase in human population, as well as an increase in affluence [often expressed as gross domestic product (GDP)/capita] will increase environmental impacts. Technology can result in increased environmental pressures (for example, the use of heavy machinery in agriculture), while other technological advances can actually result in a reduction in environmental impacts (for example, the use of renewable energy resources).

a)

b)

Figure 1.11 Resources which require time to accumulate are called stocks and include fossil fuels (a), while flows are resources which do not run out if used sustainably such as sunlight (b).

Source: (a) Sergio Russo/CC BY-SA 2.0.

is sustainable in the long run, it seems that there are strong incentives for us to radically change the way in which we grow, process, transport, consume, and recycle food. Yet, for many reasons, this is fundamentally difficult. This book aims to introduce you to the kinds of change that would be needed, as well as the reasons why such change is hard to realize. It is organized into three parts, each of which focuses on a specific dimension of the challenge of food sustainability: the environment, society, and governance. We will discuss each in turn, starting with challenges related to the environment.

 Pause and think

According to the ecological footprint approach, we are currently exceeding the carrying capacity of planet Earth. If this is correct, how is it possible that our population has not yet collapsed, similar to the reindeer example? Include the concept of stocks and flows in your answer. Next, provide arguments on whether or not our population will collapse in the future if we continue business as usual.

1.3 What are the key environmental challenges related to food production?

Imagine a field of potatoes in Germany, a rice paddy in the Philippines, or a beef cattle farm in Brazil (see Figure 1.12). These areas once consisted of natural vegetation (a temperate forest in Germany and tropical rainforests in the Philippines and Brazil). This change in land use to supply human needs has been important in producing sufficient food for growing populations, but has also resulted in a variety of adverse side effects on the environment.

For example, the removal of natural vegetation will have resulted in lower biodiversity, compared to the original habitat (*Chapter 2: Biodiversity*). During

crop growth, artificial fertilizers are usually applied to increase total production, and some of these will inevitably end up in the adjacent environment, potentially causing eutrophication of the water (*Chapter 3: Pollution*). Furthermore, pesticides are used to protect the crops against pests, including fungi, insects, and bacteria (*Chapter 3: Pollution*). As with the nutrients, some of these pesticides will spill over to the surrounding environment, potentially having negative effects on non-target organisms and further impacting on biodiversity.

Water might be pumped from ancient aquifers to provide sufficient water to the crops and cattle,

a)
b)
c)

Figure 1.12 Different land use alterations: (a) a field with potatoes in Germany; (b) a rice paddy terrace in the Philippines; and (c) beef cattle in Brazil.

Source: (b) Ori/Wikimedia Commons/CC BY-SA 3.0; (c) Scott Bauer/U.S. Department of Agriculture.

sometimes depleting water reserves for future generations (see *Chapter 4: Water*). The bare soil after the potato harvest in Germany is at risk of erosion, while the cattle on the farm in Brazil might overgraze, resulting in the run-off of fertile topsoil during a rain event (see *Chapter 5: Soils*). Finally, cattle produce relatively large amounts of methane gas—a potent greenhouse gas—thereby directly contributing to climate change (see *Chapter 6: Climate Change* and *Chapter 7: Energy*). In Part 1 of the book, we explore the environmental issues that can be directly linked to food production.

1.4 What are the important societal challenges of food production?

In Part 2 of this book, the environmental issues described in Section 1.3 are directly linked to our needs and desires as consumers. There are strong connections between food consumption and sustainability. Arguably, the most important driver of consumption is health—we need food to survive and live healthy lives, and this need is one of the drivers of what food we choose to eat. For example, a higher intake of animal products is associated with an increased likelihood of cardiovascular disease, which would reduce expected longevity, while having outsized impacts on the environment. In *Chapter 8: Nutrition*, we explore the way dietary choices can impact health and the environment.

However, what people actually eat also depends on what food is available and affordable. The availability of food varies widely around the world. Some regions have seasonal and exotic foods available all year round (think of the strawberries in a European supermarket in December). Other regions lack foods with key nutrients (for example, high-protein foods in many of the poorest nations).

Even if a food is available, it is not always affordable, and this may constrain people, even in high-income nations. For example, the rise in the consumption of cheap fast-food over more expensive and healthy food in the US has been an important factor in high rates of obesity in some communities. In *Chapter 9: Food Security*, we look at the dynamics of food availability and affordability worldwide. In *Chapter 10: Food Aid*, we focus specifically on ways to improve food security for lower-income groups around the world.

Even if high-quality, healthy, and affordable food is available, there is still no guarantee that people will eat it. Food has many meanings beyond its nutritional function, and food decisions can be heavily influenced by cultural and social norms. For example, throughout the world, increasing wealth often goes hand in hand with an increasing demand in meat consumption, as meat is widely considered to be a 'high-status' food. At the same time, in high-income countries, there has been a sharp rise in demand for organic and locally produced foods, driven by a rise in health and environmental concerns in public

...urse. In *Chapter 11: Consumption*, we look at what drives these consumption choices and discuss possible approaches to making food consumption more sustainable such as minimizing waste and maximizing recycling, food labelling, and pricing food for externalities.

1.5 How can we fix our problems? The role of governance in overcoming sustainability challenges

Against the backdrop of environmental and societal challenges outlined in Parts 1 and 2, we can begin to appreciate the complex ways in which food is produced, processed, distributed, and consumed. *Chapter 12: Food Systems* introduces you to a way of thinking that incorporates these complexities and looks at food holistically—the food systems approach. It will help you to understand not only where your food comes from, by looking at food supply chains (see Figure 1.13), but also how these supply chains are impacted by wider social, economic, and political systems. This then helps you to begin thinking about how to make these systems more sustainable.

Governance—the theme of Part 3 of this book—constitutes the use of power to make laws and policies and to influence the behaviour of people and organizations to solve problems like food sustainability. It is often associated with the state and the formal government of a country, but, in fact, governance can involve a wide range of actors, from the decentralized self-governance of village councils to the strict hierarchies of personalized dictatorships. The globalized nature of food supply chains and systems means that food sustainability can only be secured if cooperation at the global level is combined with decentralized and locally embedded forms of food governance (see *Chapter 13: Governance*). This is one of the reasons why food sustainability is often seen as a **wicked problem**—a problem that changes continuously, is fundamentally complex and unpredictable, and, as a result, is incredibly difficult to solve. As such, it is one of the most intractable collective action problems we know of, that is, it is one of the key problems that we, as humans, want to fix collectively, if only we could get our act together.

Making food systems sustainable is therefore not only a technical challenge, but also a deep political one. Different stakeholders in food systems all have different interests, and balancing them at the same time as pushing the stakeholders to behave more sustainably complicates food governance considerably. Moreover, the real benefits of food sustainability materialize in

a)

b)

Figure 1.13 The production of your chocolate bar (a) starts at cocoa plantations, for example in Peru (b), and moves through a long supply chain of transportation, processing, packaging, and distribution processes before you can eat it. Understanding this chain, however, requires us to look at how this supply chain is influenced by everything, from land governance in Peru to the determinants of chocolate consumption in your country.

Source: (a) Lee McCoy/CC BY-ND 2.0; (b) Ti galan/Wikimedia Commons/CC BY 3.0.

the long run for future generations, while the costs are paid by people (and politicians) today. For example, overfishing of the oceans will be detrimental to future generations, but it has direct monetary and health benefit to those living today. How can we ensure that the interests of these future people are protected, if balancing the interests of people living today is already such a challenge? *Chapter 14: Collective Action* will introduce the theory of collective action problems and demonstrate how it can be used not only to understand why food sustainability is so difficult to achieve, but also to identify possible ways to achieve it.

Pause and think

We argue that achieving food sustainability constitutes a wicked problem due to the dynamic nature and complexity of the issue. Another wicked problem is nuclear proliferation, in that the players are unpredictable and there are many different views and constraints when looking for solutions. Think of other wicked problems humanity is facing and the characteristics that make them wicked problems.

1.6 Aim of the book

As with many other complex global challenges, making food systems sustainable defies easy solutions. At the same time, there is an urgency to this challenge, especially with the increasing global population and growing affluence. Even if it is difficult to envision comprehensive solutions to enhance global food systems, there may be ways of finding small-scale solutions to parts of the problem that, cumulatively, help us produce sufficient good-quality food for our large and growing population, far into the future. With this in mind, the aim of this book is to present students, policymakers, researchers, and other interested readers with state-of-the-art knowledge of the main dimensions of food sustainability. However, please keep in mind that this is a rapidly changing and highly dynamic field of study. For example, over the period of preparing this book, Jair Bolsonaro was elected to power in Brazil on a platform of agricultural expansion in the Amazon, after decades of policy intervention to minimize deforestation. Likewise, a company

highlighted in a case study on urban agriculture went bankrupt (it might have been environmentally sustainable, but apparently not economically).

Overall, we hope to inspire more attention on the subject, as well as to facilitate the identification of innovative transdisciplinary approaches that bring us closer to the ultimate goal of making food sustainable.

Connect the dots

Take a piece of paper and a pen, and draw a Venn diagram (use a search engine if you do not know what this is). Use a circle each for the environment, the society, and the economy. Describe the different intersections in the Venn diagram (for example, when only the environment and the economy overlap) and what this means for sustainability.

● QUESTIONS

1.1 Why do we express human pressure on the environment as an ecological footprint, rather than talking about carrying capacity?

1.2 Why has a Malthusian catastrophe not occurred?

1.3 Give two further examples of resource stocks and two further examples of resource flows.

1.4 Define sustainability, and explain why, based on the concept of ecological footprints, humanity is currently not using resources sustainably.

1.5 What is the difference between reductionism and holism in scientific research?

● FURTHER READING

Global Footprint Network. Available from: http://www.footprintnetwork.org/ecological_footprint_nations/ [accessed 4 November 2018] **(A website which allows you to explore your own ecological footprint.)**

Jackson, T. 2011. *Prosperity Without Growth: Economics for a Finite Planet*. London: Routledge. **(An overview of how wealth and prosperity may change in a world where scarcity of resources dominate.)**

Rockström, J., Steffen, W., Noone, K, Persson, A., Chapin, F.S. 3rd, Lambin, E.F., Lenton, T.M., Scheffer, M., Folke, C., Schellnhuber, H.J., Nykvist, B., de Wit, C.A., Hughes, T., van der Leeuw, S., Rodhe, H., Sörlin, S., Snyder, P.K., Costanza, R., Svedin, U., Falkenmark, M., Karlberg, L., Corell, R.W., Fabry, V.J., Hansen, J., Walker, B., Liverman, D., Richardson, K., Crutzen, P., Foley, J.A. 2009. A safe operating space for humanity. *Nature* **461**, 472–475. **(A landmark paper bringing together, in a highly structured way, the key environmental challenges we face as humanity.)**

● REFERENCES

Boivin, N.L., Zeder, M.A., Fuller, D.Q., Crowther, A., Larson, G., Erlandson, J.M., Denham, T., Petraglia, M.D. 2016. Ecological consequences of human niche construction: Examining long-term anthropogenic shaping of global species distributions. *Proceedings of the National Academy of Sciences of the United States of America* **113**, 6388–6396.

Burger, J.R., Fristoe, T.S. 2018. Hunter-gatherer populations inform modern ecology. *Proceedings of the National Academy of Sciences of the United States of America* **115**, 1137–1139.

Ehrlich, P.R., Holdren, J.P. 1971. Impact of population growth. *Science* **171**, 1212–1217.

Food and Agriculture Organization of the United Nations (FAO). 2009. *The State of Food and Agriculture*. Rome: FAO.

Jantsch, E. 1972. *Technological Planning and Social Futures*. London: Cassell/Associated Business Programmes Ltd.

Odum, E.P., Barrett, G.W. 2005. *Fundamentals of Ecology*, 5th ed. Belmont, CA: Thomson Brooks/Cole.

Petit, J.-R., Jouzel, J., Raynaud, D., Barkov, N.I., Barnola, J.-M., Basile, I., Bender, M., Chappellaz, J., Davis, M., Delaygue, G. 1999. Climate and atmospheric history of the past 420,000 years from the Vostok ice core, Antarctica. *Nature* **399**, 429–436.

Rockström, J., Steffen, W., Noone, K., Persson, A., Chapin, F.S., Lambin, E., Lenton, T.M., Scheffer, M., Folke, C., Schellnhuber, H.J., Nykvist, B., de Wit, C.A., Hughes, T., van der Leeuw, S., Rodhe, H., Sorlin, S., Snyder, P.K., Costanza, R., Svedin, U., Falkenmark, M., Karlberg, L., Corell, R.W., Fabry, V.J., Hansen, J., Walker, B., Liverman, D., Richardson, K., Crutzen, P., Foley, J. 2009. Planetary boundaries: exploring the safe operating space for humanity. *Ecology and Society* **14**, 32.

Steffen, W., Richardson, K., Rockström, J., Cornell, S.E., Fetzer, I., Bennett, E.M., Biggs, R., Carpenter, S.R., de Vries, W., de Wit, C.A., Folke, C., Gerten, D., Heinke, J., Mace, G.M., Persson, L.M., Ramanathan, V., Reyers, B., Sorlin, S. 2015. Sustainability. Planetary boundaries: guiding human development on a changing planet. *Science* **347**, 1259855.

United Nations (UN). 2015. *The World Population Prospects: 2015 Revision*. New York, NY: UN.

Young, O.R., Steffen, W. 2009. The earth system: sustaining planetary life-support systems. In: Folke, C., Kofinas, G., Chapin, F. (eds). *Principles of Ecosystem Stewardship*. New York, NY: Springer, pp. 295–315.

Food and Environment

What are the main causes of biodiversity loss, and what does food production have to do with it? How do pesticides cause intersex in frogs? What are the consequences for freshwater resources when we grow asparagus in a desert? Why is China losing large areas of productive agricultural land? And are biofuels a good way to produce renewable energy resources?

These questions are some of the themes explored in *Part 1: Food and Environment*, which covers the impact of food on the environment and, in particular, natural resources.

In *Chapter 2: Biodiversity*, we highlight the key services biodiversity provides to humanity. Next, we focus on the flow of energy in ecosystems and the disruption of this flow by humans. After laying the groundwork, we examine the key threats of food production to biodiversity. Finally, we explore different ways in which we can produce sufficient food while maintaining biodiversity.

Chapter 3: Pollution explores the impact of pesticides and artificial fertilizers—two important agrochemicals of the Green Revolution—on the environment. We discuss the diverse ways in which pesticides impact organisms and different approaches in studying these impacts. We also show how the run-off of fertilizers into small streams can ultimately cause large-scale effects in far-away locations.

Chapter 4: Water introduces the hydraulic infrastructure developed to support agriculture. Next, we explore the way in which freshwater resources are exploited, which is often unsustainable. We stress the importance of international collaboration in water management, as freshwater resources often cross political borders.

In *Chapter 5: Soils*, we first demonstrate the alarming rate at which we degrade soils. This is especially alarming, as most food is produced in soils. We focus on the key processes that cause soil degradation and explore ways in which this can be minimized in the future.

Chapter 6: Climate Change explores the way in which climate change is impacted by food production, but also how food production is impacting climate change. We analyse key contributors to climate change, including land use change and cattle husbandry.

Finally, in *Chapter 7: Energy*, we focus specifically on the use of energy in food systems. The transition of food systems to renewable energy resources provides unique challenges; yet, it might also provide solutions to reduce the dependence on fossil fuels.

Image credit: Sergey Molchenko/Shutterstock.com

Biodiversity

What are the impacts of food production on biodiversity?

Thijs Bosker, Ellen Cieraad, and Krijn Trimbos

Chapter Overview

- The variety of life on Earth is called biodiversity, and it consists of genetic, species, and ecosystem diversity. Biodiversity is important in many ways, both in the services it provides to humanity and in its own right.

- Today, biodiversity is being lost at an alarming rate; current extinction rates are estimated to be 100- to 1000-fold higher than natural extinction levels.

- There are five key threats to global biodiversity, many of which can be directly linked to food demand and growing affluence of the human population. These threats are habitat fragmentation, the introduction of invasive species, pollution, climate change, and the overexploitation of ecosystems.

- Through agriculture, humanity uses a large portion of all the energy captured by terrestrial plants, thereby limiting the amount of energy which is available for other species.

- With agriculture dominating global landscapes, it also provides important habitats for organisms that depend on these types of landscapes, including animals and plants.

- Conservation efforts are key to maintaining biodiversity in and around agricultural landscapes. This will be even more important with the predicted growth in food demand over the coming decades. Examples of biodiversity conservation efforts in agriculture include agri-environmental schemes and nature-inclusive agriculture.

Introduction

The dodo, buffalo, and passenger pigeon went extinct through overhunting (see Figure 2.1a). Populations of Atlantic cod have collapsed due to overfishing and habitats of orangutans are exchanged for palm oil plantations (see Figure 2.1b). These are just a few examples of the direct conflict between biodiversity and humanity's need for food. Biodiversity loss is taking place at an alarming pace and is estimated

a)

b)

Figure 2.1 An example of (a) overexploitation of buffalos. Here we see a pile of American buffalo skulls in the mid 1870s. Buffalos were hunted to near extinction for a variety of reasons, one being to reduce competition for cattle. (b) A palm oil plantation, which has replaced pristine rainforest in large parts of Indonesia and Malaysia, at the cost of biodiversity.

Source: (b) Pizzaboy1/Wikimedia Commons.

to be 100- to 1000-fold higher than natural background extinction rates [these are calculated, based on fossil records, and give an estimated average species extinction (Pereira et al., 2010)].

Habitat loss, invasive species, pollution, climate change, and overexploitation are key threats to biodiversity, and all are driven by the ever-expanding human population and our increasing affluence (see *Chapter 1: Introduction*). The majority of these threats can be directly linked to food production, a topic which is explored in this chapter. We cover four themes in this chapter. First, we discuss the importance of biodiversity and explore how we can value biodiversity (Section 2.1). Next,

we illustrate how the human demand for food is affecting food web dynamics and biodiversity, looking at habitat loss, invasive species, and overexploitation of fish stocks (Section 2.2). Third, we explore conservation efforts related to maintaining biodiversity, with a special focus on food systems (Section 2.3). This is important, as agricultural landscapes constitute some 40% of the terrestrial surface (Foley et al., 2005) and can harbour many different species that we value or which are important in supporting food production. Finally, we discuss active conservation strategies employed to conserve biodiversity at different spatial scales (Section 2.4).

2.1 What is biodiversity, and why is it important?

Life on Earth is diverse and abundant, ranging from single-celled bacteria to extremely complex organisms (see Box 2.1). Scientists estimate there to be a staggering 8.7 million species on Earth, of which only 1.2 million species are currently catalogued (Mora et al., 2011). However, it is important to realize that biodiversity goes beyond counting species.

2.1.1 How is biodiversity defined?

The UN has defined **biodiversity** as the variety of life on Earth, which includes all organisms, species, and populations. It consists of the diversity within species (genetic diversity), the diversity among species (species diversity), and the diversity of ecosystems (ecosystem diversity) (UN, 1992).

BOX 2.1

Structuring life: how Linnaeus created order

The Swedish botanist and zoologist Carl Linnaeus (1707–1778) revolutionized the way organisms are grouped together, based on the similarity of their characteristics (the field of taxonomy). He systematically divided organisms within different taxonomic ranks. Until recently, the shape and characteristics (morphological characteristics) of organisms were predominantly used to group organisms. With the development of deoxyribonucleic acid (DNA) sequencing, scientists now have an additional tool available to group organisms, based on genetic similarity.

Based on both genetic and taxonomic classifications, scientists have developed a tree of life. At the base of the tree of life, there are large *Domains*, which include many different species with generic similar characteristics (for example, organisms which contain multiple cells or organisms which consist of only one cell). The further we move up the tree, the more unique the characteristics become, eventually resulting in species. To illustrate this, we will now take a journey through the tree of life, starting at the Domain level, moving through each level in turn, to finally reach the taxonomic classification of human (*Homo sapiens*) at the species level. The classification of humans at each step is highlighted in blue in the following list and is also shown in Figure A.

- At the base of the tree of life, we have three Domains— Eukarya (multicellular organisms), Bacteria (single-cell organism), and Archaea (a specific group of single-cell organisms which were previously included in the domain Bacteria).

- The domains are divided into Kingdoms. For example, the domain Eukarya includes the kingdoms of Plantae (plants), Animalia (animals), and Fungi (and several others).

- Within the Kingdom of Animalia, a variety of Phyla are differentiated, for example Mollusca (invertebrates with a shell and a muscular foot), Arthropoda (invertebrates with an exoskeleton, including insects and crabs), and Chordata (animals with a notochord, a flexible rod made out of material similar to cartilage).

- Within the phylum of Chordata, a variety of Classes of organisms can be distinguished, including the classes of Agnatha (jawless fish), Amphibia, Aves (birds), Reptilia (reptiles), and Mammalia (mammals).

- Within the class of Mammalia, we differentiate several Orders, including, among others, Carnivoria (large canines, including bears, dogs, and cats), Cetecia (aquatic mammals such as dolphins and whales), and Primates.

- Within the order Primates, we can distinguish different Families, including Lemuridae (lemurs, as found in Madagascar), Cercopithecidae (the Old World monkeys, including baboons), and *Hominidae* (the great apes).

- Within the family of Hominidae, several Genera are found, including the *Pan* (chimpanzees and bonobos), *Gorilla*, *Pongo* (orangutan), and *Homo* (modern humans and their close relatives).

- Within the genus of *Homo*, there were several species, all of them extinct, except our human species (*Homo sapiens*).

Many additional (sub-)classifications can be used, such as subphylum, or subfamily, or subspecies. However, the basic principle remains in the same organizational system, that of the tree of life. When using scientific names for species, the convention is to provide the genus and species name (with only the genus name capitalized), and to italicise both (e.g. *Homo sapiens*). To aid readability, we have not included the scientific names of animals in this textbook.

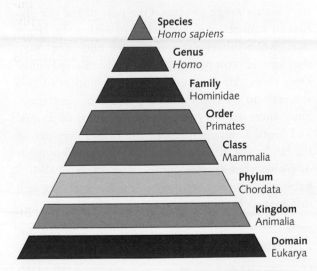

Species	*Homo sapiens*
Genus	*Homo*
Family	Hominidae
Order	Primates
Class	Mammalia
Phylum	Chordata
Kingdom	Animalia
Domain	Eukarya

Figure A Example of the formal classification of humans (*Homo sapiens*), going from the domain at the bottom, all the way up to the species level.

In practice, this means that biodiversity can be subdivided into three levels. At the base is genetic diversity, which is the amount of diversity in genetic information in a species. Next, we have species diversity, or the number of species (often referred to as species richness). Finally, at the highest level, there is ecosystem diversity.

Pause and think

How do different breeds of dogs increase biodiversity? And what about different types of habitats, for example a forest, a savannah, or an urban environment? If you have limited funds, on what level of biodiversity (i.e. genetic, species, or ecosystem) would you focus your efforts and why?

An **ecosystem** consists of a community of organisms that interact with one another and with their environment of non-living matter and energy, for example sunlight, water, and nutrients. Most commonly, we define ecosystems as distinct areas such as a forest, a lake, or the tundra. However, there is no upper or lower boundary on the size of an ecosystem. Technically speaking, this makes planet Earth one large ecosystem, but at the other end of the scale, you can buy an enclosed, fully functional ecosystem that fits in the palm of your hand (for example, http://www.eco-sphere.com). Ecosystems comprise two parts: (1) the **biotic components**, including all living and once living organisms, and (2) the **abiotic components**, including all non-living chemical and physical parts of our environment. These two systems are highly interconnected.

Ultimately, almost all life on Earth, and therefore almost all biodiversity, depends on energy captured from the sun and the cycling of nutrients. In return, plants and other organisms perform important roles in the global chemical cycles. In this chapter, we focus on the biotic part of ecosystems, including the linkage between biodiversity and the energy flow within ecosystems, and how it is possible that all this life on Earth is maintained. In other chapters, we focus on important chemical cycles (and disturbances within these cycles related to agriculture), including the

nutrient cycles (see Box 3.6 in *Chapter 3: Pollution*), water cycle (see Figure 4.2 in *Chapter 4: Water*), and carbon cycle (see Box 6.3 in *Chapter 6: Climate Change*).

2.1.2 How can life be so diverse?

Earth comprises many environments with widely varying physical conditions, for example: fresh and salt water bodies, forests, deserts, mountain, and lowland environments. In each of these environments, organisms occupy and use resources, including food, space, light, heat, and water. Organisms also compete with each other for these resources.

A leading theory to explain the diversity of species is the niche theory. An **ecological niche** is the position and role of a species within its environment, including how it reproduces, how it meets its need for water, food, and shelter, how it avoids predators, and what kind of environmental conditions it can tolerate (for example, the optimal temperature range). The niche theory assumes that, by specialization or adaptation towards a specific niche, organisms reduce competition and allow for coexistence. This adaption can, but does not always, result in speciation—the formation of new species.

A niche may already result in speciation if only one of the components (abiotic or biotic) is different. For example, imagine a population of squirrels in a forest, in which there are trees which produce nuts with a range of sizes. One part of the population of squirrels can specialize in eating the larger nuts, while others eat the smaller nuts. This way, they avoid competing over medium-sized nuts. Over time, this can result in the evolution of two species—an outcome of the process of speciation. Since there are very many combinations of abiotic and biotic components, and these also vary across the landscape, there is an almost unlimited number of niches that species can occupy. In general, if a habitat is more heterogenous, more species (with different niches) can coexist.

2.1.3 How is life on Earth supported?

Virtually all life on Earth depends on energy captured from the sun. **Autotrophs** (also called producers) are

Pause and think

Niche theory can help to describe the process of speciation—the formation of new species. With niche theory in mind, explain how human-dominated landscapes, such as agricultural areas or cities, can contribute to species diversity. Link this to biotic and abiotic factors that are present in human-dominated landscapes.

organisms that capture energy and combine this with inorganic components in their environment to produce complex organic compounds (for example, lignin and cellulose), which contain chemical energy. Autotrophs are at the base of the **food chain**, a linear network of organisms that feed on each other (see Figure 2.2a). A food chain is divided into **trophic levels**, which indicate the position an organism occupies in the chain. In reality, communities of organisms are not a set of independent food chains, but complex

a)

Food Chain

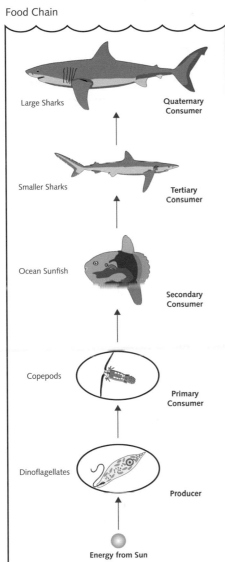

b)

Food Web

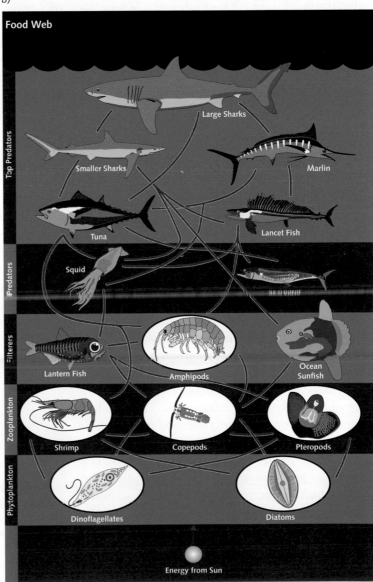

Figure 2.2 An example of a food chain (a) and a food web (b).

Source: udaix/Shutterstock.com

interactions between different organisms, creating a network called **food webs** (see Figure 2.2b).

Autotrophs, at the base of a food web, have a trophic level of 1. **Heterotrophs**, or consumers, are organisms which are not able to produce their own organic compounds, and therefore depend on the energy captured by autotrophs. A herbivore is an organism that consumes autotrophs and has a trophic level of 2. A carnivore that feeds on herbivores is termed a secondary consumer and has a trophic level of 3, and a carnivore that feeds on secondary consumers is called a tertiary consumer (trophic level 4), and so on. (See Figure 2.2 for a schematic of how this works in practice.)

The key driver of food web dynamics is the energy that is captured by autotrophs. The vast majority of autotrophs are photoautotrophs, which capture solar energy and transform this into sugars (which are simple organic compounds) using photosynthesis (See Box 2.2 for a notable exception—chemoautotrophs.) Photoautotrophs include plants, algae, cyanobacteria, and phytoplankton (phytoplankton consists of a variety of microorganisms, including bacteria and plants). The main characteristic shared by these organisms is the presence of chlorophyll (cyanobacteria do not have chlorophyll, but they have similarly functioning pigments) in their structures, which is used to capture solar energy. Solar energy is then transformed into chemical energy (i.e. organic carbon structures such as glucose).

The amount of chemical energy captured by autotrophs is called the **gross primary production (GPP)**, which is expressed in mass of carbon produced per unit area per year (for example, grams of organic carbon per m^2 per year). Part of this energy is used by autotrophs for their own metabolism. This is called respiration and involves the conversion of the chemical energy (for example, energy captured in glucose) into available energy to run cellular processes. The **net primary production (NPP)** is the GPP minus the rate at which autotrophs use some of the acquired chemical energy for respiration. This can be thought of as the 'free' energy for new biomass or reproduction (see Figure 2.3).

The ratio between the NPP and GPP (that is, how much of the captured energy measured as the GPP is transformed into the NPP) varies across different ecosystems types and depends on climatic (for example, temperature, rainfall) and geographical (for example, latitude and altitude) factors (Zhang et al., 2009). The global average over 2000–2003 was estimated to be approximately 0.52, meaning that 52% of the GPP is transformed into NPP (Zhang et al., 2009). Several factors, such as the availability of water and nutrients, cause the maximum NPP to differ dramatically between ecosystems (see Figure 2.4). For example, the NPP of algal beds and reefs is 20 times higher than that in open oceans. Similarly, the NPP of a tropical rainforest is 3.5 times higher than that of a temperate grassland.

To find the contribution of different ecosystems to global NPP, the surface area of each ecosystem type also needs to be taken into account. For example, the open ocean makes up 65% of the surface of the Earth. Even though productivity is relatively low in this system, its overall contribution to global NPP is high because of its vast area, contributing nearly 25% of the total. In contrast, the NPP per m^2 in algal beds and reefs is relatively high, but with only 0.1% of the Earth's surface covered by algal beds and reefs, the overall contribution of algal beds and reefs to global NPP is only slightly more than 1%.

Some of the chemical energy captured in biomass by plants is subsequently consumed by herbivores. On average, only 10–20% of the chemical energy is transferred from one tropic level to the next; the rest is lost as low-quality heat and waste (i.e. undigested parts) products to the environment (see Figure 2.5).

In your experience

Think about the different components of your diet. Link these components to different trophic levels in a food web. If you would start including more components at a lower or higher trophic level in your diet, how would this affect energy flows and how would this, in turn, impact biodiversity?

BOX 2.2

Life in the dark

A less well-known type of autotroph is the chemoautotroph. In oceans, below a depth of 200 m, limited-to-no sunlight penetrates, making it impossible for photoautotrophs to photosynthesize. However, chemoautotrophs can sometimes carve out an existence. Chemoautotrophs get their energy from chemical processes, rather than sunlight, and have been found around openings in the deep ocean floor, out of which heated water flows, called hydrothermal vents. Near these vents, the water is rich in dissolved minerals and supports chemoautotrophic bacteria. These bacteria can capture energy through a complex chemical reaction and form the foundation of this deep ocean food web (see Figure A).

a)

b)

Figure A An example of an ecosystem near deep ocean vents. These systems depend on chemoautotrophs, which convert chemical energy to organic components.

Source: (a) OAR/National Undersea Research Program (NURP), NOAA; (b) NOAA Okeanos Explorer Program, Galapagos Rift Expedition 2011.

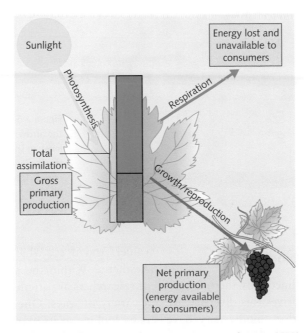

Figure 2.3 Schematic representation showing the difference between gross primary productivity (GPP) and net primary productivity (NPP).

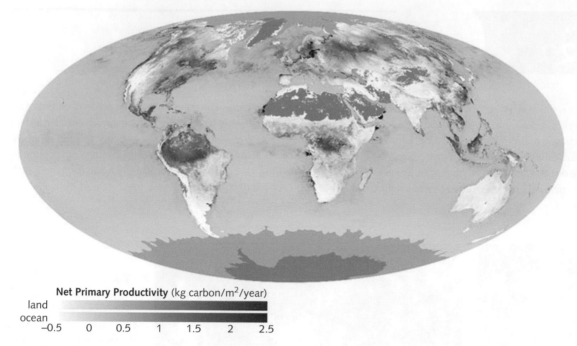

Net Primary Productivity (kg carbon/m^2/year)

land

ocean

-0.5   0   0.5   1   1.5   2   2.5

Figure 2.4 Global differences in net primary production of terrestrial and aquatic ecosystems, based on satellite data.
Source: © NASA Earth Observatory; NASA Earth Observatory.

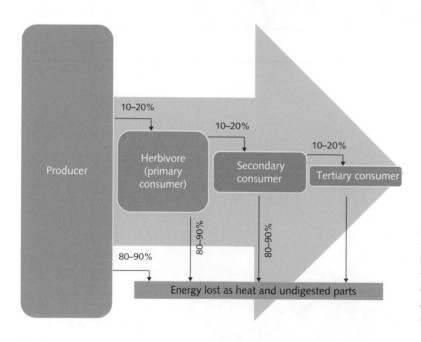

Figure 2.5 Energy flows in a food chain, with the arrow showing an increase in trophic level. In the conversion of energy between trophic levels, only 10–20% of the chemical energy is transferred to the next trophic level and much is lost as low-quality heat and undigested parts.

2.1.4 What is the value of biodiversity?

Now that we have briefly outlined why there is such a diversity of life, and what drives this diversity, we move to the next question: does biodiversity have value? And if so, what kind of value? Monetary? Aesthetic? Within the scientific community, groups of researchers are attempting to value biodiversity in a variety of ways. One way of valuing biodiversity is by looking at its extrinsic value. This is the value that organisms or ecosystems provide to animals and human beings. **Ecosystem services** are a key example of the extrinsic value of biodiversity, services, and benefits that people obtain from ecosystems and that are provided by the natural environment. Classic examples are bees pollinating our crops or earthworms recycling nutrients. There are four broad categories of ecosystem services:

- **Provisioning:** for example, the production of food and water
- **Regulating:** for example, the control of climate and disease
- **Supporting:** for example, nutrient cycles and crop pollination, and
- **Cultural:** for example, spiritual and recreational benefits.

Researchers have tried to express the value of these services in monetary units. For example, Costanza et al. (2014) estimated the combined value of ecosystem services in 2011 to be 125–145 trillion US dollars per year. For comparison, according to the World Bank, the global economic output in 2011 was 73 trillion US dollars per year. This value is, as acknowledged by Constanza et al. (2014), very high and is argued by these authors as a *'useful way to highlight, measure, and value the degree of interdependence between humans and the rest of nature'.*

Biodiversity contributes to ecosystem services, and the loss of biodiversity, be it genetic, species, or ecosystem diversity, can therefore lead to a direct loss in ecosystem services. For example, 70% of the main crops grown globally depend on pollination by insects such as bees (Klein et al., 2007). This ecosystem service is valued at 153 billion US dollars per year globally (Gallai et al., 2009). Losing this service can therefore have serious impacts for humans. Recently, there have been dramatic declines in some populations of bees, raising the possibility of losing this vital ecosystem service (one potential cause has been the use of pesticides, as discussed in Section 3.2.2 in *Chapter 3: Pollution*). Similarly, land use changes between 1997 and 2011 have been estimated to have caused a total monetary loss of between 4.3 and 20.2 trillion per year

(Costanza et al., 2014). This highlights the importance of protecting biodiversity to ensure future ecosystem services.

Pause and think

Based on the concept of ecosystem services, we can put price tags on different components of biodiversity. As a result, some components might have a higher price tag, compared to others. Can you think of any problems this could cause for biodiversity conservation?

However, we do not have to value biodiversity for its extrinsic usefulness only; it may also have value regardless of how it affects humans or ecosystems. Such value is referred to as the intrinsic value of biodiversity. Many people value knowing that pandas exist in China or the Amur leopard in the Himalayas, even though they may never see them face-to-face and they provide limited services to humanity. If we argue that biodiversity has intrinsic value, discussions about the moral obligations that humans have to protect biodiversity become important (Markku, 1997). We will discuss ethics in relation to food in *Chapter 15: Summary* in Food controversy 15.1.

Regardless of how we value biodiversity, it plays a key role in **ecosystem functioning**. Ecosystem functions are ecological processes that control the fluxes of energy, nutrients, and organic matter in the environment (Cardinale et al., 2012). Examples of ecosystem functioning are diverse and include the cycling of nutrients in food webs and the production of plant tissue by capturing sunlight. Changes in biodiversity can affect ecosystem functioning and ultimately ecosystem stability (Hautier et al., 2015). For example, a recent study demonstrated that biodiversity increases the resistance of ecosystem productivity to climate extremes (Isbell et al., 2015). A global synthesis revealed biodiversity loss as a major driver of ecosystem change (Hooper et al., 2012), potentially resulting in loss of ecosystem function and, with this, ecosystem services.

2.2 How is food production related to biodiversity loss?

Even though biodiversity has both intrinsic and extrinsic value, as explained in Section 2.2, we are losing it at an alarming rate. This section focuses on how human activities related to food production affect biodiversity. The main threats to biodiversity can be remembered by using the acronym HIPCO:

- Habitat loss and fragmentation
- Invasive species
- Pollution
- Climate change
- Overexploitation.

It is important to realize that population growth and increasing use of resources are a *driver* of the other threats—the more our human population grows, and the more resources individuals use, the larger the threats become (see *Chapter 1: Introduction*). In other chapters, we discuss impacts of pollution (see *Chapter 3: Pollution*) and climate change (see *Chapter 6: Climate Change*) on biodiversity and the environment in general. In the remainder of this chapter, our focus is on the impacts of habitat loss and fragmentation, invasive species, and overexploitation on biodiversity. At the end of this chapter, we come back to the important driver of increasing human population and resource use.

2.2.1 Habitat loss and fragmentation

Humans now dominate the use of the terrestrial ecosystem. In 1700 AD, approximately 5% of global ice-free land area was used by humans. This increased to 10% in 1800, 26% in 1900, and 55% in 2000. At the same time, the area of wild and semi-natural vegetation (for example, residential woodlands) has sharply decreased—in 1700, nearly 50% of the ice-free land surface was covered by wildlands, and by 2000, only 25% remained (Ellis et al., 2010). Currently, croplands and pastures cover approximately 40% of the terrestrial surface of our planet (Foley et al., 2005). Deforestation of natural vegetation for agricultural production still occurs in many regions of the world, including the Amazon, the USA, the European Union, and Indonesia (see Case study 2.1 on palm oil plantations in Indonesia).

As humans are transforming natural habitats into pastures and croplands, they also dominate the use of NPP within these systems. We can express this proportion as **human appropriated net primary production (HANPP)** (see Box 2.3), which is an

CASE STUDY 2.1

Palm oil: an economic success story, but a biodiversity nightmare

What do lipstick, margarine, chocolate, biofuel, shampoo, ice cream, cookies, and pizza dough have in common? All of them can, and often do, contain palm oil as an ingredient. Palm oil is an edible vegetable oil derived from oil palms (see Figure A). Compared to animal and other plant-based oils, it is relatively cheap, and smooth and solid at room temperature. The total area of oil palm plantations has increased sharply over the last decades—from 36,000 km² in 1961 to 100,000 km² in 2000 and 170,000 km² in 2012 (Koh and Wilcove, 2008; Pirker et al., 2016). To put this in perspective, 170,000 km² is over twice the surface area of Austria. The majority of palm oil production occurs in Indonesia and Malaysia (Gaveau et al., 2014), with a total revenue of $40 billion in 2012 (*Scientific American*, 2012).

There is, however, a growing global concern about the sustainability of oil palm production. One key issue is the replacement of tropical rainforests for industrial oil palm plantations. Many of these areas are biodiversity hotspots, containing significant levels of biodiversity (see Section 2.3). For example, it has been estimated that in 1973, 75.7% of Borneo (558,060 km² of Borneo's area of 737,188 km²) was covered by rainforest (Gaveau et al., 2014) (see Figure B.a). By 2010, the forested area had declined by 168,493 km² (or 30.2% of the forested area present in 1973; see Figure B.b).

a)

b)

Figure A (a) An oil palm plantation in Indonesia. (b) An oil palm bearing fruits.

Source: (a) Achmad Rabin Taim/CC BY 2.0; (b) tgerus/Flickr/CC BY-SA 2.0.

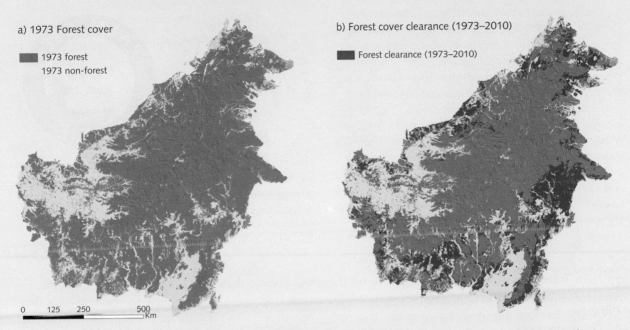

a) 1973 Forest cover

■ 1973 forest
  1973 non-forest

b) Forest cover clearance (1973–2010)

■ Forest clearance (1973–2010)

0   125   250   500
                        Km

Figure B Four decades of forest persistence, clearance, and logging on Borneo. (a) Forest (dark green) and non-forest (grey-white) in 1973, and (b) forest loss during 1973–2010 in red. Approximately 40% of the red area has been turned into industrial oil palm plantations.

Source: Gaveau DLA, Sloan S, Molidena E, Yaen H, Sheil D, Abram NK, et al. (2014) Four Decades of Forest Persistence, Clearance and Logging on Borneo. *PLoS ONE* 9(7): e101654. © 2014 Gaveau et al.

To put this in perspective, this is four times the size of the Netherlands or two-thirds of the surface area of the UK. Of the 168,493 km² of tropical rainforest lost between 1973 and 2010, a total of 64,943 km² was converted into industrial oil palm plantations. Of the remaining intact forest, 42–58% is designated to be logged or to undergo land use change (Gaveau et al., 2014).

The switch from rainforest to oil palm plantation significantly affects biodiversity in several ways. It reduces the habitat available for, and may drive the demise of, critically

endangered species such as the Bornean orangutan and the Asian elephant. A recent study found that, between 1999 and 2015, more than half of the population of orangutans on Borneo has been lost, with the most severe declines when their natural habitat was removed (Voigt et al., 2018).

In addition, oil palm plantations are species-poor communities, compared to primary forests. On average, they contain fewer than half of the vertebrate species (Fitzherbert et al., 2008) and have few forest species (Danielsen et al., 2009). Across many studies, the number of bird, lizard, and mammal species was lower in oil palm plantations, compared to forests. Finally, the monoculture of oil palm trees on the plantations (see Figure A.a) creates lower habitat heterogeneity and niche availabilities than in the primary forest. This, at least partly, drives the lower biodiversity in these plantations. This is interesting, as oil palm plantations and associated production can benefit from the ecosystem services provided by the biodiversity found in adjacent tropical rainforests. For example, the presence of large birds of prey can be an effective way to combat rats on plantations.

Because of the increased societal concern about oil palm plantations, several initiatives have started to inform the public about products which contain palm oil. For example, a group of stakeholders, including oil palm producers, consumer goods manufacturers, banks/investors, and environmental and social non-governmental organizations (NGOs), have started the Roundtable on Sustainable Palm Oil (RSPO). The RSPO has developed '*a set of environmental and social criteria which companies must comply with in order to produce Certified Sustainable Palm Oil (CSPO). When they are properly applied, these criteria can help to minimize the negative impact of palm oil cultivation on the environment and communities in palm oil-producing regions*'. One key criterion is that '*no primary forests or areas which contain significant concentrations of biodiversity (e.g. endangered species) or fragile ecosystems, or areas which are fundamental to meeting basic or traditional cultural needs of local communities (high conservation value areas), can be cleared*'. Producers can obtain certification and use a logo on their products (see Figure C.a).

Other organizations promote products that do not contain palm oil. For example, the International Palm Oil Free Certification Accreditation Programme aims to '*enable consumers, who wish to avoid palm oil for allergy, dietary or ethical reasons, to see at a quick glance if a product is palm oil free*'. Producers who meet the certification criteria can put a logo on their products (see Figure C.b).

An additional concern is that many palm oil plantations have been developed in areas previously covered by peatland forests. Once logged, the areas need to be drained, which results in high greenhouse gas emissions to the atmosphere. A more detailed discussion on the stakeholders involved in this global food system and the advantages and disadvantages of certification of products can be found in Case study 12.2 in *Chapter 12: Food Systems*.

a)   b)

Figure C Two examples of trademark logos to raise consumer awareness: (a) products which contain more sustainably produced palm oil, and (b) palm oil-free products.

Source (a) © Roundtable on Sustainable Palm Oil; (b) © Palm Oil Free Certification Accreditation Programme (POFCAP).

indicator for the impact of human land use on the available biomass in ecosystems (Haberl et al., 2007).

HANPP has roughly doubled over the last century. In 2007, HANPP was estimated at 24%. This means that approximately one quarter of the energy captured by terrestrial plants was used by humanity. HANPP is projected to increase to 29–44% by 2050 (Krausmann et al., 2013). HANPP differs dramatically between regions. In many parts of Europe and Asia, HANPP is very high and close to 100% of the potential NPP. In other regions, HANPP is relatively low; for example, in most of the Amazon basin, HANPP is less than 10% of the potential NPP (Haberl et al., 2007).

The overall increase in HANPP has direct effects on biodiversity, because in many agricultural areas, the available natural habitats and niches are often less diverse than in natural areas. For example, review studies found a significantly lower species richness in oil palm plantations, compared to the original vegetation (see Case study 2.1). Second, if humans use a large amount of the available NPP, less energy is available for other organisms, which may result in lower biodiversity (Haberl et al., 2004).

Land use change not only changes the energy dynamics, but it also affects the biodiversity within agricultural lands. In some regions, the conversion of natural habitats into cultivated landscapes has resulted in a change in the type of species: from species that were associated with the natural habitat, to agrobiodiversity. **Agrobiodiversity** is defined as the variety of animals, plants, and microorganisms that are used directly (crops, livestock) or indirectly (non-harvested species that support agro-ecosystems

BOX 2.3

HANPP: human appropriated net primary production

Imagine a piece of tropical rainforest in the Amazon, which is being removed and replaced by a soybean plantation. This is one of the many ways humans are transforming natural habitats to pastures and cropland for food production.

The natural (or potential) vegetation within an ecosystem has a potential NPP (NPP_0; see Figure 2.4). NPP_0 depends on a variety of biotic and abiotic factors, including temperature, nutrient load, water availability, and soil characteristics. The NPP_0 of a piece of tropical rainforest is relatively high, as can be seen in Figure 2.4. When humans clear this natural vegetation for agriculture, we replace it with our crops (soybean plants in our example). This now has become the actual vegetation within the system, with a different NPP production (the actual NPP, or NPP_{act}). Part of this NPP_{act} is harvested by the farmer (NPP_h; in our example, the beans), while some of the vegetation remains on the field (NPP_t; the remainder of the plant).

In addition, the full potential of the NPP might have changed due to the land use change (NPP_{luc}). For example, the tropical rainforest from our example has an overall higher NPP, compared to a field of soybean. NPP_{luc} is calculated by subtracting NPP_{act} from NPP_0. HANPP (the amount of NPP within the ecosystem used/affected by humans) then is the sum of NPP_h and NPP_{luc} (Figure A). HANPP can also be calculated by using the following formula: HANPP = $NPP_0 - NPP_{act}$.

It is important to realize that NPP_{act} can be both lower and higher than NPP_0. For example, NNP_{act} can be greater than NPP_0 when energy crops are being planted in an area with a low NPP_0 or by using fertilization and irrigation in marginal habitats. For example, in some deserts, the use of fertilizers and irrigation allows for agriculture where plants would never survive on their own (Figure B); due to the technology used, NPP_{act} in these cases is higher, compared to NPP_0.

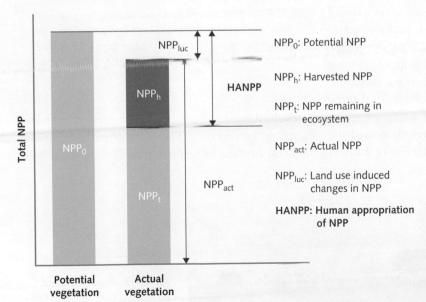

NPP_0: Potential NPP

NPP_h: Harvested NPP

NPP_t: NPP remaining in ecosystem

NPP_{act}: Actual NPP

NPP_{luc}: Land use induced changes in NPP

HANPP: Human appropriation of NPP

Figure A A schematic representation on how to quantify human appropriation of net primary production (HANPP), in relation to the potential NPP of the original vegetation.

Source: Niedertscheider M, Kuemmerle T, Müller D, Erb KH. Exploring the effects of drastic institutional and socio-economic changes on land system dynamics in Germany between 1883 and 2007. *Glob Environ Change*. 2014 Sep; 28:98-108. Copyright © 2014 The Authors. Published by Elsevier Ltd.

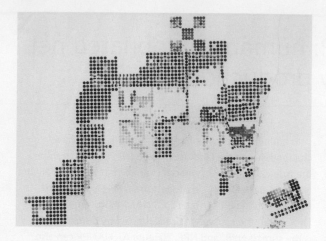

Figure B In a few cases, the actual NPP of a system (NPP_{act}) is higher, compared to the NPP of the natural vegetation (NPP_0). Growing crops in a desert using fertilizers and irrigation is one such example. This picture shows crop circles in the middle of the Sahara Desert.

Source: NASA Visible Earth.

and biocontrol) to support agricultural ecosystems. Agrobiodiversity provides important ecosystem services such as crop pollination, pest control, soil biota maintenance, and carbon dioxide (CO_2) capturing.

Land use change (i.e. turning natural vegetation into agricultural lands) is one way to produce more food to meet humanity's growing demands. An alternative is agricultural intensification, with the goal of producing higher yields. Agricultural intensification is implemented in different ways around the globe, but common practices include water management (including irrigation and drainage), simplification of crop rotation schemes, growing monocultures of highly profitable crops, mechanization of harvesting, and the use of artificial fertilizers and pesticides.

These intensification methods have decreased the heterogeneity of agrobiodiversity (see Figure 2.6). As

Connect the dots

The availability of cheap and readily available energy in the form of fossil fuels has been a driving force in agricultural intensification. Think about the sandwich you might be eating for lunch, or the bowl of rice that you will eat for dinner, or the banana you are having for a snack. Where and how do you think energy is used in the production process? How is energy used during the agricultural production? And how much in the processes afterwards (such as food processing and packaging)? We will explore the use of energy in agriculture in *Chapter 6: Climate Change* and *Chapter 7: Energy*.

a result, there has been a great reduction in overall farmland biodiversity. There are often fewer species of bees, beetles, hoverflies, true flies, spiders, birds

a)

b)

Figure 2.6 Two different approaches to growing crops: (a) intercropping different crops (coffee and tomatoes), while (b) in monoculture, only one crop is grown (soybeans).

Source: (a) Neil Palmer (CIAT)/Flickr/ CC BY-SA 2.0; (b) José Reynaldo da Fonseca/Wikimedia Commons/CC BY-SA 3.0.

CASE
STUDY
2.2

How intensification is resulting in the loss of the Dutch national bird

The Netherlands is the European stronghold of the black-tailed godwit (Figure A) where 40% of the entire European population is estimated to breed. Until the Middle Ages, black-tailed godwits were probably not very abundant. During and after the Middle Ages farmers transformed bogs, swamps, and fens into grasslands to support dairy farming. This proved to be an ideal habitat for the black-tailed godwit. In the early twentieth century, godwits initially profited from agricultural intensification, as the usage of artificial fertilizers increased food availability and habitat quality.

However, with further agricultural intensification, population numbers stagnated in the late 1960s and then started to decline. Declines in the population were driven predominantly by the lowering of groundwater tables, increased sowing of fast-growing grass species, and higher fertilizer usage. For example, lower groundwater tables negatively affect the availability of prey items for godwit adults and chicks during the breeding season, leading to less body fat and food reserves and ultimately higher mortality of godwit chicks.

Agricultural intensification also resulted in earlier and more frequent mowing during spring. This new mowing regime decreased food abundance in the freshly mown fields, and since the timing of breeding of black-tailed godwits has remained the same, chick mortality through starvation

Figure A The black-tailed godwit in a Dutch meadow.

Source: Krijn Trimbos.

increased. Earlier mowing also reduces the cover for chicks, significantly increasing the risk of chick predation.

As a result of these pressures, the Dutch population of black-tailed godwits has declined significantly from 120,000 breeding pairs in the 1960s to approximately 30,000 today. This decline is also observed in other European countries. Consequently, the IUCN has categorized the species as *near threatened*.

(see Case study 2.2), plants, and mammals in these systems (Stoate et al., 2009).

2.2.2 Invasive species

We now move to the second threat on the HIPCO list that is covered in this chapter—**invasive species** (also called alien or non-native species). These are '*organisms (usually transported by humans) which successfully establish themselves in, and then overcome otherwise intact, pre-existing native ecosystems*' (UN, 2016). Human activities, such as agriculture and aquaculture, can promote the spread of invasive species

(Kolar and Lodge, 2001). Invasive species can have an impact on biodiversity in many ways, including decreasing the abundance of native species through competition or predation (Blackburn et al., 2004; McGeoch et al., 2010).

Many species have been introduced accidentally. For example, the ballast water used by large ships to balance their loads in the water is often loaded and unloaded by ships within or near ports. Marine organisms can be pumped into the ships by accident when they load ballast water, and in this way, these organisms can be transported from port to port worldwide. In some cases, these organisms establish themselves and thrive in the new environment, becoming an invasive species. This is thought to have happened with the spread of zebra mussels, which are native in Europe but are now found in the USA and Canada where they are classified as an invasive species and clog the waterways and water pipes of cities. The Ballast Water Management Convention, adopted in 2004 and enforced from 2017, aims to prevent the spread of harmful aquatic organisms from one region to another by establishing standards and procedures for the management and control of ships' ballast water and sediments.

In other cases, invasive species have been intentionally introduced for food production and economic gain. An example of this is the Nile perch in Lake Victoria in East Africa. The Nile perch, which can grow to more than 2 m and weigh up to 200 kg (see Figure 2.7), was introduced into the lakes for commercial exploitation in the 1950s, and became an important export product (Njiru et al., 2008). However, Lake Victoria is also home to over 500 species of cichlids, an endemic group of fish species. After the introduction of the Nile perch, the population sizes of native cichlid species started to decline rapidly. The decline was initially attributed to the introduction of the Nile perch into the lake. However, more recently, a variety of other factors have also been linked to the decline (see Box 2.4).

2.2.3 Overexploitation

A final threat covered in this chapter is overexploitation of plant and animal stocks for food production, which has resulted in the reduction, and even extinction, of many species. Famous extinctions include the dodo, American buffalo (see Figure 2.1a), and passenger pigeon.

Although most terrestrial plants and animals are managed (more) sustainably these days, we are still facing a great challenge related to the exploitation of fish stocks at a global level. In 2009, the United Nations' FAO estimated that 57% of all global fish stocks are fully exploited and 30% are overexploited, leaving a small 13% which are not fully exploited. As the demand for fish is growing at a global level, fishing pressures are increasing further. For example, marine fish landings in 2012 were estimated to be 80,000,000 tonnes, an increase of 400%, compared to the 1950s.

Fisheries tend to first remove the large fish from an ecosystem, which are often at a high trophic level.

Figure 2.7 The Nile perch was introduced in Lake Victoria and has caused direct and indirect effects on the native fish species.
Source: smudger888/CC BY 2.0

If this is done unsustainably (which it often is), the total landings of fish at the higher trophic levels start to diminish. As society's demand for fish continues, fish at lower trophic levels will subsequently be targeted—resulting in the term '*fishing-down-the-food-web*' (Pauly et al., 2002). Removing species at the top of the food web (top predators) like this, not only reduces biodiversity, but can also result in additional unforeseen adverse consequences (see Box 2.5). These issues pose specific problems in governance called 'collective action problems', which are discussed in Case study 14.2 in *Chapter 14: Collective Action*.

BOX 2.4

Do not blame it all on the Nile perch

The introduction of the Nile perch has undoubtedly contributed to the declines in naturally occurring fish (mainly cichlids) populations in Lake Victoria. However, several other factors have likely contributed to the declines. As Nile perch fisheries grew over time, more people moved to live by the lake to work there (Figure A.a). Forests around the lake were cut down for firewood, homes, and agricultural plots (Figure A.b). The lack of forest resulted in increased sediment run-off into the lake (see Section 5.2.1 in *Chapter 5: Soils*). This soil erosion increased the turbidity of the water (making it murkier).

Cichlids in Lake Victoria have undergone rapid speciation over the last 10,000 years. Since this is a relatively short period for evolution processes, many of the cichlid species can still interbreed. However, under natural circumstances, this is prevented because they select their mates using visual cues and only select those of their own species. As the water turbidity increased, the fish had difficulties distinguishing their mates. This resulted in interbreeding between species, thereby losing some of the unique genetic diversity within the lake. The introduction of the Nile perch, as well as land use change and pollution, have all contributed to the decline in diversity. This shows the complex interplay between different threats and biodiversity.

a)

b)

Figure A There has been a significant increase in population around Lake Victoria (a), and many work in the fishing industry. This increase has resulted in significant land use change, transforming forests to agricultural plots, for example to keep cattle (b).

Source: (a) Sarah McCans/Flickr/CC BY 2.0, (b) Marc Veraart/Flickr/CC BY-ND 2.0.

BOX 2.5

Unforeseen cascades

Sometimes overexploitation can result in unforeseen changes in food webs and ecosystems. One example is the shark fisheries found along the east coast of the USA (Myers et al., 2007). Over the last few decades, an increased demand for shark fins in China has resulted in a global surge in shark fishing. As a result, many shark populations are now overexploited and even depleted; along the coast of North Carolina, populations decreased by over 99% since 1972.

When the numbers of top-predators decrease, the pressure on their prey reduces, resulting in increased prey populations. In North Carolina, the population of cownose ray has increased by at least an order of magnitude since the mid 1970s. The diet of this species largely depends on commercially important shellfish such as bay scallops. As the population of cownose rays increased, the levels of these bivalves declined, significantly impacting the local scallop industry.

2.3 **How to conserve our biodiversity?**

Having covered three main threats to biodiversity (habitat loss, invasive species, and overexploitation), we now discuss how biodiversity can be conserved. In practice, there are two levels at which biodiversity conservation takes place: the 'species approach' and the 'ecosystem approach'. The **species approach** focuses on areas which hold particular species, for instance rare species or **keystone species**. Keystone species are species that have a large impact on their environment relative to their abundance, thereby playing a key role in maintaining the structure and functioning of ecosystems. By protecting keystone species, other species generally are also protected indirectly.

Species conservation tries to stabilize declining species populations to prevent extinction. Conservation of a species starts with gathering data on the population trends of that species; is the population size increasing, declining, or stable? This is generally done with monitoring programmes to estimate the number of individuals or density of a certain species over time. The most complete compilation of global population trends across vertebrate taxa is contained in the annually updated *Living Planet Report*. The average population abundance of the monitored species was reduced by a staggering 60% between 1970 and 2014.

Together with knowledge on the species' ecology and the identification of the key threats that contribute to its decline, policymakers can devise appropriate conservation strategies for a given species. Strategies can be aimed at reducing the threats, increasing the population (for example, by captive breeding and release programmes or translocations) or by enabling conditions that enhance the effectiveness of conservation (including increasing public awareness or introducing policy measures).

The **ecosystem approach** targets conservation at the level of communities, habitats, or entire ecosystems, rather than individual species. This approach assumes that conservation money is better spent on an entire ecosystem than individual species, as this holds the possibility of preserving many species simultaneously. It also has the added advantage that it can help maintain specific ecosystem services within a certain area. At the largest scale, the ecosystem approach may focus on **biodiversity hotspots**. As there are limited funds for protection of biodiversity, a focus on hotspots aims to optimize biodiversity conservation by protecting areas with exceptionally high biodiversity (see Figure 2.8).

A total of 35 hotspot regions have been identified. While covering only 2.3% of the world's terrestrial systems, these hotspots contain a staggering 50% of vascular plant and 42% of amphibian, reptile, bird, and mammal species (Mittermeier et al., 2011). For instance, the Mediterranean Basin is a hotspot that contains an estimated 22,500 plant, 497 bird, and 216 mammal species. Another example is the Tropical Andes, which hosts 30,000 plant, 1728 bird, and 1095 amphibian species (Mittermeier et al., 2011).

a)

b)

Figure 2.8 The tropical rainforest in Borneo (a) and the coral reefs in the Carribean Sea (b) are two examples of biodiversity hotspots.
Source: (a) Dukeabruzzi/Wikimedia Commons/CC BY-SA 4.0; (b) Jerzy Strzelecki/Wikimedia Commons/CC BY 3.0.

Pause and think

In 2013, Timothy Lavin wrote a column entitled *'Why I hate pandas and you should too'*. In this column, he concludes that *'Darwinism isn't for crybabies. And conservation requires making tough choices. Pandas had a pretty good run for 3 million years. All that money is better spent on preserving diverse habitats rather than on a single hopeless species.'* What do you think? Is this a fair statement?

One way to conserve biodiversity is through protected areas and reserves which can be chosen and managed using the approaches listed previously. Normally, protected areas are purchased by governments, municipalities, private individuals, or conservation organizations. Most often, limited to no agricultural activities are allowed in these areas. The United Nations Environment Programme (UNEP) estimates that almost 15% of all terrestrial and inland water areas were protected in 2014. In contrast, only 3.4% of the global oceans are protected, lagging behind the protection within territorial waters (although there are plans to drastically increase this).

2.3.1 How to conserve biodiversity within agricultural landscapes?

Protected areas benefit a wide range of species. However, the large majority of our planet's surface remains unprotected. Our cultural and agricultural landscapes (crop and rangelands), including traditional farmlands, selectively logged forests, and lightly grazed grasslands, as well as commercial fishing grounds, host a significant amount of global biodiversity.

Despite the magnificent biodiversity present on Earth, some 90% of calories, protein, fat, and weight around the world are provided by 94 plant species (including plant and animal food sources) (Khoury et al., 2014); moreover, around 90% of all domestic livestock production involves only 15 mammal and bird species. However, the production of these commodities is reliant on many thousands of other species that provide ecosystem services. These include the insect and bird species involved in pollination and those species (including frogs, wasps, ladybugs, worms, and many more) that provide

biological control by feeding on crop pests (see Section 3.3 in *Chapter 3: Pollution*). Biodiversity in the soil, including small bugs, fungi, and microbes, is crucial in maintaining the soil's health and the world's carbon and nitrogen cycles on which our production crops depend.

When focusing on agricultural lands, there is an ongoing debate over whether the best way of conserving biodiversity is achieved through **land sharing**, where the objectives of conserving biodiversity and wildlife-friendly food production are integrated in the same landscape, or through **land sparing** where land is segregated. In this latter approach, high-yield agriculture is combined with the protection of natural habitats elsewhere. Particularly in already existing agricultural landscapes, this method can include taking agricultural land out of production and turning it into nature reserves. Such actions are often highly controversial and can meet fierce opposition by local farmers and communities who see themselves deprived of their livelihood and, maybe more importantly, a way of life (see Food controversy 2.1).

While agricultural intensification may allow for land sparing in some places, it does not always lead to improved conservation situations (Phalan et al., 2011). It is not always practical either, as worldwide, small-holder farms embedded in a cultural and natural landscape are more commonplace and contribute more to global food production than large-scale farming (Tscharntke et al., 2012). One important consideration is that biodiversity in productive lands is important to maintain, because it performs crucial functions and ecosystem services for food production. This has been one of the founding principles of nature-inclusive agriculture (see Case study 2.3).

The ecosystem services that all these species provide for food production may be best maintained in a landscape where agriculture is practised in such a way that it enhances biodiversity. Practices, landscape configurations, and solutions will differ by region. Local farmers and the broader community are often incorporated in conservation efforts, for instance by providing education, setting up and maintaining monitoring projects, and providing funding for conservation projects.

In many cases, farmers are concerned that protecting biodiversity will come at a financial cost.

FOOD CONTROVERSY 2.1

Intensification, extensification, or rewilding . . . Implications for biodiversity, society, and food production

There is an ongoing debate among stakeholders, including scientists, NGOs, and politicians, on the best strategy to conserve biodiversity in agricultural landscapes. Maintaining biodiversity in agricultural landscapes is important, as non-agricultural species (think bees and flies as pollinators, but also earthworms and birds) provide many of the ecosystem services that positively affect agricultural yield, and because most biodiversity occurs outside already protected areas. The proposed paths forward can differ considerably, ranging from building industrial high-rises for intensive pig farming to permaculture (see Case study 2.3) or turning marginal agricultural lands back to nature (a concept called rewilding).

At the heart of this debate is the question of how to produce sufficient food for our growing population while maintaining biodiversity. Imagine we need to produce a certain amount of food within a country to meet demand, but at the same time, we want to protect biodiversity within the landscape. There are two extreme strategies to reach both these goals. The first is land sparing, in which high-yield, intensive agriculture on a relatively small area of land is combined with nature areas. The second—land sharing—is more extensive, nature-friendly agriculture, in which the yields per unit area are lower, but biodiversity is higher. In both of these scenarios (see Figure A), we can meet food demands, while at the same time stimulating biodiversity.

So why is this a food controversy? They both seem reasonable solutions to achieve the same goal—enough food and protection of biodiversity. The problem is that either solution will affect the livelihood of the local communities, and some forms of agriculture create a strong response among certain public groups. For example, if we move towards land sparing, one option would be to return certain agricultural lands back to nature areas, a concept which has recently been coined **rewilding**. This method is popular among many conservationists but can result in strong, negative emotions, especially among the rural communities. Agricultural lands, sometimes in use for centuries and in use by the same family for many generations, are returned to nature. The farming community feels this directly impacts their livelihoods and their communities. At the same time, with food demand predicted to continue to rise, other agricultural areas will need to make up for the loss in agricultural yield in the no-longer productive area.

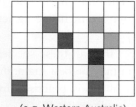
(e.g. Western Australia)

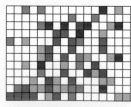
(e.g. Northern Europe)

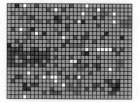
(e.g. Coto Brus, Costa Rica)

Coarse grain and abrupt change
('Land sparing')

Fine grain and spatial continuity
('Wildlife-friendly farming')

Figure A A schematic representation of land sparing and land sharing (or wildlife-friendly farming). Although the debate has polarized, there are also intermediate solutions, which include a combination of land sharing and land sparing.

Source: Based on Fischer, Joern & Brosi, Berry & Daily, Gretchen & R Ehrlich, Paul & Goldman, Rebecca & Goldstein, Joshua & B Lindenmayer, David & Manning, Adrian & A Mooney, Harold & Pejchar, Liba & Ranganathan, Jai & Tallis, Heather. (2008). Should Agricultural Policies Encourage Land Sparing or Wildlife-Friendly Farming. *Frontiers in Ecology and The Environment - FRONT ECOL ENVIRON.* 6. 380-385. 10.1890/070019. By permission of John Wiley and Sons.

One example of a rewilding project that resulted in heated public debate was the removal of a polder on the border of the Netherlands and Belgium. A polder is a form of land reclamation from a body of water. The Dutch, and several other countries, have used this form of reclamation over the last centuries to substantially increase their productive land area. In 2008, it was decided to permanently flood the polder, to provide more space for the local estuarine and river ecosystems. This meant, in effect, that it was compulsory for the local landowners (mainly farmers) to sell their land to the government. This resulted in a major uproar, which involved many politicians, public organizations, and NGOs. Several times, the decision was overturned, but the most recent verdict, by the Dutch supreme court, decided to approve the rewilding project (see Figure B).

Rewilding as a concept might be relatively new, but similar processes have happened many times in history. A good example is the displacement of Maasai from their traditional herding grounds in favour of the establishment of national parks. In these cases, there was limited or no compensation for the local community, and related conflicts still continue today (Goldman, 2011).

a)

b)

Figure B The Hertogin Hedwigepolder (a) in the Netherlands is currently agricultural land but will, in the future, potentially be rewilded and joined with the adjacent nature reserve (b).

Source: (a) Daniel van der Ree/Wikimedia Commons/CC BY-SA 3.0; ; (b) Rijkswaterstaat / Joop van Houdt.

Integrated Conservation and Development Projects (ICDPs) link the safeguarding of species to financial opportunities. Agri-environmental schemes (AES) are an example of ICDPs (see Box 2.6); the aim of these schemes is to protect biodiversity in agricultural landscapes. If this protection results in the loss of productivity, the farmer gets a compensation. While short-term losses may be associated with some biodiversity conservation practices, others can be profitable, and the view that biodiversity also provides benefits (or indeed is essential) to farmers is gaining traction.

Ultimately, whether land sharing or land sparing (or often a combination between the two) is more productive within a certain region depends on a variety of factors, including the ecological, social, and economic situation.

 Pause and think

As discussed in Food controversy 2.1, there is an active debate on the best way to protect biodiversity while increasing food production to meet global demand. As the human population will continue to grow and get richer, we need to make some important decisions on how to continue to produce food and conserve biodiversity at the same time. Do you think land sparing or land sharing is the best option to conserve biodiversity? Provide arguments to defend your stance.

CASE
STUDY
2.3

Nature-inclusive agriculture: a viable option to conserve biodiversity in agricultural systems?

By Dr Eveline Stilma, http://www.innoplant.nl

Modern agriculture is characterized by high-yield monocultures which are driven by high inputs of agrochemicals and energy. As we explored in this chapter, this affects biodiversity. One of the key challenges facing future agriculture is finding a way to protect biodiversity, while maintaining high levels of food production. Here we will explore the concept of nature-inclusive agriculture (land sharing), with a focus on permaculture and polyculture (polyculture is the reverse of monoculture where many plants are grown together, rather than one).

Nature-inclusive agriculture is based on a resilient food and ecosystem, in which biodiversity provides key ecosystem services. **Ecosystem resilience** refers to an ecosystem that is able to withstand environmental pressures and can recover quickly from any disturbances. One example of nature-inclusive agriculture that has gained momentum over the last decades is permaculture, which was developed by Bill Mollison and David Holmgren in 1978. Permaculture farms are designed to be self-sustaining in energy, water, and food. Importantly, the food production system on a permaculture farm is inspired by nature, a concept called **biomimicry**. Biomimicry can apply to products and structures (for example, Velcro tape is based on hooks found on certain seeds), but in the context of permaculture, biomimicry is the imitation of natural systems within food production.

In permaculture systems, many different plant and animal species are grown together. For example, nitrogen-fixing trees can be used for fertilization (see Box 3.6 in *Chapter 3: Pollution*), while their branches are sold as firewood. Pigs are used to eat fallen apples from the ground and clear the soil against harmful fungi and bacteria which can develop in decomposing fruits. Chickens can be used for insect and weed control. Many permaculture farmers use plants that are adapted to the local environment and are resistant to local diseases.

With the growing momentum of nature-inclusive agriculture, studies are now investigating the benefits these systems may hold for biodiversity. For example, in many tropical mountainous regions, coffee is grown beneath the shade of large trees (see Figure A). Often multiple species from

Figure A An example of a shaded coffee plantation in Mexico.

Source: Photo: 'Cafetal', by Glenn Jampol, Finca Rosa Blanca Coffee Plantation Resort.

adjacent tropical forests are present, as is the case in certain regions in Mexico. Research has shown that some species do better in the plantation, compared to the pristine forest (for example, beetles thrive), and some are not impacted (for example, bats), while for other species, there was less biodiversity in the plantation, compared to the forest (for example, frogs) (Pineda et al., 2005). More importantly, these coffee plantations connect the remaining forest fragments, thereby forming a '*highly functional resource for the preservation of biodiversity that serves as a complement to but not a substitute for cloud forest in this notably modified landscape*' (Pineda et al., 2005). In addition, using nitrogen-fixing trees in the plantations can reduce the need for artificial fertilizers, and their roots hold the soil together and protect the area from erosion (see *Chapter 5: Soils*), in turn potentially benefitting biodiversity.

To date, permaculture is mainly applied on a small scale, for example a small farm or a private garden, and few farmers can make a living off a permaculture farm. While upscaling and professionalization of permaculture is currently rare, some of the approaches used in permaculture have been more broadly adapted into agriculture. For example,

a)

b)

Figure B An example of (a) polyculture in a greenhouse in the Netherlands, in which different crops are grown in rows (intercropping) and (b) chickens in an orchard in Australia.

Source: Polyculturen in de praktijk; SIGN/LNV.

polyculture is getting more common (see Figure B.a), and some farmers now use chickens to control weeds and insects; the animals clear the ground, fertilize it, and, in addition, produce eggs and meat (see Figure B.b).

Further research is needed on nature-inclusive agriculture with the aim to designing farming systems with increased system resilience that require reduced external inputs, while maintaining conventional yields.

BOX 2.6

How can we save biodiversity in agricultural landscapes?

The most widely implemented forms of land sharing in agricultural landscapes are agri-environmental schemes (AES) and organic farming. AES were designed to preserve biodiversity and culture and to reduce environmental risks associated with modern farming. The goal of an AES is to increase the heterogeneity of the agricultural landscape (Stoate et al., 2009). Examples of measures promoted in AES include:

- Ecological management of field margins (that is, sowing field margins with flowers) (see Figure A.a) and ditches

- A more conservative mowing and fertilization regime

- Active protection of bird nests (see Figure A.b)

- Increasing and managing green landscape elements such as hedgerows

- Taking land temporarily out of production (set aside fields), and

- Increasing water tables within the productive fields.

The effectiveness of AES in reversing negative population trends on a national and global scale has remained inconclusive. Previous research showed little to no effect on species richness on farms with AES, compared to farms without AES (Kleijn et al., 2001). However, some studies find positive effects of AES. For example, a recent study in Germany demonstrated that the introduction of hedges, which is commonly done in AES, can enhance bird diversity (Batary et al., 2010).

Another potential option to save biodiversity in agricultural landscapes is to promotoe organic farming. Organic farming is an alternative agricultural production

a)

b)

Figure A Two examples of biodiversity within an agricultural landscape: (a) field margins with flowers and (b) a lapwing bird under a nest protector.

Source: Krijn Trimbos and Landschap Noord-Holland.

method which relies on fertilizers of organic origin, rather than industrial processes, such as artificial fertilizer. In addition, artificial pesticide usage is mostly prohibited. Several studies have demonstrated the positive effects of organic farming on different species groups. Overall an increase of 12% in biodiversity has been observed—specifically, more plants, higher flower diversity, more earthworms, insects, butterflies, and some birds resided on organic farms, compared to conventional farms (Gabriel et al., 2010). For example, a study showed that organic farming increased breeding density of the skylark, a bird species which nests on the ground in agricultural landscapes (see Figure B). This was attributed to higher landscape and crop heterogeneity at these farms, which increased breeding habitat and possibly food availability throughout the breeding season for the skylark (Kragten et al., 2008).

Figure B A skylark chick in a nest on an organic farm. A study showed that organic farming increased breeding density of the skylarks due to higher landscape or crop heterogeneity at these farms.

Source: Krijn Trimbos and Landschap Noord-Holland.

2.4 Future outlook

One crucial challenge for the conservation of biodiversity is how food production can be balanced with positive outcomes for biodiversity. The amount of agricultural land needed depends on the size of the human population, food consumption patterns, and output per unit of land (Kastner et al., 2012). The ultimate driver of threats to biodiversity is human population growth and increasing affluence. While our population continues to increase (our population is estimated to reach between 9 and 10 billion humans by 2050 (see *Chapter 1: Introduction*), the potential biodiversity loss also continues to increase.

Keeping food production in pace with a growing global population is a challenge in itself, but dietary changes may be an even larger driver of agricultural land requirements in the near future (Kastner et al., 2012). Diets with a higher proportion of vegetable proteins (that is, fewer proteins from higher trophic levels) have a decreased impact on the environment (see Section 8.4 in *Chapter 8: Nutrition*). However, predictions of the foreseeable future show that dietary intakes will include an increasing consumption of meat, fish, dairy, and eggs, particularly in the developing world.

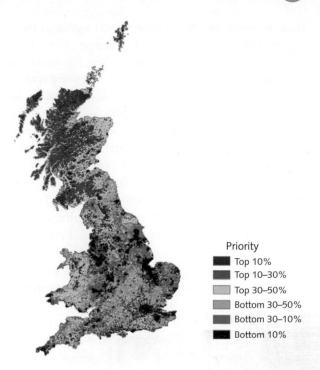

Priority
- Top 10%
- Top 10–30%
- Top 30–50%
- Bottom 30–50%
- Bottom 30–10%
- Bottom 10%

Figure 2.9 Priority map showing the top and worst locations to meet biodiversity conservation targets for Britain, based on analysis considering all major criteria (biodiversity, carbon stock retention, agriculture, and urban development).

Source: Adapted from Moilanen, A., et al. (2011), Balancing alternative land uses in conservation prioritization. *Ecological Applications*, 21: 1419–1426. Copyright © 2011, John Wiley and Sons.

In your experience

By changing diets, individuals can help protect biodiversity. For example, you can reduce your meat consumption or only buy palm oil-free products. What changes could you make to your current diet for the sake of conserving biodiversity? Would you be willing to make these changes?

While yield increases through further intensification on existing agricultural land may reduce the total land required for production, at the same time, population growth and/or dietary changes may offset these gains. It is likely that substantial additional surface area needs to be set aside for food production. In order to achieve neutral, or even positive, effects on biodiversity, we need to decide, often on a case-by-case basis, how and where different types of land use (for example, food production, urban development, and nature conservation) are best carried out.

Systematic conservation planning is one tool to help with this task; it uses spatial data to incorporate the role of habitat quality and connectivity, land, and opportunity costs. In its simplest form, it produces priority rankings which show biologically valuable areas that can be identified for conservation, and biologically low-priority areas could be identified for

alternative land uses. One such analysis in the UK showed that large areas in the north of the country were very important for biodiversity (see Figure 2.9). These regions could be appropriate for the development of nature areas, as there were much lower priorities for urban development and agricultural production, compared to the southern part of the UK (Moilanen et al., 2011).

However, in many places, trade-offs are required; if development for food production is best placed (highest expected yields) in an area where gains for biodiversity conservation would also be the largest (if areas are set aside for that purpose), then both aims cannot be simultaneously achieved. In the UK example above, conservation values and targets conflicted with agricultural and urbanization demands in many locations in the middle and south of the country. This suggests that biodiversity conservation here may have to focus on relatively small reserves, surrounded by other land uses—despite high conservation values in these regions (Moilanen et al., 2011). Recent applications of this approach also incorporate social preferences for land use and conservation to ensure that there is community support for changes, which helps create social acceptability and long-lasting solutions. Often several locations will achieve similar biodiversity conservation outcomes, creating some spatial flexibility regarding how conservation (and food production) targets can be met (Whitehead et al., 2014). Tools like these provide an insight into the potential social and policy conflicts between maintaining food security, creating space for further urban development, and protecting biodiversity.

However, as highlighted at the start of this section, it is key to realize that the projected population growth and affluence will continue to be the biggest threat to biodiversity (Crist et al., 2017). As the natural habitats of many species are shrinking, and the human population is growing, there is also an increased risk of human–wildlife conflicts (see Food controversy 2.2).

The solutions discussed in this chapter to protect biodiversity (for example, land sharing versus land sparing, permaculture, agricultural intensification, and conservation approaches) can help **minimize** future biodiversity loss but are likely not going to **stop** or **reverse** the loss. Slowing down our population growth, for example through promoting gender equality and providing effective and cheap contraception and education of girls, is likely going to be the best way to protect biodiversity (Crist et al., 2017).

Pause and think

In Food controversy 2.2, we discuss the increase in human–wildlife conflicts. Do you think farmers should have the right to protect their crops and livestock against wildlife, even if that means killing animals? Does it matter if the conflict happens to involve an endangered species (for example, wolves and tigers)?

FOOD CONTROVERSY **2.2**

Should farmers have the right to protect their livelihood?

Imagine a Maasai herdsman whose cattle is killed by a lion. Or a Indian farmer who loses a large part of his crops to elephants. Or a ranger near Yellowstone National Park who loses cattle and sheep to wolves. These types of human–wildlife conflicts happen all over the world.

In some regions, human–wildlife conflicts increase as the natural habitat of wildlife is disappearing rapidly, while the human population is expanding. This is, for example, the case in India and East Africa. In India, elephants often destroy crops, and local farmers will try to chase the animals away, sometimes with serious impacts (see Figure A). Here, it is estimated that wildlife (mainly elephants, but also tigers) kills one person a day.

In Tanzania and Kenya, the traditional herding grounds of the Maasai have long overlapped with biodiversity hotspots, including the highly productive savannahs.

Traditionally, the Maasai would protect their livestock against predators (lions, hyenas, cheetahs, and leopards) and other big grazers (elephants). However, new regulations have made killing wildlife punishable by law. With the conservation of wildlife and the increased pressure on land, conflicts are inevitable. For example, a study in south eastern Kenya found that lions killed 277 cows, goats, and sheep, with a value of almost US$35,000 between 1996 and 1999 (Patterson et al., 2004). Several NGOs now provide financial support to farmers who have lost animals to wildlife to prevent retaliations. In addition, some of these NGOs employ local Maasai as rangers to combat poaching, thereby providing an alternative livelihood within the communities.

Interestingly, in other areas in the world, wildlife is returning to reclaim its native home ranges. A good

Figure A In India, it is estimated that wildlife kills one person each day. In many cases, this happens when elephants enter agricultural fields and local farmers try to chase off the animals.

Source: AFP via Getty Images.

Figure B A wolf in Germany. Almost a century after the last wolf was killed in Germany, the wolf has made a comeback in recent years. Now wolves have also moved into the Netherlands. In both Germany and the Netherlands, there is increased social unrest and excitement about this development, causing heated public debates.

Source: Austriaca/Wikimedia Commons/CC BY-SA 4.0.

example here is the return of the wolf to areas in the USA (for example, Yellowstone National Park) and Europe (for example, in Germany and the Netherlands) (see Figure B). Due to extermination programmes in the previous centuries, wolves had disappeared from large parts of their native home-range throughout the USA and Europe. However, thanks to new conservation policies and reintroduction programmes, wolves are now making a comeback.

This reappearance of a top predator after more than a century resulted in both excitement and social unrest in the Netherlands. Many conservationists are very happy to see a top predator return to the Netherlands, while many other groups, including farmers, are concerned about their livestock. In recent years, there have been several

reports of wolves killing sheep (in some cases, this was confirmed, while in other cases, dogs had attacked and killed the sheep). For this reason, some farmers are calling for the culling of wolves entering the Netherlands in order to protect their sheep. In addition, many people have a deep-rooted fear of wolves and are worried that wolves might attack children. As wolf sightings are becoming more and more common, there is now an active debate on whether the Netherlands can support wolves and how to manage their population. This will be an important test-case for the Netherlands to see whether they can support both agriculture and biodiversity within the same landscape.

● QUESTIONS

2.1 Describe the three levels of biodiversity as used by the United Nations, and using the concept of ecological niche, explain why life on Earth is diverse.

2.2 Explain the two different ways of how we can value biodiversity, and for both ways, give an an example that was not mentioned in the main text.

2.3 The major threats to biodiversity can be remembered using the acronym HIPCO. List these major threats to biodiversity,

and give three examples of the threats which were discussed in this chapter.

2.4 One way of determining how humans dominate resource use in terrestrial systems is by determining HANPP. What does HANPP stand for, and how does it relate to the NPP from the natural vegetation?

2.5 There are two key approaches used to conserve biodiversity. Explain the difference between these two approaches.

FURTHER READING

Costanza, R., de Groot, R., Sutton, P., van der Ploeg, S., Anderson, S.J., Kubiszewski, I., Farber, S., Turner, R.K. 2014. Changes in the global value of ecosystem services. *Global Environmental Change* **26**, 152–158. (A study discussing the trends and challenges related to the value of ecosystem services, including ecosystem services provided by biodiversity.)

Díaz, S., Fargione, J., Chapin, F.S. III, Tilman, D. 2006. Biodiversity loss threatens human well-being. *PLoS Biology* **4**, e277. (A study which explains how human societies have been built on biodiversity and how the current biodiversity loss will potentially affect humanity.)

Haberl, H., Erb, K.H., Krausmann, F., Gaube, V., Bondeau, A.,

Plutzar, C., Gingrich, S., Lucht, W., Fischer-Kowalski, M. 2007. Quantifying and mapping the human appropriation of net primary production in earth's terrestrial ecosystems. *Proceedings of the National Academy of Sciences of the United States of America* **104**, 12942–12945. (This study explains how you can quantify HANPP and provides detailed maps of HANPP across the globe.)

Wilson, E.O. 2016. *Half-earth: Our Planet's Fight for Life.* New York, NY: WW Norton & Company. (A book by one of the biggest biologists of our time, proposing to set aside half of the Earth's land as human-free nature reserves to protect biodiversity.)

REFERENCES

Scientific American. 2012. The other oil problem. The world's growing appetite for cheap palm oil is destroying rain forests and amplifying climate change. *Scientific American* **307**, 10.

Batary, P., Matthiesen, T., Tscharntke, T. 2010. Landscape-moderated importance of hedges in conserving farmland bird diversity of organic vs. conventional croplands and grasslands. *Biological Conservation* **143**, 2020–2027.

Blackburn, T.M., Cassey, P., Duncan, R.P., Evans, K.L., Gaston, K.J. 2004. Avian extinction and mammalian introductions on oceanic islands. *Science* **305**, 1955–1958.

Cardinale, B.J., Duffy, J.E., Gonzalez, A., Hooper, D.U., Perrings, C., Venail, P., Narwani, A., Mace, G.M., Tilman, D., Wardle, D.A., Kinzig, A.P., Daily, G.C., Loreau, M., Grace, J.B., Larigauderie, A., Srivastava, D.S., Naeem, S. 2012. Biodiversity loss and its impact on humanity. *Nature* **486**, 59–67.

Costanza, R., de Groot, R., Sutton, P., van der Ploeg, S., Anderson, S.J., Kubiszewski, I., Farber, S., Turner, R.K. 2014. Changes in the global value of ecosystem services. *Global Environmental Change* **26**, 152–158.

Crist, E., Mora, C., Engelman, R., 2017. The interaction of human population, food production, and biodiversity protection. *Science* **356**, 260–264.

Danielsen, F., Beukema, H., Burgess, N.D., Parish, F., Brühl, C.A., Donald, P.F., Murdiyarso, D., Phalan, B.E.N., Reijnders, L., Struebig, M., Fitzherbert, E.B. 2009. Biofuel plantations on forested lands: double jeopardy for biodiversity and climate. Plantaciones de Biocombustible en Terrenos Boscosos: Doble Peligro para la Biodiversidad y el Clima. *Conservation Biology* **23**, 348–358.

Ellis, E.C., Goldewijk, K.K., Siebert, S., Lightman, D., Ramankutty, N. 2010. Anthropogenic transformation of the biomes, 1700 to 2000. *Global Ecology and Biogeography* **19**, 589–606.

Fischer, J., Brosi, B., Daily, G.C., Ehrlich, P.R., Goldman, R., Goldstein, J., Lindenmayer, D.B., Manning, A.D., Mooney, H.A., Pejchar, L., Ranganathan, J., Tallis, H. 2008. Should agricultural policies encourage land sparing or wildlife-friendly farming? *Frontiers in Ecology and the Environment* **6**, 380–385.

Fitzherbert, E.B., Struebig, M.J., Morel, A., Danielsen, F., Brühl, C.A., Donald, P.F., Phalan, B. 2008. How will oil palm expansion affect biodiversity? *Trends in Ecology & Evolution* **23**, 538–545.

Foley, J.A., DeFries, R., Asner, G.P., Barford, C., Bonan, G., Carpenter, S.R., Chapin, F.S., Coe, M.T., Daily, G.C., Gibbs, H.K., Helkowski, J.H., Holloway, T., Howard, E.A., Kucharik, C.J., Monfreda, C., Patz, J.A., Prentice, I.C., Ramankutty, N., Snyder, P.K. 2005. Global consequences of land use. *Science* **309**, 570–574.

Gabriel, D., Sait, S.M., Hodgson, J.A., Schmutz, U., Kunin, W.E., Benton, T.G. 2010. Scale matters: the impact of organic farming on biodiversity at different spatial scales. *Ecology Letters* **13**, 858–869.

Gallai, N., Salles, J.-M., Settele, J., Vaissière, B.E. 2009. Economic valuation of the vulnerability of world agriculture confronted with pollinator decline. *Ecological Economics* **68**, 810–821.

Gaveau, D.L.A., Sloan, S., Molidena, E., Yaen, H., Sheil, D., Abram, N.K., Ancrenaz, M., Nasi, R., Quinones, M., Wielaard, N., Meijaard, E. 2014. Four decades of forest persistence, clearance and logging on Borneo. *PLoS One* **9**, e101654.

Goldman, M.J. 2011. Strangers in their own land: Maasai and wildlife conservation in Northern Tanzania. *Conservation and Society* **9**, 65.

Haberl, H., Erb, K.H., Krausmann, F., Gaube, V., Bondeau, A., Plutzar, C., Gingrich, S., Lucht, W., Fischer-Kowalski, M. 2007. Quantifying and mapping the human appropriation of net primary production in earth's terrestrial ecosystems. *Proceedings of the National Academy of Sciences of the United States of America* **104**, 12942–12945.

Haberl, H., Schulz, N.B., Plutzar, C., Erb, K.H., Krausmann, F., Loibl, W., Moser, D., Sauberer, N., Weisz, H., Zechmeister, H.G., Zulka, P. 2004. Human appropriation of net primary production and species diversity in agricultural landscapes. *Agriculture, Ecosystems & Environment* **102**, 213–218.

Hautier, Y., Tilman, D., Isbell, F., Seabloom, E.W., Borer, E.T., Reich, P.B. 2015. Anthropogenic environmental changes affect ecosystem stability via biodiversity. *Science* **348**, 336–340.

Hooper, D.U., Adair, E.C., Cardinale, B.J., Byrnes, J.E.K., Hungate, B.A., Matulich, K.L., Gonzalez, A., Duffy, J.E., Gamfeldt, L., O'Connor, M.I. 2012. A global synthesis reveals biodiversity loss as a major driver of ecosystem change. *Nature* **486**, 105–129.

Isbell, F., Craven, D., Connolly, J., Loreau, M., Schmid, B., Beierkuhnlein, C., Bezemer, T.M., Bonin, C., Bruelheide, H., de Luca, E., Ebeling, A., Griffin, J.N., Guo, Q.F., Hautier, Y., Hector, A., Jentsch, A., Kreyling, J., Lanta, V., Manning, P., Meyer, S.T., Mori, A.S., Naeem, S., Niklaus, P.A., Polley, H.W., Reich, P.B., Roscher, C., Seabloom, E.W., Smith, M.D., Thakur, M.P., Tilman, D., Tracy, B.F., van der Putten, W.H., van Ruijven, J., Weigelt, A., Weisser, W.W., Wilsey, B., Eisenhauer, N. 2015. Biodiversity increases the resistance of ecosystem productivity to climate extremes. *Nature* **526**, 574–577.

Kastner, T., Rivas, M.J.I., Koch, W., Nonhebel, S. 2012. Global changes in diets and the consequences for land requirements for food. *Proceedings of the National Academy of Sciences of the United States of America* **109**, 6868–6872.

Khoury, C.K., Bjorkman, A.D., Dempewolf, H., Ramirez-Villegas, J., Guarino, L., Jarvis, A., Rieseberg, L.H., Struik, P.C. 2014. Increasing homogeneity in global food supplies and the implications for food security. *Proceedings of the National Academy of Sciences of the United States of America* **111**, 4001–4006.

Kleijn, D., Berendse, F., Smit, R., Gilissen, N. 2001. Agri-environment schemes do not effectively protect biodiversity in Dutch agricultural landscapes. *Nature* **413**, 723–725.

Klein, A.M., Vaissiere, B.E., Cane, J.H., Steffan-Dewenter, I., Cunningham, S.A., Kremen, C., Tscharntke, T. 2007. Importance of pollinators in changing landscapes for world crops. *Proceedings of the Royal Society B-Biological Sciences* **274**, 303–313.

Koh, L.P., Wilcove, D.S. 2008. Is oil palm agriculture really destroying tropical biodiversity? *Conservation Letters* **1**, 60–64.

Kolar, C.S., Lodge, D.M. 2001. Progress in invasion biology: predicting invaders. *Trends in Ecology & Evolution* **16**, 199–204.

Kraglen, S., Trimbos, K.B., de Snoo, G.R. 2008. Breeding skylarks (*Alauda arvensis*) on organic and conventional arable farms in The Netherlands. *Agriculture, Ecosystems & Environment* **126**, 163–167.

Krausmann, F., Erb, K.H., Gingrich, S., Haberl, H., Bondeau, A., Gaube, V., Lauk, C., Plutzar, C., Searchinger, T.D. 2013. Global human appropriation of net primary production doubled in the 20th century. *Proceedings of the National Academy of Sciences of the United States of America* **110**, 10324–10329.

Markku, O. 1997. The moral value of biodiversity. *Ambio* **26**, 541–545.

McGeoch, M.A., Butchart, S.H.M., Spear, D., Marais, E., Kleynhans, E.J., Symes, A., Chanson, J., Hoffmann, M., 2010. Global indicators of biological invasion: species numbers, biodiversity impact and policy responses. *Diversity and Distributions* **16**, 95–108.

Mittermeier, R.A., Turner, W.R., Larsen, F.W., Brooks, T.M., Gascon, C. 2011. Global biodiversity conservation: the critical role of hotspots. In: Zachos, F.E., Habel, F.C. (eds). *Biodiversity Hotspots: Distribution and Protection of Conservation Priority Areas*. Heidelberg: Springer, pp. 3–22.

Moilanen, A., Anderson, B.J., Eigenbrod, F., Heinemeyer, A., Roy, D.B., Gillings, S., Armsworth, P.R., Gaston, K.J., Thomas, C.D. 2011. Balancing alternative land uses in conservation prioritization. *Ecological Applications* **21**, 1419–1426.

Mora, C., Tittensor, D.P., Adl, S., Simpson, A.G.B., Worm, B. 2011. How many species are there on Earth and in the ocean? *PLoS Biology* **9**, e1001127.

Myers, R.A., Baum, J.K., Shepherd, T.D., Powers, S.P., Peterson, C.H. 2007. Cascading effects of the loss of apex predatory sharks from a coastal ocean. *Science* **315**, 1846–1850.

Njiru, M., Kazungu, J., Ngugi, C.C., Gichuki, J., Muhoozi, L. 2008. An overview of the current status of Lake Victoria fishery: opportunities, challenges and management strategies. *Lakes & Reservoirs: Research & Management* **13**, 1–12.

Patterson, B., Kasiki, S., Selempo, E., Kays, R. 2004. Livestock predation by lions (*Panthera leo*) and other carnivores on ranches neighboring Tsavo National Park, Kenya. *Biological Conservation* **119**, 507–516.

Pauly, D., Christensen, V., Guenette, S., Pitcher, T.J., Sumaila, U.R., Walters, C.J., Watson, R., Zeller, D. 2002. Towards sustainability in world fisheries. *Nature* **418**, 689–695.

Pereira, H.M., Leadley, P.W., Proença, V., Alkemade, R., Scharlemann, J.P., Fernandez-Manjarrés, J.F., Araújo, M.B., Balvanera, P., Biggs, R., Cheung, W.W. 2010. Scenarios for global biodiversity in the 21st century. *Science* **330**, 1496–1501.

Phalan, B., Onial, M., Balmford, A., Green, R.E. 2011. Reconciling food production and biodiversity conservation: land sharing and land sparing compared. *Science* **333**, 1289–1291.

Pineda, E., Moreno, C., Escobar, F., Halffter, G. 2005. Frog, bat, and dung beetle diversity in the cloud forest and coffee agroecosystems of Veracruz, Mexico. *Conservation Biology* **19**, 400–410.

Pirker, J., Mosnier, A., Kraxner, F., Havlík, P., Obersteiner, M. 2016. What are the limits to oil palm expansion? *Global Environmental Change* **40**, 73–81.

Stoate, C., Baldi, A., Beja, P., Boatman, N.D., Herzon, I., van Doorn, A., de Snoo, G.R., Rakosy, L., Ramwell, C. 2009. Ecological impacts of early 21st century agricultural change in Europe—a review. *Journal of Environmental Management* **91**, 22–46.

Tscharntke, T., Clough, Y., Wanger, T.C., Jackson, L., Motzke, I., Perfecto, I., Vandermeer, J., Whitbread, A. 2012. Global food security, biodiversity conservation and the future of agricultural intensification. *Biological Conservation* **151**, 53–59.

United Nations (UN). *United Nations System-Wide Earthwatch. Biodiversity. Invasive Species*. Available at: http://www.un.org/earthwatch/biodiversity/invasivespecies.html [accessed 22 October 2018].

United Nations (UN). 1992. *Convention on Biological Diversity*. New York, NY: UN.

Voigt, M., Wich, S.A., Ancrenaz, M., Meijaard, E., Abram, N., Banes, G.L., Campbell-Smith, G., d'Arcy, L.J., Delgado, R.A., Erman, A., Gaveau, D., Goossens, B., Heinicke, S., Houghton,

M., Husson, S.J., Leiman, A., Sanchez, K.L., Makinuddin, N., Marshall, A.J., Meididit, A., Miettinen, J., Mundry, R., Musnanda, N., Nurcahyo, A., Odom, K., Panda, A., Prasetyo, D., Priadjati, A., Purnomo, R.A., Russon, A.E., Santika, T., Sihite, J., Spehar, S., Struebig, M., Sulbaran-Romero, E., Tjiu, A., Wells, J., Wilson, K.A., Kühl, H.S. 2018. Global demand for natural resources eliminated more than 100,000 Bornean orangutans. *Current Biology* **28**, 761–769.e765.

Whitehead, A.L., Kujala, H., Ives, C.D., Gordon, A., Lentini, P.E., Wintle, B.A., Nicholson, E., Raymond, C.M. 2014. Integrating biological and social values when prioritizing places for biodiversity conservation. *Conservation Biology* **28**, 992–1003.

Zhang, Y., Xu, M., Chen, H., Adams, J. 2009. Global pattern of NPP to GPP ratio derived from MODIS data: effects of ecosystem type, geographical location and climate. *Global Ecology and Biogeography* **18**, 280–290.

3

Pollution

How are food systems related to environmental pollution?

Thijs Bosker and Martina G. Vijver

Chapter Overview

- Since the 1950s, public awareness of the potential long-term consequences of pollutants has increased. This has resulted in a shift in paradigms, which has had great implications on how we observe the world around us and how we think about minimization of pollution of our environment.

- Food production is inherently related to environmental pollution, for example through pesticides and artificial fertilizers entering the environment. Pollution from food production is often diffuse, which makes both prevention and removal difficult.

- The application of fertilizers to agricultural fields can also result in the release of nutrients into the environment. The excess input of nutrients into aquatic systems can lead to a cascade of effects, for example shifts in food webs and hypoxic dead zones in estuaries.

- Pesticides that enter the environment can have a variety of unintended effects on non-target organisms, ranging from the disruption of the endocrine system to the decline of populations.

- Scientists conduct both field and laboratory studies to better understand the impact of pollutants. This information can be used to minimize the risk of pollutants in the environment.

- Research is conducted to find innovative ways of producing food without (or with minimal) environmental impact, but some of these inventions are highly controversial, including new types of pesticides and genetically modified crops.

Introduction

How can the use of artificial fertilizers lead to dead zones in estuaries? Are pesticides responsible for the decline of bee and bird populations? Are they also linked to decreased fertilization rates and increased intersex in amphibians? And how do we quantify and assess the impact of these compounds? These

a)

b)

Figure 3.1　(a) Light pollution from greenhouses near Luttelgeest (the Netherlands) seen from approximately 7 km distance and (b) a ghost net in the ocean (right).

Source: (a) IIVQ / Tijmen Stam/Wikimedia Commons/CC BY-SA 4.0; (b) Mstelfox/Wikimedia Commons/CC BY-SA 4.0..0.

are some of the questions scientists are trying to answer when investigating the impact of agricultural pollutants.

Two groups of chemicals which have been directly linked to environmental pollution are pesticides and artificial fertilizers. We have been using increased quantities of these agrochemicals since the Green Revolution (see *Chapter 1: Introduction*), and they are two driving forces behind the significant increase in agricultural yields (another important driver has been the use of energy, which will be discussed in *Chapter 7: Energy*). But while pesticides and artificial fertilizers have been instrumental in feeding the world, they can also cause environmental pollution. However, pollution related to food is by no means restricted to the use of fertilizers and pesticides. Other examples include light pollution associated with greenhouses (see Figure 3.1a), ocean pollution caused by fishing nets lost at sea (so-called ghost nets) (see Figure 3.1b), and air pollution related to the release of greenhouse gases such as methane from cattle manure or the use of fossil fuels.

The focus in this chapter, however, is on environmental pollution caused by pesticides and artificial fertilizers. First, Section 3.1 gives a brief historical background on public perceptions of environmental pollution and relates this to a shift in environmental awareness among the general public. Section 3.2 introduces key concepts related to environmental pollution and associated fields of study. Finally, Section 3.3 focuses on the potential impact of fertilizers and pesticides on the environment. We provide several case studies on the impact of pollutants on ecosystems, including the relationship between dead zones and fertilizers, and the impact of pesticides on organisms.

3.1 When and why did we start worrying about environmental pollution?

Environmental pollution has become a regular topic on the news, ranging from local issues, such as light pollution (see Figure 3.1a), to global issues such as the release of greenhouse gases that cause climate change. **Pollution** is defined by the UN as the presence of substances (for example, chemicals or plastics) or energy (for example, light or heat) in the environment, resulting in undesirable environmental effects (UN, 2016) (see Box 3.1).

BOX 3.1

What is the difference between a xenobiotic chemical, a contaminant, and a pollutant?

In this chapter, you will see a variety of terms to describe foreign compounds in the environment, but not all of them classify as pollutants. For clarification, these are the definitions used in this chapter to describe foreign compounds in the environment:

- Xenobiotic compound: a compound, typically a synthetic chemical, which is foreign to the body or an ecological system

- Contaminant: a compound in the environment which is found at *elevated concentrations* above the natural background level for the area and for the organism

- Pollutant: a compound in the environment whose nature, location, or quantity produce undesirable environmental effects. This means that a xenobiotic compound is always a contaminant (as it is a foreign compound and therefore always above natural background levels). However, not all contaminants are xenobiotic compounds. Copper, for example, is found in all organisms and ecological systems, and therefore not a xenobiotic compound. It becomes a contaminant if levels exceed normal background levels. If this results in undesirable effects on a system or organism, copper is classified as a pollutant.

In your experience

Think of environmental pollution over which there is concern within your own community. What is the source of this pollution? Who should be responsible for dealing with this pollution? Is there anything you could do to minimize this pollution?

Although there is a long history of environmental awareness and conservation efforts (see Box 3.2 for a short description on historical environmentalism), our current awareness of environmental pollution can be traced back to the 1950s and 60s. Here we highlight three examples which raised global environmental awareness: the sudden outbreak of neurological disorders in Japan, the publication of the book *Silent Spring*, and a picture of Earth taken from outer space. These events, along with many others, have shaped our current perspective on the environment.

3.1.1 How a community was affected by mercury poisoning

In the 1950s and 60s, local medical practitioners in Minamata (Japan) noticed a sharp increase in specific neurological disorders, including general muscle weakness, loss of peripheral vision, and damage to hearing and speech. Even more worrying, there were also severe cases which included insanity, paralysis, and even death (Harada, 1995). Over time, ever greater numbers of children were born with neurological defects. These events raised alarm bells, and through further investigation, a link was found between the discharge of industrial waste water, released by a nearby chemical factory, and the neurological disorders. This waste water contained the highly poisonous chemical methylmercury ($[CH_3Hg]^+$).

Methylmercury has two important characteristics that made it a direct problem to the communities around the factory. First of all, organisms are able to accumulate methylmercury from their environment, a process called **bioaccumulation** (a net result of uptake and elimination). Organisms can have great difficulties eliminating specific chemicals from their system,

BOX 3.2

Environmentalism in history

There is a long history of environmental awareness, dating back many centuries. Early Arabic writing on medicine dating back to the sixth century dealt with environmental pollution, relating it to human diseases and illnesses (Gari, 2002). One of the first documented cases of environmental protection in Europe dates back to 676 AD. Cuthbert, a monk in the monastery of Lindisfarne on Holy Island (off the north east coast of England), issued two laws protecting eider ducks on the island (Jones-Walters and Čivić, 2013). In some cases, environmental awareness was directly linked to religion; hundreds of Hindus in India died in 1731 when trying to protect sacred trees which were needed by the Maharaja of Jodhpur for fuel (Veeraraj, 1990).

Most of these early environmental issues were local, affecting people in the direct vicinity (Kula, 2013). With growing human populations and the rise of industrialization, the magnitude and complexity of environmental issues increased rapidly, often causing effects on a much larger scale (Kula, 2013). As a response, environmental awareness about environmental degradation increased among the public, fuelled by influential authors such as John Muir, Henry Thoreau, Rachel Carson, and Aldo Leopold. This increased awareness resulted in the formation of several important NGOs, largely funded by the public through memberships or fundraising. Well-known international examples still active today include the Nature Conservancy (founded in 1951), World Wildlife Fund (founded in 1961), Friends of the Earth (founded in 1969), and Greenpeace (founded in 1971).

---

as is the case for methylmercury. As a result, the level of methylmercury increases with exposure duration and with an organism's higher trophic status in a food chain, a process called **biomagnification** (see Box 3.3).

The problem in Minamata was the result of local fishermen collecting fish and shellfish for consumption. These fish and shellfish had been exposed to methylmercury and had biomagnified this compound in their tissues. Because fish and shellfish were an important food source, and humans are at the highest trophic level, the concentration of methylmercury in local citizens increased through biomagnification. It is important to realize that high levels of a chemical in an organism do not necessarily imply adverse health effects. After all, our bodies are full of essential (bio)chemical compounds in high amounts needed for normal functioning.

This brings us to the second important characteristic of methylmercury—it is a highly toxic compound. Methylmercury affects the normal functioning of the nervous system (neurotoxin). The combination of biomagnification and the neurotoxic effects of methylmercury caused Minamata disease. Although tragic, the events in Minamata did raise awareness regarding pollution of the environment and the subsequent impact on human health.

3.1.2 Will there still be birds whistling in spring?

'*These sprays, dusts, and aerosols are now applied almost universally to farms, gardens, forests, and homes–nonselective chemicals that have the power to kill every insect, the "good" and the "bad," to still the song of birds and the leaping of fish in the streams, to coat the leaves with a deadly film, and to linger on in soil–all this though the intended target may be only a few weeds or insects. Can anyone believe it is possible to lay down such a barrage of poisons on the surface of the earth without making it unfit for all life? They should not be called "insecticides," but "biocides".*'

Rachel Carson in her book *Silent Spring* (1962)

What if pesticides decimate bird populations? Will there still be birds whistling in spring? Will spring be the same? These questions were of great concern to Rachel Carson, a scientist employed by the US Fish and Wildlife Service. In 1962, she wrote a book *Silent Spring*, which is widely considered as one of the most influential books in environmental history. It highlights the potential impacts of pesticides on the environment. At that time, DDT was a common insecticide used all over the world. DDT is highly

BOX 3.3

What is the difference between bioaccumulation and biomagnification?

Bioaccumulation refers to the accumulation of a xenobiotic compound within an organism and is a net result of uptake and elimination. Here, elimination includes elimination of the compound back to the environment, storage of the contaminant in tissues where it cannot cause harm, and detoxification of the compound by changing the structure. The rate of accumulation depends on a variety of factors unique to the compound, including its size and lipophilicity (ability to accumulate in fatty tissues). In addition, the total accumulation of a compound in an organism depends on the environmental concentration; an equilibrium (or steady state) is formed between the concentration in the organism and the environment.

In specific cases, **biomagnification** can occur. This is caused by the inability of organisms to eliminate compounds from their bodies. Organisms at the base of the food pyramid accumulate xenobiotic compounds from the environment. Animals at the next trophic level feed on these organisms and thereby also accumulate the compound in their bodies. However, they are unable to excrete the compound, and as a result, they concentrate or magnify the compound. When these organisms are subsequently consumed by another predator higher in the food chain, the concentration increases even further (Figure A). This results in biomagnification

across trophic levels—the increase in the concentration of the contaminant with an increase in trophic level (Figure A).

This increase in concentration can potentially lead to serious adverse health effects. Well-known examples of chemicals which biomagnify can be found in the persistent organic pollutant group (see Box 3.4) such as dichlorodiphenyltrichloroethane (DDT) or polychlorinated biphenyls (PCBs). Other chemicals, like metals, hardly show biomagnification (or only under specific conditions such as mercury when present as the organic compound methylmercury).

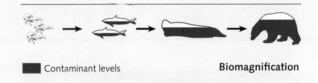

■ Contaminant levels   **Biomagnification**

Figure A Schematic illustrating the concept of biomagnification—the concentration of a xenobiotic compound (for example, in grams of xenobiotic compound/kg of body weight) increases with an increase in trophic level because of the inability of organisms to excrete the compound.

Source: © WWF.

effective against insects, as it affects the functioning of neurons in invertebrates, causing paralyses, ultimately leading to death. Mammals are less sensitive to this pesticide; this is why DDT became one of the most widely used insecticides in the world, resulting in a Nobel prize for the Swiss inventor Paul Müller. Unfortunately, DDT turned out to be a **persistent organic pollutant (POP)** (see Box 3.4), with many unwanted and unexpected side effects.

Research in Europe and the USA in the 1950s and 60s revealed that DDT and some of its metabolites

(a daughter compound produced during biological, chemical, or physical transformation from the original compound) caused eggshell thinning in birds, including in the iconic, and nationally important, American bald eagle (see Figure 3.2). Due to the thinner eggshells, the eggs broke or cracked when the birds sat on them during incubation. This led to population declines in a variety of bird species. For example, in Long Island, NY (USA), the population of osprey, a bird of prey species which feeds mainly on fish, was severely impacted. The number of young

BOX 3.4

Why do we find herbicides in the Arctic?

POPs, or **persistent organic pollutants**, are a group of xenobiotic chemicals which can cause environmental degradation due to four key characteristics:

1. They are very persistent in the environment (that is, they take a long time to break down)

2. They are highly toxic, causing a variety of effects on organisms

3. They are lipophilic, meaning that they like to accumulate in the fatty tissues of organisms, potentially resulting in biomagnification

4. They can be transferred around the globe through a process called atmospheric deposition, in which a POP attaches to particles in water droplets and is transported through the atmosphere. This transport is mainly towards colder regions.

This combination of characteristics results in an accumulation of these contaminants in organisms and potential toxic effects. Many POPs (including pesticides such as DDT) accumulate in polar regions due to atmospheric transfer. Studies in the Canadian Arctic have revealed high levels of POPs in the tissues of Arctic animals, especially in top predators such as polar bears (Braune et al., 2005). In addition, a study on indigenous communities in the Canadian Arctic, including the Inuit, revealed high levels of POPs in their bodies, a result of biomagnification (Van Oostdam et al., 1999). These high levels are linked to their traditional diets, which contain relatively large quantities of animal products, including fish and seals (Van Oostdam et al., 1999).

Figure 3.2 Two adult American bald eagles and a chick in Seedskadee National Wildlife Refuge in Wyoming, USA.

Source: Tom Koerner/USFWS (CC BY 2.0).

per nest surviving to fledgling age decreased by 74–94% (Spitzer et al., 1978), causing a sharp decline in the overall population.

Silent Spring raised public awareness of how humans can impact the environment around them. It has been instrumental in the ongoing efforts to minimize the input of POPs to the environment. In 2001, almost 40 years after *Silent Spring* was published, the Stockholm Convention was signed, with the goal to reduce the input of 12 POPs globally. These 12 POPs, named **the dirty dozen**, including compounds such as PCBs, dioxins, and pesticides, including DDT. In 2004, the convention came into effect, but several key countries did not ratify the agreement, including the USA, Italy, and Israel. However, as DDT is a POP, it will continue to be detected in the environment decades long after it has been banned. For example, in European and US surface waters, DDT and its metabolites are still found in detectable levels, even though it was banned in the 1970s. Currently (2018), DDT is still in use, as it is a very effective insecticide against the malaria mosquito.

 Pause and think

DDT is still one of the most effective insecticides against *Anopheles* mosquitos, which carry malaria. Every year, hundreds of millions of people are affected by malaria, resulting in an estimated 500,000 casualties. Knowing the environmental risk of DDT, and the health risk of malaria, do you think the use of DDT to prevent malaria outweighs the environmental impact?

3.1.3 A lonely planet and its shifting paradigms

On Christmas evening, 1968, the crew of Apollo 8 circled around the moon and took one of the most iconic pictures of our planet—Earthrise (see Figure 3.3). The relatively small size of our planet and the colour contrast with the lifeless moon made many people realize that our planet has finite resources and a finite ability to deal with pollutants (Westbroek, 2012).

Due to events like these, public opinion on environmental pollution has undergone a fundamental paradigm shift. Historically, the **dilution paradigm** was common practice. This paradigm can be summed up as: *the solution to pollution is dilution*. In other words, the belief was that if we contaminate a large body of water (for example, with POPs) or the atmosphere (for example, with greenhouse gases), the self-regulatory principles of our planet will dissolve any potential harm. Due to the increasing public understanding of the risk of pollution, a new paradigm emerged—the **boomerang paradigm**. Within this paradigm, the pollutant does not disappear when we emit or discharge it into the environment. It will, at some point and at some time, resurface, just like a boomerang. This important shift in how we think about our environment has great implications on how we observe the world around us and how we think about minimization of pollution of our environment.

The increase in environmental awareness has resulted in new areas of study, including the fields of environmental chemistry (determining the fate of chemicals in the environment) and environmental toxicology (quantifying ecological effects of

Figure 3.3 Earthrise over the surface of the moon. Picture taken during the Apollo 8 Mission in 1968.

Source: NASA.

chemicals). In addition, governments have developed regulatory bodies and agencies to prevent, minimize, and combat environmental pollution. This resulted in new concepts and definitions related to environmental pollution and new approaches to study impacts.

In your experience

The combined impact of individual consumers can significantly contribute to the overall load of pollutants in the environment. One example is the increase in plastic waste in the environment. Consider how your current lifestyle is contributing to environmental pollution and what you could do to minimize your own contribution to environmental pollution.

3.2 How is agriculture linked to environmental pollution?

Imagine the large smokestacks of a factory or all the sewage effluent we have to clean (see Figure 3.4). Both of these examples could result in environmental pollution and have a clearly defined discharge point to the environment. We call this **point-source pollution**. As the discharge location of the pollutants can be clearly identified, **front-of-pipe solutions** (or front-end solutions) can often be used. These solutions prevent pollution by removing the pollutant before it is released into the environment. Examples include scrubbers in coal-fired power plants, which can remove most mercury or sulfur, or sewage treatment facilities that remove most of the solids, nutrients, and pathogens.

a)

b)

Figure 3.4 Two examples of point-source pollution: a smokestack (a) and a wastewater outfall (b).
Source: (a) © Jorge Royan / http://www.royan.com.ar / CC BY-SA 3.0; (b) United States Department of Agriculture.

In contrast, **non-point source** (or diffuse) **pollution** comes from many different sources simultaneously, for example metal run-off from urban environments and sediment run-off from logged forests (see Figure 3.5). Non-point source pollution is diffuse and often occurs over relatively large areas. In addition, different mixtures of chemicals can occur at different locations, and there is no clear emission route which can be identified. **End-of-pipe solutions** are often used for non-point source pollution, in which pollutants that have already been formed and released into the environment are cleaned up. In general, this is very time-consuming and costly.

As an alternative, we can try to prevent or minimize non-point source pollutants from entering the environment in the first place. For example, agricultural fields often feature buffer strips (see Figure 3.6). As the name implies, these strips form a buffer at the edges of agricultural fields to prevent or minimize the movement of agrochemicals to adjacent surface waters or non-target fields. In addition, they can provide a habitat for organisms, including natural predators against pests which may be a problem for the farmer, which increases the self-regulatory capacity of the ecosystem (see Box 2.6, *Chapter 2: Biodiversity*).

Figure 3.5 Land run-off in Kahoolawe, Kuheia (Hawaii, USA) pouring into the sea, as an example of non-point source pollution. Soil is eroded, and, with the soil, come nutrients and agrochemicals, which can accumulate in the environment.
Source: Forest & Kim Starr/CC BY 3.0.

Figure 3.6 An example of a buffer strip of multiple rows of trees and shrubs, which creates a buffer to protect a creek in Iowa (USA) from the agrochemicals used on the adjacent agricultural fields.
Source: Lynn Betts / Photo courtesy of USDA Natural Resources Conservation Service.

Agricultural pollution can generally be labelled as non-point source pollution, as the discharge of pollutants happens over a diffuse area and spreads rapidly via adjacent (drainage) ditches and streams. The movement of both nutrients and pesticides into the surrounding environment are two key examples of non-point source pollution resulting from agriculture.

3.2.1 How do fertilizers cause pollution?

Plants require a wide variety of nutrients to grow, with phosphorus (P), nitrogen (N), and potassium (K) being needed in the highest quantities. Plants can only take up specific forms of these nutrients. Phosphorus is most readily available as phosphates (PO_4^{3-}). Nitrogen forms available to plants include nitrates (NO_3^{2-}), ammonia (NH_3), and ammonium (NH_4^+). Potassium is often provided as potash, which is a mixture of potassium minerals that includes potassium chloride (KCl) and potassium sulfate (K_2SO_4). All three of these nutrients are essential for plant growth.

However, at the start of the nineteenth century, it became evident that the available nitrogen for plant growth was insufficient to support the growing human population. This situation could potentially lead to a Malthusian catastrophe (see *Chapter 1: Introduction*). For this reason, there was a desperate need for a source of readily accessible nitrogen. With the advancement of chemistry, two German scientists, Fritz Haber and Carl Bosch, developed a method (called the **Haber–Bosch method**) to transform the readily available N_2 gas in the atmosphere into the forms of nitrogen available to plants, resulting in artificial fertilizers. Because of its immense implications (unlimited access to nitrogen for food production), this process has been termed the biggest technological discovery of the twentieth century and resulted in the Nobel Prize for both Haber and Bosch in 1931. Despite its enormous benefits, one major downside of the Haber–Bosch process is the high amount of energy it demands—1–2% of all global energy is used for the production of artificial fertilizers (see Section 6.2.1 in *Chapter 6: Climate Change*).

Unlike nitrogen, there is no chemical process to create large-scale phosphorus fertilizers, but it can be mined. Phosphorus mining happens in areas with phosphate deposits (locations where the soil contains high levels of phosphates), for example in Northern Africa (for example, Morocco and Algeria), Australia, Florida (USA), and West Africa (see Figure 3.7).

The invention of the Haber–Bosch process and the ability to mine phosphate deposits resulted in the production of artificial fertilizers on an industrial scale. This has been one of the biggest successes of the Green Revolution, allowing for a large reservoir of key plant nutrients needed to increase overall agricultural output. The way nutrients behave at the global level is depicted in **biochemical cycles** (see Box 3.5).

Both the nitrogen and phosphorus cycles have been seriously affected by the human need for fertilizers, with enormous impacts on ecosystem health. One direct consequence of the increase in fertilization has been the loss of biodiversity. Only a few plant species can tolerate the high levels of nutrients applied to agricultural fields (see Section 2.1.2 in *Chapter 2: Biodiversity*, concept of ecological niche). This has resulted in the increase in, and even domination by, the few plant species that prefer (or tolerate) higher levels of nutrients. Subsequently, animals that feed on these plant species may be favoured, allowing them to outcompete other species. This impacts the species at the next trophic level and ultimately can lead to shifts in a food web, again affecting biodiversity. As a result, many meadows, which once supported a highly diverse range of species, are now dominated by only a few. This can have knock-on effects on other species higher in the food chain (see Case study 2.2 and Box 2.5 in *Chapter 2: Biodiversity*).

Figure 3.7 A phosphate mine in Togo.
Source: Alexandra Pugachevsky, CC BY-SA 3.0.

In addition, not all nutrients that are applied to agricultural fields are taken up by plants, and a significant amount can be lost to the environment. For example, it is estimated that only 30% of the nitrogen available to plants is taken up on Dutch dairy farms (Daatselaar et al., 2015). Some of the applied nitrogen is emitted into the atmosphere. Through atmospheric transport, this nitrogen can be deposited in distant terrestrial systems, a process called **nitrogen deposition**. Nitrogen deposition is linked to a variety of sources, including the increased use of artificial fertilizers and the combustion of fossil fuels. This has resulted in an

BOX 3.5

The biochemical cycles vital to plant growth

The nitrogen cycle

The nitrogen cycle is an important biochemical cycle and includes chemical (for example, nitrogen gas in the atmosphere) and biological components (for example, nitrogen in plants or animals) (see Figure A). The majority of nitrogen is found as a gas (N_2) in the atmosphere. In fact, N_2 makes up 79% of our atmosphere (as a comparison, oxygen (O_2) makes up 21%, and CO_2 0.0038–0.0041%). This atmospheric nitrogen is not directly available to plants, but it can be captured by

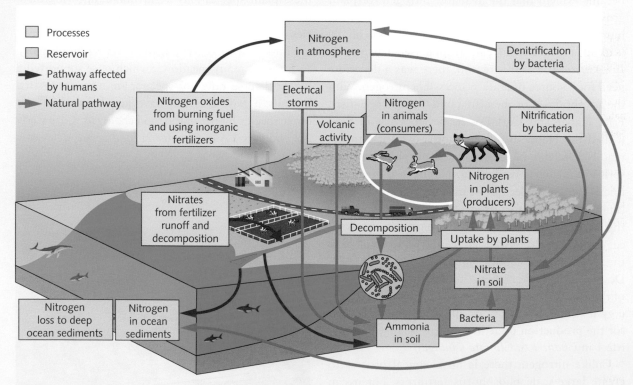

Figure A Simplified diagram of the global nitrogen cycle. Red boxes and lines indicate key human impacts on the cycle.

Source: Reproduced with permission from Miller, G. and Spoolman, S. (2012). *Living in the Environment*. 17th ed. Brooks/Cole.

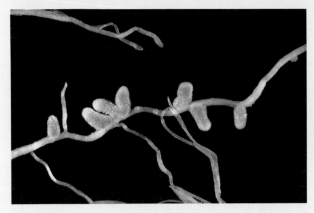

Figure B A root nodule on a legume, which contains bacteria from the Rhizobia family, which can fix atmospheric nitrogen.

Source: Ninjatacoshell/Wikimedia Commons/CC BY-SA 3.0.

Figure D Intercropping of spinach and beans (a legume).

Source: Polyculturen in de praktijk, SIGN/LNV.

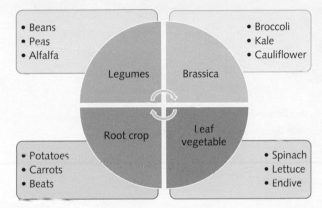

- Beans
- Peas
- Alfalfa

Legumes

Brassica

- Broccoli
- Kale
- Cauliflower

Root crop

Leaf vegetable

- Potatoes
- Carrots
- Beats

- Spinach
- Lettuce
- Endive

Figure C Example of a crop rotation which includes a legume for natural nitrogen fertilization.

Source: Thijs Bosker.

bacteria which transform it into forms of nitrogen that are available to plants (termed **N-fixation**). It is recycled through the decomposition of biological waste such as animal and plant tissues and excrements.

N-fixation normally occurs within soils by free-living bacteria. A select group of plants, which include legumes (for example, cloves, alfalfa, soybean, beans, and lentils), have developed a more efficient way to benefit from bacterial N-fixation. These plants have small nodules on their roots (see Figure B). Bacteria from the *Rhizobia* family live within these nodules and can fix atmospheric nitrogen into nitrogen available to plants. The host plant can use part of this fixed nitrogen for growth. In return, the host plant provides sugars to the bacteria, making this a true symbiosis (both organisms involved have a direct benefit from the cooperation).

Legumes are often used in agriculture to enhance soil fertility. They are used in crop rotations, ensuring that, every 3–5 years, the nitrogen status of agricultural soils is improved, thereby increasing the long-term yields within a farming system (see Figure C). As an alternative, some farmers conduct intercropping, in which two or more crops are produced on the same plot, one of which is a legume (see Figure D). The legume will fix the nitrogen and will make it available to the other crops (the exact process is still not completely understood but is likely related to root decomposition of the legumes).

Through the development of the Haber–Bosch process, humans have found a way to fix N_2 from the atmosphere and make it available to plants. The massive application of artificial fertilizers containing nitrogen has resulted in huge changes in the nitrogen cycle (see Figure A and main text).

The phosphorus cycle

Similar to the nitrogen cycle, the phosphorus cycle includes chemical (for example, phosphorus in rocks) and biological components (see Figure E). In natural systems, the main source of phosphorus comes from the weathering of rocks through erosion and the river transport of phosphorus down to the oceans. After phosphorus has been taken up within a food web, it can be recycled through the production of animal or plant waste, similarly to that with nitrogen.

Through the run-off of rain across the land and rivers, phosphorus is transported to our oceans where some of it reaches deep-sea sediments. Through the rock cycle, phosphorus can be recycled, reappearing at the surface at the earth where it can be mined. However, this happens over very long timescales (millennia to millions of years), and for this reason, phosphorus is considered a non-renewable resource.

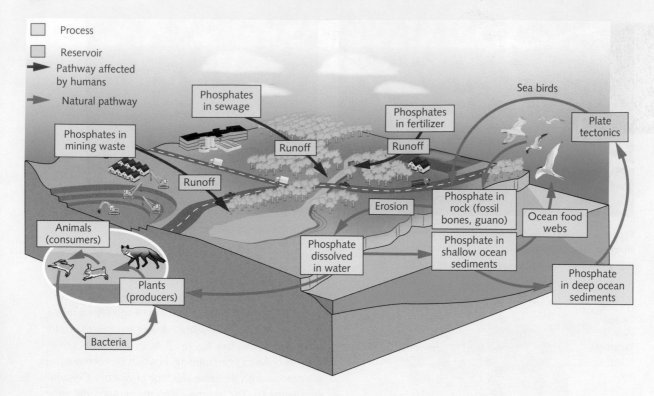

Process

Reservoir

Pathway affected by humans

Natural pathway

Phosphates in mining waste

Phosphates in sewage

Runoff

Runoff

Phosphates in fertilizer

Runoff

Sea birds

Plate tectonics

Animals (consumers)

Erosion

Phosphate in rock (fossil bones, guano)

Ocean food webs

Plants (producers)

Phosphate dissolved in water

Phosphate in shallow ocean sediments

Phosphate in deep ocean sediments

Bacteria

Figure E Simplified diagram of the global phosphorus cycle. Red boxes and lines indicate key human impacts on the cycle.

Source: Reproduced with permission from Miller, G. and Spoolman, S. (2012). *Living in the Environment*. 17th ed. Brooks/Cole.

through direct run-off, especially on rainy days (see Figure 3.8). The flow of nutrients to water bodies can result in **eutrophication**—the process by which a body of water is enriched with nutrients resulting in excessive plant and algal growth. As in terrestrial systems, only few aquatic plant species do well under conditions of increased nutrient availability. As a result, many other species will be outcompeted, resulting in a loss of biodiversity. On a larger scale, there can be even more dramatic impacts, resulting in areas deprived of almost any species, called dead zones (see Case study 3.1).

3.2.2 Pesticides and environmental impacts

Farmers have a long history in battling pests to protect their crops, as pests can have severe impacts on yields. An illustration is the outbreak of potato blight

Figure 3.8 Run-off from an agricultural field, which can result in eutrophication of surface waters.

Source: Lynn Betts/USDA Natural Resources Conservation Service

CASE STUDY 3.1

How nutrients can lead to dead zones

Aquatic systems are highly interconnected; they include small headwater streams, flowing on to connect to medium streams, which connect to larger rivers, ultimately discharging into estuaries. The combined area for which all water drains to one single point is called **watershed (or drainage basin)**. Huge quantities of water exit the watershed through this single point. For example, the river Rhine discharges approximately 2 million litres of water per second, the Mississippi River 16 million litres per second, and the Amazon River a staggering 200 million litres of water per second.

Now imagine the loss of nutrients from an agricultural field into a small headwater stream, resulting in a small increase in nutrients in the stream. As the water from all the small headwater streams in a watershed combines into medium streams and rivers, the total load of nutrients also increases. This results in a large influx of nutrients into the discharge points of the watershed—estuaries. This increase in nutrients stimulates the growth of primary producers, most notably algae. When the influx of nutrients reaches a critical point, massive algal blooms can occur. This typically happens in spring and summer when the water warms up. Some of these algal blooms are so voluminous they can be observed from space using satellite images (see Figure A).

During daytime, algae have a net oxygen production, meaning they produce more oxygen than that they use for respiration. However, at night, their oxygen production is negligible as they cannot photosynthesize, but they continue to use oxygen for respiration, resulting in a net deficit of oxygen production. In addition, the number of bacteria which decompose decaying algal matter increases during the algal blooms; these bacteria also require oxygen for respiration. As a result, there are daily fluctuations in oxygen levels, with higher levels during the day and lower levels during the night. At the peak of the bloom, this can result in zones with very low amount of oxygen (hypoxic) or even without any oxygen at all (anoxic).

When oxygen levels in the water drop below 25% saturation, the level of oxygen for many organisms is too low to survive, including most shellfish, invertebrates, and fish (see Figure B). The result are **dead zones**—large areas with can sustain only very low biodiversity due to the lack of oxygen. Since the 1960s, the number of dead zones has increased rapidly; they have been reported in 400 systems, affecting more than 245,000 km² (Diaz and Rosenberg, 2008). Estuaries and near-shore coastal environments are especially at risk of dead zones, significantly impacting the local ecology, but also local fisheries.

Figure A Satellite image of an algal bloom on the coast of England.

Source: NASA.

Figure B Fish kill in August 2003 due severe hypoxia—near anoxia—in Greenwich Bay (Narragansett Bay, Rhode Island, USA).

Source: Chris Deacutis/Integration and Application Network (CC BY 2.0).

in Ireland. Potato blight is caused by the oomycote (a fungus-like organism) *Phytophthora infestans*, which causes potatoes to rot from the inside. Potatoes were traditionally the food of the poor; they contain a relatively high amount of energy and have relatively high yields for the amount of effort needed by the farmer.

In the 1840s, a severe outbreak of potato blight in Ireland resulted in a near-complete loss of the potato yield. A variety of factors made the potato production in Ireland vulnerable to *P. infestans*. One of them is the clonal reproduction of potatoes, resulting in a genetically identical crop which is vulnerable to diseases. Second, potatoes were grown in lazy bed systems, with minimal crop rotation. The result of the crop failures was dramatic—a large proportion of the population of Ireland depended on potatoes, and when the yields failed, approximately 0.8 to 1.5 million people died due to hunger (Cousens, 1960; Mokyr and Ograda, 1984).

Massive famines resulting from pest-related crop failures are now rare. This is not because pests have disappeared; in fact, *P. infestans* is still an important pest which farmers need to deal with when growing potatoes. Due to advancements in agricultural science, there are now a variety of ways to protect crops by combating pests, including breeding pest-resistant crops, manually removing weeds or sick plants, and crop rotation.

In addition, synthesized pesticides are a key method for protecting crops and ensuring optimal yields. Much

like artificial fertilizers, pesticides have played a major role in ensuring food security, by helping to control vectors of disease (Matthews, 2007). Pesticides are used in agriculture to protect crops against a variety of pests, including fungi (fungicides), insects (insecticides), weeds (herbicides), and rodents (rodenticides).

In your experience

Next to their use in agriculture, pesticides are also commonly used by the general public. For example, people use herbicides for their lawns, insecticides to get rid of ants in their back yard, and poisons against rats and mice. In contrast to the use of pesticides by farmers, these activities are poorly regulated and monitored. Think of a situation in which a pesticide is used by the public, and consider what can be done to minimize the associated risks.

The global success of pesticides is based on their effectiveness in protecting crops against pests, their ease of use (see Figure 3.9), and promotion by governments and institutes. For example, in the USA, the use of pesticides on 21 important crops (including wheat, soybean, potatoes, and corn) was monitored between 1960 and 2008. Usage increased from 88 million kg of active ingredients (the actual compound in the pesticide which is biologically active) in 1960 to 234 million kg in 2008 (Fernandez-Cornejo et al., 2014).

a)

b)

Figure 3.9 Two different ways in which pesticides are commonly applied to agricultural fields: (a) by using tractors and (b) by manual application.

Source: (a) Lite-Trac/Wikimedia Commons/CC BY-SA 3.0; (b) T. R. Shankar Raman/Wikimedia Commons/CC BY-SA 4.0.

Similar to fertilizers, pesticides are inadvertently released into the surrounding environment when applied to agricultural fields. This can happen when a pesticide treatment is followed by a rain event, whereby a pesticide leaches into the soil and moves into ground and surface waters. It can also happen when pesticides are sprayed and there is drift towards adjacent fields due to wind. For this reason, pesticides are ubiquitous in the environment, and especially in surface waters. A study based on 4000 European monitoring sites found that pesticides are a major risk factor for aquatic life, including fish, invertebrates, and algae (Malaj et al., 2014).

Pesticides have a variety of effects on organisms (see Box 3.6 on how scientists assess the effects on organisms). The most extreme cases lead to direct mortality on non-target species. When direct mortality events occur, the effects are very visible, as dead or dying organisms can be observed. This causes strong negative responses in the general public, and therefore there has been considerable action to prevent these events from occuring.

In addition to lethal impacts, pesticides can have a variety of other sublethal effects on organisms, including adverse impacts on growth, development, and behaviour. In fact, nearly every system in an organism can be affected by pesticides. In a recent example, concerns have been raised about the possibility that pesticides, among other compounds, impact the endocrine system of organisms (see Case study 3.2).

As a response, most countries have developed regulations to minimize the risk of pesticides on the environment. Scientific research on impacts forms the foundation of an **Environmental Risk Assessment** (or ecological risk assessment)—a structured approach to quantitatively or qualitatively estimating the risk related to a specific (mix of) pollutant(s). This information is subsequently used by risk managers in a process called **Environmental Risk Management** to determine the course of action. Importantly, the decision not only depends on the environmental risk, but also other factors are taken into account. Economic, social, or legal constraints are important factors which are also considered by risk managers to determine the course of action and whether or not the use of a pesticide should be banned or restricted. As a result, different stakeholders actively try to influence the risk assessment, including environmental NGOs and the chemical industry. In 2017, the French prime minister Emmanuel Macron highlighted that this could potentially affect the independent decision-making process: '*We have endured too much pressure in recent years on these questions, too many hidden interests and industry expertise that is in no way scientific expertise.*' (Reuters, 2017).

Connect the dots

In *Chapter 13: Governance*, we will focus on the governance of environmental issues and how stakeholders influence these processes. As preparation, consider the key stakeholders who are active in debates around pesticide regulations. What are the roles of these different stakeholders in the debate? Can you think of ways in which they are trying to influence the decision-making process? Do you think that all of these stakeholders should be allowed to take part in this process?

BOX 3.6

How do we assess whether a pesticide can cause adverse impacts on organisms?

As a result of exposure to a pesticide, undesirable or adverse impacts on organisms can occur. **Acute toxicity** impacts occur almost immediately after exposure to a pollutant (often a high dose), while **chronic toxicity** refers to effects that manifest after a longer period of exposure to a pollutant (often to lower doses). Observed impacts can be very diverse but can be grouped into two categories: **lethal**, that is resulting in mortality events, or **sublethal**, in which case the exposed organism does not die but has lower fitness (for example, lower reproductive output, behavioural changes, or reduced immune function).

To study the impact of pollutants, a variety of methods can be used. The two most common approaches are **laboratory studies** and **field studies**. In laboratory studies, organisms are exposed under controlled laboratory conditions to a (mix of) pollutant(s) of interest in an experimental research design (see Figure A.a). In field studies, the impacts of a (mix of) pollutant(s) are studied using both observational and experimental studies directly within ecosystems (see Figure A.b). Both approaches have clear advantages and disadvantages, and, when possible, a combination of both laboratory and field studies should be used to assess pollutant impacts.

a)

b)

Figure A   Example of a laboratory and field study on environmental pollution.

Source: Thijs Bosker.

CASE STUDY 3.2

How pesticides impact hormone levels

The endocrine system is one of the most important systems in organisms—it regulates growth, development, behaviour, metabolism, and reproduction, to name but a few key processes. Hormones are the key compounds in the endocrine system, which are excreted by glands and travel in blood to distant target organs. They are signalling substances, which means they bind to receptors, after which a signal is transduced, resulting in a response. In many cases, amplification happens—one hormone binds to a receptor, resulting in the production of hundred of thousands of proteins. These proteins, for example enzymes, cause a response in the organism. Because of this amplification, the quantities of hormones needed to elicit a response are low, and most often levels of hormones are in the parts-per-trillion range. To put this in perspective, one part per trillion is approximately the same as one drop of water in an Olympic-size swimming pool.

The endocrine system is important in maintaining the steady state within organisms, called homeostasis. This happens through a complex series of feedback loops, which control the number of hormones within the blood of organisms. The endocrine system has remained remarkably unaffected by evolution. For example, most vertebrate organisms share the same hormones—testosterone has a similar structure in the mouse, fish, and human. The function of hormones within organisms can change, however. For example, in humans, thyroid hormones are important in growth, while in amphibians, they are essential for metamorphosis.

Some pesticides turn out to have the ability to affect the endocrine system and thereby disrupt homeostasis in organisms. And as hormones are signalling substances, even small amounts of pesticides in the environment can cause serious impacts on organisms, including humans. One compound which has been linked to endocrine disruption is atrazine. It is one of the most widely used herbicides in the USA (see Figure A) and is found in the environment (Solomon et al., 1996), as well as in drinking water (Benotti et al., 2009).

Research by Tyrone Hayes demonstrated that exposure to low levels of atrazine resulted in feminization of male frogs (Hayes et al., 2002; Hayes et al., 2003; Hayes et al., 2006). Follow-up studies have found that exposure to atrazine turned male tadpoles into females with impaired fertility (Hayes et al., 2010). In addition, atrazine has been linked to mammary and prostate cancers in laboratory rats, hypothesizing the link between atrazine

exposure and cancer in humans (Fan et al., 2007). However, this research has sparked a lot of debate, as other studies, some of which were funded by pesticide companies, showed no or limited effects of the pesticide on organisms (Kloas et al., 2009; Renner, 2008; Van der Kraak et al., 2014).

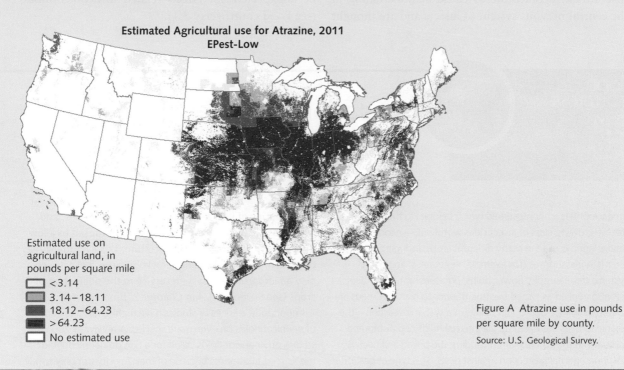

Estimated Agricultural use for Atrazine, 2011
EPest-Low

Estimated use on agricultural land, in pounds per square mile
- < 3.14
- 3.14 – 18.11
- 18.12 – 64.23
- > 64.23
- No estimated use

Figure A Atrazine use in pounds per square mile by county.
Source: U.S. Geological Survey.

One problem with minimizing the risk of pesticides in the environment is that the overall impacts or risks of pesticides are often very difficult to quantify. This is especially true for sublethal effects and mixtures of chemicals found in the environment. Since pesticides can have a wide variety of sublethal effects, and they are often present in mixtures, there is a near-endless combination of scenarios to test in order to quantify direct and indirect adverse effects on ecosystems. This makes it extremely costly and difficult to screen all (combinations of) pesticides on potential sublethal impacts prior to approval. As a result, there is significant uncertainty about the risk of pesticides, often resulting in conflicting reports in the media. Some media report on the enormous pressure that pesticides have on the environment, while others focus on the progress made in reducing environmental impacts and how the use of pesticides in intensive agriculture is vital to sustain high yields. Due to the conflicting

reports, it is hard for the general public to understand the potential risk of pesticides on the environment.

Due to the growing awareness of the potential side effects of some pesticides, the chemical industry is developing new pesticides to replace them. The intention is to develop pesticides which degrade faster in the environment or which have more targeted modes of action (that is, are more selective). For example,

Pause and think

Acute toxicity is often linked to lethal effects, while chronic toxicity is commonly linked to sublethal effects. However, other scenarios are also possible. Can you think of an example of an acute exposure resulting in a sublethal impact? And an example of a chronic exposure resulting in a lethal impact?

the chemical industry has developed neonicotinoids as an alternative to organophosphate insecticides. Organophosphates affect nerve functioning in many different animals (including humans, insects, and fish) and are therefore labelled as **non-selective pesticides**. In contrast, neonicotinoids cause malfunctioning in the central nervous system of insects and are thought to be much less toxic to other groups of organisms. Neonicotinoids are chemically similar to nicotine, a natural insecticide produced by tobacco plants. These pesticides are referred to as **selective pesticides**. However, recent studies have shown that there might also be unexpected side effects related to neonicotinoids (see Food controversy 3.1).

FOOD CONTROVERSY **3.1**

Imidacloprid: riskier than expected?

Imidacloprid is a neonicotinoid which belongs to the fastest-growing class of insecticides worldwide—neonicotinoids. Imidacloprid is used in many different crops, including crops grown in glasshouses (for example, vegetables) and open systems (for example, flower bulbs, potatoes, and sugar beets). It is also applied as a seed treatment; prior to planting, seeds are coated by imidacloprid. This results in a systemic reaction—the neonicotinoids are taken up via internal fluids and distributed throughout the whole plant (including the pollen and nectar), ensuring optimal protection against pests. This type of pesticide application is presumed to be environmentally friendly, as the chemical is applied directly to the target crop, minimizing the risk of run-off to the environment.

Imidacloprid (and neonicotinoids, in general) acts as a neurotoxin in insects, which results in paralysis of the insects, and ultimately death. It binds much more strongly to receptors in insects, compared to other groups of animals, making it effective against insects while there are limited impacts on other groups of animals. This is confirmed by laboratory studies; insects appear to be the most sensitive groups, while amphibians, fish, and mammals, are much less so. Based on this type of laboratory data, an Environmental Risk Assessment was conducted, and risk managers deemed that neonicotinoids are relatively safe to use. As of 2016, neonicotinoids are registered for use on hundreds of field crops in over 120 different countries.

However, many follow-up studies, both in the laboratory and in the field, have since found potential adverse effects of imidacloprid. The European Academies Science Advisory Council (EASAC) performed a large-scale analysis of available field and laboratory studies and concluded that the continuous preventive usage of imidacloprid (for example, in seed coating) caused major impacts on biodiversity in agricultural areas. In particular, honey bees seem to be affected (see Figure A.a). Many other species of wild bees, bumblebees, butterflies, and hoverflies are also very vulnerable when exposed to a low dose of neonicotinoids. For agriculture, this eventually may create a dramatic cascading effects because these insects play an important ecosystem service in the pollination of crops (see Section 2.1.4 in *Chapter 2: Biodiversity*).

Importantly, a series of studies have highlighted the impact of neonicotinoids on even more species. A review of 200 studies on neonicotinoids called for a ban on neonicotinoid use as seed treatments because of their toxicity not only to insects, but also to other wildlife, including birds and other invertebrates (Mineau and Palmer, 2013). A recent study in the Netherlands on 15 different insectivorous bird species found a strong decline in overall population (Hallmann et al., 2014) (Figure A.b). This negative trend could be statistically related to the amount of imidacloprid concentrations, as determined in surface water. The risk might be even higher for aquatic invertebrates, as a recent study found that they are extremely sensitive to neonicotinoids, even when exposed to relatively low doses (Vijver et al., 2017). As a result, imidacloprid and other neonicotinoids have received significant media attention.

However, there have been heated debates about the adverse impacts of imidacloprid on non-target organisms. The study linking declining Dutch bird populations to imidacloprid has been questioned by other scientists (Ter Braak, 2014). Next to the scientific debate, other key stakeholders also actively participate in this discussion. For example, Bayer CropScience, one of the leading producers of imidacloprid, states the following on their website: '*With hundreds of studies conducted and their demonstrated safe use on farmland across the country, we know more about the safe use of neonics to honey bees than any other pesticide. New studies continue to confirm their safety to bees and other pollinators when used appropriately.*'

a)

b)

Figure A Two examples of species on which imidacloprid potentially has adverse impacts: (a) the European honey bee (*Apis mellifera*) and (b) the Eurasian skylark (*Alauda arvensis*).

Source: (a) Luc Viatour / https://Lucnix.be / CC BY-SA 3.0; (b) Daniel Pettersson/CC BY-SA 2.5.

In contrast, Kirsten Thompson (a scientist at Greenpeace Science) was quoted as follows: '*Even if neonicotinoids are used exclusively inside greenhouses, they contaminate the surrounding environment. Their use may be restricted in principle, but the danger they pose to bees and other wildlife is not. Only a full ban would protect bees, other pollinators and wildlife from neonicotinoids.*'

The European Union (EU) has approved a ban on three of the main neonicotinoids in the field, including imidacloprid. It is unclear what will happen when the ban comes into force by the end of 2018. Will organophosphates be used again? Is this a worse alternative? Or are other options available? Does the ban need to be extended to usage in greenhouses as well? This debate will continue to unfold over the next years.

3.3 **Future outlook**

With the projected growth in the human population, the demand of food will continue to grow in the foreseeable future, as will the use of agrochemicals. Effectively minimizing environmental pollution in food production is therefore of great concern. Over recent decades, enormous strides have been made in producing food in a more environmentally friendly way, while maintaining high production volume (Foley et al., 2011).

To further minimize the impacts of pesticides, there is a focus on stimulating the production and use of more environmentally friendly chemicals and pesticides, including pesticides with shorter lifetimes (faster degradation) and pesticides which are more targeted to a specific pest and with minimal needless emissions to, and side effects on, the environment. However, even these new generations of pesticides can have potentially unwanted and unexpected side effects (see Food controversy 3.1). Because of this, many others want to move away from pesticides and artificial fertilizers completely, and move to organic agriculture, in which these agrochemicals cannot be used. Instead, fertilization relies on organic fertilizers, including compost, legumes in crop rotations, and animal manure. However, there are intense debates on whether organic agriculture can actually feed humanity. A study directly comparing the yields in conventional and organic agriculture found that, on average, yields are 8% lower in organic, compared to conventional, agriculture (Muller et al., 2017).

Regardless of this discussion, many techniques which have been developed or employed on a large scale in organic agriculture are now commonly applied in conventional agriculture as well. For example, biological pest control is an alternative to pesticides and commonly used in organic agriculture to control pests and diseases. In biological pest control, a pest is controlled by using other organisms, for example natural enemies. Ladybugs are important predators on aphids, which themselves are an important pest organism (see Figure 3.10a). Farmers can purchase ladybugs from commercial companies and introduce them in their crops to control aphid populations. Another example are parasitoid wasp species, which, as part of their life cycle, deposit eggs within another organism (see Figure 3.10b). These eggs hatch in the host organism, providing a direct source of food to the developing larvae, ultimately killing the host. The introduction of natural enemies has been especially successful in greenhouses; in these enclosed environments, the introduced predators cannot escape to the adjacent environment, resulting in effective control of pest organisms.

a)

b)

Figure 3.10 Two examples of natural enemies predating on pest insects: (a) a ladybug eating a daphnid, and (b) a parasitoid wasp placing an egg into a gypsy moth caterpillar.

Sources: (a) John Flannery/CC BY-SA 2.0; (b) Agricultural Research Service/United States Department of Agriculture.

Plant breeding is also a successful strategy to reduce environmental pollution. Plant breeding has a long history, dating back many centuries. The goal of plant breeding is to create crops with specific desired traits such as increased yields. Plant breeding has been used to develop crops which are resistant against pests and diseases (such as potato blight), reducing the need for pesticides. This is beneficial for both the environment and the farmer (as less pesticides usage means a reduction in production cost). Although this is a successful approach, it also takes a considerable amount of time to select and create crops with the desired traits. Now that we have entered the age of molecular biology, genetic manipulation of crops and animals provides numerous new opportunities to battle pests or increase nutrient uptake (see Box 3.7).

Of course, exciting as these pathways may seem, they also come with new challenges. There are fierce scientific and public debates about the potential environmental impacts of GM crops on the environment and human health (see Box 3.7). Also, there are ongoing debates about whether organic agriculture is indeed a viable alternative to conventional agriculture and about the harmfulness of pesticides to the environment. This highlights the continued need for independent scientific research in this highly dynamic field.

Connect the dots

In *Chapter 8: Nutrition*, we talk about health issues, one of which is VAD (see Box 3.7). Do you think the use of GM crops is warranted in the battle against VAD? And what about using genetic modification to develop crops that have a longer shelf-life to combat food waste, an issue discussed in *Chapter 11: Consumption*?

BOX 3.7   Genetically modified organisms

Genetic modification involves the direct manipulation of an organism's genetic material by using biotechnology. Most commonly, this involves the introduction of genetic material from one organism into the genetic material of another organism. In agriculture, this technique has gained popularity over traditional plant breeding as a way to introduce new, favourable traits into crops. One example is Bt-crops, in which genes from *Bacillus thuringiensis* (a bacterium) are genetically engineered into the genetic material of plants such as corn. During reproduction, *B. thuringiensis* produces a crystal protein which acts as a natural insecticide, killing a variety of different insects, including caterpillars and nematodes. In Bt-crops, the gene responsible for the crystal protein is incorporated into the plant DNA, allowing the plant to also produce the protein, resulting in a built-in insecticide.

Genetically modified (GM) crops with a large variety of different traits are now on the market, including crops with a longer shelf-life to reduce the waste of food, increased nutritional value, or the ability to grow under drier conditions or in saline environments (see Box 4.5 in *Chapter 4: Water*).

Another example of a GM crop is golden rice, which has been genetically modified to produce vitamin A. Vitamin A deficiency (VAD) is a major health issue in many developing countries, and especially in children. The WHO estimates there are 250 million pre-school children struggling with VAD. Every year, approximately 250,000–500,000 children with VAD become blind. About half of these children will die within

12 months of losing their sight. Most vitamin A is found in animal products, including eggs and meat, which often is not available in sufficient quantities in developing countries. As an alternative source scientists have developed golden rice, which produces beta-carotene, which can be converted into vitamin A in animal bodies. Beta-carotene has a distinct red-orange colour, resulting in the yellow colour which gives golden rice its name (see Figure A). As rice is a staple food for many in developing countries, the goal is to battle VAD in this way. At the moment there are still debates on whether golden rice is an effective way to reduce VAD in children.

The most common and controversial GM crops, however, are Roundup™-ready crops. These crops have been genetically modified to be resistant to the herbicide glyphosate, the active ingredient in Roundup™ (a tradename used by Monsanto, one of the world's leading agrochemical and agrobiotechnical companies). Glyphosate is a herbicide which is effective against many grasses and other plants. Roundup™-ready crops can be sprayed with glyphosate, killing weeds, but without affecting the crop.

There are huge controversies surrounding the use of GM crops, and the debate is highly political and highly polarized.

Figure A Golden rice is an example of a genetically engineered crop to battle VAD. It produces beta-carotene, which gives it a yellow colour. Beta-carotene can be transformed into vitamin A.
Source: International Rice Research Institute (IRRI)/CC BY 2.0.

Activists, for example, have regularly destroyed GM crops, and the industry is blamed for controlling markets, exploiting developing countries, and even driving farmers to suicide (see Food controversy 9.1 for more details).

● QUESTIONS

3.1 Describe the difference between the dilution paradigm and the boomerang paradigm, and place this in historical context.

3.2 Define biomagnification and bioaccumulation, and explain the difference between the two concepts.

3.3 Explain how dead zones develop in estuaries.

3.4 Endocrine disruptors are a group of environmental pollutants of emerging concern. Explain what they are and how they affect the endocrine system of organisms.

3.5 Environmental risk assessment is a process in which the risk of pesticides is quantified. Explain how these data are used by risk managers and what additional data they use to decide whether the use of pesticides should be allowed, banned, or restricted.

● FURTHER READING

Carson, R. 1962. *Silent Spring*. Boston, MA: Houghton Mifflin. **(Classic study by Rachel Carson on the risk of pesticides to our natural environment. She highlights the impact of DDT on birds and how this might result in a spring without birds whistling.)**

Diaz, R.J., Rosenberg, R. 2008. Spreading dead zones and consequences for marine ecosystems. *Science* **321**, 926–929. **(This paper describes the growing global issue of dead zones all across the globe and how this will impact ecosystems and human systems.)**

Nature. *GM Crops: Promise And Reality*. Available at: https://www.nature.com/news/specials/gmcrops/index.html [accessed 19 October 2018] **(An excellent website by *Nature*, providing unbiased and clear insights and background information on the highly controversial application of GM crops globally.)**

● REFERENCES

Benotti, M.J., Trenholm, R.A., Vanderford, B.J., Holady, J.C., Stanford, B.D., Snyder, S.A. 2009. Pharmaceuticals and endocrine disrupting compounds in US drinking water. *Environmental Science & Technology* **43**, 597–603.

Braune, B.M., Outridge, P.M., Fisk, A.T., Muir, D.C.G., Helm, P.A., Hobbs, K., Hoekstra, P.F., Kuzyk, Z.A., Kwan, M., Letcher, R.J., Lockhart, W.L., Norstrom, R.J., Stern, G.A.,

Stirling, I. 2005. Persistent organic pollutants and mercury in marine biota of the Canadian Arctic: an overview of spatial and temporal trends. *Science of The Total Environment* **351–352**, 4–56.

Cousens, S.H. 1960. Regional death rates in Ireland during the great famine, from 1846 to 1851. *Population Studies* **14**, 55–74.

Daatselaar, C.H.G., Reijs, J.R., Oenema, J., Doornewaard, G.J., Aarts, H.F.M. 2015. Variation in nitrogen use efficiencies on Dutch dairy farms. *Journal of the Science of Food and Agriculture* **95**, 3055–3058.

Diaz, R.J., Rosenberg, R. 2008. Spreading dead zones and consequences for marine ecosystems. *Science* **321**, 926–929.

Fan, W.Q., Yanase, T., Morinaga, H., Ondo, S., Okabe, T., Nomura, M., Komatsu, T., Morohashi, K.I., Hayes, T.B., Takayanagi, R., Nawata, H. 2007. Atrazine-induced aromatase expression is SF-1 dependent: implications for endocrine disruption in wildlife and reproductive cancers in humans. *Environmental Health Perspectives* **115**, 720–727.

Fernandez-Cornejo, J., Nehring, R.F., Osteen, C., Wechsler, S., Martin, A., Vialou, A. 2014. *Pesticide Use in US Agriculture: 21 Selected Crops, 1960–2008*. Economic Information Bulletin Number 124, United States Department of Agriculture, Economic Research Service.

Foley, J.A., Ramankutty, N., Brauman, K.A., Cassidy, E.S., Gerber, J.S., Johnston, M., Mueller, N.D., O'Connell, C., Ray, D.K., West, P.C., Balzer, C., Bennett, E.M., Carpenter, S.R., Hill, J., Monfreda, C., Polasky, S., Rockstrom, J., Sheehan, J., Siebert, S., Tilman, D., Zaks, D.P.M. 2011. Solutions for a cultivated planet. *Nature* **478**, 337–342.

Gari, L. 2002. Arabic treatises on environmental pollution up to the end of the thirteenth century. *Environment and History* **8**, 475–488.

Hallmann, C.A., Foppen, R.P.B., van Turnhout, C.A.M., de Kroon, H., Jongejans, E. 2014. Declines in insectivorous birds are associated with high neonicotinoid concentrations. *Nature* **511**, 341–343.

Harada, M. 1995. Minamata disease—methylmercury poisoning in Japan caused by environmental pollution. *Critical Reviews in Toxicology* **25**, 1–24.

Hayes, T., Haston, K., Tsui, M., Hoang, A., Haeffele, C., Vonk, A. 2003. Atrazine-induced hermaphroditism at 0.1 ppb in American leopard frogs (*Rana pipiens*): laboratory and field evidence. *Environmental Health Perspectives* **111**, 568–575.

Hayes, T.B., Collins, A., Lee, M., Mendoza, M., Noriega, N., Stuart, A.A., Vonk, A. 2002. Hermaphroditic, demasculinized frogs after exposure to the herbicide atrazine at low ecologically relevant doses. *Proceedings of the National Academy of Sciences of the United States of America* **99**, 5476–5480.

Hayes, T.B., Khoury, V., Narayan, A., Nazir, M., Park, A., Brown, T., Adame, L., Chan, E., Buchholz, D., Stueve, T., Gallipeau, S. 2010. Atrazine induces complete feminization and chemical castration in male African clawed frogs (*Xenopus laevis*). *Proceedings of the National Academy of Sciences of the United States of America* **107**, 4612–4617.

Hayes, T.B., Stuart, A.A., Mendoza, M., Collins, A., Noriega, N., Vonk, A., Johnston, G., Liu, R., Kpodzo, D. 2006. Characterization of atrazine-induced gonadal malformations in African clawed frogs (*Xenopus laevis*) and comparisons with effects of an androgen antagonist (cyproterone acetate) and exogenous estrogen (17 beta-estradiol): support for the demasculinization/feminization hypothesis. *Environmental Health Perspectives* **114**, 134–141.

Jones-Walters, L., Čivić, K. 2013. European protected areas: past, present and future. *Journal for Nature Conservation* **21**, 122–124.

Kloas, W., Lutz, I., Springer, T., Krueger, H., Wolf, J., Holden, L., Hosmer, A. 2009. Does atrazine influence larval development and sexual differentiation in *Xenopus laevis*? *Toxicological Sciences* **107**, 376–384.

Kula, E. 2013. *History of Environmental Economic Thought*. London: Routledge.

Malaj, E., von der Ohe, P.C., Grote, M., Kühne, R., Mondy, C.P., Usseglio-Polatera, P., Brack, W., Schäfer, R.B. 2014. Organic chemicals jeopardize the health of freshwater ecosystems on the continental scale. *Proceedings of the National Academy of Sciences of the United States* **111**, 9549–9554.

Matthews, G.A. 2007. Pesticides and agricultural development. In: Matthews, G.A. (ed) *Pesticides: Health, Safety, and the Environment*. Oxford: Blackwell Publishing Ltd, pp. 1–28.

Mineau, P., Palmer, C. 2013. *The Impact of the Nation's Most Widely Used Insecticides on Birds*. The Pains, VA: American Bird Conservancy.

Mokyr, J., Ograda, C. 1984. New developments in Irish population history, 1700–1850. *Economic History Review* **37**, 473–488.

Muller, A., Schader, C., El-Hage Scialabba, N., Brüggemann, J., Isensee, A., Erb, K.-H., Smith, P., Klocke, P., Leiber, F., Stolze, M., Niggli, U. 2017. Strategies for feeding the world more sustainably with organic agriculture. *Nature Communications* **8**, 1290.

Renner, R. 2008. Atrazine effects in *Xenopus* aren't reproducible. *Environmental Science & Technology* **42**, 3491–3493.

Reuters. 2017. *EU pesticide debate needs more 'independent expertise': Macron*.

Solomon, K.R., Baker, D.B., Richards, R.P., Dixon, D.R., Klaine, S.J., LaPoint, T.W., Kendall, R.J., Weisskopf, C.P., Giddings, J.M., Giesy, J.P., Hall, L.W., Williams, W.M. 1996. Ecological risk assessment of atrazine in North American surface waters. *Environmental Toxicology and Chemistry* **15**, 31–74.

Spitzer, P.R., Risebrough, R.W., Walker, W., Hernandez, R., Poole, A., Puleston, D., Nisbet, I.C.T. 1978. Productivity of ospreys in Connecticut-Long-Island increases as DDE residues decline. *Science* **202**, 333–335.

Ter Braak, C. 2014. *Het verband tussen gif en vogelafname is niet aangetoond*. NRC Handelsblad: Amsterdam.

UNdata. n.d. *Glossary: Pollution*. http://data.un.org/Glossary.aspx?q=pollution [accessed: 22 October 2018].

Van der Kraak, G.J., Hosmer, A.J., Hanson, M.L., Kloas, W., Solomon, K.R. 2014. Effects of atrazine in fish, amphibians, and reptiles: an analysis based on quantitative weight of evidence. *Critical Reviews in Toxicology* **44**, 1–66.

Van Oostdam, J., Gilman, A., Dewailly, E., Usher, P., Wheatley, B., Kuhnlein, H., Neve, S., Walker, J., Tracy, B., Feeley, M., Jerome, V., Kwavnick, B. 1999. Human health implications of environmental contaminants in Arctic Canada: a review. *Science of The Total Environment* **230**, 1–82.

Veeraraj, A. 1990. God is green. *International Review of Mission* **79**, 187–192.

Vijver, M.G., Hunting, E.R., Nederstigt, T.A.P., Tamis, W.L.M., van den Brink, P.J., van Bodegom, P.M. 2017. Postregistration monitoring of pesticides is urgently required to protect ecosystems. *Environmental Toxicology and Chemistry* **36**, 860–865.

Westbroek, P. 2012. *De ontdekking van de aarde*. Uitgeverij Balans: Amsterdam.

Water

How does agriculture impact freshwater resources?

Paul F. Hudson

There are two spiritual dangers in not owning a farm. One is the danger of supposing that breakfast comes from the grocery, and the other that heat comes from a furnace.

A Sand County Almanac, Aldo Leopold (1949)

Chapter Overview

- Freshwater resources are essential to human sustenance and sustainable food production. The exploitation of freshwater resources requires extensive hydraulic infrastructure to ensure a stable supply of water to crops and safety from flooding.

- A large proportion of agriculture is dependent upon pumping freshwater from water supplied by dams, reservoirs, and underground aquifers.

- Excessive pumping of groundwater from aquifers to support irrigation lowers the water table, a major driver of land subsidence. Subsidence is a significant problem along low-lying rivers and deltas because it increases flood risk and coastal erosion.

- Dams are heavily utilized by agriculture to support irrigation and flood control. This can create stable hydrologic conditions; however, they also impact rivers and biodiversity by degrading habitat variability and the storage of large volumes of river sediment, with adverse downstream consequences.

- There have been a large number of dams removed in Europe and North America, a trend which is accelerating. In other regions, dams are still being built, which remains a controversial topic.

- Dykes (levees) significantly restrict the natural 'flood pulse' and degrade floodplain ecosystems. However, they also facilitate the development of agriculture in regions near rivers and streams.

- Governmental institutions are necessary to design and implement appropriate policy and should especially utilize an integrated watershed management framework (IWM) to sustainably manage freshwater resources.

Introduction

Asparagus is a water-intensive crop, but one of the largest areas of asparagus cultivation is the Atacama Desert of southern Peru, one of the driest places on Earth, with an average rainfall of only 1 mm per year. This raises the question of how sustainable this actually is. To address the question requires consideration of economics, environmental policy, and hydrologic processes, especially related to the utilization of water resources. The ability of humans to exploit water resources for cultivation enables reliable food production in what are otherwise marginal or non-arable settings (see Figure 4.1). This exploitation and the rapidly increasing global population are two key reasons for the fourfold increase in agriculture-related water use over the past century.

Asparagus production in Peru represents an extreme case of water resource exploitation. The water used in this case can only be supplied because of high-land waters from the Ica River basin being diverted into canals and excessive reliance upon groundwater (water held under the surface in the pores of soil or the fractures of rock) (Hepworth et al., 2010). Diversion of highland waters has adverse consequences for aquatic ecosystems, as well as local communities who live in the highlands and have, for centuries, depended on the water resources to sustain their livelihoods. In the desert lowlands, where the crops are intensively cultivated by large agricultural enterprises, withdrawal of groundwater occurs at a rate much higher than that of its replenishment and is therefore unsustainable.

Agriculture for food production covers over a third of Earth's ice-free land surface and accounts for a staggering 70% of the freshwater utilized by humans (Foley et al., 2005). However, the continued intensification of food production needed to feed our growing global population and increasing wealth is dependent upon

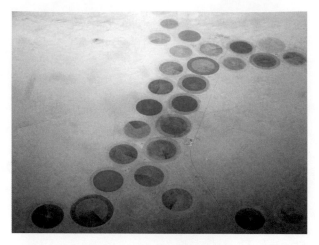

Figure 4.1 Pivot irrigation in the Sahara desert in Libya is an example of unsustainable use of freshwater resources for the production of crops in a desert.
Source: Futureatlas.com/CC BY 2.0.

freshwater resources that are increasingly vulnerable to global environmental change, especially in riverine lowlands and deltas. In this chapter, we provide an overview of key environmental issues related to freshwater resources and agriculture, with a focus on groundwater and surface water resources. The chapter first looks at how freshwater is distributed across the world (Section 4.1). Next, fundamental characteristics of hydrologic processes are reviewed (Section 4.2). Key examples of the environmental impacts of agriculture are given in Section 4.3. Finally, the role of international organizations and governance of water resources is reviewed in Section 4.4. As a whole, the chapter takes a watershed or drainage basin perspective, beginning with upland groundwater, river, and stream hydrology, and finally environmental change as related to agriculture in coastal deltaic environments where food production is increasingly concentrated.

4.1 How is freshwater distributed?

Water exists as a solid (ice), liquid, or gas, and its distribution and organization are governed by the global **hydrologic cycle** (see Figure 4.2a). The vast majority of water on our planet is saline (salt) water and is found in our oceans and seas (see Figure 4.2b).

Freshwater accounts for only 2.5% of Earth's water budget.

Most freshwater exists in a solid form (69%), as ice (see Figure 4.2b). Ice is primarily stored in thick polar glaciers (ice caps) and mountain glaciers,

a)

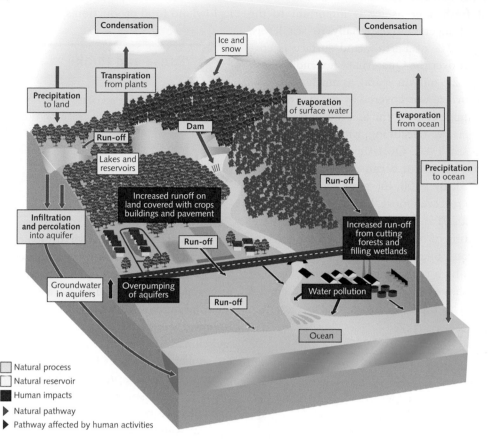

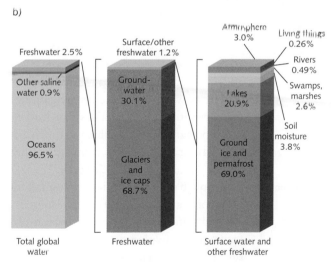

b)

Figure 4.2 (a) The hydrologic cycle within the watershed concept, illustrating several forms of agricultural influences on terrestrial pathways, and (b) its proportion within different states.

Source: (a) Reproduced with permission from Miller, G. and Spoolman, S. (2012). *Living in the Environment*. 17th ed. Cengage; (b) © NASA. Adapted from from Water in Crisis: A Guide to the World's Fresh Water Resources edited by Peter H. Gleick (1993). © Oxford University Press, USA.

as well as permafrost. The storage of substantial amounts of freshwater as ice is important to the global water budget, and especially to sea levels. Today, about 10% of Earth's land surface is covered by ice, but over the Quaternary (extending back approximately 2.6 million years), as much as 30% of Earth's land surface was covered by extensive glacial ice. Due to climate change, the amount of freshwater stored as ice has been declining, with significant consequences to sea levels. Over the twentieth century, sea levels rose by 16 cm because of climate change (see Figure 4.3). By 2100, the global mean sea level is projected to rise even more, between 36 and 64 cm (Church et al., 2013). The leading cause of sea level rise is the heating of ocean water, which expands the volume the water takes up, raising the sea level. The other factor is high rates of ice melt during the past 50 years, which is unprecedented over the recent 2.6 million years. This rapid rise in sea level increases the vulnerability of exploding coastal populations and its lowland food production systems.

Pause and think

Both the Netherlands and Bangladesh are home to major deltas—areas where rivers drain into the sea. Consider a scenario in which there is a 75 cm rise. What would be the consequences to humans, nature, and agriculture in both of these countries?

Atmospheric moisture represents an important transitory state in the global hydrologic cycle, between evaporative and condensation processes. Atmospheric water resources are very dynamic and respond rapidly to changes in temperature and wind patterns. While the total amount of water stored in the atmosphere is minor, water vapour is vital to Earth's energy flux and climate change, as it scatters and reflects incoming sunlight and stores heat emitted from Earth (see Section 6.1 in *Chapter 6: Climate Change*). Water enters the atmosphere through **evapotranspiration**—the collective process referring to evaporation from surficial water bodies, soil, and the surface of plant leaves. Although a natural process, evapotranspiration can impact freshwater resources and agriculture. For example, in dry environments, evaporative loss of water stored within large reservoirs (for example, lakes or storage ponds) can approach 10% annually and, if unaccounted for, can impact downstream flow, as well as water available for irrigation. In addition, improper irrigation practices can lead to high rates of evapotranspiration, which can drive soil salinization, having serious impacts on agricultural yields (see Section 5.2.3 in *Chapter 5: Soils*).

Although only accounting for a small fraction of the water on Earth, the freshwater resources available in a liquid state have important physical properties that facilitate the exploitation of freshwater for agriculture (see Box 4.1). As a whole, approximately 31% of Earth's freshwater is stored as either surface waters, such as lakes, wetlands, and rivers, or as groundwater (see Figure 4.2b). Groundwater is generally considered to be of higher quality than surface water. This is because the downward movement of water through soil and rock—**infiltration**—results in the removal of

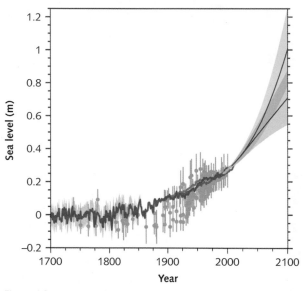

Figure 4.3 Historic and projected global sea level variation from 1700 to 2100. The data are a compilation of paleo sea level, tide gauge, and altimeter data sets from the Fifth Assessment Report (AR5) of the United Nations Intergovernmental Panel on Climate Change (IPCC). The central estimates and likely ranges for projections of global mean sea level rise are for low RCP2.6 (blue) and high RCP8.5 (red) emission scenarios. All data relative to pre-industrial values.

Source: IPCC, 2013: Climate Change 2013: The Physical Science Basis. Contribution of Working Group I to the Fifth Assessment Report of the Intergovernmental Panel on Climate Change [Stocker, T.F., D. Qin, G.-K. Plattner, M. Tignor, S.K. Allen, J. Boschung, A. Nauels, Y. Xia, V. Bex and P.M. Midgley (eds.)].

BOX 4.1

Key properties of water which are important for agriculture

Freshwater resources in a liquid state are easily exploited for agriculture because of several key characteristics. First, water exists as a liquid across a large temperature range (0–100°C), enabling it to be utilized for agriculture in cool and warm environments alike. Second, considerable freshwater is stored in the approximate top 10 to 30 cm of soil as soil moisture, which is directly available to crops. Third, because water has a high surface tension, it is stored within the soil matrix or can move upwards through the soil by a process called **capillary flow**. Briefly, capillary water is able to flow in small narrow spaces within the soil, even upwards against gravitational forces. Capillary flow is important to agriculture because it provides water to plant roots from deeper ground layer long after a rainfall event has passed, enabling crop cultivation in seasonally dry environments.

many physical and chemical pollutants. This natural cleansing of water, a valuable ecosystem service (see Section 2.1.4 in *Chapter 2: Biodiversity*), enables much groundwater to be utilized for agriculture and human consumption, with little or no treatment. Indeed, a common approach to municipal water treatment in developed and developing nations is the creation of artificial wetlands to foster water infiltration.

Soil moisture is the most important type of water in the hydrologic cycle (see Figure 4.2a) for sustainable food production, because of being directly available to plants (see Box 4.1). Soil moisture can be replenished by precipitation (rain, snow, sleet, etc.). The net amount of soil moisture depends on several hydrologic processes, including infiltration, run-off, and evapotranspiration, which redistribute water supplied by precipitation.

When the amount of water stored in the soil is less than its storage potential there is a soil moisture deficit. Importantly, the impact of climate change is expected to result in greater variability in soil moisture and in greater annual soil moisture deficits. By 2100 it is predicted that many land surfaces will experience a reduction in soil moisture (see Figure 4.4), including key agricultural regions such as south-eastern and western Australia, the US western Great Plains, South Africa, east China, and western Europe. The reduction in soil moisture will result in increased frequency and severity of agricultural drought. Agricultural drought refers specifically to extended periods of dry soil and is driven by a combination of reduced seasonal precipitation, higher rates of evaporation, and land change. Soil moisture projections are important to consider because of increasing future reliance upon irrigation to support food production, especially from groundwater resources, but also from surface waters. In the next sections, we will discuss groundwater and surface water resources in more detail and then highlight specific threats.

4.1.1 Groundwater resources

The largest source of available freshwater—by far—is groundwater, accounting for approximately 30% of Earth's freshwater resources (see Figure 4.2b). Groundwater is stored in underground **aquifers**, which are porous and permeable bodies of rock or sedimentary deposits with an ability to both store and transmit water (see Box 4.2). Aquifers represent a bountiful water resource for agriculture, providing high-quality water that can be readily utilized for crop irrigation.

In your experience

Do you know where your drinking water comes from? If you don't, try to find out. Do you know if it was cleaned before you drink it? If so, how was it cleaned? And how much do you pay for your drinking water? Do you think you pay a fair price?

Δ Soil Moisture

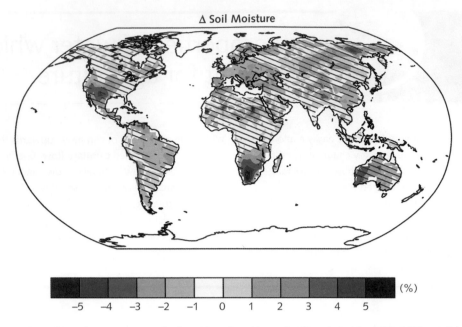

(%)

−5   −4   −3   −2   −1   0   1   2   3   4   5

Figure 4.4 The percentage change in average annual soil moisture (top 10 cm of soil) projected for 2018–2100, relative to a 20-year reference period (1986–2005). The change is modelled for a moderate emission scenario (that is, RCP4.5). The hatched areas indicate where the change is small relative to natural variability (<1 standard deviation), whereas the stippled areas indicate where the change is large (>2 standard deviations), compared to natural variability, with 90% of the models in agreement.

Source: IPCC, 2013: Climate Change 2013: The Physical Science Basis. Contribution of Working Group I to the Fifth Assessment Report of the Intergovernmental Panel on Climate Change [Stocker, T.F., D. Qin, G.-K. Plattner, M. Tignor, S.K. Allen, J. Boschung, A. Nauels, Y. Xia, V. Bex and P.M. Midgley (eds.)].

BOX 4.2

What is the difference between confined and unconfined aquifers?

Groundwater is stored in confined and unconfined aquifers with variable physical properties. **Confined aquifers** (also called artesian aquifers) are separated from the surface by an **aquiclude**, which is a dense sedimentary layer of clay or rock that is relatively impermeable (see Figure A). The aquiclude prevents the transfer of water and other materials between layers. From an environmental perspective, an aquiclude provides a protective layer from overlying pollution sources that can infiltrate (seep) into the soil from contaminant spills or excessive use of agricultural pesticides and fertilizers. By comparison with confined aquifers, **unconfined aquifers** lack an aquiclude (see Figure A). Unfortunately, the lack of an aquiclude makes unconfined aquifers highly susceptible to contamination. Chemical fertilizers and pesticides utilized

in large-scale mechanized agriculture (see Section 3.2 in *Chapter 3: Pollution*) may pollute unconfined aquifers.

In confined aquifers, deeper water is under greater pressure than shallower water, because of confining rock layers. This is important to farmers when using groundwater from confined aquifers for irrigation. Due to the pressure, groundwater from confined aquifers flows freely upwards, without using pumps. Because unconfined aquifers lack an aquiclude the water is not under pressure. However, because the water table of unconfined aquifers is at shallower depths (perhaps <200 m below surface), deep wells are not required. A substantial amount of water is stored in shallow unconfined aquifers. The main sources of drinking water for Amsterdam and The Hague (the

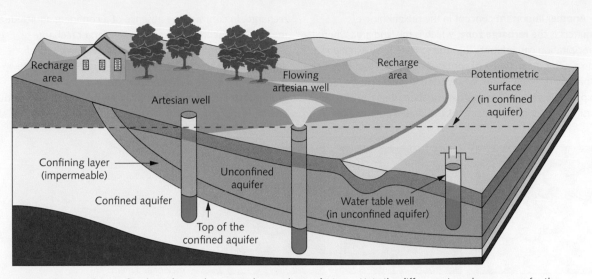

Figure A Confined and unconfined aquifers and associated groundwater features. Note the difference in recharge zones for the confined and unconfined aquifers.

Source: Adapted from U.S. Geological Survey. https://water.usgs.gov/edu/earthgwaquifer.html

Figure B Extensive sand dunes along the Dutch coast provide natural filtration and storage for precipitation and also riverine (river) water (which artificially helps to recharge the aquifer), providing a reliable source of freshwater for large metropolitan areas, including Amsterdam and The Hague (in photo). Land use activities within the aquifer recharge zone are strictly regulated to reduce pollution.

Source Thijs Bosker.

Netherlands), for example, are the extensive sand dunes along the Dutch coast (see Figure B).

There are several key properties which will allow aquifers to recharge (that is, refill with water). Two key physical controls on aquifers are porosity and permeability, which refers to the ability of an aquifer to store and transmit water. **Porosity** refers to the space between rock or sedimentary particles and is expressed as a percentage of the total volume (see Figure C). **Permeability** is related to the 'connectedness' of individual pore spaces and refers to the rate of water flow through the aquifer (see Figure C). For example, the permeability of clay (<0.13 cm per hour) is very slow, while that of coarse sand (>25.4 cm per hour) is very rapid (see Table A).

Another important concept in the replenishing of aquifers is the **recharge zone**, which is the land area where precipitation can infiltrate into the soil and rock to replenish the aquifer. A limitation of confined aquifers is that they have a narrow recharge zone. While artesian wells provide readily exploitable water resources for irrigation, they can also be easily overexploited because of the limited potential for recharge. In contrast, the absence of a confining aquiclude means that unconfined aquifers have larger recharge zones than confined aquifers. Rainfall over any part of an unconfined aquifer, therefore, can directly contribute to aquifer recharge. Conservation and management of groundwater often include regulations on the land use above aquifer recharge zones.

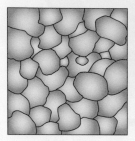
Low porosity
Low permeability

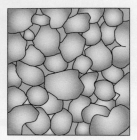
High porosity
Low permeability

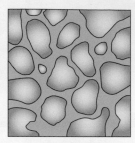
High porosity
High permeability

Figure C Porosity and permeability rates of rock or sediment.
Source: Food & Agricultural Organization (FAO) of the United Nations.

Table A Permeability and soil texture*

Permeability class	Permeability (cm per hour)	Textural class (median grain size)
Very slow	<0.13	Clay (0.004 mm)
Slow	0.13–0.5	Sandy clay, silty clay
Moderately slow	0.05–2.0	Clay loam, sandy clay loam, silty clay loam
Moderate	2.0–6.3	Very fine sandy loam, loam, silt loam, silty clay loam, silt (0.03 mm)
Moderately rapid	6.3–12.7	Sandy loam, fine sandy loam
Rapid	12.7–25.4	Sand, loamy sand
Very rapid	>25.4	Coarse sand (1.0–0.5 mm)

* See soil texture diagram in Chapter 5, Modified from: *Nature Education*.

Freshwater supplied by underground aquifers is vital to international food production systems. Large agricultural regions, such as southern Spain, western and south-eastern Australia, the Great Plains of North America, and even the perennially wet Mekong delta, rely heavily upon groundwater resources to generate high agricultural yields. An unsustainable high rate of groundwater pumping results in the **water table**, the upper level of the aquifer, to lower (decline in height). This is crucial because continued withdrawal of groundwater then requires larger pumps and greater amounts of energy, which is especially a problem to small farmers. The excessive withdrawal of groundwater in southern Peru associated with asparagus, for example, is reducing the level of the water table by as much as 0.5 m per year. The high rate of groundwater withdrawal has resulted in an overall lowering of the water table by 9.5 m between 1990 and 2008.

An important issue facing many of Earth's large aquifers, such as the Ogallala Aquifer of the American Great Plains (see Case study 4.1), is that they were primarily supplied during cooler and wetter conditions over the last 2.6 million years (the Quaternary). Thus, a pressing concern, especially in dry environments, is whether contemporary precipitation patterns will provide adequate rainfall for aquifer recharge in areas of high irrigation. The situation is expected to become increasingly dire over the twenty-first century. Recharge projections for many aquifers used for agriculture predict a reduction in annual precipitation, further diminishing aquifer recharge (Jiménez Cisneros et al., 2014).

CASE
STUDY
4.1

The decline of the Ogallala aquifer, a vital groundwater resource in the USA

The Ogallala aquifer is an enormous and complex aquifer that spans some 450,000 km² across the western American Great Plains (see Figure A.a). The natural vegetation of the western Great Plains was an extensive shortgrass prairie, but now the region is more known for grain cultivation. For this reason, the region is also known as the American breadbasket, as it accounts for 20% of total American food production (Konikow, 2013). The region is not known for its rainfall, however, and 54% of the land directly above the Ogallala aquifer is used for agriculture. This is possible because of intensive irrigation to support the growing of crops, with 85% of the agriculture relying

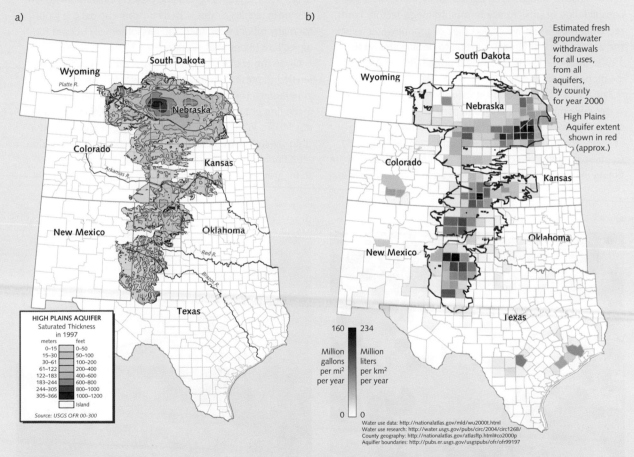

Figure A (a) The Ogallala aquifer in the western Great Plains of the USA, and (b) groundwater withdrawal rates (fresh water, all sources) by county in 2000.

Source: (a) USGS OFR 00-300; (b) U.S. Geological Survey.

directly upon irrigation from Ogallala groundwater (see Figure A.b).

The volume of the Ogallala aquifer is immense, and contains enough water to cover all 50 US states with 0.5 m of water. In the mid 1900s it was estimated to store some 1.7 quadrillion (1,700,000,000,000,000) litres of water, approximately the volume of one of the US Great Lakes. This was great news to farmers following the epic droughts of the 1930s and 1950s, and the Ogallala seemed like an unlimited resource. But the tremendous volume of groundwater reportedly stored in the Ogallala aquifer masked problems that were already beginning.

As a response to drought and the Dust Bowl era of the 1930s the Ogallala was intensively pumped to provide agriculture with a stable water supply. Initially the water table was close to the surface and, for a minor investment, small farmers could easily dig wells. But post-World War II saw the rise of industrialized agriculture, with considerable consolidation of small family farms into mega-factory farms, requiring more intensive exploitation of groundwater. While the water table was already starting to decline around the 1930s, after the 1940s, it rapidly dropped. By 1980, some locations had dropped by 30 m.

The 1980s saw temporary stabilization of the Ogallala's water table due to the improvement of irrigation technologies, specifically **centre pivot irrigation**, resulting in a very characteristic agricultural landscape (see Figure B). Centre pivot irrigation consists of a rotating sprinkler, and the timing and volume of water delivery are much more precise than irrigation by flooding. Nevertheless, the 1980s were a brief interruption to the larger trend of aquifer depletion, as high rates of Ogallala depletion have continued into the 2000s.

The problem with irrigation in the western Great Plains is that it is dry and is projected to become drier. The average annual precipitation across the aquifer, from south to north, is 350–500 mm per year. Evaporation rates from south to north, however, range from 1554 to 2667 mm per year. This large annual deficit is made worse because most of the rainfall occurs during spring and summer when evaporation rates are the highest. As a result, there is minimal aquifer recharge. Consequently, over the past five decades, about one-third of the total volume of the Ogallala has been depleted, and if current trends continue, by 2050, an additional one-third will be lost. To make things worse, climate change projections over the American Great Plains are worrying. Simulations under a 2.5°C warming scenario projects a minimum 20% decline in recharge rates, a likely outcome across the globe for many other unconfined aquifers intensively utilized for agriculture (Bates et al., 2008).

Given the above challenges, conservation is desperately needed to preserve the Ogallala aquifer for future generations. Unfortunately, the existing approach to managing groundwater is inherently flawed. Groundwater regulations in the USA are political, being determined on a state-by-state basis (presenting a collective action problem; see *Chapter 14: Collective Action*). And within a single state, groundwater laws are far from being in accord with

a)

b)

Figure B (a) Crop circles on the American Great Plains created by centre pivot irrigation and pumping of the Ogallala aquifer. NASA satellite photo of distinctive pattern of crop circles, Finney County, Kansas. The small and large irrigated plots are 800 and 1600 m in diameter, respectively. (b) Centre pivot irrigation and hay cultivation, Texas panhandle. The system delivers precise amounts of water over a specific area but encounters high losses due to high evapotranspiration rates.

Sources: (a): NASA; (b): Source: P.F. Hudson, 2002.

basic laws of hydrology. Texas, for example, has a 'right to capture' law, which can be freely translated into *the law of the biggest pump*. This means that a landowner can pump as much groundwater for use or sale, without consideration of adverse impacts on adjacent property owners or the environment.

Importantly, it is not food production only that is impacted by the depletion of aquifers. In dry regions, natural groundwater seepage sustains springs and associated riparian ecosystems (riparian zones refer to the interface between the river/stream and the terrestrial system and originate from the Latin word *ripa* meaning river bank). When the water table drops, water to sustain aquatic species dependent upon spring flow is also lost, which causes negative impacts to both ecosystems and biodiversty.

In addition to groundwater depletion, there are two other important environmental degradations associated with unsustainable pumping of groundwater. First, the lowering of the water table drives land **subsidence,** the sinking (lowering) of the land surface. Land subsidence is especially a problem in lowland coastal areas, because minor changes in elevation increase flood risks from rising sea levels and storm surge events (see Section 4.3).

Second, freshwater aquifers located along the coastal zone are especially vulnerable to **salt water intrusion,** which is caused by overpumping and sea level rise (Jiménez Cisneros et al., 2014). When groundwater withdrawal is not balanced with recharge, denser saline groundwater displaces freshwater and degrades the water quality. This is especially the case in highly porous and permeable sands (see Figure C in Box 4.2), which enable rapid seepage and displacement of freshwater.

Groundwater is considered contaminated by saltwater when sodium chloride (salt) concentrations exceed 250 mg/L (the EU standard). The tolerance of different crops varies widely, for example fruit trees have lower tolerance (200–600 mg/L), compared to tomatoes (1400 mg/L) and wheat (3000 mg/L). To prevent the crossing of tipping points, whereby recovery is not possible, requires expensive management and mitigation measures (Mazi et al., 2013). Such a scenario is of great concern along coastal Holland, the centre of agricultural production for the Netherlands, a country that has among the world's highest agricultural export value. Unless expensive management measures are implemented, saltwater intrusion and aquifer degradation in the Netherlands will become more severe with climate change and sea level rise (de Vries, 2007).

4.1.2 Watersheds and surface waters

Although they account for a meagre 1.2% of Earth's total freshwater resources, surface waters are the second major source of freshwater, behind groundwater (see Figure 4.2b). Surface waters include rivers, springs, wetlands, and lakes. Earth's surface waters are organized by **watersheds** (or **drainage basins**). A watershed is a topographically defined area of land that directs all run-off to a single outlet via a network of rivers (see Figure 4.5). The quantity of **discharge** [volumetric rate of streamflow, usually expressed in cubic metres per second (m³/s)] from a watershed is primarily determined by the amount of precipitation falling above the watershed and the size of the watershed. The size (area) of a watershed is determined by the **drainage divide,** a topographic border defined by elevation that encloses the watershed, except for a single point where the water discharges (see Figure 4.5). Headwater tributary

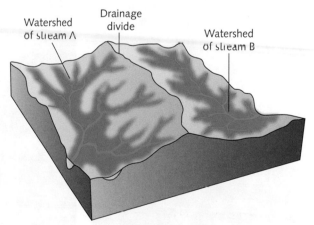

Figure 4.5 A watershed is an area of land that directs all run-off to a single outlet via a hierarchical network of rivers. The boundaries between watersheds are drainage divides, which are topographic (elevation) borders.

streams are often located within hilly or mountainous terrain. Downstream, small headwater tributaries join to form larger rivers, which eventually become large enough to form wide valleys able to support agricultural activities. While some watersheds drain into local 'closed basins', such as lakes or even inland seas, most watersheds drain into the ocean.

Connect the dots

In *Chapter 3: Pollution*, we discussed the occurrence of dead zones in estuaries. Explain how the usage of artificial fertilizers in a watershed can contribute to the development of these dead zones.

4.2 How have humans shaped our landscape to have access to water?

Having access to sufficient water for agriculture was vital during the development of human societies. This is captured in the concept of **hydraulic civilization**, which refers to ancient agrarian societies oriented around the management of water resources by government institutions (see Figure 4.6). Historically, the concept was applied to the Middle East and Asia (Butzer, 1976), and in the New World to pre-Columbian civilizations (that is, before 1492) of the Andes and Central Mexico (Mann, 2005).

As with the past, human reliance upon intensive agriculture for food production continues to depend upon substantial **hydraulic infrastructure**—engineering structures that modify landscapes and water resources. The major types of hydraulic infrastructure include dams in the upper and middle reaches of watershed, and dykes (levees), canals, and pumping and irrigation systems in fluvial (river) lowlands and deltas to distribute waters to crops and to protect fluvial lowlands from flooding. Because of their extensive size and design sophistication, hydraulic infrastructure requires robust government institutions for planning, construction, and management.

Unfortunately, hydraulic infrastructure is associated with considerable environmental impacts. In this section we explore several challenges related to

a)

b)

Figure 4.6 Two examples of hydraulic civilization: (a) an aqueduct in Cusco, Peru, developed by the Incas, and (b) an aqueduct near Nimes, France, developed by the Romans.

Source: (a) Rainbowasi/CC BY-SA 2.0; (b) Emanuele/Flickr/CC BY-SA 2.0.

BOX 4.3

How hydrologists study watersheds

Intensive agriculture impacts watershed hydrology, for example the streamflow in a river. The streamflow is supplied by both surface run off (also called overland flow) and **baseflow**, which is groundwater that seeps into a river channel. Because land use change from natural to agriculture land can alter run off, it can also impact streamflow. Hydrologists examine the impacts of different agricultural practices by analysing the shapes of **hydrographs**, plots of time (x-axis) and discharge (y-axis) (see Figure A).

A single-event hydrograph illustrates the response of a watershed to an individual precipitation event and reveals the influence of land cover on hydrologic processes (see Figure A). Under natural land cover, such as a forest or natural grassland, a single-event discharge hydrograph is associated with a relatively smooth rising slope and declining slope, revealing that forests or grasslands reduce run-off. However, when the natural cover is replaced by agriculture, the run-off rates increase and baseflow declines. This results in several fundamental changes to the hydrograph, including: (i) higher flood peaks (more stream power), (ii) steeper rising and declining slopes, (iii) less baseflow, and (iv) reduced hydrologic **lag time**. The latter refers to the time between the rainfall event and the peak discharge (see Figure A).

Overall, a change to agriculture results in a more 'flashy' hydrograph, a qualitative expression to refer to the increased variability in land drainage from an individual rainfall event. There are two problems caused by the change in the hydrograph in Figure A; the first is greater stream power, which can result in erosion of the channel bed. The second

problem is related to reduced baseflow, which results in transformation of a stream with year-round flow (perennial flow) to streamflow occurring immediately after rainfall/run-off events and then drying up (ephemeral flow). In many cases, transformation means that rivers have insufficient water to support aquatic ecosystems.

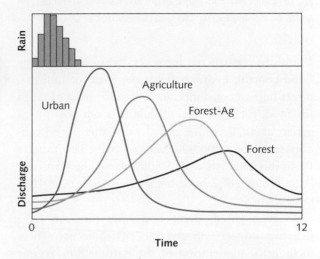

Figure A A continuum of discharge hydrographs over a 12-hour period, from natural to agricultural to urbanized settings, generated by a discrete rainfall event. Comparison of the curves illustrates the hydrologic response of landscapes to land change, with natural (forest) and urbanized settings representing the extreme end-points. The hydrographs change from 'smooth' for the forest conditions to 'flashy' for agriculture and urban conditions..

hydraulic management of water resources for food production, to illustrate its environmental impacts on river systems and freshwater resources. We begin by reviewing the impacts of dams and reservoirs on water flow. Second, we consider agriculture in fluvial lowlands, including the impacts of dyke systems on floodplains. Finally, we move further downstream and examine unique types of environmental impacts caused by agricultural activities in large river deltas.

4.2.1 Dams and reservoirs

Dams and reservoirs (see Figure 4.7) are essential to sustainable food production, providing stable water resources for irrigation and enabling agricultural activities to expand upon marginal lands (Biermans et al., 2011; Zarfl et al., 2015). Over half of Earth's global run-off is regulated by dams, with most dams located in watersheds with heavy agriculture.

a)

b)

Figure 4.7 Dams and reservoirs come in a variety of sizes and designs: (a) a small run-of-the-river dam and associated reservoir in India and (b) a high arch hydroelectric dam and reservoir in France.

Source: (a) Vijayanrajapuram/Wikimedia Commons/CC BY-SA 4.0. (b) Global Ref/Wikimedia Commons/CC BY-SA 3.0.

Dams are constructed for several reasons, including flood control, hydropower, water uses for industry, direct human consumption, and cultivation of crops for food, fuel, and fibre. While it is not possible to arrive at an accurate global tally, there are probably more than 200,000 dams globally. The Yangtze Basin in China has over 50,000 dams, the USA over 75,000 dams of 2 m or higher, and India over 4147 large dams.

As highlighted, dams and reservoirs can have negative impacts on riverine environments. These include: (i) moderation of the natural flow regime (see Box 4.3), (ii) reducing downstream sediment loads to riparian and coastal environments due to sediment trapping behind dams, and (iii) fragmenting watersheds, so that aquatic organisms are unable to migrate along the riparian corridor (upstream and downstream). In most instances, the environmental flow downstream of a dam is less variable, compared to natural flowing rivers (see Box 4.4).

While adverse impacts are well understood by scientists, this has not prevented the construction of new dams, especially large dams. Indeed, an additional 3700 'mega dams' are either planned or under construction, mainly in Asia and South America (Zarfl et al., 2015). Large dams are also planned for the Congo River, long considered among the most pristine large rivers on Earth. The construction of new dams is highly controversial because of prospective impacts on unique biotic and abiotic riverine environments (see Food controversy 4.1).

Connect the dots

Hydropower is a source of renewable energy (see *Chapter 7: Energy*), which combats climate change (see *Chapter 6: Climate Change*). Give arguments for and against the use of hydrodams, and link this to both climate change and other environmental impacts. In addition, hydrodams can have social impacts. What are examples of positive and negative social impacts of hydrodams?

In other areas, there is a trend towards the removal of dams. Dam removal is occurring at an accelerated pace in Europe and the USA. Since the 1970s, 1150 US dams have been removed, although the regional pattern is very uneven (Foley et al., 2017) (see Box 4.5). In the Pacific Northwest, dams are being removed in the intensively dammed Columbia River basin of Washington and Oregon, as well as in California, with 12 dam removals in 2014 (https://www.americanrivers.org). The pattern of dam removal is also uneven across Europe, with Spain, France, and the UK in the lead. Overall, 3450 dams have been removed, but hundreds of thousands of dams, weirs, and small dam-like blockages continue to fragment Europe's rivers (Dam Removal Europe, 2018). The European champion in dam removal is France, which has some 90,000 obstacles or structures in its waterways, with 70,000 of these being either dams or weirs. Since 1996, France has removed 2100 structures, including several larger dams.

BOX 4.4

Environmental flows and the impacts of dams

The amount of streamflow in a river varies both spatially along the river because of tributary inputs and temporally with seasonal changes in precipitation. Different flows support different abiotic and biotic functions (see Figure A). The concept of environmental flows is embedded in modern 'integrated' approaches to river basin management. **Environmental flows** refer to managing river streamflow, so that human, geomorphic (especially erosional processes), and ecological functions are maintained. Essentially, establishing environmental flows for a river requires a hydrologist to determine how much water can be allocated for agriculture, such as for irrigation, while still sustaining the physical and aquatic habitat of the river.

Establishing environmental flows for a river means that a hydrologist must determine the relationships between different streamflow levels with associated aquatic ecological zones. These can be unique to each river (Whiting, 2002). Fine sediments such as silt are transported downstream by low-magnitude discharge events which happen often, perhaps ten times per year (see Figure A). Such 'flushing' flows are important, because they prevent pore spaces of the channel bed from becoming 'clogged' with fine sediment, which leads to habitat degradation. The health of aquatic organisms living in the river sediment, for example, is heavily dependent upon accessing food within the top 5 cm to 20 cm of the channel bed.

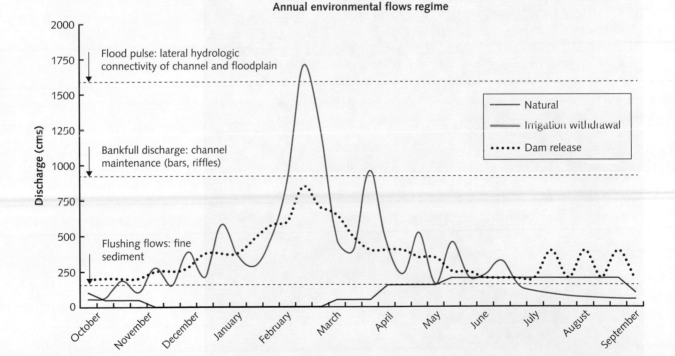

Figure A This annual hydrograph shows the varying flows in a river basin over a year and relates it to different biophysical functions and human uses. In temperate regions, the hydrological or water year usually runs from 1 October to 30 September because of the seasonality of precipitation and streamflow. The hydrograph includes withdrawal of water for irrigation and typical discharge release patterns downstream of a hydroelectric dam. The units for discharge are in cubic metres per second (m³/s).

Source: © the author.

Larger discharge events move and reorganize the river's macro features such as large rocks and river banks. Such features are associated with varying stream velocities associated with different fish habitats, among other organisms. The **bankfull discharge** refers to the amount of discharge (streamflow) that fills up the channel but does not flow overbank (not a true flood event) (see Figure A). Because it is mostly associated with channel maintenance, it is a key concept in river channel formation and management.

Flood pulses rhythmically occur about once per year (see Figure A) in large humid zone rivers such as the Amazon (Junk et al., 1989). Flood waters seep into low features along the floodplain and infill floodplain water bodies, remaining inundated for months (see Figure B). Flood pulses are essential to the overall environmental health of riparian corridors, because they can connect rivers, which drive an exchange of nutrients, water, fine sediment (silt/clay), and organisms between rivers and floodplains. Flooding is essential to floodplain biogeochemistry, including the sequestration of phosphorus and nitrogen so that they are not flushed to coastal waters. Additionally, the life cycles of many fish species are dependent upon floodplain water bodies for feeding and spawning.

In most instances the streamflow downstream of a dam is less variable, as low flow periods are higher and high flow periods are lower. Importantly, critical high-flow pulses and bankfull discharge events are often reduced, if not eliminated downstream of dams. During warm and dry periods (summer months), hydropower releases can resemble a 'sawtooth' pattern (see dotted line on Figure A). Irrigation demands are greatest during dry seasons (see red line on Figure A), which increases stress on the river, as there are low flows during that period.

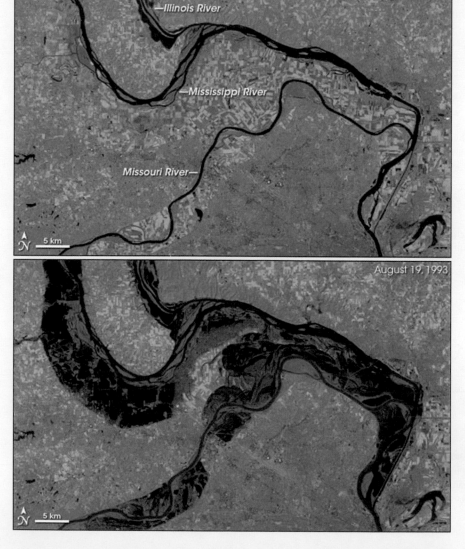

Figure B The middle Mississippi River near St Louis in the USA before (top) and during (below) the massive 1993 flood that breached the dikes (levees).

Source: NASA Earth Observatory.

FOOD
CONTROVERSY
4.1

Dam construction

Dams are strongly linked to agriculture, and for many rivers they will remain as part of the hydrologic framework for the foreseeable future. Nevertheless, dams remain a highly controversial topic in watershed management (Jørgensen and Renöfält, 2013). In addition to environmentalists, other groups can also be negatively impacted by dam construction. For example, indigenous peoples whose livelihoods were (or will be) disrupted by impounded flows, upstream landowners who lose flooded property, and recreational interests that depend upon natural flows may oppose the construction of new dams. However, overall, dam proponents are much more powerful than dam opponents. Proponents of dams include formidable interests such as federal agencies, large construction companies, utility companies, downstream developers, and property owners (including farmers) who benefit from flood control.

Because of the polarized position of stakeholders when it comes to dams, there are often fierce debates. A recent controversial example was associated with the planned construction of the Sivens Dam on the Tescou River in France. The Tescou River is a tributary to the Atlantic draining the Garonne River in southwestern France. Construction began in 2011, and the reservoir would have flooded 13 hectares of critical wetland and aquatic habitat, but would have benefitted only a small number of farmers. The issue pitted farmers against environmentalists and caused clashes (see Figure A), that resulted in an unfortunate fatality.

Construction of the Sivens Dam was halted in 2014 by utilizing a powerful EU policy instrument: the EU Water Framework Directive (WFD). The WFD requires all EU nations to develop plans for the management of rivers (EC, 2000) and includes specific ecological and hydrologic targets. Because the dam would have degraded such hydrologic and ecologic habitat, the French government ultimately had to abandon construction of the dam.

Figure A Riots erupted between environmentalists protesting against the construction of the Sivens Dam in the Tescou River in France and police, resulting in a fatality.

Source: Davel4444/Wikimedia Commons/CC BY-SA 4.0.

BOX 4.5

The art of dam removal

While much is known about the impact of dams on aquatic environments, the science of dam removal, however, is in its infancy. The removal of the Elwha River dams in coastal Washington represents a vital case study for the development of procedures and protocols in removing larger dams and in monitoring their post-dam geomorphic and ecological response. The Elwha River dam removal project started in 2011 and is the largest dam removal project in the world (US Geological Survey, 2015). The removal of the two dams occurred in two phases and included the initial removal of the 30-m high Elwha River Dam, followed by a phased removal of the 64-m high Glines Canyon Dam (see Figure A.a).

The key motivation in removing the Elwha River dams was recognition of the cultural value of the annual salmon migration to resident Native Americans, which has been a controversy for many decades. Removal of the dams abruptly released 7.1 million m³ of sediment into downstream riparian and coastal zones (see Figure A.b). Such a dramatic change to the physical environments raises questions about sediment management, so that associated shallow marine environments are not degraded. Additionally, the upstream riparian restoration is expected to take decades. Nevertheless, the early results are impressive, as salmon are returning to the headwater zone for the first time in nearly a century.

a)

b)

Figure A The Elwha River dam removal project in coastal Washington (USA) in 2011, the largest dam removal project in the world. (a) Destruction of the 64-m Glines Canyon Dam on the Elwha River. (b) Sediment plume at the mouth of the Elwha River, Washington (USA), following removal of two upstream dams.

Source: U.S. Geological Survey.

While downstream degradation of aquatic ecosystems is often seen as the motivating factor for removal, dams are usually removed because they are no longer economically viable or because of liability concerns. Many older reservoirs have infilled with sediments and provide less hydropower than originally projected. Also, older dams were often constructed of different materials and specifications (standards) and considered hazardous. The cost to maintain, repair, and upgrade dams to modern government safety standards often exceeds economic returns. Small dam owners, which are often local-scale power companies, are increasingly opting to remove dams. Thus, an ironic twist is that the same rationale for dam construction—economics—is also the justification for their removal.

4.2.2 Fluvial lowlands and agricultural systems

Fluvial lowlands along large river valleys and deltas have very high fertility and have been used for agriculture for thousands of years. The earliest large hydraulic civilizations settled along lowland rivers to exploit plentiful water and fertile soils. In addition to the Nile, classic examples include the lower Indus River basin and Mesopotamia and the floodplain lands between the lower Euphrates and Tigris Rivers in modern-day Iraq. The bounty of resources provided by the location, including access to water for consumption and irrigation and fisheries for food and trade, as well as the riverine transportation network, provided the only viable location for extended settlement in an otherwise arid environment (Butzer, 1976). Traditional subsistence floodplain agriculture remains important in developing nations (see Case study 4.2).

As discussed in Section 4.2.1, the seasonal flooding of a river floodplain—the flood pulse (see Box 4.3)—is an important process for riparian ecosystem health. In addition, it also reduces flooding in downstream areas, as water is dispersed over the floodplain and temporarily stored until the flood wave passes. However, the annual flood pulse can also cause considerable crop damage (see Figure 4.8). An inherent conflict exists

Figure 4.8 Flooding and sediment dispersal of the Yazoo River along the lower Mississippi valley, resulting in inundation and damage to agriculture (soybean fields).

Source: U.S. Department of Agriculture.

CASE STUDY 4.2

Subsistence floodplain agriculture in the Amazon

Subsistence floodplain agriculture is widely practised along lowlands in rural Africa, Asia, and Central and South America. Along the lower Amazon River in South America, small floodplain communities practise subsistence agriculture and are sustained by both floodplain farming and floodplain fishing (WinklerPrins and Barrios, 2007). In Figure A, we can see that local communities are often established along the river banks atop natural levees, which are a type of floodplain landform comprised of flood sedimentary deposits. From the river bank, moving down the levee back slope, the sediment decreases in size, from sand to clay and becomes wetter. The sandy soils at the levee crest support settlement and cultivation of fruit trees and manioc, the latter being a major source of carbohydrates (starch) and essential to the daily diet. Lower down the slope, the silty-clayey soils support cultivation of crops such as maize, beans, melons, and cucumbers. Finally, the wet clayey soil at the levee base, the transition to the flood basin, is utilized mainly as pasture for seasonal cattle grazing.

Large floodplains provide more than fertile soil and water for food production, they are also essential fisheries. One of the most important Amazon fish is the pirarucu (*Arapaima giga*), a very large fish that grows up to 3 m in length (see Figure B). The complex assemblage of floodplain lakes in the Amazon represents a hybrid natural-managed resource. Pirarucus move into the floodplain lakes during flooding. The fish remain in the floodplain for an essential part of their life cycle, especially for spawning and feeding on small fish, nuts, fruits, and small floodplain animals.

Management is essential to the survival of floodplain fisheries (McGrath et al., 2007). Due to its high commercial value, the pirarucu had been in decline for decades, with a precipitous drop after 1950 with the development of larger private fishing enterprises. Top-down federal approaches to manage the fisheries were ineffective because agencies failed to engage with local communities who do most of the fishing. Federal agencies also lacked human and physical resources to

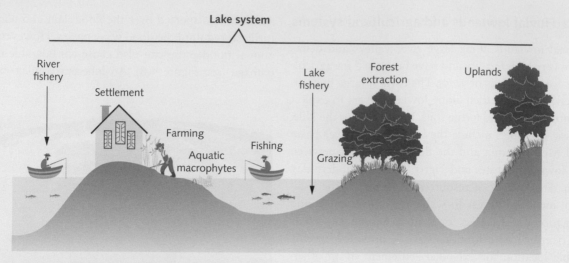

Figure A A model of a subsistence floodplain food system along the lower Amazon River, showing settlement and farming upon natural levees that slope from the river bank to the floodplain lake fishery.

Source: Adapted with permission from McGrath, D.G., de Castro, F., Futemma, C. et al. *Human Ecology* (1993) 21: 167. https://doi.org/10.1007/BF00889358. Copyright © 1993, Plenum Publishing Corporation.

Figure B A pirarucu (*Arapaima giga*), an important staple in the Amazon.

Source: pirarucu/Flickr/CC BY-ND 2.0.

enforce fisheries policies over the vast and remote Amazonian floodplain. A more recent community-driven (bottom-up) approach is attributed to a rebound in fish stocks (this is a good example of what the economist Elinor Ostrom's work argued for in order to deal with collective action problems; see *Chapter 14: Collective Action*).

The essence of the community-oriented approach is that access to the floodplain lake is controlled by local residents, but with communication with federal organizations. Federal and state organizations provide key information as to the broader health of the fishery resource. Federal organizations also have placed a moratorium on the use of gill net fishing, requiring traditional methods only by local inhabitants. Since locals are incentivized to conserve their fishery, they take measures to reduce degradation such as restricting cattle to specific locations and planting fruit trees to provide food for the pirarucu. Perhaps most importantly, local inhabitants take inventory of fish stocks and control access to the fishery. This is important, as controlling access provides a sense of ownership and ensures that their local knowledge of the hydro-ecosystem is integrated within the broader management framework.

between floodplain stakeholders—floodplain inundation reduces downstream flood levels and is essential to riparian ecology, while damaging agricultural lands.

Large floodplain agricultural enterprises also need lots of freshwater and protection from flooding. Freshwater is usually supplied from groundwater pumping of shallow aquifers. Flood control often comes from one of the oldest human inventions: the humble—millennial old—earthen dyke.

Dykes are constructed to control flooding, providing an opportunity to develop floodplains for agriculture and other human activities (for example, settlement, industry). This phenomenon can be succinctly stated as: **development follows the dyke** (White, 1945). Flood control infrastructure is not actually constructed to protect floodplains from flooding; it is to protect human activities (mostly agriculture). Of course, the development of a dyke system involves more than earthen dykes, but also a complex assemblage of pumps, canals, ditches, and drains that must be designed, constructed, and maintained by effective government organizations (see Box 4.6).

BOX 4.6

That sinking feeling: agriculture and ground subsidence

Along lowland floodplains and deltas, accelerated land subsidence (or sinking of the land) is one of the most significant environmental impacts related to agriculture, substantially increasing flood risk and increasing the vulnerability of humans and food production systems. Making land suitable for agriculture requires drainage and pumping of groundwater, which abruptly initiates land subsidence (Allison et al., 2016; FAO, 1988; Van de Ven, 2004).

The environment to which we refer consists of enormous basins of peat, which is plant material slowly decaying in anaerobic (absence of oxygen) conditions with a high water table at or near the ground surface. But once dykes are built and groundwater is pumped for drainage, the water table lowers

and organic matter comes in contact with oxygen (aerobic conditions) and starts to rapidly decompose (Galloway et al., 1999). The organic carbon in peat and soil is converted to carbon dioxide (CO_2). This results in the loss of organic material, and therefore subsidence. Biological activity by fungi, bacteria, and microflora further contributes to decomposition, resulting in rapid degassing of CO_2 from organic soil and peatlands. This loss of organic mass then also increases soil porosity.

The combined impact of increased porosity and water table lowering (by drainage) increases the overlying stress on soil, which further drives compaction and subsidence. A common result is a subsidence bowl (see Figure A), a large depression basin that is lower than the surrounding

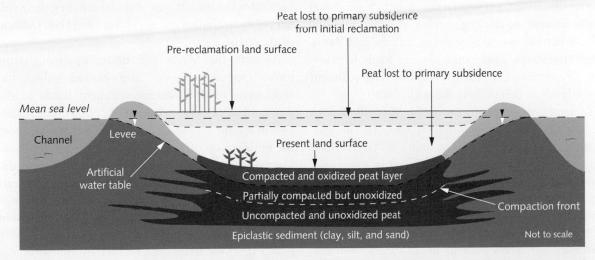

Figure A Depression basin created by subsidence of peatlands as a result of human-induced oxidation caused by diking (levees) and drainage to support intensive agriculture.

Source: U.S. Geological Survey.

landscape. This can generate a type of positive feedback, requiring additional drainage and pumping to prevent flooding, but which drives further subsidence (Allison et al., 2016). The process occurs most rapidly within the first decades to a hundred years, and then declines to a steady state (Van Asselen et al., 2011).

Subsidence basins within interior delta plains often result in the formation of new shallow lakes. In coastal regions, it often results in loss of land. This occurs because as deltas subside, waves and tides drive saline marine water further inland. This increases the salinity of wetlands, killing plants and exposing the soils to waves, driving erosion. In the Netherlands (see Figure B), subsidence began around 1000 years ago as peatlands were dyked, drained, and reclaimed for agriculture and energy. Such processes formed the Zuidplaspolder in the Rhine delta, the lowest place in Europe at 7 m below sea level (Van de Ven, 2004). Extensive subsidence also can be seen in lowland Indonesia and Malaysia. In these regions, subsidence has been caused by extensive oil palm cultivation to support a global processed food industry, among other human uses (see Case study 2.1 in *Chapter 2: Biodiversity*) (Erkens, 2015; FAO, 1988).

Figure B A dyke along the River Rotte (left, with windmill) near Zevenhuisen (approximately 5 km North East of Rotterdam), with deep subsidence of an adjacent polder (field on right). Historically, windmills were used to drain the polder. The difference in elevation between the dyke and polder is approximately 2.5 m.

Source: http://www.mooizuidplas.nl/wp-content/uploads/2013/07/IMG_0555.jpg.

4.2.3 Deltas, agriculture, and environmental change

Many of the Earth's large deltas are especially vulnerable to global environmental change (Cruz et al., 2007). There are three main impacts: climate change (precipitation, temperature, winds/storms), sea level rise, and land use change. Land use change includes conversion of natural vegetation cover to agricultural lands, as well as the engineering of large hydraulic infrastructure, specifically dykes, canals, and dams. The latter traps upstream sediment loads in reservoirs, restricting the redistribution of river sediment in wetlands in deltas (Syvitski et al., 2009).

Coastal erosion problems in the Mississippi delta (see Figure 4.9) represent all three impacts. The Mississippi delta is losing almost half a hectare of land per hour, the highest rate of land loss in the world. This is largely caused by a reduction in sediment supply. The specific causes are: (i) sediment trapped in upstream reservoirs (which supply irrigation to agricultural lands) are no longer transported to coastal wetlands, (ii) dykes prevent flood waters from dispersing sediment to delta wetlands, which would

reduce subsidence, and (iii) high rates of sea level rise (for example, see Figure 4.3) and land subsidence.

The Asian mega deltas, supporting a population over 500 million and located at the mouth of 11 large river basins, have also undergone substantial degradation (Cruz et al., 2007). Because the Asian mega deltas are important to global food production, it highlights the importance of understanding their vulnerability to climate change and sea level rise (Allison et al., 2016).

As with the Mississippi basin, upstream dams have trapped sediment and caused substantial reductions to downstream sediment loads of the Asian mega deltas (Syvitski et al., 2009). The reduction in sediment loads reduces wetland aggradation, the building up of wetland by sediment deposition. Continued floodplain and delta wetland aggradation is required to keep up with **relative sea level rise**, which is defined as the summation of absolute sea level rise plus local ground subsidence. In addition, significant amounts of groundwater are pumped up for irrigation, further complicating the issue (see Case study 4.3).

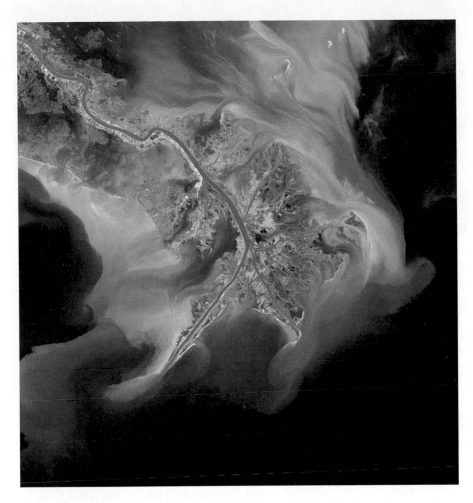

Figure 4.9 The sediment plume of the Mississippi delta in the northern Gulf of Mexico. The lower Mississippi annually transports some 225×10^{-6} tons of sediment per year to the northern Gulf of Mexico, which represents about 50% reduction from pre-dam sediment loads of the early 1950s.

Image Source: NASA.

CASE STUDY 4.3

The Mekong under threat

The vulnerability with Asian mega deltas represents a direct threat to regional—and global—food security (Redfern et al., 2012). The premier example occurs in the Mekong Delta, one of the earliest regions of rice cultivation. The Mekong is the third largest delta on Earth, and as of 2000 agriculture extended across 85% of the region, the majority of which is used by smallholder farmers who cultivate rice and increasingly fish and shellfish.

The development of Mekong delta agriculture relies upon extensive hydraulic infrastructure (see Figure A) developed over several hundred years. Flood control is provided by some 20,000 km of dykes. Many dykes are comprised of local earthen materials that are easily eroded with rising sea

levels. While they are often breached and in need of repair (see Figure A.a), the extensive network of dykes results in an almost complete disconnection of the flood pulse with the deltaic wetlands. Other threats include the impacts of climate change, aquifer degradation, and ground subsidence.

Climate change and sea level rise are already responsible for declines in annual rice production in the Mekong delta. Increased variability in precipitation and streamflow is expected to amplify wet and dry seasons, further reducing crop yields. Tropical cyclone activity is expected to increase, which, when combined with higher sea levels, will cause more severe flooding for extended periods. During storm surges, salt water is also driven through irrigation networks

a)

b)

Figure A Hydraulic infrastructure in the Mekong Delta (a) Erosion of a local coastal dyke along lower Mekong Delta, Tran Van Thoi District, Ca Mau Province, Vietnam. (b) Groundwater pumping to support irrigated rice cultivation.

Source: Vietnam News Agency, Vietnam Ministry of Information and Communications, used with permission.

and into rice fields. The area of rice cultivation submerged each year has been increasing, ranging from 150,000 to 200,000 km^2 across the Asian mega deltas. Coastal dykes are being eroded and breached, resulting in increased submergence of agricultural fields.

Another threat confronting the deltas is the unsustainable use of groundwater (from both confined and unconfined aquifers). Because of their protection from surface contamination, the older confined aquifers are considered a higher-quality source of freshwater (as a bonus, they do not need to be pumped, as they are artesian) (see Box 4.2). Unfortunately, older aquifers are only recharged in a narrow zone located in the northern (upstream) portions of the delta, a wetland area that has increasingly been drained for agriculture and polluted by agricultural fertilizer (see Section 3.2.1 in *Chapter 3: Pollution*), threatening the quality of the pristine groundwater (Erban et al., 2014).

Rapid aquifer pumping has resulted in substantial lowering of the water table, driving subsidence (see Box 4.4). A 1000-km^2 regional hotspot, south of Ho Chi Min city (Vietnam), for example, is driven by groundwater pumping from over 900 deep wells, resulting in an average subsidence rate from 11 mm per year to 24 mm per year (see Figure B).

As groundwater quality degrades (see Figure C), there are public health concerns from arsenic contamination (Erban et al., 2014). Arsenic levels are highest in shallow aquifers, adjacent to river channels. Deeper (and older) aquifers are preferred, but within 10 to 20 years, the process of groundwater withdrawal results in compaction of the

confining clay layers (aquicludes). This 'squeezes' naturally stored arsenic out of clay pores, forcing it into adjacent aquifer sands and increasing arsenic levels in the aquifer. World Health Organization safe levels for arsenic in drinking water are set at below 10 µg/L, but deep wells in the Mekong delta average 20 µg/L.

Finally, increased demand for protein (see Section 8.1.1 in *Chapter 8: Nutrition*), which includes intensive fish and shellfish farming, is driving land use change (Syvitski et al., 2009). While some of this protein is consumed by the exploding populations of Asian mega cities, most is exported for consumption to Western nations. Asia produces 89% of the global supply of farmed fish, with most of the production located within its deltas. The coastline of several Asian mega deltas, such as the Mekong, Huang He, and Pearl Rivers, is increasingly dominated by aquaculture, replacing mangrove swamps and wetlands. This causes significant reductions to marine and estuarine biodiversity.

In comparison to rice cultivation, aquaculture facilities require more substantial flood control infrastructure and larger amounts of groundwater. Because of their high sensitivity to water quality fluctuation, such as salinity, alkalinity, and temperature, the hydraulic infrastructure associated with intensive fish farming must be more precise and larger. It is not surprising, therefore, that subsidence is even higher in land underlying aquaculture facilities. Along the Huang He delta in China, for example, subsidence is especially alarming, and is as high as 250 mm per year.

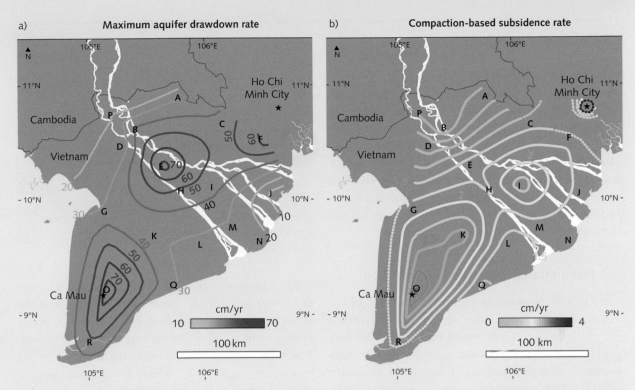

Figure B Maximum rate of water table lowering (a) and associated compaction-based subsidence (b) over the Mekong Delta. The letters correspond to specific wells (not discussed). Data for both figures are visually displayed as isolines, which represent equal rates of aquifer drawdown and subsidence, respectively.

Source: Erban, L.E., Gorelick, S.M., and Zebker, H.A. 2014. Groundwater extraction, land subsidence, and sea-level rise in the Mekong Delta, Vietnam. Environmental Research Letters 9 (8). doi:10.1088/1748-9326/9/8/084010. CC BY 3.0.

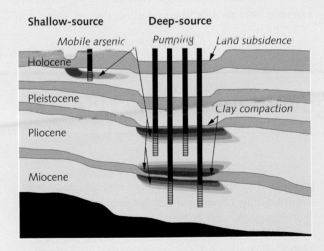

Figure C A profile of aquifers utilized for groundwater pumping in the Mekong delta. The Holocene aquifer is unconfined, whereas the older (Pleistocene and Tertiary) aquifers are confined by clay aquicludes (grey layers) that are being compacted, resulting in the 'squeezing' and movement of arsenic (red) from the clay layers into the aquifer. The lower aquifer is more than 500 m below the surface.

Source: Laura E. Erban, Steven M. Gorelick, Howard A. Zebker, and Scott Fendorf. 2013. Release of arsenic to deep groundwater in the Mekong Delta, Vietnam, linked to pumping-induced land subsidence. PNAS 110 (34) 13751-13756; doi:10.1073/pnas.1300503110.

4.3 **Political hydrology**

Protecting surface waters and groundwater from the adverse impacts of agricultural activities requires effective governmental institutions. These institutions need to implement policies that directly relate to physical processes which influence the quality and quantity of water resources. This can happen at national (see Box 4.7 for an example), international, and even global levels. Here we consider international water cooperation at a watershed level (that is, when a watershed overlaps several countries) and the global politics of water.

4.3.1 **International watershed cooperation**

Water is a vital natural and national resource. However, the distribution of water resources, both surface and groundwater, is not bound by political borders (see Figure 4.10). Water resources that cross borders require unique governmental frameworks to assure equitable and sustainable water management. To get a sense of the scale of the issue, some 263 watersheds cross international political borders, while over 300 aquifers flow under international borders (Rubio, 2001). Watersheds which cross international political borders house 40% of the world's population and cover 44% of land (Rubio, 2001). Clearly, shared management of water resources looms large,

especially in view of projected water shortages due to climate change.

Pause and think

In what watershed do you live? Is this an international or a national watershed? Which key institutes govern the watershed? And which stakeholders use the water resources in the watershed?

By comparison with groundwater policies, surface water policies are much more evolved, and the importance of a watershed perspective to water management is becoming the international standard. A notable example occurs with the EU's WFD (see Food controversy 4.1), which links landscapes to water, channels to floodplains, and rivers to the coast. Although the plans have been unevenly implemented between western and eastern European nations, the attempt should be lauded for its scope and vision.

But not all large international basins are so well coordinated. Nations within the Nile basin, for example, have a long and undistinguished history of unsigned agreements regarding the fair allocation of water for

BOX 4.7

Water governance in the USA

In the USA, major federal policies, such as the Clean Water Act (CWA) (1972) and the Endangered Species Act (1973), have helped to preserve and restore surface waters and associated environments. An important dimension of the CWA is the notion of 'physical integrity', which refers to the preservation of the abiotic and biotic dynamics of surface waters. Major federal institutions, such as the US Fish and Wildlife, are squarely engaged with issues such as environmental flows, so that waterways provide a suitable habitat for aquatic ecosystems, especially when it involves endangered species. Other policies provide financial incentive to farmers to preserve waterways.

In the USA, this was necessary because of the tremendous pace at which wetlands were being degraded and lost. Of the US wetlands lost between the 1950s and 1970s, 87% were lost because of their conversion to agricultural lands. The Swampbuster Bill, a provision on the 1985 Food Security Act, was passed by the US Congress to arrest this trend. The Swampbuster Bill extends financial benefits to farmers if they do not convert wetlands or erodible soils upon their lands for agriculture. The provision is credited with a sharp reduction in the loss of wetlands to agriculture, from 95,000 hectares in 1985 to 11,000 hectares by 1997.

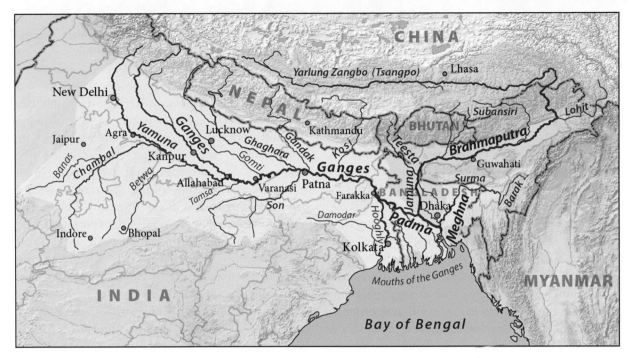

Figure 4.10 Effective management of the Ganges-Brahmaputra-Meghna River basin would require international coordination between five nations: China, Nepal, India, Bhutan, and Bangladesh.

Source: Pfly/Wikimedia Commons/CC BY-SA 3.0.

hydropower, irrigation, and conservation. Ethiopia's recent construction boom of large hydropower dams has caused tension with Egypt, threatening to collapse the Nile Basin Initiative. In North America, freshwater extraction for floodplain irrigation along the Rio Grande, a 1765-km border between the USA and Mexico and the world's longest international riparian border, is regulated by an outdated 1944 treaty. The 1944 treaty is fundamentally flawed because the volume of water Mexico is required to supply was established during a wetter climate. In essence, Mexican farmers in the region are under constant stress because they have to release water for Texan farmers. This is a problem especially during dry periods, which are increasingly frequent and prolonged, when irrigation for agriculture is most critical.

4.3.2 Beyond the basin: the global perspective

The conversion of lands has dramatically increased over the past three centuries, especially over the twentieth century (see Figure 4.11). Detailed monitoring by government agencies of several watersheds reveals

that such changes impact freshwater resources by decreasing soil moisture and evapotranspiration and increasing run-off and river flows. Not only are rivers impacted by the increased streamflow, but the water quality also degrades, associated with flushing of

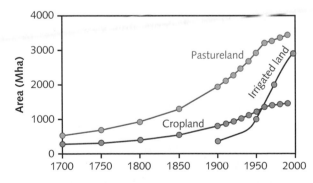

Figure 4.11 Global change in the area of pastureland and cropland between 1700 and 2000, as well as irrigated land area between 1900 and 2000.

Source: Adapted from Scanlon, B. R., I. Jolly, M. Sophocleous, and L. Zhang (2007), Global impacts of conversions from natural to agricultural ecosystems on water resources: Quantity versus quality, Water Resour. Res., 43, W03437, doi: 10.1029/2006WR005486. By permission of John Wiley and Sons.

pesticides and salts that degrade downstream aquatic and riparian ecosystems.

While irrigated lands represent a small proportion of overall agricultural lands, their impact on water resources is considerable. Internationally, irrigated land areas increased over the twentieth century, with a sharp rate of increase after the 1950s, from about 100 Mha to 280 Mha (see Figure 4.11). Over 18% of croplands are irrigated, though the increased yields are responsible for approximately 40% of the total global food production (Scanlon et al., 2007).

The global trajectory of land conversion associated with agriculture and irrigation is projected to continue, with international variation. While some developed nations are undergoing a reduction in agricultural lands associated with demographic shifts (rural to urban migrations), a rapid conversion of forests and grasslands for agriculture continues in developing nations, especially in South America and sub-Saharan Africa (Scanlon et al., 2007).

Globalization and increased recognition of human vulnerability to stressed freshwater supplies have stimulated new ways of thinking about water (this is similar to the trade concepts highlighted in Box 6.5 of *Chapter 6: Climate Change*). A **water footprint** refers to the total volume of freshwater used to produce a good or service. The water footprint classifies water into three types, specifically green, blue, and grey water (Hoekstra et al., 2011; Mekonnen and Hoekstra, 2010). Green water is supplied by precipitation (for example, see Figure 4.3) and is directly available to cultivated plants. Blue water is supplied

from a water body, such as a lake, a river, or an underground aquifer, via a hydraulic infrastructure for utilization in industry, consumption, or irrigation. Grey water is used to transfer pollutants such as industry waste, urban run-off, and sewage.

Comparing footprints, beef has a whopping footprint of 15,410 L per kilogram (L/kg), although 94% of it is green water. Beef is the highest of any food product, and much higher than other livestock, including sheep (10,400 L/kg), pork (6000 L/kg), and poultry (4300 L/kg). To return to green asparagus introduced at the start of the chapter, it has among the highest global water footprint of any crop, at 2370 L/kg. However, the footprint does not tell us where the water is actually used or whether that use is sustainable; it could be that products with a high water footprint are grown in regions with large water resources. In the absence of this link, application of the water footprint for the development of sustainable freshwater resources policy therefore is yet to be realized (Perry, 2014).

In your experience

How much water do you use every year? To find out go to http://waterfootprint.org/en/resources/interactive-tools/personal-water-footprint-calculator/personal-calculator-extended/ and calculate your annual water footprint. How big is your impact? And what part of this footprint is related to your food consumption? In what way could you reduce your water footprint?

4.4 Future outlook

Fortunately, sustainable management approaches have been developed and are being implemented in some cases. A key management approach is **integrated watershed management** (IWM), embracing a watershed perspective. In contrast to traditional approaches whereby management was oriented around singular purposes (for example, navigation, agriculture, flood control), IWM is an adaptive approach considering a comprehensive array of functions, including:

1. Land cover regulation in the upper basin to reduce downstream run-off and pollution
2. Implementation of environmental flows strategy
3. Making space for flood waters along select riparian reaches
4. Modifying engineering structures to improve ecological functions
5. Stakeholder involvement, and

6. Policy and management aligned across different governmental scales (that is, local, regional, national, international).

The Rhine delta, degraded for centuries by human activities, was unsustainable from the standpoint of projected changes in climate and sea level rise (for example, see Figure 4.2). Large floods in 1993 and 1995 raised awareness of the delta's vulnerability to projected climate change and sea level rise scenarios. IWM concepts are being applied to climate-proof and restore the Dutch Rhine in an internationally acclaimed case study referred to as 'Room for the River.'

An encouraging trend is that many smaller dams and flow obstructions are increasingly being removed in Europe and North America. Results thus far are promising, and in many instances, the natural flow and sediment regime are recovering, as are the associated aquatic and riparian ecosystems.

The concept of IWM has been around for about 25 years, a sufficient period to test and refine various facets of the strategy across a range of physical and human–environmental landscapes. Implementation of IWM requires effective policies and institutions to plan, design, maintain, and monitor hydraulic infrastructure that influence hydrologic and associated ecological processes. On a large scale this is occurring with the EU's Water Framework Directive which requires European river basins to reach specific targets of hydrologic and ecological health. Successful implementation of such an ambitious plan in Europe—or elsewhere—requires alignment of agricultural and environmental stakeholders with an interest in sustaining freshwater resources for future generations.

● QUESTIONS

1. Why is groundwater management important for sustainable agriculture and flood safety in large deltas and coastal lowlands?

2. Why should agriculture be practised from a drainage basin framework?

3. How is streamflow influenced by agricultural practices and related hydraulic infrastructure?

4. Why is ground subsidence of special concern in deltas with intensive agricultural activities and large populations?

5. How can concepts of integrated watershed management be utilized to manage the consequences of global climate change?

● FURTHER READING

Leopold, A. 1949. *A Sand County Almanac*. Oxford and New York, NY: Oxford University Press. **(Among the first and most important environmental books written for a popular audience to raise awareness of environmental problems caused by mismanagement of water and soil resources.)**

Reisner, M. 1986. *Cadillac Desert: The American West and its Disappearing Water*. New York, NY: Viking Press. **(The highly acclaimed book that chronicles the politics and economics of large dam and irrigation projects across western USA and their environmental impacts. Written for a popular audience.)**

Stegner, W. 1954. *Beyond the Hundredth Meridian. John Wesley Powell and the Second Opening of the West*. New York, NY: Penguin. **(The classic Pulitzer Prize-awarded text that chronicles exploration of western USA and the vulnerability of its water resources to drought; highly scholarly and written for a popular audience.)**

Syvitski, J.P.M., Kettner, A.J., Overeem, I., Hutton, E.W., Hannon, M.T., Brakenridge, G.R., Day, J., Vörösmarty, C., Saito, Y., Giosan, L., Nicholls, R.J. 2009. Sinking deltas due to human activities. *Nature Geoscience* **2**, 681–686. **(A key scientific article that outlines different human causes of delta subsidence, with a focus on Asian mega deltas.)**

● REFERENCES

Allison, M.A., Törnqvist, T., Amelung, F., Dixon, T.H., Erkens, G., Stuurman, R., Jones, C., Milne, G., Steckler, M., Syvitski, J.P.M., Teatini, P. 2016. Global risks and research priorities for coastal subsidence. *Eos* **97** Available at: https://doi.org/10.1029/2016EO055013 [accessed 4 November 2018].

Bates, B.C., Kundzewicz, Z.W., Wu, S., Palutikof, J.P. (eds). 2008. *Climate Change and Water. Technical Paper of the Intergovernmental Panel on Climate Change*. Geneva: IPCC Secretariat.

Biermans, H., Haddeland, I., Kabat, P., Ludwig, F., Hutjes, R.W.A., Heinke, J., Von Bloh, W., Gerten, D. 2011. Impact of reservoirs on river discharge and irrigation water supply during the 20th century (Figure 3, Cumulative capacity of global dam storage over the 20th century). *Water Resources Research* **47**, W03509.

Butzer, K.W. 1976. *Early Hydraulic Civilization in Egypt: A Study of Cultural Ecology*. Chicago, IL: University of Chicago Press.

Church, J.A., Clark, P.U., Cazenave, A., Gregory, J.M., Jevrejeva, S., Levermann, A., Merrield, M.A., Milne, G.A., Nerem, R.S., Nunn, P.D., Payne, A.J., Pfeffer, W.T., Stammer D., Unnikrishnan, A.S. 2013. Sea level change. In: Stocker, T.F., Qin, D., Plattner, G.-K., Tignor, M., Allen, S.K., Boschung, J., Nauels, A., Xia, Y., Bex, V., Midgley, P.M. (eds). *Climate Change 2013: The Physical Science Basis. Contribution of Working Group I to the Fifth Assessment Report of the Intergovernmental Panel on Climate Change*. Cambridge and New York, NY: Cambridge University Press, pp. 1137–1216.

Collins, M., Knutti, R., Arblaster, J., Dufresne, J.-L., Fichefet, T., Friedlingstein, P., Gao, X., Gutowski, W.J., Johns, T., Krinner, G., Shongwe, M., Tebaldi, C., Weaver, A.J., Wehner, M. 2013. Long-term climate change: projections, commitments and irreversibility. In: Stocker, T.F., Qin, D., Plattner, G.-K., Tignor, M., Allen, S.K., Boschung, J., Nauels, A., Xia, Y., Bex, V., Midgley, P.M. (eds). *Climate Change 2013: The Physical Science Basis. Contribution of Working Group I to the Fifth Assessment Report of the Intergovernmental Panel on Climate Change*. Cambridge and New York, NY: Cambridge University Press, pp. 1029–1136.

Cruz, R.V., Harasawa, H., Lal, M., Wu, S., Anokhin, Y., Punsalmaa, B., Honda, Y., Jafari, M., Li, C., Huu Ninh, N. 2007. Asia. Climate Change 2007: impacts, adaptation and vulnerability. In: Parry, M.L., Canziani, O.F., Palutikof, J.P., van der Linden, P.J., Hanson, C.E. (eds). *Contribution of Working Group II to the Fourth Assessment Report of the Intergovernmental Panel on Climate Change*. Cambridge: Cambridge University Press, pp. 469–506.

Dam Removal Europe. Available at: https://damremoval.eu/ [accessed 4 November 2018].

de Vries, J.J. 2007. Groundwater. In: Wong, T.H.E., Batjes, D.A.J., de Jager, J. (eds) *Geology of the Netherlands*. Amsterdam: Royal Netherlands Academy of Arts and Sciences, pp. 295–315.

Erban, L.E., Gorelick, S.M., Zebker, H.A. 2014. Groundwater extraction, land subsidence, and sea-level rise in the Mekong Delta, Vietnam. *Environmental Research Letters* **9**, 084010.

Erban, L. E., Gorelick, S. M., Zebker, H. A., Fendorf, S. 2013. Release of arsenic to deep groundwater in the Mekong Delta, Vietnam, linked to pumping-induced land subsidence. *Proceedings of the National Academy of Sciences of the United States* **110**, 13751–13756.

Erkens, G. 2015. Delta subsidence: an imminent threat to coastal populations. *Environmental Health Perspectives* A204–A209.

European Commission (EC). 2000. Water Framework Directive. Directive 2000/60. *Official Journal of the European Communities*. Available at: https://eur-lex.europa.eu/resource.html?uri=cellar:5c835afb-2ec6-4577-bdf8-

756d3d694eeb.0004.02/DOC_1&format=PDF [accessed 8 November 2018]

Foley, J.A., DeFries, R., Asner, G.P., Barford, C., Bonan, G., Carpenter, S.R., Chapin, S., Coe, M.T., Daily, G.C., Gibbs, H.K., Helkowski, J.H., Holloway, T., Howard, E.A., Kucharik, C.J., Monfreda, C., Patz, J.A., Prentice, I.C., Ramankutty, N., Synder, P.K. 2005. Global consequences of land use. *Science* **309**, 570–574.

Foley, M.M., Magilligan, F.J., Torgersen, C.E., Major, J.J., Anderson, C.W., Connolly, P.J., et al. 2017. Landscape context and the biophysical response of rivers to dam removal in the United States. *PLoS One* **12**, e0180107.

Food and Agricultural Organization of the United Nations (FAO). 1988. *Nature and Management of Tropical Peat Soils*. FAO Soils Bulletin, 59. Rome: FAO.

Galloway, D., Jones, D.R., Ingebritsen, S.E. 1999. *Land Subsidence in the United States*. United States Geological Survey, Circular 1182. Available at: https://pubs.usgs.gov/circ/circ1182/pdf/circ1182_intro.pdf [accessed 8 November 2018].

Hepworth, N.D., Postigo, J.C., Guemes Delgado, B., Kjell, P. Drop by Drop. 2010. *Understanding the Impacts of the UK's Water Footprint Through a Case Study of Peruvian Asparagus*. London: Progressio, CEPES and Water Witness International.

Hoekstra, A.Y., Chapagain, A.K., Aldaya, M.M., Mekonnen, M.M. 2011. *The Water Footprint Assessment Manual: Setting the Global Standard*. London: Earthscan.

Jiménez Cisneros, B.E., Oki, T., Arnell, N.W., Benito, G., Cogley, J.G., Döll, P., Jiang, T., Mwakalila, S.S. 2014. Freshwater resources. In: Field, C.B., Barros, V.R., Dokken, D.J., Mach, K.J., Mastrandrea, M.D., Bilir, T.E., Chatterjee, M., Ebi, K.L., Estrada, Y.O., Genova, R.C., Girma, B., Kissel, E.S., Levy, A.N., MacCracken, S., Mastrandrea, P.R., White, L.L. (eds). *Climate Change 2014: Impacts, Adaptation, and Vulnerability. Part A: Global and Sectoral Aspects. Contribution of Working Group II to the Fifth Assessment Report of the Intergovernmental Panel on Climate Change*. Cambridge and New York, NY: Cambridge University Press, pp. 229–269.

Jørgensen, D., Renöfält, B.M. 2013. Damned if you do, dammed if you don't: debates on dam removal in the Swedish media. *Ecology and Society* **18**, 18.

Junk, W.J., Bayley, P.B., Sparks, R.E. 1989. The flood pulse concept in river-floodplain systems. In: Dodge, D.P. (ed). *Proceedings of the International Large River Symposium (LARS)*. Special Publication of Canadian Fisheries and Aquatic Sciences, pp. 110–127.

Konikow, L.F. 2013. *Groundwater depletion in the United States (1900–2008): U.S. Geological Survey Scientific Investigations Report 2013–5079*. Available at: http://pubs.usgs.gov/sir/2013/5079 [accessed 8 November 2018].

Leopold, A. 1949. *A Sand County Almanac: With Other Essays on Conservation from Round River*. New York, NY: Oxford University Press.

Mann, C.C. 2005. *1491: New Revelations of the Americas Before Columbus*. New York, NY: Knopf.

Mazi, K., Koussis, A.D., Destouni, G. 2013. Tipping points for seawater intrusion in coastal aquifers under rising sea level. *Environmental Research Letters* **8**, 014001.

McGrath, T., Almeida, O.T., Merry, F.D. 2007. The influence of community management agreements on household economic strategies: cattle grazing and fishing agreements on the lower Amazon floodplain. *International Journal of the Commons* **1**, 101–121.

Mekonnen, M.M., Hoekstra, A.Y. 2010. *The Green, Blue and Grey Water Footprint of Crops and Derived Crop Products.* Value of Water Research Report Series No. 47. Delft: UNESCO-IHE.

Perry, C. 2014. Water footprints: Path to enlightenment, or false trail? *Agricultural Water Management* **134**, 119–125.

Redfern, S.K., Azzu, N., Binarima, J.S. 2012. *Rice in Southeast Asia: Facing Risks and Vulnerabilities to Respond to Climate Change. Building Resilience for Adaptation to Climate Change in the Agriculture Sector.* Proceedings of a Joint Food Agricultural Organization and Organization for Economic Co-operation and Development. Available at: http://www.fao.org/fileadmin/templates/agphome/documents/climate/Rice_Southeast_Asia.pdf [accessed 8 November 2018].

Rubio, M. 2001. *Internationally Shared (Transboundary) Aquifer Resources Management: Their Significance and Sustainable Management.* Paris: UNESCO International Hydrological Programme, Non-Serial Publications in Hydrology.

Scanlon, B.R., Jolly, I., Sophocleous, M., Zhang, L. 2007. Global impacts of conversions from natural to agricultural ecosystems on water resources: quantity versus quality. *Water Resources Research* **43**, W03437.

Syvitski, J.P.M, Kettner, A.J., Overeem, I., Hutton, E.W., Hannon, M.T., Brakenridge, G.R., Day, J., Vörösmarty, C., Saito, Y., Giosan, L., Nicholls, R.J. 2009. Sinking deltas due to human activities. *Nature Geoscience* **2**, 681–686.

United States Geological Survey. 2015. *Elwha River Dam Removal—Rebirth of a River.* United States Geological Survey, Fact Sheet 2011–3097. Available at: https://pubs.usgs.gov/fs/2011/3097/pdf/fs20113097.pdf [accessed 8 November 2018].

van Asselen, S. Karsenberg, D., Stouthamer, E. 2011. Contribution of peat compaction to relative sea-level rise within Holocene deltas. *Geophysical Research Letters* L24401.

van de Ven, G.P. 2004. *Man-Made Lowlands: History of Water Management and Land Reclamation in the Netherlands,* 4th ed. Utrecht: Uitgeverij Matrijs.

White, G.F. 1945. *Human Adjustment to Floods.* Department of Geography Research Paper no. 29. Chicago, IL: University of Chicago Press.

Whiting, P.J. 2002. Streamflow necessary for environmental maintenance. *Annual Review of Earth and Planetary Science* **30**, 181–206.

WinklerPrins, M.G.A., Barrios, E. 2007. Ethnopedology along the Amazon and Orinoco Rivers: a convergence of knowledge and practice. *Revista Geográfica* **142**, 111–129.

Zarfl, C., Lumsdon, A.E., Berlekamp, J., Tydecks, L., Tockner, K. 2015. A global boom in hydropower dam construction. *Aquatic Sciences* **77**, 161–170.

5

Soils

What are the impacts of agriculture on soils?

Paul F. Hudson, Peter Houben, and Thijs Bosker

'[. . .] *Soils constitute the foundation for agricultural development, essential ecosystem functions and food security and hence are key to sustaining life on Earth.*'

68th UN General Assembly, 2015

Chapter Overview

- Soil development involves both abiotic (non-living) and biotic (living) processes, and it takes centuries to produce a well-developed soil horizon. For this reason, soils are considered a non-renewable resource.

- Soil is a vital natural resource essential for food production and provides a range of ecosystem services. About a third of global soil resources already have been degraded by human activities, mostly due to modern agricultural practices.

- Key causes of soil degradation are water and wind erosion, pollution, and salinization. There is a significant risk that soil degradation will reduce future crop yields.

- Current threats and future agricultural demands require effective management of soil resources. New techniques and policies will be necessary to ensure sustainable use of soil resources, both on a local and on a global scale.

Introduction

Why is the frequency of sand storms increasing in certain parts of China (see Figure 5.1a)? Why do rivers in Madagascar have the colour of chocolate (see Figure 5.1b)? And why are there deep channels in agricultural fields in many parts of the world such as Australia (see Figure 5.1c)? The answer to all three questions has to do with the rapid disappearance of soils resulting in severe **land degradation**.

a)   b)   c)

Figure 5.1 Three examples of effects of soil degradation: (a) an increase in sand storms in China; (b) the Bombetoka Bay in Madagascar turning chocolate brown because of soil erosion; and (c) a trench (gully) in an agricultural field near Adelaide, Australia, caused by water erosion.

Source: (a) Photo by Se Hasibagen/CC BY 2.0; (b) Photo by oledoe/Flickr/CC BY-SA 2.0; (c) John Coppi, CSIRO/CC BY-SA 3.0.

About 95% of the world's food production comes from the soil (the remaining 5% comes from glass greenhouses and fisheries). It is easy to appreciate that soil is among Earth's most vital natural resources, next to fresh air and freshwater. However, we are losing this precious resource at an alarming rate (see Figure 5.2). Earth has already lost a third of its soil (UNCCD, 2017), mostly due to modern agricultural practices. There has been a considerable increase in loss over the past century. We now lose 120,000 km² of agricultural land every year due to soil degradation (Rickson et al., 2015), about half the size of the UK. The current global rate of soil loss from both direct and indirect human activities is unsustainable,

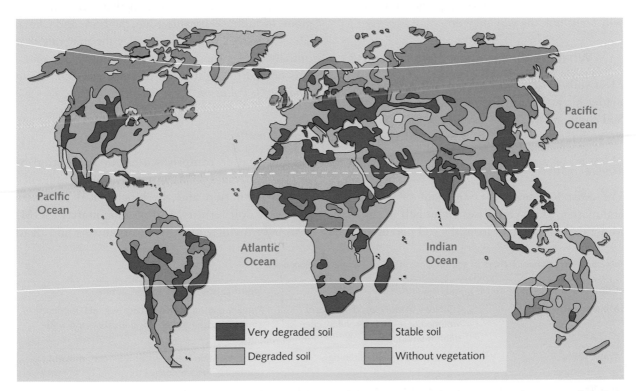

Figure 5.2 Soil degradation on a global scale. The majority of Earth's lands utilized for agriculture are either degraded or are very degraded.

Source: Philippe Rekacewicz, UNEP/GRID-Arendal (http://www.grida.no/resources/6338).

occurring at a much greater rate than its natural formation and replenishment (Amundson et al., 2015).

In your experience

About 95% of the world's food production comes from soil. Given the alarming rate of soil degradation, how might your diet and monthly food budget change if the current rate of soil loss continues for the next century? Do you think the impacts are equal around the world?

In this chapter, we examine soil resources in relation to agriculture and human–environment interactions, beginning with a review of the fundamental concepts of soil properties and soil formation (Section 5.1). Next, we explore how soils are impacted by agriculture (Section 5.2), and how this impacts agriculture in return (Section 5.3). Finally, we consider how to manage our soils more sustainably, including by looking to traditional agricultural societies, which can sometimes be more sustainable than modern mechanized agriculture (Section 5.4).

5.1 What are soils?

As soil sits between the solid Earth and the gaseous atmosphere, soil is the medium where complex exchanges occur between minerals, atmospheric gases, water, and living and non-living organisms. Important functions of soil include, among others: nutrient (re-)cycling (see Box 3.5 in *Chapter 3: Pollution*), water management (see *Chapter 4: Water*), carbon sequestration, and habitat for soil organisms. In this section, we look at the basic ingredients of soils and why soils look the way they do.

5.1.1 What are the key ingredients of soils?

Most soils have the same basic components: mineral matter, air, water, and organic material (see Table 5.1). Soil components can be thought of as '*ingredients to the soil*' (Schaetzel and Thompson, 2015).

Mineral matter

The most fundamental component of soil is mineral matter, which comprises about half of soil mass. Mineral matter derives from the **soil parent material**

which may have an origin from weathered rocks or sedimentary deposits (e.g. sand dunes, or river deposits on floodplains and deltas). A key soil parent material is **loess**, which is a wind-blown deposit of mineral matter, primarily silt (0.0625–0.004 mm). Loess accumulated over thousands of years during the Pleistocene and comprises about 10% of Earth's surface. Loess, however, is the soil parent material for some of the world's most productive agricultural regions, including central Europe, the USA, and China. The soil parent material provides the minerals that initially define the soil chemistry, including nutrients needed for cultivation of crops (see Box 3.5 in *Chapter 3: Pollution*).

Organic matter

Organic matter comprises a minor proportion of overall soil mass (5%) but has a large influence on soil formation and the biochemical processes related to soil fertility. **Soil organic matter (SOM)** is usually divided into living organisms and dead

Table 5.1 Typical characteristics of soil components of an average developed soil

Component (% of soil mass)	External influence	Significance related to agriculture
Mineral matter (50%)	Lithology (rock type), climatic and weathering regime, secondary chemical neoformation of minerals	Primary control on soil fertility (e.g. pH, nutrients) and soil texture (determining aeration and water budget)
Water (25%)	Climate, topography, lithology	Plant-available water, water retention, drought resistance, field drainage
Air (20%)	Climate, topography, lithology, soil fauna	Key for maintaining soil life, oxidation
Organic matter (5%)*	Climate, vegetation, soil fauna	Key for soil fertility, water storage, carbon storage

* Organic matter further differentiated by humus (80%), living organisms (10%), and roots (10%).

or decaying organic matter (Schaetzl and Thompson, 2015). As much organic matter is on the surface of soil, the organic matter in soil decreases as you move deeper through the soil. Normally, soil organic matter accumulates at or near the soil surface.

Living organisms include plant roots, animals, fungi, and microorganisms. During soil formation plant roots are essential, as they help to physically churn the soil and increase soil **porosity** (the space between individual rock or soil particles, expressed as a percentage of soil volume). Animals living in the soil directly mix and churn the soil as well, and increase porosity and **infiltration** (the downward percolation or movement of water). Earthworms, for example, have an influential role in soil development within temperate and tropical climates (see Box 5.1). In addition to mixing soil materials, earthworms add considerable microbial contributions

through their waste because they directly ingest soil particles.

Microorganisms include a broad range such as algae, protozoa, and bacteria. A single gram of soil, the weight of a paper clip, may contain some three billion microorganisms, vital to a range of soil processes and functions. Microorganisms play an underappreciated, vital role in processing organic waste products as part of nutrient cycling, especially as they release and fixate vital soil nutrients essential to plant growth.

Two important categories of dead organic matter are litter and humus. **Litter** refers to dead organic matter at the soil surface that has not decomposed (sometimes called duff) and may include twigs, grasses, and leaves (see Figure 5.3a). Within wet and cold regions, a thick (around 20–40 cm) litter layer (O-horizon) often accumulates atop soils. In contrast, in wet tropics, litter does not have time to

BOX 5.1  Darwin's earthworms

Next to his groundbreaking work on evolution, Charles Darwin conducted a 40-year (!) study on a down-to-earth topic: earthworms. He published his work in 1881 in a book entitled *The Formation of Vegetable Mould through the Action of Worms, with Observations on their Habits.* In the book, he describes earthworms as an '*unsung*

creature which, in its untold millions, transformed the land as the coral polyps did the tropical sea'. Darwin estimated that, on average, every acre (4046 m²) in his experimental plot contained 53,767 earthworms (see Figure A).

Darwin's research included a 29-year experiment, in which he measured the rate at which a stone is buried by earthworm-burrowing activities. He linked this research to Stonehenge, arguing that earthworm activity resulted in the partly buried stones at Stonehenge (see Figure B).

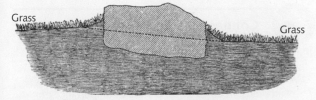

Figure B A diagram of a fallen stone at Stonehenge from Darwin's book on earthworms. Darwin conducted a 29-year experiment on the influence of earthworms on the sinking of stones and related this to partly buried stones at Stonehenge.

Source: Charles Darwin: *The formation of vegetable mould, through the action of worms, with observations of their habits.* London: John Murray, 1881.

Figure A Earthworms are vital in shaping our soils, and they are present in soils in large numbers.

a)

b)

Figure 5.3 Dead organic matter consisting of two important components: (a) litter, which is made of dead organic matter which is not decomposed such as leaves and twigs; and (b) humus (dark top part in figure), which remains after the decomposition of leaves and other plant material by organisms.

Source: (b) Wulf Grube/Wikimedia Commons/CC BY-SA 3.0.

accumulate on the soil surface because of the high rate of biological decomposition. As such, tropical soil usually have low amounts of organic material (Stocking, 2003).

The most important organic component of soil is **humus** (see Figure 5.3b). Humus is the chemically stable remains of organic matter after decomposition of plant organic matter through microbial activities (Jobbágy and Jackson, 2000; Schaetzl and Thompson, 2015). It is primarily organic carbon that is important, as more carbon is stored in the soil than the atmosphere and biosphere combined (Jobbágy and Jackson, 2000). Because of the tremendous amount of carbon stored within topsoil, conserving soil humus is important to managing Earth's carbon budget (Lal, 2004) (see Box 5.2).

Connect the dots

In Case study 2.1 of *Chapter 2: Biodiversity*, we discussed deforestation and the increase in palm oil plantations. In Box 5.2, we discuss the importance of soils in the carbon cycle (for more details on this cycle, see Box 6.3 in *Chapter 6: Climate Change*). Can you link deforestation to changes in the carbon cycle and the impacts on climate change? This will be a focus of *Chapter 6: Climate Change* (specifically Section 6.2.1)?

Soil water

Soil water (also called soil moisture or green water) is water within soil and is stored within soil pores in the upper approximately 1 m. Ninety per cent of the world's agricultural production is dependent upon soil containing sufficient water (Amundson et al., 2015). Variability in soil moisture occurs over different temporal scales, including after a rain event, seasonally, and even over longer periods in response to regional climate change (i.e. drought and wet periods).

The loss of soil moisture can significantly reduce overall yields. In the absence of rain or irrigation, soil moisture will be lost due to evapotranspiration. If this continues, the **wilting point** is reached. The wilting point refers to a critical level of soil moisture, after which plant roots can no longer take in water from the soil, resulting in wilting of the plants, which reduces the yields and can even cause plants to die (see Figure 5.4). The soil wilting point differs according to the moisture demands of each specific crop type and the ability of plant roots to take up water.

To prevent soils from reaching the wilting point, an annual **soil water budget** can be developed, in which evapotranspiration is linked to rainfall (see Figure 5.5). The soil water budget helps to determine when a soil moisture deficit is likely to occur.

BOX 5.2

Soils and climate change

As important as fertile soil is to food production, it is also vital to Earth's carbon cycle and the global climate system (Montgomery, 2007). Naturally, a lot of attention is paid to atmospheric carbon, as it is here that heat is trapped and temperatures go up (see Box 6.3 in *Chapter 6: Climate Change*), but it is worth noting that three times as much carbon is stored in the top metre of soil. Organic carbon is primarily transferred to soil with the decomposition of plant litter (dead plant matter). In the soil, carbon may be immobilized for a period ranging from a few years to thousands of years, until it is disturbed and released.

Since the advent of agriculture some 12,000 years ago, 133 billion tonnes of soil carbon has been lost—about 8% of total soil carbon reserves (Sanderman et al., 2017). And about two thirds of this lost carbon likely ended up in the atmosphere. Much of soil carbon loss occurred with land change—conversion of natural vegetation to industrial agriculture, especially after about the 1950s. Not surprisingly, these losses occurred in nations where industrial agriculture has been most intense, including western Europe, the USA, China, and Australia. Following suit, nowadays, a significant amount of carbon is released in developing countries when forests are replaced by agriculture (see Figure A; see Box 6.3 in *Chapter 6: Climate Change*).

Figure A Land use change, for example the conversion of forests to agricultural land, as seen here in Madagascar, can result in massive release of carbon from the soil to the atmosphere, thereby directly contributing the climate change.

Source: Cunningchrisw/CC BY-SA 4.0.

Figure 5.4 An example of wilting corn in east Kenya. With the absence of rain and an infrastructure for irrigation, soil moisture levels reached the wilting point, resulting in significant loss of yields.

Source: giro555/Flickr/CC BY-SA 2.0.

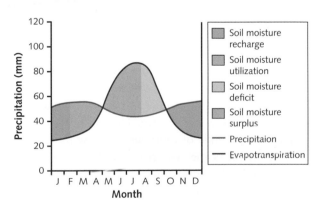

Figure 5.5 Annual soil water budget. The ability of plants to obtain sufficient water is related to the ratio of evapotranspiration and precipitation with deficit periods resulting in a wilting point that requires irrigation to support cultivation.

This makes soil water budgets a useful tool for assessing the viability of potential crop types for cultivation and helps to develop irrigation strategies.

Soil air

The final component of soil is air, which fills the pore spaces that are not occupied with water. Soil air differs from atmospheric air; because of the decomposition of organic matter, soil air is high in CO_2 and low in O_2. Eventually CO_2 seeps through the pore space and escapes into the atmosphere, a process referred to as soil respiration. Indeed, this latter function can be directly monitored by scientists who study the carbon cycle and the influence of different agricultural practices on atmospheric CO_2 levels.

5.1.2 How do we describe soils, and why can they look so different?

While the proportion of individual components (that is, mineral matter, air, water, and organic material) in soil are fairly consistent, minor differences have large influences on **pedogenesis** (the process of soil formation) and soil dynamics. In addition, soils look different because they reflect Earth's wide range of environmental conditions such as spatial variability in climate, vegetation, fauna, and human activities. The given configuration of these environmental controls brings about a place-specific combination of processes that eventually form soil.

Examination of a soil profile (within a trench; see Figure 5.6) reveals that the soil is organized into horizons (see Figure 5.6). **Soil horizons** are distinctive layers within the soil that are defined by soil colour, texture, structure, and chemistry. Soil horizons develop because soil-forming processes change with depths below the soil surface. The soil horizon usually found a couple of centimetres below the surface is referred to as the A-horizon is often marked by a dark colour because high amounts of decomposed soil organic matter which derives from biological organisms and activity. By contrast, the lower soil horizon (C) is largely devoid of soil life, except for occasional worm or animal burrows or tree roots.

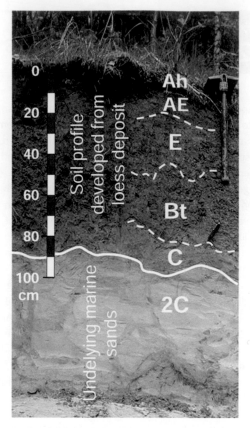

Figure 5.6 Soil profile developed from loess in Wetterau, Germany. The bottom layer (2C) is sandstone from marine coastal deposits. The subsequent horizons have all developed from loess and make this very suitable for agriculture.

Source: P. Houben.

Instead, the lower soil horizon is often characterized by the presence of chemicals that were translocated by soil water moving up or down in the soil. Soil horizons inform soil scientists about the environmental processes that made the soil, as well as current soil functioning and its linkages to human activities. The specific combination of soil horizons makes up a soil profile (see Figure 5.6).

A soil horizon is defined by soil texture, structure, chemistry, and colour. **Soil texture** is an important soil property defined by the relative proportion of sand-, silt- (dust-sized, floury), and clay-sized particles. The combination of these particles defines the texture class (see Figure 5.7). Soil texture influences the downward redistribution of clay, organic material, and dissolved matter by percolating water. Sandy soils allow high rates of infiltration after a

a)

b)

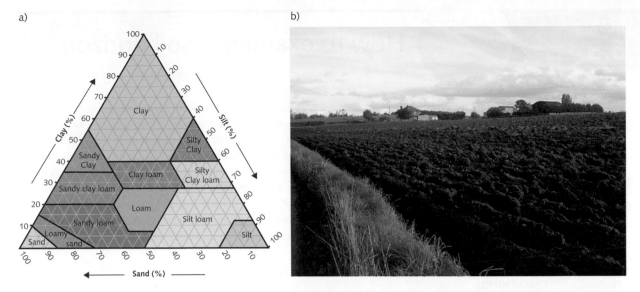

Figure 5.7 (a) Soil texture diagram, which plots the proportion of sand, silt, and clay within soil onto a single texture class. (b) A loamy soil, here in Lancashire in the UK, associated with high agricultural productivity.

Source: (a) Mikenorton/Wikimedia Commons/CC BY-SA 3.0; (b) Alan Murray-Rust / Lancashire Loam / CC BY-SA 2.0.

rainfall event, resulting in rapid loss of soil moisture and the leaching of soil nutrients. In contrast, clayey soils retain soil moisture following a rainfall event. The ideal soil texture for agriculture is a loamy soil, which has the optimal combination of water, nutrient, and organic matter retention. Globally, loam soils are associated with some of the most agriculturally productive regions.

Soil structure refers to how soil particles are clumped together into naturally formed peds, which range from approximately 1 mm to approximately 30 mm. Soil that is loosely clumped together allows for sufficient infiltration so that run-off is not problematic.

Soil chemistry is usually considered from two main criteria: soil pH and soil cation exchange capacity (CEC), which relates to the ability of soil to hold on to, and attract, essential nutrients. Soil pH defines whether a soil is acidic or basic (alkaline). In acidic soils (which have a low pH), the soil water is high in hydrogen (H^+) ions, which can replace soil nutrients. The nutrients are now no longer bound to the soil and can be lost to deeper

horizons. Alkaline soils (which have a high pH) are unable to either dissolve or release nutrients and are therefore less fertile. The ideal soil acidity is around pH 5.5 to 7.0.

Of all soil properties, **soil colour** is the most diagnostic, and helps to understand the processes of soil formation and the suitability of different soil types for cultivation. Soil colour provides insight into soil quality. Reddish soils, for example, are dominated by metallic ions in the soil, especially iron and aluminium. A black or brown soil horizon is often associated with the storage of soil humus, an indication of soil fertility. Light-grey or white splotches in the soil can indicate a problem with salinization, an indication of degraded soil quality, which is not good for agriculture. These colours can be seen in Box 5.3, Figure A.

Soil scientists can classify soils based on the combination of soil horizons (see Box 5.3). Globally, there are thousands of soil types. Soil scientists examine specific soil properties in soil profiles to map the agricultural potential of different soils across the landscape.

BOX 5.3

How to examine a soil horizon

Soil scientists use a letter system to describe soil horizons. Each of these letters describes a specific soil horizon, based on soil texture, chemistry, structure, and colour. There are two major groups of horizons. The first are **solum** (surface and subsoil) horizons (O, A, E, and B horizons), which are horizons that have undergone soil formation. The second are the parent material horizons (C and R horizons), which have undergone limited or no soil formation.

Table A provides details for each of these horizons. Cultivation occurs mainly within the A horizon, whereby deeper tapping roots make use of the soil water retained in B, and even C, horizons. Not all horizons have to appear in each soil, and sometimes the sequence of soil horizons can differ. Combined, the soil horizons make up a soil profile. The most basic soil profile would show a sequence of A–B–C (–R) horizons. In addition, many subcategories are present to further characterize soils (see Figure A)

Table A Description of basic horizons within an idealized soil profile

Horizon	Characteristics
O	Uppermost soil horizon comprising only intact or partly decomposed organic matter referred to as litter (twigs, leaves, duff); usually formed in environments with low rates of decomposition
A	Horizon formed at the soil surface (or below the O horizon, if present). Contains humus, hence has a darker colour (brown, grey, blackish) than the underlying horizon. Often contains evidence of agricultural activities
E	A horizon which has lost a significant part of the mineral and/or organic matter; appears coarser-grained and lighter (ashy) than the underlying horizon
B	A horizon which accumulates organic matter, clay, oxides, and nutrients from above horizons; in consequence, the B horizon is often darker and higher in clay than overlying horizons
C	Dominated by broken-up original parent material that is not very chemically altered; has very low levels of nutrients
R	Unaltered parent material (bedrock)

(Modified from Schaetzl and Thompson, 2015).

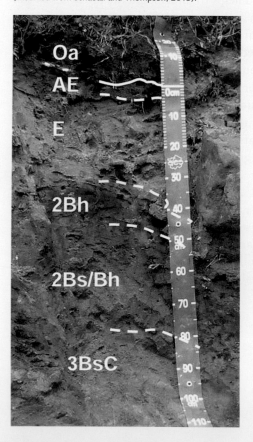

Figure A Soil profile displaying a typical sequence of soil horizons. Note the presence of an organic litter (Oa) horizon and the thick leached elluvial (E) horizon. Location: Eifel Mountains near Gerolstein, Germany.

Source: P. Houben.

5.2 How does agriculture impact soils?

Although soil is fundamental to Earth's living surface and it provides essential ecosystem services (see Box 5.4), the overall health of soils is deteriorating (see Figure 5.2). The alarming magnitude of global soil degradation over the twentieth century is of great concern to food security and to a range of interrelated environmental and societal issues. This is especially true in developing nations, in which slash-and-burn agriculture is still commonly practised (see Case study 5.1). The loss of beneficial soil functions, often termed **soil degradation**, occurs primarily in two ways, including a reduction in soil quality and a physical loss of soil by erosion (UNCCD, 2017).

In your experience

Box 5.4 gives examples of key ecosystem services provided by soils. Which of these services do you use regularly? Could you go without these services in the future, given the high rate of soil degradation?

Soil erosion results in a physical loss of soil and is triggered by wind- and water-driven processes.

Large-scale agriculture often results in bare soil being exposed to run-off (overland flow), which can accelerate soil erosion and the loss of fertile topsoil where crops are cultivated (see Figure 5.8). **Soil quality** refers to soil functioning it its natural capacity, which includes its ability to sustain plant and animal productivity while maintaining its hydrologic and biogeochemical roles within the broader landscape. Soil quality can decline due to physical and chemical changes, such as loss of soil nutrients or salinization, which reduce crop yields.

The tragedy is that, by far, the greatest loss of soil is caused by the very type of land use for which it is most needed—agriculture (UNCCD, 2017) (see Box 5.5 to explore hydroponics—agriculture without soils). Agriculture imposes a myriad range of abrupt changes to soil properties and environmental processes that can degrade soil productivity, as well as increase soil vulnerability to erosion by water and wind. The severity of agricultural impacts can be appreciated by considering that over 40% of Earth's natural landscapes have been converted to agricultural purposes, especially for food production. Indeed, human landscapes are among Earth's largest biomes (Foley et al., 2005).

a)

b)

Figure 5.8 Soil erosion by (a) overland flowing water (run-off) and (b) the formation of erosional gullies in the land, which rapidly removes fertile topsoil and reduces agricultural yields.

Source: (a) U.S. Department of Agriculture, Natural Resources Conservation Service; (b) Rodney Burton / Soil erosion, Wigborough, Somerset / CC BY-SA.2.0.

Indigenous farming and soils

Traditional forms of agriculture have developed over centuries in association with natural environmental rhythms. They continue to be widely practised by indigenous peoples, particularly in developing nations. Traditional agriculture includes a range of models such as farming, pastoralism, aquaculture, and agroforestry (Hall, 2001).

Common components of traditional farming practised in its 'pure indigenous' form include reliance on manual labour, simple tools, small plots, and a local-scale orientation that relies upon community interactions and organizations. Additionally, such systems often rotate crops and intercrop to take advantage of nutrient exchange. For example, over thousands of years, the indigenous peoples of the Americas practised a system of intercropping with corn and legumes, as the latter has nitrogen-fixating roots (see Box 3.6 in *Chapter 3: Pollution*) that helped to replenish the high nitrogen uptake of corn.

Traditional agriculture also has reduced reliance on artificial fertilizers and is considered more sustainable from the perspective of biodiversity and natural resource preservation, including soil resources (Stocking, 2003). This is important, because globally some 375 million indigenous peoples occupy over 20% of Earth's land area, representing over 75% of Earth's biodiversity (Erni, 2015).

Swidden (slash-and-burn) cultivation is an ancient farming system that continues to be widely practised throughout many rural areas of the humid and semi-arid tropics of Africa, Asia, and Latin America (Lal, 2004). While there are variations to swidden cultivation (see Figure A), the basic model is that a plot of natural forest is cut down with crude implements, such as machetes and axes, usually before the beginning of the wet (rainy) season. The cut plant biomass is allowed to dry and then burnt on site, which provides a pulse of nutrients to the soil, especially carbon, nitrogen, and phosphorus.

The newly created agricultural plot is used to cultivate a variety of crops for several years (Denevan, 2001; Doolittle, 2001). After crop yields decline substantially after a few years, the plots are abandoned, called the fallow stage (see Figure B). The fallow stage is designed to allow the lost nutrients from cultivation to recover. The duration of the fallow stage varies, depending upon food demands, as related to available land, economics, and population density. In areas with low population densities, the fallow period is generally of long duration, from 10 to 30 years, allowing abandoned plots to revert back to forest. The fallow stage in regions with high population densities, however, is shorter, from 3 to 10 years (Rossi et al., 2010).

The alteration of inherent soil–vegetation relations disrupts the natural process of nutrient cycling and result in degradation in soil quality, especially in a loss of nutrients (Rumpel et al., 2006). Soil quality degradation is more severe with intense (high-temperature) burning that incinerates all organic matter, rather than allowing for storage (Lehmann, 2006). Additionally, soil degradation increases with slope steepness, which increases run-off and can generate excessive erosion in the form of deeply entrenched gullies (see Case study 5.2).

Figure A Slash-and-burn agriculture, as seen here in Brazil, still occurs in many developing regions.

Source: Alzenir Ferreira de Souza/CC0.

Figure B Field plots in different stages of the swidden (slash-and-burn) farming cycle. Sierra Madre Oriental, San Luis Potosi, Mexico.

P.F. Hudson, 2001.

BOX
5.4

What ecosystem services do soils provide?

Soils are vital to food production, as they provide key ecosystem services to humanity (for a detailed discussion on ecosystem services, see Section 2.1.4 in *Chapter 2: Biodiversity*). Soil ecosystem services can be divided into provisioning, supporting, regulating, and cultural services (see Table A).

The **provisioning role** of ecosystem services relates to products derived from soils that have marketable value. This includes clay use for Chinese ceramics (see Figure A.a), the utilization of soils in food production (Lal, 2004), and peat utilized for fuel. Medicinal uses include soil-based antibiotics, such as streptomycin, a common soil microbe.

The **supporting role** refers to soils providing the basis in which other natural and anthropogenic systems are maintained. Examples include the support that soil provides to crops. Additionally, soil supports plants and agriculture by providing a stable supply of nutrients.

The **regulating role** sees soils as a primary influence on the rates and magnitude of other natural processes. Here examples include flood regulation, soil erosion control, filtration and cleansing of groundwater, and regulating soil moisture to plants. The latter is an especially important role to agriculture in seasonally dry landscapes. Additionally, soils help to regulate Earth's long-term climatic regime by storing carbon.

At the intersection of soil providing a **cultural role** and a supporting role is the role of food in society where some food production systems highly valued by society are strongly associated with specific soil types. Additionally, cultural heritage is literally stored in soil—excavation is commonly used to unearth artefacts and architecture which have been preserved in soils (see Figure A.b). The importance of this is appreciated by the discipline of geoarchaeology, at the nexus of soil science, sedimentology, and archaeology.

a)

b)

Figure A Two examples of ecosystem services soils can provide: (a) a Chinese ceramic vase made from clay, in which soil provides a provisioning role; and (b) an excavation during an archaeological expedition, in which soil provides a cultural role.

Source: (a) Thomas Quine/Flickr/CC BY 2.0; (b) Smithsonian Institution/CC BY-SA 3.0.

Table A Examples of different types of soil ecosystem services

Service	Type of service	Main soil component and/or property	Example
Food	Supporting, cultural	Whole soil	Cultivation, ethnobotany
Fibre	Supporting	Whole soil	Cotton, jute
Topsoil	Supporting	Clay, humus	Gardening topsoil
Fuel	Provisional, supporting	Organic matter	Peat, corn for ethanol
Carbon sequestration	Regulating	Clay, humus	Top soil
Water purification	Provisioning, supporting, regulating	Whole soil	Municipal sewer systems
Climate	Regulating	Clay, humus	Carbon sequestration
Nutrient cycling	Regulating, supporting	Clay	P, N
Habitat for organisms	Supporting	Whole soil	Burrowing animals and insects (for example, prairie dog)
Flood abatement	Regulating	Whole soil	Infiltration and water storage
Source of pharmaceuticals and genetic resources	Provisioning, supporting	Clay, organic material	
Foundation for human infrastructure	Supporting, cultural	Sand	Non-subsiding
Construction materials	Provisioning	Sand	Non-subsiding foundation materials

(Based on NRCS 2017.)

BOX 5.5

Hydroponics: agriculture without soils

Hydroponic agriculture refers to the growing of crops outside the soil (see Figure A.a). In hydroponic systems, the nutrients are supplied in solutions (most commonly artificial fertilizers) and the soil is replaced by another supporting medium, for example rockwool (see Figure A.b). The nutrients can be very precisely managed and can even be provided to individual plants, using drip irrigation. Hydroponic systems are often closed systems, in which nutrients and water are recycled for optimal use.

One of the best known examples of large-scale hydroponics can be found in some of the greenhouses in the Netherlands (see Figure B.a) (note that not all greenhouses use hydroponics; some do grow crops in soil). The focus in this region is on the production of vegetables [cucumbers, tomatoes (see Figure B.b), and bell peppers], as well as ornamental plants and flowers that have a high export value. These greenhouse crops are cultured under computerized and optimized conditions, using state-of-

the-art techniques. Under these conditions, yields can be extremely high—tomatoes have an average yield of around 50 kg per m², bell peppers 25 kg per m², and cucumbers a staggering 65 kg per m². By comparison, in 2011–12, the average yield of these crops in Almeria, Spain, another key production area for these vegetables, was 9.6 kg per m² for tomatoes, 9.1 kg per m² for cucumbers, and 6.9 kg per m² for bell peppers (Valera et al., 2016). A high wheat production is 1.5 kg per m².

However, it is very important to realize that hydroponics are only effective for a very limited group of crops. The key staple crops that form the basis of most of our meals (for example, wheat, corn, and rice) are not usually grown using hydroponics. Because of the high production cost of staples, hydroponics is not economically viable. However, several space agencies, such as the European Space Agency, are experimenting with growing these crops on hydroponics, a vital prerequisite if humans want to ever colonize other planets (Page and Feller, 2013).

a)

b)

Figure A Food can be grown without soil. Nutrients are provided in a solution, often using drip irrigation (a) and compounds, such as rockwool (b), are used as supporting medium.

Source: (a) Borisshin/Wikimedia Commons/CC BY-SA 4.0; (b) Achim Hering/Wikimedia Commons/CC BY 3.0.

a)

b)

Figure B (a) The Netherlands has some of the highest hydroponic food production in the world, mostly in greenhouses. (b) Vegetables, such as tomatoes, are a popular crop.

Source: (a) Wolk9/CC0; (b) Goldlocki/CC BY-SA 3.0.

Pause and think

Currently, 95% of our food is produced in soils. In Box 5.5, we discussed hydroponic systems, in which food is produced without soils. As soils are not required, they can even be placed on rooftops (see Case study 7.1 in *Chapter 7: Energy*). Do you think that hydroponics is a viable alternative to feed our global population? List some of the practical benefits and constraints of these systems.

5.2.1 Soil erosion by water

Modern mechanized agriculture causes soil erosion by water, which can be several orders of magnitude greater than natural soil erosion (Montgomery, 2007). This is due to two important reasons. First, landscapes covered with natural vegetation, such as grasslands and forests, have considerable **surface roughness** (see Figure 5.9). This refers to the roots, twigs, and litter which cover the soil and which buffer the shear stress generated by run-off,

a)

b)

Figure 5.9 Surface roughness can impact erosion. Natural vegetation, such as a forests (a), often have high surface roughness, consisting of trees, roots, twigs, and litter, which greatly reduces soil erosion by water. In contrast, agricultural lands, such as potato fields (b), have a considerable amount of bare soil and low surface roughness.

Source: (a) Mickaël Delcey/Wikimedia Commons/CC BY-SA 4.0.

as these physical structures interrupt run-off paths. Natural forests and grasslands therefore have low rates of soil erosion (Renard et al., 1997). But across landscapes dominated by agriculture, considerable bare soil exposed for much of the year represents less resistance to overland flow, resulting in higher rates of run-off. Much of this run-off ends up in the aquatic system, which can cause significant degradation of downstream aquatic environments (see Case study 5.2).

CASE STUDY 5.2

Madagascar: linking erosion to watershed degradation

Erosion of fertile topsoil not only causes a threat to the productivity of the agricultural lands, but it can also cause significant impacts on watersheds. Madagascar is the fourth largest island in the world, and agriculture is still the mainstay of the economy. As over 70% of the population works in the agricultural sector.

Over the last century, there has been extensive logging in Madagascar that has resulted in the replacement of large areas of tropical rainforest by agriculture, especially tavy, a form of slash-and-burn or swidden (see Case study 5.1 for more details). In short, an area of rainforest is cut and burnt, and the area is planted with rice. After 2 years of production, the field is left fallow (no crop), and after 4–6 years, the process is repeated. After 2–3 of these cycles, the soil is exhausted of nutrients and the farmers leave. Now the fields are replaced by shrubs and grasses, which, on hills, are not sufficient to hold the soil in place after rainfall. This leads to high rates of erosion of the top soil (see Figure A).

More than 160,000 km^2 of land have been affected in Madagascar, an area equivalent of four times the Netherlands, or two-thirds of the UK, significantly reducing the productivity of the land. This has directly affected the livelihoods of 4 million.

Once eroded, soils may be flushed into streams whereby they are deposited in the channel bed and adjacent floodplain, causing degradation of the riparian ecosystem. Additionally, the magnitude and frequency of flooding can increase downstream because the channel capacity is reduced from infilling, which also compromises important infrastructure (for example, bridges, pipelines, flood control).

Figure A An example of massive trenches (called gullies) caused by erosion in Madagascar.

Source: Volker Häring/Wikimedia Commons/CC BY-SA 3.0.

The lifespans of dams and reservoirs are commonly reduced because of high rates of reservoir sedimentation that ultimately reduce water storage capacity and hydropower generation.

Coastal and marine environments, including coral reefs, can be adversely impacted by high rates of sedimentation derived from the eroded soil. This is especially the case when viewing these issues from a watershed perspective (see Section 4.1.2 in *Chapter 4: Water*), because of the indirect consequences of upstream (hill slope) erosion to downstream environments. In Madagascar, eroded soil is transported to rivers, such as the Betiboka River which empties in the Bombetoka Bay, causing extensive degradation to the associated coastal ecosystem.

The vegetation along the Madagascar coastline is naturally dominated by mangrove forests, providing a key habitat and nursery ground for many fish species, as well as crustaceans, sea turtles, birds, and dugongs (a sea mammal like a manatee). In addition, coral reefs are located along the northwest coast of Madagascar. The mangrove forests provide an important function in that they trap river-borne sediment that would otherwise be deposited atop the coral reefs. Unfortunately, between 1990 and 2010, over 20% of Madagascar's mangrove forests were lost, mainly due to agriculture, aquaculture, coastal development, and the demand for firewood (Jones et al., 2016). Without the protective buffer of the mangrove forests, the coral reefs, which are biodiversity hotspots (see Section 2.3 in *Chapter 2: Biodiversity*), are severely degraded by the high rate of sedimentation.

Figure B A satellite image of the Betsiboka River entering the Bombetoka Bay. The brown colour of the water is caused by erosion of fertile topsoil.

Source: NASA.

Figure C Madagascar has lost >20% of its mangrove forests between 1990 and 2010. Mangroves provide a key habitat to many species, including fish, sea turtles, and dugongs, and trap sediment, thereby protecting coral reefs.

Source: Smiley.toerist/Wikimedia Commons/CC BY-SA 3.0.

Figure 5.10 Impeded soil drainage caused by soil compaction along tractor wheelings in the Thames River Valley, England (winter 2014).
Source: Environment Agency.

Figure 5.11 Gully development in the Sierra Madre Oriental of eastern Mexico.
Source: P.F. Hudson, 2001.

A second way that mechanized agriculture increases run-off and soil erosion is by soil compaction. Compaction prevents rainwater from infiltrating into the soil, thereby increasing the risk of run-off and erosion. Precipitation that would have been stored in the soil and seeped into rivers as base flow (see Box 4.3 in *Chapter 4: Water*) now flows overland as run-off (see Figure 5.10).

Soil erosion by flowing water can occur in three main ways: sheet wash, rills, and gullies. Erosion by **sheet wash** refers to the incremental removal of fine layers (approximately 1 mm thick) of topsoil, especially soil particles that were already dislodged by raindrop impact (Renard et al., 1997). Sheet wash slowly degrades soil quality and reduces soil fertility, thereby lowering agricultural yields. Because sheet wash does not result in significant physical change to the landscape, it can be difficult to detect, thereby occurring for decades before management intervention. In extreme cases, sheet wash results in near-complete removal of the organic rich soil A-horizon.

By comparison with sheet wash, **rills** are concentrated paths of run-off and produce much higher rates of soil erosion. Rills are usually approximately 10–50 cm wide and approximately 1–10 cm deep (see Figure 5.8b). Agricultural fields are especially susceptible to rill development when they are bare (see Figure 5.9b). This frequently occurs at the onset of the rainy season, before crops have grown, or after the crops have been harvested and the litter removed from the fields.

Rills are usually short-lived, sealing over with subsequent rainfall events or agricultural activities (e.g. ploughing). Without intervention, however, a network of interconnected rills develops that concentrates run-off, resulting in higher rates of soil erosion. The concern

here is the development of a **gully**. In contrast to rills, gullies are semi-permanent erosional features that range approximately from 1 to 10 m in width and depth (see Figure 5.11). This is a larger issue in mechanized agriculture, because the treads and furrows from the machinery (tractors) themselves often contribute to rill formation. While sheet wash and rills can be managed with contour ploughing, buffer strips, and no-till farming, the onset of gullies is a more difficult (and costly) type of land degradation problem to manage.

5.2.2 Soil erosion by wind

Dry regions are especially prone to soil loss from wind erosion and a decline in soil quality by salinization. In contrast to water erosion, wind erosion is more of a problem on flat and expansive dry landscapes. A classic example is the Dust Bowl of the USA, which refers to a period in the 1930s during which over approximately 350,000 km² of the High Plains were impacted by drought and severe winds events. Wind-driven soil erosion increases exponentially with decreasing ground cover, a major concern with modern mechanized agriculture because it leaves the soil bare and exposed to wind (see Figure 5.9.b).

A vegetation cover of about 50% is required to prevent excessive soil erosion by wind. During the Dust Bowl events, the fine-grained humus-rich prairie topsoil was selectively eroded by wind, a process known as **deflation**, decreasing the average topsoil depth by 20–80 cm. The subsequent dust storms were often referred to as 'black rollers' or 'black blizzards' (see Figure 5.12), because of the large amounts of airborne, blackish soil organic matter (humus), causing massive reduction in soil fertility for cultivation.

Figure 5.12 Wind erosion of dark humus topsoil near Dodge City, Kansas during the era of the American Dust Bowl in 1935. Because of the large amount of blackish soil organic matter, these clouds were referred to as **black rollers** or **black blizzards**.

Source: Special Collections, Wichita State University.

Worryingly, dust bowls are now happening on a large scale in China (see Figure 5.1a and Figure 5.13). In 2000, 1.91 million km^2 in China alone were affected by wind erosion (Wang et al., 2006). This is 20% of the surface area of China, and roughly the same area as Mexico. In these degraded soils, the carbon and nitrogen contents have reduced by up to 66% and 73%, respectively (Wang et al., 2006). A prime location where this occurs is the fertile Loess Plateau (see Case study 5.3).

5.2.3 Salinization of agricultural lands

Land degradation by soil **salinization** is often a problem in agricultural settings that rely upon irrigation (UNCCD, 2017). Salinization occurs by transfer of dissolved salts with upward moving capillary water, from lower horizons to upper soil horizons (see Figure 5.14). While soil salinization occurs across a range of climate types, it is especially acute in dry environments.

Figure 5.14 Salinization (indicated by white colouring) of soils in the Grand Valley, Colorado, a dry environment with large-scale modern agriculture heavily reliant upon irrigation water from the Colorado River.

Source: Davies, 2017.

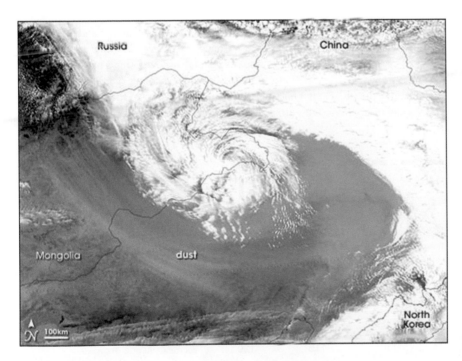

Figure 5.13 A dust bowl in China, causing massive loss of fertile topsoil and degradation of agricultural lands.

Source: NASA image courtesy Jeff Schmaltz, MODIS Rapid Response team.

CASE
STUDY
5.3

Losing and restoring the soil resource—the example of the Chinese Loess Plateau

Loess consists of wind-blown deposits of silt that accumulated over thousands of years during the Pleistocene (ice ages during the past 2.6 million years). Soil developed upon loess is among the most fertile for agriculture. Loess soils support the world's most productive grain-growing regions, such as Ukraine, south-western Russia, north-central USA, and central Europe. The largest loess region is the Chinese Loess Plateau. Loess deposits 50–100 m in thickness cover an area of around 500,000 km² in north-central China (see Figure A). The superb soil properties of loess have long been recognized by the earliest farmers on Earth. The Loess Plateau is said to be the cradle of Chinese civilization, with evidence of the oldest millet cropping dating back to 10,000 years.

Loess consists predominantly of floury particles referred to as silt (see Figure 5.7). The formation of the upper portions of the Chinese Loess Plateau occurred primarily during the past 2.6 million years. Ice ages promoted glacial erosion and physical weathering. Strong winds resulted in the erosion and transport of the silt, with subsequent deposition resulting in the accumulation of thick loess deposits. The soils that developed upon loess are rich in organic matter and enriched with nutrients. Water availability and warm air temperatures in summer enabled biotic activity to thrive, resulting in optimized water and air exchange and nutrient cycling.

Unfortunately, the Chinese Loess Plateau is now known as one of the world's most prominent examples of extreme soil degradation. Originally, the Loess Plateau had rolling topography with extended forests that promoted a very active soil life, while also protecting the soil from erosion. After the natural vegetation was converted into agriculture, large-scale erosion occurred. The Loess Plateau now is marked by flat-topped, elongated ridges (representing the remains of the original plateau) that are deeply dissected by gullies and river valleys (see Figure B).

Since 1958, the Chinese government has implemented a variety of environmental conservation policies to protect the Loess Plateau (Cao, 2011). The first measures focused on erosion control, flood control, and enhanced freshwater availability (Chen et al., 2007). Erosion control has included the construction of numerous artificial terraces to safeguard hill slopes from erosion, a traditional method to control water and sediment flows across sloping fields on vulnerable surfaces (see Figure C).

Despite clear achievements, it is almost impossible to restore soil quality, and thus beneficial functions of the original soil resource. With respect to the nature of a healthy soil, any 'modern' soil is a soil in its infancy, or just slightly weathered loess sediment. These soils lack organic material (humus) that helps to bind particles together and resist run-off, and thus will remain susceptible to erosion by run-off events, with sparse vegetative cover. This is because natural soil-forming processes require a couple of hundred years, if not millennia, to create soil aggregates that prevent surface run-off, gullying, and, as a consequence, local flooding.

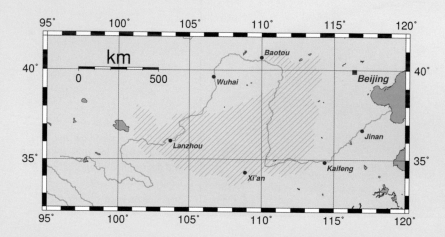

Figure A The Chinese Loess Plateau extends out over 500,000 km² in north-central China.

Source: CSIRO/CC BY-SA 4.0.

The Chinese Loess Plateau is an important example of a landscape that has been stripped of its valuable natural soil resource. The process is irreversible on the broader Loess Plateau scale, and only a small portion has been restored.

Figure B Large part of the fertile loess soil has been lost due to water and wind erosion, leaving deep gullies and river channels.

Source: Vmenkov/Wikimedia Commons/CC BY-SA 3.0.

Figure C Terraces have been promoted on hillsides. This is a successful measure promoted by the Chinese government to prevent erosion through water run-off.

In dry environments, the salinization process is often driven by improper irrigation practices, which is made more severe because of the high rates of surface evapotranspiration. The salt is either provided naturally by the parent material or irrigation water. Soils degraded by salinization are referred to as sodic soils, in reference to the high concentration of sodium (Na^+). At a minimum, salinization results in diminished crop yields, but with continued salt concentration, it eventually results in field abandonment. The subsequent loss of supporting vegetation increases the susceptibility of the soil to wind and water erosion (see Figure 5.15).

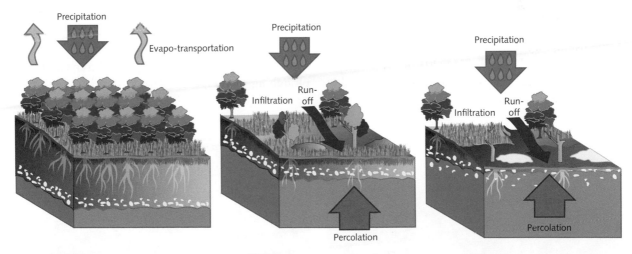

Before clearing
Most water is used where it falls. The system is in balance.

After clearing
Saline groundwater rises and is concentraed at the surface by evaporation. Vegetation growth is affected.

Later
Accumulation of salt at the surface kills protective plant cover. The land is open to erosion.

Figure 5.15 Schematic diagram of salinization caused by changes in land cover in dry environments.

Salinization impacts developed and developing nations. But it is especially acute in Africa where large irrigation projects in the 1960s and 1970s, funded by international development agencies, were inappropriately implemented. After a couple of decades, the initial increase in crop yield stalled and then declined with increased Na⁺ concentration in soil. Salinization has already resulted in the abandonment of millions of hectares of irrigated lands. As globally irrigation is most common in dry environments, it is estimated that some 10–48% of irrigated lands are impacted by salinization (FAO, 1997). While new irrigation technologies hold promise for reducing salinization, of great concern is that the area of irrigated lands have doubled over the past two decades (UNCCD, 2017).

5.3 How will soil degradation impact agriculture?

Agricultural production has increased tremendously over the last century, with the rise of the Green Revolution (see *Chapter 1: Introduction*). For example, there has been a long-term increase in cereal production, especially since the 1950s. This increase is, in a large part, due to a tremendous increase in cultivated land (FAO and ITPS, 2015). Nevertheless, annually, some 7.6 million tonnes of cereal production are lost because of soil erosion associated with the large-scale industrial agricultural system. By 2050, it is estimated that the annual reduction in cereal production due to soil erosion will total over 250 million tonnes, a yield equivalent to removing 1.5 million km² of arable land from production. If this were a country, it would be the nineteenth largest country on the planet, right after Mongolia.

The issue of soil degradation and crop declines varies regionally, depending upon the natural setting and landscape history in relation to modern agricultural practices (Bakker et al., 2007; FAO, 2015). Most African nations experience a net annual decline in soil quality, as nutrient loss exceeds its annual replenishment. Soil acidity is also associated with a decline in crop yield, especially in tropical regions that have been deforested. Salinization, the accumulation of salt in the upper soil horizons, reduces crop yields and, in extreme cases, prevents land cultivation. Globally, an area equal to the size of Brazil is impacted by human-generated salinization.

In North America and the EU, historic reductions in crop yields were estimated at 4% per 10 cm of topsoil loss, which reduces rooting depth and available soil moisture. Considerable variability exists, with crop yields in northern Europe expected to remain relatively stable. However, Southern Europe is expected to experience crop yield reduction because of soil erosion, especially in regions that are under intensive cultivation such as Mediterranean Spain, Greece, and Italy.

A major problem with modern mechanized agricultural practices is that it reduces the content of SOM, which is considered a reduction in soil quality. This is mostly because of repeated soil ploughing (see Figure 5.16), which literally turns

Figure 5.16 Many farmers plough their field before sowing a new crop. By ploughing, the A horizon is turned upside down.
Source: (a) © The International Livestock Research Institute/Wikimedia Commons (CC BY 2.0).

the micro-habitat of soil life upside down through tilling. The reduction largely occurs because the crop litter is removed (for example, straw production following cereal harvesting) and is therefore not allowed to decompose and feed back into the nutrient cycling. A reduction in SOM from 5% to 3% may reduce the ability of soils to capture and store nitrogen and phosphorus by 20% (a process called **sequestration**), which, in turn, reduces crop yields (Lal and Moldenhauer, 1987). In addition, the loss of soil humus makes topsoil more vulnerable to erosion.

An important motivation for **no-till agriculture**, an approach that does not annually till the soil, is to 'leave the litter in the field' (see Figure 5.17). Such approaches have been shown to increase soil moisture retention by reducing evaporative losses and increase soil organic matter within 5–10 years. Additionally, the litter protects the soil surface from flowing water (run-off) and wind, thereby reducing the rates of soil erosion. This is also true for the wind erosion occuring in China—switching to no-till agriculture can greatly reduce wind erosion and associated dust storms (Wang et al., 2006). No-till agriculture also offers benefits for climate change, as it reduces the release into the atmosphere of CO_2 stored in the soil.

Why then is this technique not applied everywhere and by every farmer? Because leaving plant litter on

Figure 5.17 No-till soy cultivation, with the soil surface protected by wheat residue. The crop litter reduces soil erosion by raindrop impact and run-off and also reduces soil moisture loss by evaporation.

Source: Tim McCabe / Photo courtesy of USDA Natural Resources Conservation Service.

the field also leaves pests and diseases on fields, endangering follow-up crops. The unwanted side effect of no-till or less-till farming thus can require a significantly increased use of pesticides. In consequence, many farmers would rather accept losses in soil fertility for the benefit of saving a more intense application of costly and polluting pesticides (see Section 3.2.2 in *Chapter 3: Pollution*). This raises important questions regarding industrial agriculture and soil quality (see Box 5.6).

BOX 5.6

Industrial agriculture and soil quality

A large proportion of industrial food production occurs in the form of mono-culture, an agricultural system whereby farmers intensively cultivate a single crop (see Figure A, and Case study 2.3 in *Chapter 2: Biodiversity* for a discussion on polyculture as an alternative to mono-culture). Mono-culture is an especially efficient agricultural system for crop planting, harvesting, irrigation, and in

the application of fertilizers and pesticides. Mono-culture agricultural systems also streamline costs and lowers the opportunity costs of producers, in that they do not need to gain knowledge of multiple crops. However, mono-cultures also make soil more vulnerable to the loss of organic carbon by degassing and erosion. Additionally, mono-culture systems reduce the input of organic matter into the soil, thereby reducing soil biodiversity and diminishing key micro- and macro-organisms that contribute to healthy soils.

The adverse consequences to soil caused by intensive monocrop agriculture are widely considered unsustainable and can result in reduced crop yields. Global cereal yields are estimated to level off by 2030, although demand will continue with increased population and dietary shifts. Thus, a question that is increasingly being asked concerns whether Earth—and humanity—has reached 'peak soil', that is considering the tremendous degradation of soil resources, will we be able to sustain food production to support a projected global population of 9.5 billion in 2050?

a)

b)

Figure A Examples of mono-cultures: (a) Wheat production in the US Midwest and (b) a potato field in France.
Source: (a) © United States Department of Agriculture; (b) Olivier Bacquet/Flickr (CC BY 2.0).

5.4 Future outlook

The increasing demand for agriculture to support food, fuel, and fibre for a rapidly growing population means that soil resources will continue to be under great pressure (see Food controversy 5.1). Over 50% of agricultural lands carry soils that are degraded to varying degree, a condition that directly impacts 1.5 billion of the global population. An additional type of soil loss occurs directly through construction activities such as urbanization (UNCCD, 2017). Between 1970 and 2000, for example, some 43,000 km² of Earth's surface were urbanized (an area the size of Denmark), resulting in irreversible soil loss (Amundson et al., 2015). The remaining soils should be managed as a valuable non-renewable natural resource (see, for example, Lehmann and Kleber, 2015), as it requires centuries to recover by natural processes of soil formation.

The greatest global environmental threat to soil resources continues to be land use change (UNCCD, 2017). Nowadays, land conversion of unused or 'underused' land often comes along with the implementation of industrial farming schemes, which may cause reduced soil quality, increased soil erosion by water or wind, and loss of associated ecosystem services on large scales. This is especially problematic in small-scale communities in lesser developed nations (see Case studies 5.1 and 5.2). Populations within such nations that continue to practise traditional forms of agriculture are

FOOD CONTROVERSY 5.1

Land grabbing: who owns the soil for growing crops?

China's population has doubled since 1960, and the overall affluence in China has increased rapidly over the last decades. This has resulted in massive mechanization of Chinese agriculture, but with some serious negative impacts on soils (see Case study 5.3). Over 20% of agricultural land in China is degraded due to wind erosion, and in addition, there are deep concerns regarding soil pollution.

One recent development has been large-scale land acquisition, or **land grabbing**. Land grabbing refers to the buying or leasing of large pieces of land (most often defined as areas >1000 ha) by companies and governments. Leasing involves the large-scale renting of land for a specific period, often over multiple decades. Over 70% of all land grabbing occurred in only 11 countries, predominantly in East Africa (including Ethiopia, Kenya, Tanzania, and the Democratic Republic of Congo) and South East Asia (including Indonesia, Papua New Guinea, and the Philippines) (Anseeuw et al.,

2012) (see Figure A). By 2011, at least 560,000 km² of African land was involved in land grabs, eight times the size of the Republic of Ireland and 4.8% of Africa's total agricultural area (Anseeuw et al., 2012). This figure is likely to have grown in recent years. The biggest investor in large-scale land acquisition is China, followed by Saudi Arabia and Brazil.

The purchase of land is highly controversial. Local people can lose access to land and resources, increasing the risk of food insecurity (Cotula et al., 2009). A large proportion of the rural population in sub-Saharan Africa depends on land for their food security and livelihood (Cotula et al., 2009). This makes land grabbing a very sensitive and controversial issue. In addition, there is a risk of corruption, especially when there is a lack of transparency around the deals (Cotula et al., 2009).

One notorious case occurred in Madagascar where the South Korean company Daewoo Logistics attempted to lease 13,000 km² of land from the government (Cotula et al.,

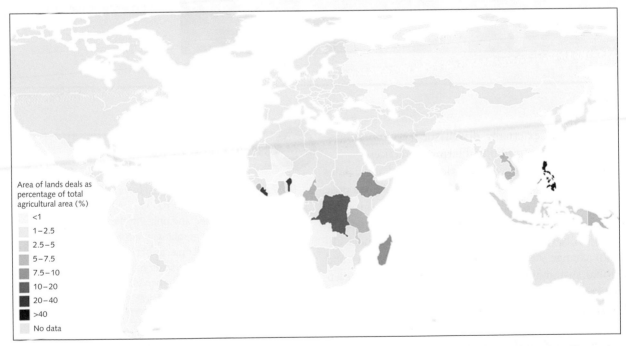

Area of lands deals as percentage of total agricultural area (%)

- <1
- 1–2.5
- 2.5–5
- 5–7.5
- 7.5–10
- 10–20
- 20–40
- >40
- No data

Figure A A global map showing that large-scale land acquisition mainly happens in the Global South and is concentrated in Africa and South East Asia. The darker the colour, the higher the percentage of the total agricultural area in that specific country which has been involved in land deals.

Source: Matthias Engesser/Centre for Development and Environment (CDE).

2009). The company wanted to grow maize and palm oil for export to South Korea and other countries (*Financial Times*, 2008). This resulted in widespread protests, and some condemned the deal as '*neo-colonialism*' (BBC, 2009). It also led to public anger against the government (see Figure B) and contributed to the resignation of the former president Marc Ravalomanana. In the face of this public pressure, the deal was then cancelled by the new government (BBC, 2009).

Although this land grab failed, between 2004 and 2009, over 8000 km² of land was grabbed in Madagascar alone,

involving an investment of almost 80 million USD. The increasing expansion of biofuels (see *Chapter 7: Energy*) has also played a role in land demand. For example, in 2008, a 4525-km² biofuel project in Madagascar was approved. In this case, GEM Biofuels, with its head office on the Isle of Man (UK), has secured a 50-year lease (Cotula et al., 2009). We can expect these trends to continue with ongoing increases in demand from population growth, rising affluence, and biofuel targets (see Food controversy 7.1 in *Chapter 7: Energy*).

Figure B　A protest in Antananarivo, the capital of Madagascar, partly fuelled with public anger against a large land deal between the government and the South Korean company Daewoo Logistics.

Source: fanalana_azy/Flickr (CC BY 2.0).

vulnerable to externally driven socio-economic pressures to abandon traditional farming practices.

In addition to soil degradation by agricultural activities, the projections of future climate change pose another threat. This is the case especially in semi-arid regions where annual soil moisture budgets are marginally viable and projected to decline under climate change scenarios. Given this, food security and global sustainability are intricately linked to the effective management of Earth's soil resources. While the current trajectory of soil resource management does not look good (Amundson et al., 2015), there are pockets of optimism (see Case study 5.4). Much is already understood about the process of land degradation and soil erosion, which can be mitigated by

 Pause and think

Land grabbing is defined as the buying or leasing of pieces of land larger than 1000 ha. What does this land grabbing do to other land prices in the region? And do you consider it land grabbing when wealthy foreign individuals buy smaller pieces of land for personal use (for example, to build a vacation home)? If you do not consider it land grabbing, when would the buying or leasing of land tip into a land grab?

strategies such as non-till agriculture, formation of terraces on hillsides, buffer strips, and crop rotation (see Box 3.6 in *Chapter 3: Pollution*).

CASE STUDY 5.4

Regenerating soils: the *Terra preta* approach

One approach to increasing food production in areas with poor soil fertility is to remake the soil. **Terra preta**, or dark earth soils (see Figure A), also rely upon the burning of organic matter and can be seen as a variation of slash-and-burn agriculture. In *terra preta*, the soils are anthropogenically enriched by burning plant material, food waste, and other organic detritus (Woods and Denevan, 2007). The carbon content of *terra pretta* soils can be as high as 9%, which contrasts with <1% in adjacent undisturbed soils. The dark earth soils of *terra preta* stand in strong contrast to adjacent nutrient-poor soils (see Figure A).

A single hectare of metre-deep *terra preta* soil is estimated to sequester 250 tons of carbon, far higher than adjacent undisturbed soils at 100 tons of carbon (see Figure B). In addition to *terra preta* having high amounts of organic matter and microbial activity, they also have elevated levels of important plant nutrients such as phosphorus, nitrogen, and potassium (Glaser et al., 2001).

Terra preta with the highest soil organic matter and carbon content develop by low-temperature, slow burning of plant material in oxygen-poor shallow soil environments, rather than rapid incineration at high temperatures. The lower temperatures leave considerable amounts of partially burnt organic residue, which is termed biochar, in the shallow soil subsurface. The importance of this is that it stimulates microbial activity. Thus, *terra preta* represents slash-and-char, rather than slash-and-burn.

While they have been documented to exist across an international range of settings, the heart of *terra preta* research is in the Brazilian Amazon. Some *terra preta* sustained high population densities for centuries, and some have existed for several thousand years (Woods and Denevan, 2007), and fields may range in size from about 1 ha to 350 ha. As a whole, in the Amazon basin, the extent of this human-made soil is estimated to be about 18,000 km². Although it is a small proportion of the Amazon, it was sufficient to significantly increase

a)

b)

Figure A Two soil profiles developed in the same parent material in the Amazon basin: (a) *terra preta* soil profile (approximately 100 cm deep) showing pronounced dark organic, rich A-horizon; (b) a typical Amazon soil profile (approximately 120 cm deep) developed upon deeply weathered tertiary surface, displaying an oxidized colour with relatively low organic matter in the A-horizon.

Source: Glaser, B., Haumaier, L., Guggenberger, G. et al. Naturwissenschaften (2001) 88: 37. Copyright © 2001, Springer-Verlag.

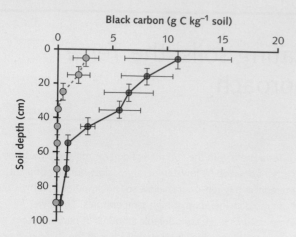

Figure B Black carbon in *terra preta* soils (black dots) in the Amazon is much higher than in a typical Amazon soil profile (open dots).

Source: Glaser, B., Haumaier, L., Guggenberger, G. et al. Naturwissenschaften (2001) 88: 37. Copyright © 2001, Springer-Verlag.

food production for local indigenous communities. Crop yields in *terra preta* are at least double those of adjacent unmodified soils.

Wim Sombroek, an internationally renowned Dutch soil scientist, is credited with rediscovering *terra preta* and developing the systematic study of its characteristics in relation to fertility and human activities (Mann, 2007). Sombroek recognized that the cultural dark earth soils in the Brazilian Amazon had similarities to the plaggen soils on his own family farmstead in his native Netherlands. Plaggen soil has an anthropogenically fortified topsoil horizon (see Figure C). In many cases, plaggen soil was created over hundreds of years from the constant dumping of ashes, heather, and stable droppings.

Terra preta now is regarded as a proven model of sustainable agriculture and soil management (Glaser et al., 2001; Lehmann, 2006). The increased study of dark earth soils (in general) and the verification of their much higher food production and carbon storage have led to innovative proposals to utilize biomass from industrial-scale agriculture to create biochar to replenish depleted soils in developed nations (see, for example, Lehmann, 2006; Woolf et al., 2010). While it has yet to be proven economically viable, some scholars maintain that large-scale biochar storage in soils could potentially store enough carbon to help mitigate global climate change (Lehmann, 2007; Woolf et al., 2010).

Figure C Photo of plaggen soil profile in the Netherlands. Note the sharp colour contrast between dark organic material at the surface and oxidized sandy soil at the base (podzol).

Source: © European Union, 1995-2018.

Pause and think

At the start of Section 5.4, we highlighted that it takes centuries for soils to form. However, in Case studies 5.3 and 5.4, we give examples of soil restoration activities. Think back to the definition of a renewable and non-renewable resource (see Section 1.2.1 in *Chapter 1: Introduction*). Do you think that soils are renewable or non-renewable resources? Provide arguments based on the content of this chapter which justify your answer.

To address current challenges to soil resources and to sustainably secure their beneficial functions require an understanding of how soil forms, functions, and degrades. Studying soil has focused on developing effective management and production strategies as related to agriculture. Nevertheless, soil science is now considered as a truly transdisciplinary field that spans a wide range of natural and social sciences. Soil science is included as an academic programme within agricultural and forestry studies at universities, in addition to government agencies and international institutions, such as the FAO of the United Nations. Many universities also contribute to sustainable agricultural practices by working with government agencies and directly with farmers.

These institutions have developed systematic methodologies and procedures to describe the nature of a soil and study how various environmental processes and agricultural practices influence soil quality and soil degradation. Again, major efforts historically were directed towards optimizing agricultural production (that is, crop yields). In recent decades, however, emphasis has oriented towards understanding the interaction of soils across a broader range of environmental and ecosystem processes, including the role of soils in global environmental change (Lal, 2004; Woolf et al., 2010; Lehmann and Kleber, 2015).

A key international initiative of the FAO is its 2015 Sustainable Development Goals (SDGs), an update to its Millennium Development Goals (see *Chapter 1: Introduction*). In addition to the SDGs highlighting the importance of environmentally oriented land management for climate change (SDG no. 13), the FAO also specifically champions agriculture in the interest of sustainable soil management. Food security should occur in accord with practices that preserve and respect small-scale (and often indigenous) agriculture, while enhancing biodiversity and preserving soil and water resources (SDG no. 2). A pressing concern in sub-Saharan Africa and tropical Latin America is the threat of declining biodiversity caused by the conversion of forests to large-scale industrial agriculture. Inappropriate practices threaten to further increase desertification, land degradation, and soil erosion, and thereby reduce food security (SDG no. 15).

A vital lesson learnt after decades of monitoring and research is that the most effective way to preserve soil resources is to increase the storage of soil organic matter, which inherently means promoting the biological activities of soil life. Retention and replenishment of soil organic matter improve soil structure and soil moisture, and serve as the basis for a range of soil ecosystem services.

In sum, global food security will continue to rely upon large-scale industrial agriculture. In this scenario, a key soil conservation goal would be to maintain, or even improve, physical and chemical qualities of soil to support ecosystem services. While Earth's natural soil humus stocks take millennia to recover, improvements in the physical and chemical quality of soils can occur within about a decade. Implementing large-scale measures to enhance soil quality should be a primary goal of government institutions that oversee the management of agriculture, soil resources, and food and water security.

QUESTIONS

5.1 What is the importance of humus to soil fertility?

5.2 What are the properties used to define soil horizons?

5.3 How does salinization result in land degradation?

5.4 Why does *terra preta* result in higher agricultural yields?

5.5 How is soil organic matter important to ecosystem services?

FURTHER READING

Amundson, R., Berhe, A.A., Hopmans, J.W., Olson, C., Sztein, A.E., Sparks, D.L. 2015. Soil and human security in the 21st century. *Science* **348**, 1261071. **(An important recent study that relates soil to food production and its vulnerability to a range of social challenges, particularly in the developing world.)**

Montgomery, D.E. 2007. *Dirt: The Erosion of Civilizations*. Berkeley, CA: University of California Press. **(A landmark book that relates societal problems—from prehistoric to modern times—to soil resources.)**

United Nations Convention to Combat Desertification (UNCCD). 2017. *Global Land Outlook*, 1st ed. Bonn: UNCCD. **(A recent compendium and global synthesis of problems caused by land change, especially related to food production and soil resource management.)**

● REFERENCES

Amundson, R., Berhe, A.A., Hopmans, J.W., Olson, C., Sztein, A.E., Sparks, D.L. 2015. Soil and human security in the 21st century. *Science* **348**, 1261071.

Anseeuw, W., Boche, M., Breu, T., Giger, M., Lay, J., Messerli, P., Nolte, K. 2012. *Transnational Land Deals for Agriculture in the Global South*. Bern/Montpellier/Hamburg: CDE/CIRAD/GIGA.

Bakker, M.M., Govers, G., Jones, R.A., Rounsevell, M.D.A. 2007. The effect of soil erosion on Europe's crop yields. *Ecosystems* **10**, 1209–1219.

BBC News. 2009. *Madagascar Leader Axes Land Deal*. Available at: http://news.bbc.co.uk/2/hi/africa/7952628.stm [accessed 30 April 2018].

Cao, S. 2011. Impact of China's large-scale ecological restoration program on the environment and society in arid and semiarid areas of China: achievements, problems, synthesis, and applications. *Critical Reviews in Environmental Science and Technology* **41**, 317–335.

Chen, L., Wei, W., Fu, B., Lü, Y. 2007. Soil and water conservation on the Loess Plateau in China: review and perspective. *Progress in Physical Geography* **31**, 389–403.

Cotula, L., Vermeulen, S., Leonard, R., Keeley, J. 2009. *Land Grab or Development Opportunity? Agricultural Investment and International Land Deals in Africa*. London/Rome: IIED/FAO/IFAD. ISBN: 978-1-84369-741-1

Davies, J. 2017. The business case for soils. Comment. *Nature* **543**, 309–311.

Denevan, W.M. 2001. *Cultivated Landscapes of Native Amazonia and the Andes*. Oxford: Oxford University Press.

Doolittle, W.E. 2001. *Cultivated Landscapes of Native North America*. Oxford: Oxford University Press.

Erni, C. 2015. Introduction. In: Erni, C. (ed). *Shifting Cultivation, Livelihood and Food Security New and Old Challenges for Indigenous Peoples in Asia*. Rome: Food & Agricultural Organization of the United Nations, pp. 3–40.

Financial Times. 2008. *Daewoo to Cultivate Madagascar Land for Free*. Available at: https://www.ft.com/content/6e894c6a-b65c-11dd-89dd-0000779fd18c [accessed 30 April 2017].

Foley, J.A.N., DeFries, R., Asner, G.A.P., Barford, C., Bonan, G., Carpenter, S.R., Chapin, F.S., Coe, M.T., Daily, G.C., Gibbs, H.K., Helkowski, J.H., Holloway, T., Howard, E.A., Kucharik, C.J., Monfreda, C., Patz, J.A., Prentice, I.C., Ramankutty, N., Synder, P.K. 2005. Global consequences of land use. *Science* **309**, 570–574.

Food and Agricultural Organization of the United Nations (FAO). 1997. *Irrigation Potential in Africa: A Basin Approach*. Food and Agricultural Organization of the United Nations, FAO Land and Water Bulletin 4. Rome, FAO.

Food and Agricultural Organization of the United Nations (FAO). 2015. *International Year of Soils: Healthy Soils for a Healthy Life*. Available at: http://www.fao.org/soils-2015/en/ [accessed 5 October 2017].

Food and Agricultural Organization of the United Nations (FAO), Intergovernmental Technical Panel on Soils (ITPS). 2015. *Status of the World's Soil Resources—Main Report*. Rome: FAO and ITPS.

Glaser, B., Haumaier, L., Guggenberger, G., Zech, W. 2001. The 'Terra Preta' phenomenon: a model for sustainable agriculture in the humid tropics. *Naturwissenschaften* **88**, 37–41.

Hall, M. 2001. *Farming Systems and Poverty: Improving Farmers' Livelihoods in a Changing World*. Rome/New York, NY: Food & Agricultural Organization of the United Nations/World Bank.

Jobbágy, E., Jackson, R. 2000. The vertical distribution of soil organic carbon and its relation to climate and vegetation. *Ecological Applications* **10**, 423–436.

Jones, T., Glass, L., Gandhi, S., Ravaoarinorotsihoarana, L., Carro, A., Benson, L., Ratsimba, H., Giri, C., Randriamanatena, D., Cripps, G. 2016. Madagascar's mangroves: quantifying nation-wide and ecosystem specific dynamics, and detailed contemporary mapping of distinct ecosystems. *Remote Sensing* **8**, 106.

Lal, R. 2004. Soil carbon sequestration impacts on global climate change and food security. *Science* **304**, 1623–1627.

Lal, R., Moldenhauer, W.C. 1987. Effects of soil erosion on crop productivity. *Critical Reviews in Plant Sciences* **5**, 303–367.

Lehmann, J. 2006. Black is the new green. *Nature* **442**, 624–626.

Lehmann, J. 2007. A handful of carbon. Commentary. *Nature* **447**, 143–144.

Lehmann, J., Kleber, M. 2015. The contentious nature of soil organic matter. *Nature* **528**, 60–68.

Mann, C.C. 2007. A few words about Wim Sombroek. In: Woods, W.I., Teixeira, W.G., Lehmann, J., Steiner, C., WinklerPrins, A., Rebellato, L. (eds). *Amazonian Dark Earths: Wim Sombroek's Vision*. Dordrecht: Springer, pp. ix–xiii.

Montgomery, D.E. 2007. *Dirt: The Erosion of Civilizations*. Berkeley, CA: University of California Press.

National Resources Conservation Service Soils (NRCS). 2017. *Soil Health and Ecosystem Services*. Available at: https://www.nrcs.usda.gov/wps/portal/nrcs/main/soils/health/ [accessed 29 September 2017].

Page, V., Feller, U. 2013. Selection and hydroponic growth of bread wheat cultivars for bioregenerative life support systems. *Advances in Space Research* **52**, 536–546.

Rekacewicz, P. 2008. *Global Soil Degradation*. UNEP/GRID Arendal—From Collection: IAASTD (International Assessment of Agricultural Science and Technology for Development). Available at: http://www.grida.no/graphicslib/detail/global-soil-degradation_9aa7 [accessed 16 November 2017].

Renard, K.G., Foster, G.R., Weesies, G.A., McCool, D.K., Yoder, D.C. 1997. *Predicting Soil Erosion by Water: A Guide to Conservation Planning with the Revised Universal Soil Loss Equation (RUSLE)*. Agriculture Handbook 703. Washington, DC: United States Department of Agriculture.

Rickson, R.J., Deeks, L.K, Graves, A., Harris, J.A.H., Kibblewhite, M.G., Sakrabani, R. 2015. Input constraints to food production: the impact of soil degradation. *Food Security* **7**, 351–364.

Rossi, J.P., Celini, L., Mora, P., Mathieu, J., Lapied, E., Nahmani, J., Ponge, J.F., Lavelle, P. 2010. Decreasing fallow duration in tropical slash-and-burn agriculture alters soil macro invertebrate diversity: a case study in southern French Guiana. *Agriculture, Ecosystems and Environment* **135**, 148–154.

Rumpel, C., Chaplot, V., Planchon, O., Bernadou, J., Valentin, C., Mariotti, A. 2006. Preferential erosion of black carbon on steep slopes with slash-and-burn agriculture. *Catena* **65**, 30–40.

Schaetzl, R.J., Thompson, M.L. 2015. *Soils: Genesis and Geomorphology*, 2nd ed. Cambridge: Cambridge University Press.

Stocking, M. 2003. Tropical soils and food security: the next 50 years. *Science* **302**, 5649.

United Nations Convention to Combat Desertification (UNCCD). 2017. *The Global Land Outlook*, 1st ed. Bonn: UNCCD.

Valera, D., Belmonte, L., Molina, F., López, A. 2016. *Greenhouse Agriculture in Almería: A Comprehensive Techno-economic Analysis*. Almería: Cajamar.

Wang, X., Oenema, O., Hoogmoed, W.B., Perdok, U.D., Cai, D. 2006. Dust storm erosion and its impact on soil carbon and nitrogen losses in northern China. *Catena* **66**, 221–227.

Woods, W.I., Denevan, W.M. 2007. Amazonian dark earths: the first century of reports. In: Woods, W.I., Teixeira, W.G., Lehmann, J., Steiner, C., WinklerPrins, A., Rebellato, L. (eds). *Amazonian Dark Earths: Wim Sombroek's Vision*. Dordrecht: Springer, pp. 1–14.

Woolf, D., Amonette, J.E., Street-Perrott, F.A., Lehmann, J., Joseph, S. 2010. Sustainable biochar to mitigate global climate change. *Nature Communications* **1**, 56.

Climate Change

Paul Behrens and Meredith T. Niles

Chapter Overview

- The natural greenhouse effect increases the temperature of the land, atmosphere, and ocean, and without it, Earth would be inhospitable to life.

- Anthropogenic climate change is the human-driven addition of more warming gases through human activities, which enhances the natural greenhouse effect.

- The food sector is a significant emitter of greenhouse gases, contributing between 19% and 29% of total global emissions.

- Emissions from the food sector come from many different processes along the production chain. Life cycle analysis (LCA) is one method for calculating these emissions.

- Reducing emissions of the food sector presents unique challenges when compared to other sectors such as energy and industry. Emissions are diffuse, composed of many different gases, and many of the most promising solutions are social.

- Solutions can be thought of as supply-side- and demand-side-based. Supply-side implies more technological and producer behaviour changes. Demand-side implies more consumer behaviour and social changes.

- Climate change will likely impact the production of food, creating the need for system adaptation. These adaptations can vary from low-tech (shade for animals) to high-tech (new vaccines for developing pest threats).

Introduction

How does your diet contribute to climate change? How does the clearing of a rainforest to grow food not only result in the reduction of biodiversity, but also in increased greenhouse gas (GHG) emissions? What is the natural greenhouse effect, and is it important in providing key processes that support life on Earth? How has humanity and our demand for food impacted our climate? These are some of the key questions addressed in this chapter.

This chapter is split into three main parts. First, we explain the physical process of climate processes, from the sunlight we receive on Earth to the way GHGs trap infrared radiation. Second, we look at how humans are impacting these processes, resulting in anthropogenic climate change. Next, we demonstrate how human food systems contribute to climate change and what can be done to reduce our impact. To show the complexity in finding solutions, we provide two case studies which show how counterintuitive and complex some food choices are. Finally, we explore the other side of the issue—how climate change will impact food systems as the environment changes.

6.1 How does climate change actually work?

Here we will discuss *how* the greenhouse effect **works**, not only *what* it **does**. It is important to remember that sunlight is transferring **energy** from the sun to Earth. As sunlight (scientifically, electromagnetic radiation; see Box 6.1) reaches the top of the atmosphere, it has a power of approximately 1370 W/m^2, the same power as an electric heater over every square metre (we will explore concepts of power and energy further in *Chapter 7: Energy*). This is termed the **solar constant**. However, because Earth is approximately spherical, as we move further from the equator, light is spread over more of Earth's surface area. This results in warm tropics and cold polar regions (see Figure 6.1).

As sunlight passes through the atmosphere, it is reflected and absorbed by different gases and surfaces (including clouds). The amount of light actually making it through to hit the Earth's surface is termed **insolation**. Due to reflection, absorption, and averaging over night and day, the insolation at Earth's surface reduces from 1370 W/m^2 (the solar constant) to an average of 250 W/m^2 (this, however, is highly dependent on location). We will discuss reflection and absorption, in turn.

Reflection: The amount of light reflected from the planet is called the **albedo**. If you look at a satellite picture of the Earth, you can see the bright white clouds that clad the atmosphere, and the dark blue

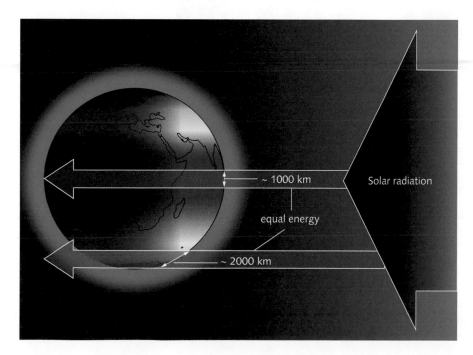

Figure 6.1 The amount of power from solar radiation hitting Earth depends on where on Earth you stand. Solar radiation hitting the equator vertically has a greater power per square metre than at the poles.

BOX
6.1

What is sunlight?

Sunlight is an everyday term for a selection of the electromagnetic radiation originating from the sun. Electromagnetic radiation is made up of a spectrum of different wavelengths, not only including the optical wavelengths (those we can see), but also infrared, X-rays, UV, gamma-rays, for example (see Figure A). This becomes important later because different wavelengths of light are absorbed by different gases in the atmosphere. Objects that radiate at a hotter temperature, such as the sun, radiate at higher frequencies and higher energies, than cooler objects (such as the Earth). For further information on the physics of these phenomena, see **Wien's law**, **blackbody radiation**, and **Stefan–Boltzmann law**.

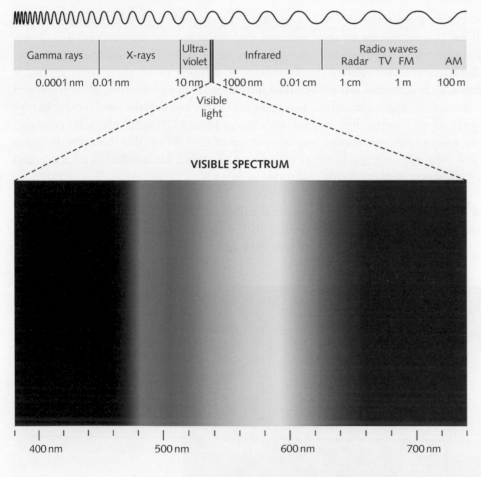

Figure A The electromagnetic spectrum. Visible light is only one part of this spectrum, which extends from gamma-rays (short wavelengths, high frequency to the left) to radio waves (long wavelengths, low frequency to the right).

Source: Peter Houben.

of the Earth's oceans (see Figure 6.2). Some surfaces reflect more light back into space than others; the whiter the surface, the greater the light reflected. Albedo is the fraction of light reflected; an albedo of 1 means 100% of the light is reflected back into space, and an albedo of 0 means that all the light is absorbed by the surface of Earth. For example, the albedo of a thick cloud is 0.7–0.8 (70–80% is reflected back out); the albedo of soil 0.2–0.5, and the albedo of forests is 0.1. Averaging all these different surfaces, the Earth has an average albedo of around 0.3.

Figure 6.2 A picture of the Earth from space. The bright white of clouds and the darker blue of oceans can be seen clearly. The different brightness results in different amounts of radiation absorbed and reflected. The ratio of this is termed the albedo.

Source: NASA.

Absorption: Absorption is the amount of electromagnetic radiation which is absorbed in the atmosphere and is key to the greenhouse effect. All surfaces radiate electromagnetic radiation, depending on their temperature. As the sun is very hot, it emits electromagnetic radiation at high frequencies, mainly in the ultraviolet (UV) range (see Box 6.1). Different gases in the atmosphere absorb electromagnetic radiation from sunlight in different ways. For example, sunlight at the top of the atmosphere contains X-rays, which would be particularly damaging for animals on the surface, causing burns and cancers. Fortunately, some gases, including ozone and O_2, absorb X-rays and prevent their transmission through to the surface (see Figure 6.3).

The atmosphere is transparent to several areas of the electromagnetic spectrum such as visible light (if it was not, we would not be able to see!), infrared, and some longer wavelengths. The sun's electromagnetic radiation emitted at these wavelengths can be absorbed by the Earth's surface, as long as it is not reflected by other surfaces such as clouds.

Given that the Earth has an average albedo of 0.3, around 70% of light is absorbed by our planet, heating the Earth up. Importantly, as the Earth itself is an object at a certain temperature, it will also emit

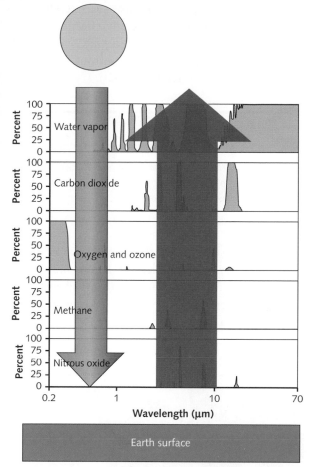

Figure 6.3 A schematic of the greenhouse effect. Starting on the left of the figure, we see sunlight at many wavelengths passing through the atmosphere to the Earth's surface (in green at the bottom of the figure). Each of the plots behind the rainbow arrow shows the absorption of energy for different GHGs at those wavelengths. As the atmosphere is generally transparent at these wavelengths, most of the light reaches the surface and warms up. Earth re-radiates this energy at a longer wavelength (infrared), seen in the red arrow on the right. As this re-radiated energy travels through the atmosphere, GHGs absorb the energy, heating up the Earth.

Source: Peter Houben.

electromagnetic radiation. However, as the Earth is not nearly as hot as the sun, the electromagnetic radiation is emitted at a much lower frequency, compared to sunlight, mainly in the infrared part of the electromagnetic spectrum (see Figure 6.3). Unlike the high-frequency radiation emitted from the sun, this lower-frequency radiation does not pass as easily through our atmosphere. Certain gases within the atmosphere absorb the outgoing electromagnetic radiation emitted by the Earth strongly, and as a result, they heat up. **Greenhouse gases (GHGs)** are the molecules which are especially effective in trapping this radiation. The dominant GHGs are carbon dioxide (CO_2), methane (CH_4), nitrous oxide, and water vapour (H_2O) (see Figure 6.3).

Pause and think

If Earth has a habitable temperature and atmosphere, why is it that planets very close (Mercury) and far away (Pluto) from the sun do not?

Figure 6.3 summarizes this process—most of the high-frequency sunlight (left of Figure 6.3) passes through the atmosphere and is absorbed by Earth. In contrast, some of the re-emitted lower-frequency radiation from Earth (right of Figure 6.3) is absorbed by gases in the atmosphere. The net result is the trapping of outgoing radiation by GHGs, heating the atmosphere up, a process called the **natural greenhouse effect**. Without this effect, Earth would probably be a cold, lifeless planet (see Box 6.2 for a calculation of the average temperature on Earth without the natural greenhouse effect).

Over the last centuries, especially since the start of the Industrial Revolution, humans have added significant amounts of extra GHGs into the atmosphere. This increases the total amount of heat-absorbing molecules in the atmosphere, and we enhance the natural greenhouse effect; this process is called **anthropogenic climate change**. Most people associate anthropogenic climate change with the emission of CO_2 from the burning of fossil fuels, but we have also increased the emission of CO_2 through other means like land use change (see the carbon cycle in Box 6.3) and have released other GHGs, including

---

BOX 6.2

What would the temperature of Earth be like without the greenhouse effect?

Why is the natural GHG effect essential for life on Earth? This can be demonstrated with a series of simple calculations. Due to the conservation of energy (see *Chapter 7: Energy*), the incoming radiation on Earth must equal the outgoing radiation from Earth. The incoming radiation is the sum of all the solar insolation over the area of the Earth facing the sun (it looks like a circle from the perspective of the sun). The amount of radiation absorbed by Earth can be calculated, using the following formula:

$$(1-A)S\pi R_e^2$$

where A is the albedo for Earth (0.3 on average), S is the solar constant (1370 Wm^{-2}) or the amount of energy hitting Earth per second, and R_e is the radius of Earth.

The amount of outgoing radiation is given by the surface area of Earth (the surface area of a sphere, $4\pi R^2$) multiplied by the formula for radiative power:

$$4\pi R_e^2 \varepsilon \sigma T_e^4$$

where epsilon (ε) is the emissivity (often taken as 1 for the sun) and sigma (σ) is the Stefan–Boltzmann constant 5.67 × 10^{-8} $Wm^{-2}K^{-4}$. T_e is the temperature of Earth in Kelvin.

Without an atmosphere, there is no natural greenhouse effect. Therefore, we equate incoming radiation with outgoing radiation:

$$(1-A)S\pi R_e^2 = 4\pi R_e^2 \varepsilon \sigma T_e^4$$

Rearranging gives an average, chilly, temperature of:

$$T_e = \sqrt[4]{\frac{(1-A)S}{4\varepsilon\sigma}} = 255°K = -18.15°C$$

The equation is very accurate for planets like Mars without an atmosphere. With the natural greenhouse effect, the Earth's atmosphere traps heat, increasing the temperature to an average of 14.6°C.

BOX 6.3

The carbon cycle

The carbon cycle is another major biogeochemical cycle, similar to the nitrogen and phosphorus cycles (see Box 3.5 in *Chapter 3: Pollution*). Carbon is found everywhere in the environment, including in soils, organisms, rocks, water, and the atmosphere. It is also the element which makes carbon-based life forms (and thus all life on Earth) possible. Carbon atoms can move between different reservoirs, for example between the soil and the atmosphere or between rocks and oceans (see Figure A).

As with all cycles, the carbon cycle can be in a relative balance, with the concentration in any one reservoir (say, the

atmosphere) remaining unchanging. Until recently, the level of CO_2 and CH_4 in the atmosphere has remained stable for a very long time (see Figure 1.2 in *Chapter 1: Introduction*). However, through human activity, we have fundamentally altered the global flows of carbon. The prime example is fossil fuel combustion, which takes large, previously unavailable underground reservoirs of carbon and, mainly through combustion, places it in the atmosphere (forming CO_2). However, there are a wide range of other important processes which also affect the carbon cycle. For example,

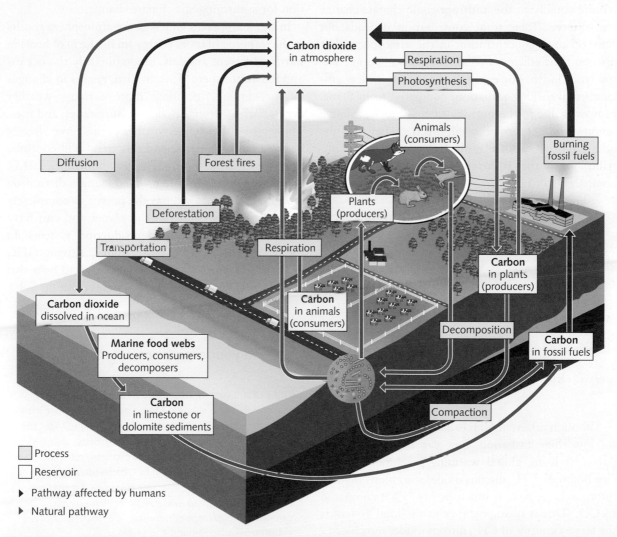

Figure A The carbon cycle showing carbon flows between different reservoirs which can be sources or sinks, depending on other environmental factors.

Source: *Environmental Science*, International Edition, 14th Edition by G. Tyler Miller Jr. and Scott Spoolman, 9781133104391.

as we have seen in *Chapter 5: Soils*, tilling soil releases the carbon stored in the soil into the atmosphere as well.

These biogeochemical cycles are highly connected. The increase in atmospheric carbon also increases the level of carbon in the oceans. This is because the levels of carbon in the oceans and the atmosphere are linked by continuously exchanged carbon. If the amount rises in one, it elevates the levels in the other. This has led to an acidification of the oceans since an increase in CO_2 levels increases the rate of processes that create acidic compounds and reduces the amount of freely available calcium. This creates issues for animals such as corals, bivalves, and snails, which use this calcium to build their tissues. This can cause knock-on effects on the entire marine food web.

As carbon is stored in all life on Earth, there are further complex dynamics and feedback effects between changes in conditions and the ability of different areas of the biosphere to store carbon. For example, as described in this chapter, elevated CO_2 levels increase temperature, which can cause stress to many plants. At the same time, the elevated levels might allow some plants to grow faster, perhaps increasing yields in some cases, as CO_2 is used in photosynthesis (although if you refer to Figure 6.14 later in this chapter, you can see that the changes are generally very negative).

CH_4 and nitrous oxide. Not all of these GHGs contribute equally to the anthropogenic climate change we observe. Their total contribution depends not only on their concentration in the atmosphere, but also on their efficiency to absorb the radiation emitted from Earth. For example, CO_2 is much less efficient in absorbing radiation, compared to methane. However, due to the burning of fossil fuels, massive amounts of CO_2 are released, much greater than the addition of CH_4. As a result, CO_2 has a greater contribution to the anthropogenic greenhouse effect, compared to CH_4.

Connect the dots

Think back to *Chapter 1: Introduction* and the concept of **sources** and **sinks**. If plants suffer more than they gain from climatic changes, then areas which were carbon sinks may turn into carbon sources—think of forests, which are usually sinks, turning into sources due to wildfires caused by elevated temperatures. Can you think of other examples in which a source is turned into a sink, or vice versa? If not, search online for some examples.

Although different GHGs have different capacities for absorbing radiation, they can be directly compared by using **global warming potentials** (GWPs) (see Box 6.4). CH_4, nitrous oxide, and chloro-/hydro-fluorocarbons have a much higher GWP, compared to CO_2. This is of importance as the food system is the largest emitter of CH_4, nitrous oxide, and chloro-/hydro-fluorocarbon to the atmosphere. Because these GHGs have a high GWP, this has significant implication for anthropogenic climate change.

In summary, extra GHGs in the atmosphere results in the trapping of more energy (in the form of heat) in Earth's different systems, predominantly the oceans and the atmosphere. This, in turn, results in changes in our climate, including more extreme weather events such as droughts, floods, hurricanes, and fires. These types of extreme weather events have already become more frequent, and this trend will continue to intensify in the future (Fischer and Knutti, 2015). Climate change, also often called climate disruption or climate breakdown, has the power to completely remake the ecosystems of the planet and can have severe effects on our food production systems. In the next sections, we explore what is driving GHG emissions in more detail (Section 6.2) and how this is impacting food production systems (Section 6.3).

In your experience

Think of the climate where you live. Have there been recent noticeable changes? Have generations older than you noticed changes? What are those changes? Has the intensity of weather events, or the frequency, changed? Once you have made a list of the changes, compare them against the reporting in the media (do a search online) and against the scientific literature summarized in the interactive map available here: http://www.carbonbrief.org/mapped-how-climate-change-affects-extreme-weather-around-the-world

BOX 6.4

Why time horizons matter: global warming potentials and politics

GWPs are defined as the average warming potential of a GHG over a specific **time horizon** (usually 20 or 100 years). This is because GHGs do not last forever in the atmosphere (in the case of CH_4, for example, it undergoes reactions to produce CO_2 over a period of decades). GWPs are always expressed as CO_2 equivalents (therefore, CO_2 has a GWP of 1 across all time horizons). For example, CH_4's effect, when averaged over 100 years, has a GWP of 28, meaning it has 28 times the warming potential of CO_2 over a 100-year period (examples for other GWPs are listed in Table A).

However, if we take the time horizon of 20 years, CH_4 has a GWP of 84. Why is this important? Agriculture has much higher emissions of CH_4 than other sectors; if we develop policies to control GHGs, it is in the sector's interest to choose the longest time horizon possible in order to underplay the importance of those CH_4 emissions. While this may seem like a dry discussion of numbers, it becomes of critical importance to agriculturally intensive countries such as America and New Zealand.

Table A Global warming potentials of selected gases

GWP values	Lifetime (years)	20-year time horizon	100-year time horizon
Methane	12.4	84	28
HFC-134a (hydro-fluorocarbon)	13.4	3710	1300
CFC-11 (chloro-fluorocarbon)	45	6900	4660
Nitrous oxide (N_2O)	121	264	265
Carbon tetrafluoride (CF_4)	50,000	4950	7350

HFC-134a and CFC-11 are both refrigerants which are now controlled due to their high GWPs. N_2O is an important part of the agricultural contribution to climate change, as it is released when microbes interact with nitrogen in soil. N_2O emissions are enhanced through fertilizer production and use. The majority of CH_4 is emitted through animal agriculture production and natural gas energy production, and through several industrial processes (IPCC, 2013).

6.2 What is driving anthropogenic climate change, and how does the food system contribute?

The human-driven, enhanced greenhouse effect is termed anthropogenic climate change (from now on, we refer to anthropogenic climate change as climate change for brevity). Climate change is a particularly difficult problem to solve, as it will require a complex set of scientific, technological, economic, social, cultural, and ethical solutions (part of the definition of wicked problems, as discussed in *Chapter 1: Introduction*). Furthermore, these solutions have to be implemented under uncertainties of their effect (we have predicted effects, but these predictions can be off due to the complexity of climate modelling). We are also trying to fix these issues, while recognizing that time is running out. The impacts

of climate change on the food system will be discussed in detail later in this chapter; here, we focus on how we got to this stage, what the global picture looks like, how the food system is contributing to climate change, and what mitigation options are available.

6.2.1 What is driving climate change?

Climate change is fundamentally a problem of the way our energy and food systems operate.[1] The size of both of these systems has expanded hugely over time, and

---

[1] There are emissions from the production of cement as well, but these are small in comparison.

they are tightly coupled with economic development and GHG emissions. As we will see in *Chapter 7: Energy*, modern society is simply not possible without the large-scale use of energy, nor is it possible without large-scale use of land for food production. GHG emissions from the energy system are dominated by the burning of fossil fuels releasing CO_2. By contrast, GHG emissions from the food system are more varied, including CH_4, N_2O, and CO_2. These emissions need to be placed in a historical context; since the start of the Industrial Revolution, industrialized countries have caused much of the existing problems, at the expense of other countries. Nowadays, other major players have emerged, most notably China. Historically, China was not a major contributor to overall GHG emissions; however, with the development in wealth in China (with 17% of the total global population), it now emits significant amounts of GHGs. This is further complicated by the role of China in global manufacturing and emissions for products consumed in high-income countries (for some of the ethical ramifications in global emission agreements, see Box 6.5).

The idea that everyone is entitled to a budget of GHGs, based on how much they are currently emitting

(the inertia concept in Box 6.5) could be seen as completely unjust, as it allows those that emit the most to continue emitting the most. However, if we share emissions by the population in each country (the equity approach in Box 6.5) emission reductions of 20% per year in high-income countries would be needed. It is quite possible that this would be technically, socially, and politically impossible. Let us make a hypothetical decision that it should be a mix, perhaps weighted as 50% by current use and 50% by per capita allowance. Now we have another problem: how do we measure these emissions? Do we count at the national level, limited to the border of the country (**production-based accounting**), or do we take into account all the emissions due to our consumption, which includes trade (**consumption-based accounting**)? If someone in Italy purchases a laptop made in China, should the emissions released in making the laptop reside with a Chinese person or with the Italian? Carbon accounting is a minefield of ethical, historical, and political power issues, which explains much of the difficulty in reaching global agreements on climate change.

If we now look at the specific activities contributing to climate change, we can break down GHG emissions

BOX 6.5
Carbon accounting

The development of energy and food systems did not happen equally around the globe; some countries underwent the Industrial Revolution very early (that is, Western Europe and the USA), while some areas have not industrialized (that is, some regions in Africa and Asia). Countries that have benefitted from using fossil fuels have added to the total GHGs in the atmosphere, at the expense of everyone else who has to deal with the resulting global climate change. This is termed a collective action problem, a topic which will be covered in depth in *Chapter 14: Collective Action*. The inequality in development and emissions becomes even more important when we consider that the countries that have emitted the most are also the countries most able to adapt to the changing climate, given their financial resources.

We know there are limits to the concentration of GHGs in the atmosphere, while maintaining a comfortable

climate. We can think of this as a carbon budget (note that we will actually have to reduce the carbon dioxide already in the atmosphere; this is discussed further in *Chapter 7: Energy*). But here comes the most important question: how do we share the responsibility of emission reductions among countries and people? There are two common divisions:

- Inertia: everyone is entitled to a budget of GHGs, based on how much they are currently emitting, with decreasing emissions over time

- Equity: the remaining budget is divided among all the people on the planet, so that lower-income countries are able to increase emissions (for a short time before reducing again) and high-income countries have to decrease emissions extremely fast.

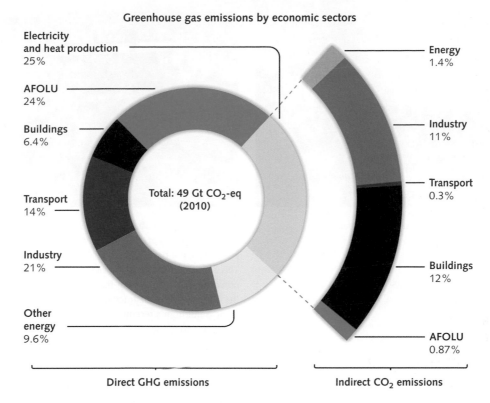

Figure 6.4 The contribution of different sectors to direct and indirect anthropogenic emissions sector. Direct emissions occur at the point of use such as the tailpipe of a car or the smokestack of an aluminium smelter. Indirect emissions are those that do not occur directly at point of use; for example, a light in an office may need a coal power plant to generate the electricity, which releases GHGs in a different location.

Source: IPCC, 2014: Climate Change 2014: Synthesis Report. Contribution of Working Groups I, II and III to the Fifth Assessment Report of the Intergovernmental Panel on Climate Change [Core Writing Team, R.K. Pachauri and L.A. Meyer (eds.)]. IPCC, Geneva, Switzerland, 151 pp. © Intergovernmental Panel on Climate Change, 2018.

by different economic sectors (see Figure 6.4). Together, electricity production, heat production, and industry make up almost 50% of emissions. The remaining emissions are from transport, buildings, and agriculture, forestry, and other land use (AFOLU).[2]

Given the focus of the book, we will look closer into AFOLU and how this contributes to climate change. In forestry, for example, trees break down CO_2 in the atmosphere and use the carbon during photosynthesis to create chemical energy (see the concept of gross and net primary production in *Chapter 2: Biodiversity*). Some of this energy is stored in the tree, for example in the trunk, roots, and leaves. When a tree is felled,

the carbon may continue to be stored (e.g. in the form of furniture or paper) or released into the environment (e.g. in a fire). The balance of CO_2 emission from planting, growing, and felling of trees gives us the emissions due to forestry. This is important because a large amount of forestry has been removed for food production. A similar process occurs in agricultural land use change.

The second component of AFOLU—other land use (OLU)—is less obvious. **Other land use** refers to the emissions of GHGs when land is being converted from the natural habitat (original land use) into a different land use (for example, clearing a mature forest in order to grow crops). By converting a forest, more agricultural production becomes available, but at the cost of the carbon which was stored in that original forest. Often the forest is simply burnt, directly releasing CO_2 into the atmosphere, as discussed

[2] For a full description of the breakdown of different emissions, the consequences of climate change, and the possible solutions, refer to the summary for policy-makers report from the Intergovernmental Panel on Climate Change, as referenced in the Further reading section at the end of this chapter.

Figure 6.5 Slash-and-burn agriculture in Madagascar.
Source: Diorit/Wikimedia Commons (CC BY-SA 3.0).

Figure 6.6 A demonstration of 'strip-till'. The tilled soil can be seen in the stripes of deep brown in the photo, between regions which have not been tilled. The plants will be grown in this strip. This is an alternative to tilling all the soil which minimizes, among other things, the amount of fertilizers which need to be applied, the fuel needed by the tractor, and soil compaction. Although this scene is not from a recently cleared forest, it is clear to see what is happening to the soil.
Source: Alandmanson/Wikimedia Commons (CC BY-SA 4.0)

previously (often called 'slash-and-burn agriculture'; see Figure 6.5).

Connect the dots

Consider how different types of land use change may impact biodiversity (see *Chapter 2: Biodiversity*), pollution (see *Chapter 3: Pollution*), water (see *Chapter 4: Water*), soils (see *Chapter 5: Soils*), and food security (see *Chapter 9: Food Security*). Think of the land near you—what is it used for, and what would be the physical impact if it changed? Can you also think of social impacts on the local community?

However, even if a forest is not burnt, a change in land use can still result in significant contributions of GHGs to the atmosphere. This is because there are large amounts of carbon stored in the soil (as described in *Chapter 5: Soils*, soil is not 'dirt', but a collection of organic and inorganic materials which contain carbon). When a forest is cleared for agriculture, a farmer may then till the soil (turning the soil over and breaking it up; see Figure 6.6). In doing so, the organic material in the soil which stores carbon is moved to the top of the soil layer. This allows carbon to move back (ventilate) into the atmosphere. The contribution of **land use change** to GHG emissions is a significant issue and, by some estimates, accounted for approximately 12.5% of total anthropogenic carbon emissions between 1990 and 2010 (Houghton et al., 2012). For example, land use change is the main source of GHG emissions for Brazil, Indonesia, and Papua New Guinea, all countries where significant amounts of pristine rainforest have been converted to agricultural lands.

6.2.2 How big is the impact of the food system on climate change?

Estimates of the total GHG emissions from the food system range significantly, but recent estimates suggest that, in 2015, up to one-third of total global anthropogenic emissions come from our food system (Niles et al., 2018). Given that food systems are changing rapidly, these emissions may be even greater today.

Figure 6.7 shows the activities, actors, drivers, and outcomes of the food system schematically. Food undergoes several steps through the cycle of (pre-)production, production, post-production, consumption, and disposal. We can break down these food steps into:

- **Pre-production**—the manufacture of agricultural inputs such as fertilizers and pesticides
- **Production,** such as:
  - the growing of crops
  - the rearing of livestock
  - the farming of fish

Food System Components, Processes, and Activities

Figure 6.7 A schematic of the activities and actors, outcomes, and drivers of the food system, from pre-production to waste and disposal.

Source: Niles, M.T., Ahuja, R., Esquivel, M.J., Mango, N., Duncan, M., Heller, M., Tirado, C. 2017. *Climate change and food systems: Assessing impacts and opportunities*. Meridian Institute. Washington, DC.

- **Post production,** including:
  - **Processing** steps such as drying, supplementation (e.g. with vitamins), grinding, mixing, etc.
  - **Distribution** in the transport and refrigeration of products from the processing point to the consumption point (the consumer)
- **Consumption**—the purchasing of food at the point of sale and how we cook food
- **Waste and disposal,** including landfill, production of fertilizers, combustion, collection of biogas, etc.

Emissions arise from all along the food production chain, either directly (e.g. cows burping; see Box 6.6) or indirectly (e.g. a supermarket freezer). However, agricultural emissions of crops and livestock are, by far, the most important by volume and make up 80–86% of the total food system emissions globally.

The number of different people, processes, and machines involved in supplying food is quite bewildering. Let us take an example. In her local London restaurant, Ms Smith orders a steak dinner, which is served to her 20 minutes later. What is the footprint of this steak dinner in terms of GHG emissions? To explore this, we will take a simplified look at how this steak is produced for Ms Smith and what GHGs are emitted along the way.

In this case, let us assume Ms Smith's steak was produced in Brazil (the second largest beef exporter in the world after the USA). Even before we arrive at the farm at which the cow was reared, the Brazilian cattle farmer had to buy animal feed, such as corn or soy, from a supplier. In growing the animal feed, fertilizers and herbicides were used by the supplier, as well as petrol and electricity in the farm machinery, all resulting in the emission of GHGs. For example,

BOX 6.6

Got Gas?

Enteric fermentation is the breaking down of carbohydrates by microorganisms in the stomachs of cattle and other similar hooved animals. These microorganisms break down food into simpler molecules that can be absorbed by the animal. One of the largest by-products of this process is CH_4, which is released in belches and flatulence. Scientists have estimated these emissions using measurements of CH_4 in a cow backpack (see Figure A). Based on this and other research, it was shown that CH_4 from enteric fermentation makes a substantial contribution to GHG emissions. For example, in the USA, these CH_4 emissions comprised approximately 3.5% of total US emissions and 45% of total US agricultural emissions in 2015 (EPA, 2017).

There are many promising solutions available that can help reduce CH_4 emission from cattle, which include advanced biotech solutions such as altering the microorganisms in the cow stomach, vaccines for CH_4 prevention, and animal selection.

Figure A The cow backpack. Developed in Argentina for the measurement of CH_4 emissions from cows. A tube is passed directly through to the stomach, which allows for the collection of CH_4 which would normally be passed through the animal.
Source: REUTERS/Marcos Brindicci.

it is estimated that around 27% of all artificial fertilizer (a major source of GHG emissions due to the high-energy Haber–Bosch process; see *Chapter 3: Pollution*) used worldwide is for growing feed for livestock (Gerber et al., 2013).

GHG emissions associated with rearing cattle are already at least equal to growing the animal feed needed to feed the animal, and we have not looked at emissions from the animals yet! However, even in terms of the impacts of this animal feed, this is not the end of the story. With the extra demand for corn for feed and biofuels, the price of corn has probably gone up. Now many corn farmers have an incentive to grow more and benefit from the higher prices. In order to do so, farmers purchase pristine forests near their land on which they grow more corn. Typically, they clear and burn the forests and till the soil, releasing CO_2 in the process (see Figures 6.5 and 6.6). In 2015, agriculture covered 37% of the Earth's surface. When new cropland is made, 80% of it comes from replacing forests (https://data.worldbank.org/indicator/AG.LND.AGRI.ZS). This specific form of land use change contributes between 6% and 18% of total global emissions (van der Werf et al., 2009). As the demand for meat grows, not only do emissions for the inputs grow, but so do emissions due to land use change.

Let us return to the cattle farmer; now that the Brazilian cattle farmer has access to animal feed, she can rear the cattle. As cattle consume feed and water, and digest it via enteric fermentation, they release large amounts of CH_4 gas by belching (see Box 6.6). As we have discussed, CH_4 is a powerful GHG, and direct emission from livestock represents approximately 8% of total global emissions. Of this, around three-quarters are from cattle stomach gas and the remainder from manure (Gerber et al., 2013). The type of animal feed is an important variable which determines total CH_4 emissions. In Food controversy 6.1, we look at how grass-fed or grain-fed cattle can impact health and environmental outcomes.

FOOD
CONTROVERSY
6.1

Grass-fed or grain-fed beef?

There has been a growing market for grass-fed beef in recent decades. In many grocery stores in Europe and the USA, grass-fed beef is now a commonly available option. Depending on what label or certification the product has, the cow may have spent some or all of its life feeding on grass. Labels indicating '100% grass-fed' mean the cow spent their entire life eating grass (see Figure A.a).

In the USA, it is very common for cattle to start their life on grass but make their way through a production system which ends in a feedlot (an intensive feed yard where animals are kept in larger numbers). In this feedlot, they typically eat more energy-intensive foods, such as corn and soy, which provide additional energy as cattle are 'finished' before slaughter (see Figure A.b).

So which is 'better'? Here we have to define what we are worried about. Are we concerned about human health? The environment? The welfare of the animals? While there has been a lot of debate about grass-fed versus grain-fed beef, it really depends on what metric you look at to make your distinction. There is clear and growing nutritional evidence that grass-fed beef is higher in healthier fatty acids like omega-3s (also found in wild salmon, for example) (Daley et al., 2010). But whether grass-fed beef is better for the environment depends on how you want to measure it.

Grass-fed cows typically take longer to reach their slaughter weight and can require a lot of land. All those extra days on grass mean they are spending more time releasing CH_4. But it also takes a lot of energy and inputs to grow corn and other cattle feed and transport them to feedlots. In addition, in grass-fed systems, perennial (year-round) pastures can store significant amounts of carbon, which would not be possible in most corn and other crop production. However, cattle in feedlots are increasingly fed by-products from other industries like the leftover mash after brewing beer, which is cutting down on landfilling and repurposing a potential waste product. So . . . which type of beef is better for the environment? A pretty complicated question, right?

One tool scientists have to answer these types of comparative questions is **life cycle assessment (LCA)**. Generally speaking, most existing LCA studies have found that intensive feedlot systems have fewer GHG emissions, but many of these do not include potential soil carbon benefits of grass systems—growing grass increases the amount of carbon sequestered in the soil itself. So, for now, more research is needed to better understand the environmental differences from a climate perspective.

a)

b)

Figure A Two types of cattle farming: (a) grass-fed, and (b) feedlot production.

Source: © United States Department of Agriculture.

Returning to the steak dinner of Ms Smith. Once grown, the cattle are transported to the slaughterhouse, slaughtered, processed, and packaged for sale. Processing and packaging require electricity, which can be produced from many sources, both carbon-intensive and renewable (see *Chapter 7: Energy* for more details). Next the packaged steaks will need to be transported to the consumer. Depending on the mode of transportation and where the food is shipped to, transportation can make up an important component of GHG emissions from the food system. In some cases, this has led to campaigns to eat food produced locally—in the UK, this has been linked to the concept of **food miles**. However, as demonstrated in Case study 6.1, it is not always clear whether one should 'eat local' or 'buy global'.

The steaks will have to be refrigerated for their entire journey to the restaurant where Ms Smith is eating. Depending on the type of food or beverage, refrigeration can be a large contributor to GHG emissions,

as 40% of all foods require refrigeration. For example, Coca-Cola (2008) estimated that 71% of its total carbon footprint is due to refrigeration. Globally, across all food, about 15% of all electricity is consumed for refrigeration (James and James, 2010).

Pause and think

When eating your next dinner, think of the journey that your food has taken to reach your plate. As you take each ingredient, look at the package to see where it was grown—was it locally or abroad? In what part of the country might it have been grown? Who might have grown the food? Where might it have been transported? And so on. When you have an idea of this, look up some of the pictures of this food production in these countries online.

CASE STUDY 6.1

Food miles

Food miles refer to the distance food travels between the place it is produced to the place it is consumed. In this chapter, we have already given the example of Brazilian beef and its consumption in Europe. Food miles was a term coined during the 1990s in the UK and began a drive to 'eat local' (see Figure A for an example shop display).

At a glance, eating locally produced food seems like it could make a significant reduction in GHGs, but this would assume that the transportation of food is a significant contributor to food emissions. We have to account for the emissions in the whole production chain, much like we described in the LCA of grass-fed versus feedlot-fed cattle earlier in the chapter (see Food controversy 6.1).

A good example is the consumption of tomatoes for the UK market, which highlights the difference in energy used to grow tomatoes in Spain, compared to the UK. Even though Spanish tomatoes must be transported across the continent, they do not need to be grown in heated greenhouses. In contrast, the transportation cost for locally produced UK tomatoes are much lower, compared to Spanish tomatoes; however, they are grown in heated greenhouses. As a result, tomatoes can be grown in a more energy-efficient and GHG emission way

in Spain, compared to the UK (DEFRA, 2007). This example shows the importance of looking at the entire production chain.

These effects have been shown for many different foods. For example, in a New Zealand study, there is evidence that, despite being on the other side of the planet, dairy production in New Zealand for the UK market is twice as efficient as production in the UK itself, even when including transport emissions.

In fact, transportation of food makes up about 11% of total GHG emissions for most food types (Weber and Matthews, 2008). These findings are possible because bulk transport by ship and truck can be very energy-efficient, when compared to local transport by car. The same amount of fuel can transport 5 kg of food over 97,000 km by a large truck, or it can transport a single consumer only 64 km (Coley et al., 2009). To confuse things further, there has been a recent trend to more air freighting of foods, which no matter what type of food, emits large amounts of carbon. Fortunately, the total amount of air-freighted food is still quite low.

In general, if you decide to drive to the nearest farmer's market to buy locally produced food, you could end up

Figure A Advertising for the purchasing of local food in order to reduce food miles in the USA.

Source: © Metcalfe's Market.

causing more emissions than if you did all your shopping at the local supermarket by foot, even if the majority of the products come from distant locations. This complexity makes it hard to make 'the right choice' for consumers wishing to make an environmentally informed decision (see *Chapter 11: Consumption* for more details).

Finally, the steak arrives at the restaurant where it is refrigerated for a short time, before being ordered by Ms Smith. The restaurant cooks on gas, so CO_2 is released until the steak is ready to be served. Estimates put the emissions of refrigeration and cooking at the point of consumption at 15% in high-income countries (Garnett, 2012).

At the end of this long process, Ms Smith is eating and starts to feel quite full; she wants to 'leave space' for a dessert and decides to leave half of her steak on the plate. The leftovers from her meal are put into a waste bin and the next day taken to the local rubbish tip where they decompose and release CH_4. Direct and indirect emissions from waste make up approximately 13% of total UK food-related emissions (Garnett, 2012).

We have just described the process for one product, and we can already see that due to the interlinked and trade-dependent world in which we live, understanding environmental impacts will always be a highly complex task (a simplified overview is shown in Figure 6.8). Gathering the emission data for all the different processes for a product is daunting. It also has to be done for each different product; a yogurt does not have the same inputs as a bag of rice. To put the challenge in perspective, just think about the number of different yogurts you can buy at the supermarket: Greek, soya, flavoured, yogurts with fruit syrup, yogurts with chocolate flakes, all of which have a different environmental footprint.

As discussed in Case study 6.1 and Food controversy 6.1, one method we can use is called LCA, which attempts to account for all emissions over the product lifetime, generally referred to as cradle-to-grave analyses. Other LCA approaches are also useful, such as **well-to-wheel** for oil, or **cradle-to-gate** for farm animals. If this sounds time-intensive, it is, often taking many days, and sometimes years, to gather the information required.

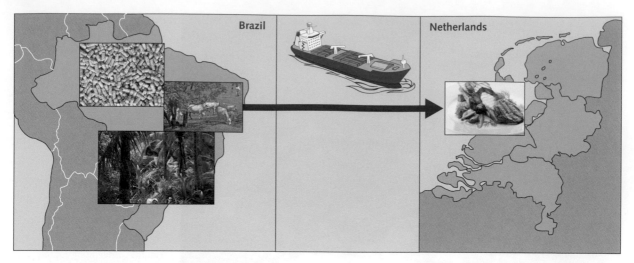

Figure 6.8 The journey from production to consumption of Brazilian beef in the Netherlands. The inset pictures on the left show a picture of the Brazilian rainforest, feed for cattle, and cattle in pasture. On the right, the inset picture shows steak on the table in the Netherlands.

6.3 What can we do to reduce the contribution of the food system to climate change?

Emissions from energy and agriculture make up the large majority (over 90%) of total GHG emissions. Additionally, as discussed previously, the world population is set to grow by another 1–2 billion people by 2050 (FAO, 2014); it is also likely to increase in wealth, which is often accompanied by increased consumption of animal products and more processed food (Tilman and Clark 2014). Here we enter an important feedback loop—if emissions grow due to changes in our food system and food demands (for example, a global increase in meat consumption), we can expect severe impacts on the environment. In turn, this is likely to result in a reduced capacity of the environment to maintain the desired output of food production.

We have two main options for dealing with the problem of climate change, commonly termed **mitigation** strategies and **adaptation** strategies. Mitigation strategies are those that attempt to reduce the stock of GHGs in the atmosphere, reducing the impacts of climate change. Adaptation strategies involve approaches that deal with the changing climate—for example, high-tech interventions like building sea walls against rising sea levels, or low-tech ones like changing the hours of activity so people do not work through the hottest hours of the day. In this section, we discuss mitigation strategies, while in Section 6.5, we will focus on adaptation strategies needed in the food system to cope with climate change.

Emission mitigation in food systems presents some unique challenges, when compared to energy emission mitigation. There are three key challenges, which we will discuss in some detail below:

1. **Food system emissions are diffuse.**

   Generally, a large amount of land is used to produce many foods, so the emissions are spread over a large area, making it non-point source pollution (see *Chapter 3: Pollution*). This is in contrast to, say a power plant, which has a single point source of emissions. Controlling the emissions from farm machinery and livestock, as well as any emissions from soil, will require many different technological solutions, some of which may be impractical, particularly in extensive systems (see the large area of land used for farming in Figure 6.9).

   This is particularly true when you think about the food product that ends up on your plate, the emissions ranged from the inputs it took to grow the food, the potential emissions associated with land

Figure 6.9 A picture of a tractor raking grass hay (left of the picture), and a combine harvester harvesting dried grass and unloading onto a tractor (right of the picture).

Source: GLF Media/Shutterstock.com.

use changes, the processing, transportation, refrigeration, and cooking energy. These emissions were diffused throughout the supply chain and cannot be easily targeted through a single change. Nevertheless, there is great potential for reducing emissions across the food system, as we detail later.

2. **Food systems emit many different types of GHG.**

In contrast to the energy system, which largely emits CO_2, the food system predominantly emits other gases. As discussed, CH_4 and N_2O are emitted in the production of fertilizers, the decomposition of manure and under certain conditions from the soil, and the burning of biomass. In addition, hydro-fluorocarbons are used as refrigerants and can be released through the improper disposal of refrigerators (see Figure 6.10).

These gases have different warming potentials, and they are accounted for in many diverse ways (see Box 6.3). These different gases may require different regulations and will certainly require different policies and instruments for their control.

3. **Many important mitigation strategies depend on social and cultural changes.**

Of course, the food we grow is influenced by what we consume, and thus how much and what types of foods people eat has an impact on GHG emissions (we discuss this at length in *Chapter 8: Nutrition*). Many of our food choices are preferences, but not always, and it is important to recognize that what and how people eat is deeply social and cultural. As a result, changing these practices involves a behaviour change in a significant portion of the population (we explore this further in *Chapter 11: Consumption*).

It is also important to recognize that dietary changes may not need to happen universally. For example, in many low-income countries, finding adequate nutrition that is culturally appropriate is still a critical problem, and these diets are relatively low in GHG emissions. Conversely, in

Figure 6.10 Thousands of refrigerators lined up for disposal after Hurricane Katrina destroyed homes in 2005. People evacuating their houses left their refrigerators without power with food inside; now there are hundreds of thousands of refrigerators needing careful disposal in the region. This is a clear example of a climate change feedback. Intense storms like Hurricane Katrina are much more likely due to climate change, which causes devastation, having the potential to release further GHGs.

Source: Don Ryan/AP/Shutterstock.

many high-income countries, there is a significant consumption of diets associated with high GHG emissions. Thus, how diets will shift and which populations are most important to make these shifts is also necessary to understand (see *Chapter 7: Energy* for further details).

Nevertheless, a growing body of scientific research is examining how different diets influence GHG emissions. Research suggests that diets that have fewer animal products, particularly beef, result in reduced GHG emissions. Ensuring that your dietary intake matches your energy output can also minimize your dietary GHG emissions (this is discussed further in *Chapter 8: Nutrition*).

At first glance, this may provide a quick solution to reducing emissions—easier than large investments in renewable energy, nuclear power, new batteries, and biofuel production facilities, etc. However, dietary shifts will likely result in negative impacts also, including for the large number of people working in the food system, and changes in the nutritional content of food, requiring education and engagement. As with all transitions, there will be winners and losers, and often many of these negative aspects are not widely discussed.

Given these challenges, the mitigation opportunities for the food system will be quite unique. We next take a look at what some of these opportunities might be and how we might be able to implement them.

6.3.1 Food system mitigation opportunities

With the three characteristics introduced previously, there are ample opportunities to reduce food system emissions. These can play a significant role in avoiding catastrophic changes in our climate. In some areas of the food system, mitigation methods are very similar to those in the energy system. For example, any part of the food system relying on electricity can use new renewable energy, rather than coal or gas. Generally, this is in post-production, for example refrigeration, cooking, and retail activities.

Likewise, transport emissions will need to be reduced, which might include electric tractors, hydrogen- or biofuel-powered ships and airplanes, or shifting transportation modes, from air- to ship-freight, as well as increasing the use of local food systems (local farming,

vertical farming, hydroponics, etc.). In addition, we may be able to divert agricultural wastes into the production of biofuels, which could then be used in these ships and planes, solving several problems in one go.

However, while these strategies are critical to making overall reductions, it is important to remember that for most countries, agricultural production (and not transportation or processing) makes up the majority of food system emissions. As a result, the type of food and the way we produce food is critical to reducing food GHG emissions. This is why it is so important that both farmers and consumers work together in reducing food system emissions.

Below we describe many of the existing and future mitigation opportunities by **supply** (section titled 'Can we supply food while emitting fewer emissions?'), which concerns farmers, food distributers, and food manufacturers, and **demand** (titled 'Can we reduce emissions through what we eat and what we do?') which concerns consumers. Supply-driven opportunities focus on ways we can produce food with fewer emissions and impacts. Demand-driven opportunities focus on what behaviours different cultures and societies can implement to reduce emissions. This supply/demand dichotomy is a common way to analyse many impacts, and economy itself, as it isolates different types of solutions. From a systems perspective, however, it is critical to look at both of these and to recognize that the two are highly connected. How farmers and others in the food production process produce food may influence what people buy; conversely, what consumers want to purchase can also influence and drive what farmers grow and food processors produce.

Can we supply food while emitting fewer emissions?

For agriculture, many of the best strategies for reducing GHG emissions (of all gas types) are also the ones that can help farmers become more efficient and save money. Farms are a good place for cutting energy costs by using renewable energy such as biodigesters, wind or solar, especially on land that may not be best for producing food (see Figure 6.11). Also, farmers can reduce CO_2 emissions on farms through using their tractors and equipment less or buying more fuel-efficient equipment.

Since most emissions from agriculture are the result of CH_4 and N_2O, strategies that can reduce these GHG

Figure 6.11 A farm in Horstedt, Germany demonstrating how many renewable energies can be integrated onto a farm through the use of a Biogas (fermenter), wind power, and solar energy.

Source: Florian Gerlach (Nawaro)/Wikimedia Commons (CC BY-SA 3.0).

implement practices that can help store CO_2 in the soil, rather than releasing it to the atmosphere. This can include planting trees, reducing tillage, and more efficient water and crop management. For animal systems, there is potential to be more efficient in the production of cattle specifically, reducing the amount of time it takes for an animal to grow, and so reducing how much CH_4 they produce (although this may concern consumers, highlighting how interconnected these approaches are). This also includes changes in animal diets; for example, recent research discovered that feeding cows certain kinds of fats can significantly reduce their CH_4 emissions (Moate et al., 2011).

Livestock farmers can also manage animal manure to reduce emissions by capturing it in a pond and covering it before its use on the field. This captured CH_4 can then be used for energy or heat by burning, reducing CH_4 emissions (while still producing CO_2), though this option is not feasible when animals are kept in pastures (see Food controversy 6.1).

We may even find new, previously unthought-of technological supply-side solutions. If we are absolutely wedded to the concept of eating meat and maintaining our current diets (see further text), we may find ways to produce synthetic substitutes. For example, instead of allowing cows to freely emit CH_4, we could avoid raising cows altogether and culture meat from the constituent proteins. In fact, researchers have already begun making such substitutes, although these artificial meats are still very expensive

emissions are much more important for agriculture. Effective practices that farmers can implement may include reducing nitrogen fertilizers, spreading fertilizers onto fields by injecting them into the soils so they do not run off or evaporate as easily (see Figure 6.12) (this is also very relevant to *Chapter 3: Pollution*, since it limits nitrogen losses). In addition, plant breeders are trying to develop crops which are more efficient in taking up fertilizers and which have a longer shelf-life (these are often genetically modified, as discussed in Box 3.7 of *Chapter 3: Pollution*). Taking these together could drastically reduce fertilizer application.

Another opportunity for farmers is through new efforts focused on 'carbon farming' whereby farmers

a)

b)

Figure 6.12 Traditional fertilizer spraying (a) versus fertilizer injection (b). The injection is highly targeted and could result in less waste and run-off.

Source: (a) Werktuigendagen Oudenaarde/Flickr (CC BY-SA 2.0), (b) David Wright / Crop Spraying near Saxby All Saints / CC BY-SA 2.0.

a)

b)

Figure 6.13 Can you fake a burger from a live animal? (a) Cultured (printed) meat, grown in cell culture, and not animals. (b) The Impossible Burger, a plant-based meat substitute that bleeds when cooking.

Source: (a) Nederlandse Leeuw / Wikimedia Commons (CC BY 3.0); (b) © Copyright 2019 Impossible Foods Inc.

(see the photo on the left of Figure 6.13). This price will likely decrease drastically, as all new technologies do, and if it follows other product revolutions, it may even become cheaper than farmed steak.

Other technological substitutions may include further replication of the taste and qualities of meat with plants. For example, the Impossible Burger is a plant-based alternative burger patty that even bleeds when cooked (see the photo on the right of Figure 6.13 and https://www.impossiblefoods.com/burger/). An overview of more supply-side interventions, impacts, challenges, and co-benefits is given in Table 6.1.

In your experience

Would you consider eating cultured burger and why? What would stop you? And conversely, why would you jump at the chance? Would you be more likely to eat cultured burger than plant-substitute burger and why?

Can we reduce emissions through what we (as consumers) eat and do?

Many consumers are simply unaware of how their food consumption activities may be impacting climate change. Compare the last time anyone you know mentioned solar, wind, or nuclear energy with the last time they mentioned the composition or size of diets with respect to the environment. Renewable energy infrastructure is, for many, easier to talk about than fundamentally altering what we eat (de Boer et al., 2016; Macdiarmid et al., 2016).

At the retail and consumer level, we all have a huge role to play in making sure that our food systems can adapt and mitigate climate change. One of the greatest opportunities to both reduce GHG emissions and help adapt our food systems to future shocks is to reduce food waste. It is estimated that globally up to one-third of all food produced is lost through the food system, which results in huge amounts of water, energy, land, and inputs (FAO, 2015). The United Nations even estimated that if food waste were a country, it would be the third largest global emitter! An overview of some demand-side interventions, impacts, challenges, and co-benefits is given in Table 6.2.

When developing strategies to shift dietary patterns, we have to consider the related industries associated with such shifts. For example, nearly one billion people globally rely on food and income from livestock production (FAO, 2018). Related to this, while it is certainly possible to thrive on plant-based diets, we have to consider an individual's nutritional needs and any changes have to be balanced to prevent malnutrition, which is still prevalent in many low-income countries (see *Chapter 7: Energy* and Payne et al., 2016).

Table 6.1 Examples of low GHG food supply interventions

Intervention	Impact area	Challenges	Co-benefits
Minimize disruption of soil in production	Reductions in CO_2 and N_2O emissions	Training of farmers and new equipment costs in wet soils may actually increase N_2O	Potential for increased yields in some conditions. Tractor use reduced (lowers energy and labour costs). Erosion benefits
Reduce fertilizer usage or increase fertilizer use efficiency	In high-income countries, fertilizers are often overapplied, without any gain in yield. New fertilizer technologies also allow farmers to apply fertilizers more efficiently	Training of farmers on the correct amount to apply. Cost of high-efficiency fertilizers.	Water quality improvements. Cost savings for farmers if they are reducing their fertilizer inputs
Crop residue burning	Residues often burnt on farms in countries like India. Reduction in CO_2 and CH_4	Training of farmers	Health benefits from air pollution
Improved grazing management of animals through pasture quality	Reducing GHGs through enteric fermentation and manure. Pasture quality can improve digestion	May require increased fertilizer application	May improve productivity
Rotational grazing (adjust grazing periods)	Improved yields and carbon sequestration of grasslands	Large adaptation for small-scale farms	Grassland management may improve
Covering manure ponds	In confinement systems, manure is often concentrated into ponds, which can cause GHG emissions. Covering the ponds, and ideally capturing their CH_4 for energy, can reduce GHG emissions	Cost of covers and CH_4 digesters. Farmer training	Reduces odours, generates energy

(Adapted from Niles, 2018)

Table 6.2 Examples of low GHG food demand interventions

Intervention	Impact area	Challenges	Co-benefits
Choose more plant-based meals	N_2O and CH_4 emissions; land use change; reduce fossil fuel use	Problem may be shifted overseas; risk that fish substitutes in diets, so increasing pressure on fish stocks	Improved health outcomes due to saturated fat intake reduction
Balance energy intake and output	Higher energy intake increases all impacts (Nelson et al., 2016)	Ensure there is no loss of nutrients. Risk of body shaming	Improved health outcomes
Do not waste food/ manage unavoidable waste properly	Less food waste may result in unnecessary production. If food waste is managed well, it can be used for bioenergy	Waste is linked to the 'consuming less' debate	Reduced energy costs in the case of biomass energy production
Eat seasonally	Tackles areas of refrigeration, transport, and spoilage	Measures to reduce air-freighted foods may clash with international development objectives	Increased awareness of seasonal agricultural production
Accept different notions of quality	Less waste means reduced production	May reduce availability of undesired food for animal feed	Greater awareness of food production activities
Eat fewer foods with low nutritional value, for example alcohol, tea, coffee, bottled water	'Unnecessary' foods are not needed in our diet	Raises major questions around free choice. These foods provide a living for many in the world	May have health improvements
Cook and store foods in energy-conserving ways	Energy use in the home	Simple to do; impacts limited, but useful	Saves money

(Adapted from Niles et al., 2018 and Garnett 2011)

Finally, behavioural change is difficult, as anyone who has just started exercising regularly can testify. There are numerous psychological and cultural barriers to behavioural change, from cravings for the smell of cooked meat to the number of vegetarian or vegan options on a restaurant menu.

6.4 What is the impact of climate change on food systems?

In the previous sections, we learned how food systems can contribute to climate change and what emission-reducing strategies may be available across the food systems. However, food systems both contribute to, and are impacted by, climate change. Looking at these impacts can seem overwhelming, as there are lots of possible changes to consider (see Figure 6.15 for an overview of some impacts). In this section, we will first explore how the various parts of the food system may be affected by climate change in the future, and second what types of adaptation strategies could help minimize these impacts.

Climate change has the potential to create new and different climates than what we are used to across up to 40% of our planet (Williams et al., 2007). It is estimated that climate change accounts for about one-third of yield variability in crops (Ray

et al., 2015), so climatic changes will have profound impacts on how we grow, process, distribute, and consume food.

Some places may actually benefit from climate change, particularly in northern latitudes as it becomes warmer, which can extend growing seasons. In fact, in the USA, the location with the largest growth in new farmers is Alaska (USDA, 2014). Expansion of the growing season may allow for additional crops to be grown.

Despite potential local gains, scientists generally agree that beyond 2030, there will be an overall negative impact on global crop yields (see Figure 6.14). These negative impacts will come from a variety of climate stresses, including higher temperatures, increasing numbers of extreme weather events such as hurricanes or droughts, changes in water availability, and the

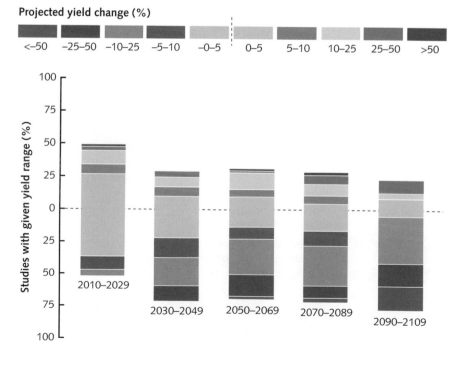

Projected yield change (%)

| <−50 | −25–50 | −10–25 | −5–10 | −0–5 | 0–5 | 5–10 | 10–25 | 25–50 | >50 |

Figure 6.14 Most studies now agree that climate change will negatively affect yield, starting in 2030 overall.

Source: Challinor, A J, J Watson, D B Lobell, S M Howden, D R Smith, and N Chhetri. 2014. "A Meta-Analysis of Crop Yield under Climate Change and Adaptation." Nature Clim. Change 4 (4): 287–91. Copyright © 2014, Springer Nature.

expansion of new weeds, diseases, and pests. Some traditional methods for controlling these problems (for example, herbicides or pesticides) may become less effective, as weeds are expected to be better at competing with crops in elevated CO_2 conditions (Varanasi et al., 2016).

Farmers will be at the forefront of dealing with these unprecedented changes and, depending on what resources and tools they have, could significantly suffer. Smallholder farmers—predominantly living in low- and middle-income countries and often producing much of their own food—will likely be the most heavily affected by climate change, including impacts on their own food security (Morton, 2007). This is especially challenging, since many of these farmers lack the resources to adapt to climate change.

Climate change will lead to the expansion of new weeds and pests, which could have negative impacts on crops and other biodiversity (see *Chapter 2: Biodiversity*). Farmers appear to be most worried about pests when considering climate change, since this will present new challenges to maintaining their yields (see Figure 6.15).

Rising temperatures can cause heat stress in animals, which decreases animal welfare and also reduces production efficiency. Additionally, animals may face new diseases in the future and require different kinds of care. Animals will also face changes because of the crops or pasture they consume, which may have yield or quality shifts (Thornton et al., 2009).

Climate change will also have significant impacts on our food distribution system, as well as on how we process, cook, consume, and dispose of food. For example, in an increasingly globalized food system, transportation and distribution have become a critical component of ensuring access to food (see *Chapter 9: Food Security* for more details). Sea-level rise could pose a significant risk to distribution (IPCC, 2011). Seven out of the top ten largest ports in the world are in China, and it is estimated that ports in Asia could be the most heavily affected by sea-level rise (Hallegatte et al., 2013). Given the volume of agricultural products that Asia imports (see Figure 6.16), this could have a huge impact not only on how China grows and consumes food, but also on places like the USA and Brazil, which send agricultural products to China (this is another example

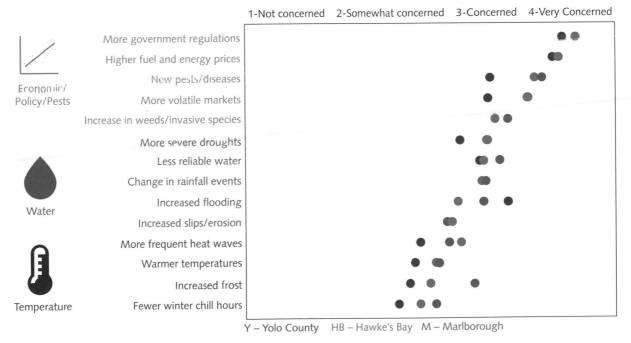

Y – Yolo County   HB – Hawke's Bay   M – Marlborough

Figure 6.15 Farmer concerns related to climate change across Yolo County, California, Hawke's Bay, New Zealand, and Marlborough, New Zealand. For additional detail on farmer studies, see Niles et al., 2013; Niles et al., 2015; and Niles et al., 2016.

Source: Niles et al., 2013, Niles et al., 2015, and Niles et al., 2016. See References for full citations.

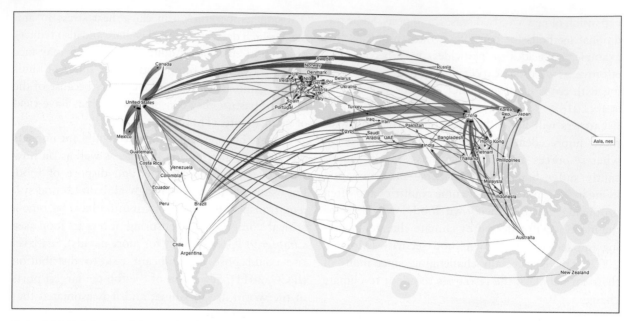

Figure 6.16 Global trade of all agricultural products, 2015. Blues indicate importers, while pinks indicate exporters. This figure highlights the complexity of how our global food system moves food and provides an indication of how global food system transportation may be influenced by future climate change.

Source: Chatham House (2018), 'resourcetrade.earth', http://resourcetrade.earth/.

of the importance of a globally connected world) (see *Chapter 7: Energy*).

New research suggests that, beyond impacts on crop yield and food distribution, climate change could also profoundly change our food's nutritional quality and impact food safety. For example, for every 10°C increase in temperatures above 10°C, bacteria growth doubles. This means that we could face increasing concerns for food safety with rising temperatures (James and James, 2010).

There is also growing evidence that the nutritional content of foods could change with climate change. Much of the potential impact could come from the very important micronutrients in our diet—things like zinc, iron, and selenium, which are all critical for human functioning (see *Chapter 8: Nutrition*). Scientists have found that many staple crops like rice, barley, and wheat, as well as legumes (such as beans), have fewer micronutrients and protein when they are grown under higher levels of CO_2 (Myers et al., 2014).

These effects may also not be confined to land; with a warming ocean, the kinds of food available for fish and seafood may change, which may influence both the availability and nutritional content of ocean species (Bermúdez et al., 2015).

6.4.1 Adaptation strategies to climate change

Since many of the climate change impacts we discussed will affect agriculture directly, we will focus on farmers and finish the section with a short discussion on transport and diet adaptations.

For crop systems, introducing crop rotations and more variety within their farming systems can help farmers to diversify their income sources and avoid losses across a single crop in the future. In the event of significant future pest outbreaks, farmers with more than one crop would be able to withstand impacts better.

Two other crop strategies which can be easily implemented are the use of cover crops (to avoid bare land which is susceptible to water and nutrient run-off) and using compost or mulch (see Figure 6.17). Both of these strategies build carbon and organic material in the soil, which allows the soil to hold more water. This, in turn, will help plants to withstand droughts.

There is also a growing focus on genetically modifying crops to withstand future climate change impacts. This has been made much easier with new technologies such as CRISPR, which allows direct editing of the genes. These potential technologies can engineer crops to be better yielding in drought

a)

b)

Figure 6.17 (a) Italian ryegrass is used as a cover crop following maize in South Africa. Maize residues, together with ryegrass, are used as winter fodder for livestock. In spring, the grass is allowed to grow without further grazing, then killed with herbicide before the following crop (maize or soyabean) is planted using a no-till planter. (b) A rural, central compost site in Germany.

Source: (a) Alan Manson/Wikimedia Commons (CC BY-SA 3.0); (b) Crystalclear / Wikimedia Commons (CC BY-SA 3.0).

conditions and increased salinity. However, most of these technologies are still emerging and are typically only available to farmers in high-income countries that can afford the additional seed costs. It is also worth noting that some countries have regulations against the use of genetically modified crops, so this strategy may be less viable in some places.

For ranchers and livestock farmers, adaptation strategies largely revolve around ensuring their animals can adapt to future heat stresses, pests, and diseases. For heat stress, farmers may need to install fans and cooling systems in indoor animal systems and plant additional trees for shade in pasture systems (see Figure 6.18). Future technological solutions may include vaccines to new pests and diseases, as well as increased veterinary medicines.

Future transportation adaptation strategies include securing shipping routes which may include moving ports

a)

b)

Figure 6.18 (a) Cattle sheltering under a tree in the Netherlands. (b) A poultry shed with an industrial fan system for cooling clearly visible at the end of the building.

Source: (a) Crystalclear/Wikimedia Commons (CC BY-SA 3.0); (b) Gualberto Becerra/Alamy Stock Photo.

to new locations or altering shipping routes. Research is only just beginning on the nutritional changes in seafood and other food crops mentioned previously. Dietary or technological substitutions may be necessary to account for future micronutrient changes.

Importantly, some adaptation strategies may also mitigate GHG emissions; thus, adopting these practices could have co-benefits for both climate change adaptation and mitigation. An example of this is the use of cover crops, which can help with soil loss and potentially add organic matter to the soil that can be beneficial in drought conditions. But cover crops can also be used in the place of synthetic fertilizers, so they may also provide mitigation benefits.

6.5 Future outlook

Avoiding extreme climate change will require a concerted effort across the entire food system, ranging from farmers to ranchers, from food processors to companies, and from wholesalers to consumers. The fact of the matter is that we will need to reach zero, or even negative, emissions by the end of this century if we wish to avoid damaging climate change (without the aid of drastically geoengineering our planet).

Importantly, there are significant mitigation opportunities to reduce food system emissions across all sectors. There are numerous options, including technological advances, production changes, and behavioural changes.

Since we are already experiencing climatic change, development of adaptation strategies will also be important. In some cases, these strategies have co-benefits, for example diversifying production and reducing other pollutants like artificial fertilizers.

We will also have to manage the changes we are already experiencing and the uncertain future changes we will see in response to current GHG levels. This will result in not only a large and deep energy transition, but also a food transition which simultaneously lowers emissions and adapts to changing conditions.

However, we will need to implement these mitigations in the near future, and on a large scale, to significantly reduce global emissions and minimize adverse effects on the environment.

● QUESTIONS

6.1 Briefly describe the process of climate change in a paragraph (this may seem difficult, but the best way to learn is to try and explain this yourself).

6.2 How are life cycle assessments used to calculate the environmental impacts from food systems, and why are they important?

6.3 In this chapter, we outlined the story of a steak, from production to consumption. Create your own story for a food product you eat often. Can you discover where the ingredients were made and how they were produced? Consider if knowing the whole process would impact your purchasing decisions.

6.4 What choices would you make when looking at Table 6.2? What difference would they make to your lifestyle?

6.5 What kinds of impacts can we expect from climate change across the food system?

● FURTHER READING

Intergovernmental Panel on Climate Change (IPCC). *Climate Change 2014 Synthesis Report Summary for Policymakers*. Available at: https://www.ipcc.ch/pdf/assessment-report/ar5/syr/AR5_SYR_FINAL_SPM.pdf [accessed 9 November 2018] **(An accessible overview of the most recent IPCC assessment, including all three working groups.)**

Jamieson, D. 2017. *Reason in a Dark Time*. Oxford: Oxford University Press. **(Describes the historic and current issues preventing substantial action on climate change. Takes political, policy, economic, and ethical perspectives. Lucidly written and comprehensively referenced.)**

Niles, M.T., Ahuja, R., Barker, T., Esquivel, J., Gutterman, S., Heller, M.C., Mango, N., Portner, D., Raimond, R., Tirado, C., Vermeulen, S. 2018. Climate change mitigation beyond agriculture: a review of food system opportunities and

implications. *Renewable Agriculture and Food Systems* **33**, 1–12. **(A comprehensive overview of all the interactions between climate change and food systems.)**

● REFERENCES

Bermúdez, R., Feng, Y., Roleda, M.Y., Tatters, A.O., Hutchins, D.A., Larsen, T., Boyd, P.W., Hurd, C.L., Riebesell, U., Winder, M. 2015. Long-term conditioning to elevated CO_2 and warming influences the fatty and amino acid composition of the diatom *Cylindrotheca fusiformis*. *PLoS One* **10**, e.0123945

Challinor, A.J., Watson, J., Lobell, D.B., Howden, S.M., Smith, D.R., Chhetri, N. 2014. A meta-analysis of crop yield under climate change and adaptation. *Nature Climate Change* **4**, 287–291.

Coca-Cola. 2008. *Our CRS Journey: Delivering on our Commitments*. Available at: https://www.unglobalcompact. org/system/attachments/1763/original/COP. pdf?1262614286 [accessed 4 November 2018].

Coley, D., Howard, M., Winter, M. 2009. Local food, food miles and carbon emissions: a comparison of farm shop and mass distribution approaches. *Food Policy* **34**, 150–155.

Daley, C.A., Abbott, A., Doyle, P.S., Nader, G.A., Larson, S. 2010. A review of fatty acid profiles and antioxidant content in grass-fed and grain-fed beef. *Nutrition Journal* **9**, 10.

de Boer, J., de Witt, A., Aiking, H. 2016. Help the climate, change your diet: a cross-sectional study on how to involve consumers in a transition to a low-carbon society. *Appetite* **98**, 19–27.

Department of Food, Environment, and Rural Affairs (DEFRA). 2007. *Comparative Life Cycle Assessment of Food Commodities Procured for UK Consumption through a Diversity of Supply Chains*. Final Report for DEFRA, Project FO0103. London: Department of Food, Environment, and Rural Affairs.

Food and Agriculture Organization of the United Nations (FAO). 2014. *State of Food Insecurity in the World 2014: In Brief*. Rome: FAO.

Food and Agriculture Organization of the United Nations (FAO). 2015. *Food Wastage Footprint and Climate Change: Global Food Loss and Waste*. Available at: http://www.fao.org/3/a-bb144e.pdf [accessed 4 November 2018].

Food and Agriculture Organization of the United Nations (FAO). 2018. *Livestock and the Environment*. Available at: http://www.fao.org/livestock-environment/en/ [accessed 4 March 2018].

Fischer, E.M., Knutti, R. 2015. Anthropogenic contribution to global occurrence of heavy-precipitation and high-temperature extremes. *Nature Climate Change* **5**, 560–564.

Garnett, T. 2012. Where are the best opportunities for reducing greenhouse gas emissions in the food system (including the food chain)? A comment. *Food Policy* **36**, 463–466.

Gerber, P.J., Steinfeld, H., Henerson, B., Mottet, A., Opio, C., Dijkman, J., Falcucci, A., Tempio, G. 2013. *Tackling Climate through Livestock: A Global Assessment of Emissions and Mitigation Opportunities*. Rome: Food and Agriculture Organization of the United Nations (FAO).

Hallegatte, S., Green, C., Nicholls, R.J., Corfee-Morlot, J. 2013. Future flood losses in major coastal cities. *Nature Climate Change* **3**, 802–806.

Houghton, R.A., House, J.I., Pongratz, J., Van Der Werf, G.R. Defries, R.S., Hansen, M.C., Le Quéré, C, Ramankutty, N. 2012. Carbon emissions from land use and land-cover change. *Biogeosciences* **9**, 5125–5142.

Intergovernmental Panel on Climate Change (IPCC). 2011. *Summary for Policymakers, Climate Change 2014: Mitigation of Climate Change*. Contribution of Working Group III to the Fifth Assessment Report of the Intergovernmental Panel on Climate Change. Cambridge and New York, NY: Cambridge University Press.

Intergovernmental Panel on Climate Change (IPCC). 2013. Stocker, T.F., D. Qin, G.-K. Plattner, M. Tignor, S.K. Allen, J. Boschung, A. Nauels, Y. Xia, V.B., Midgley, P.M. (eds). *Climate Change 2013, The Physical Science Basis*. Working Group I Contribution to the Fifth Assessment Report of the Intergovernmental Panel on Climate Change. Cambridge and New York, NY: Cambridge University Press.

James, S.J., James, C. 2010. The food cold-chain and climate change. *Food Research International* **43**, 1944–1956.

Macdiarmid, J.I., Douglas, F., Campbell, J. 2016. Eating like there's no tomorrow: public awareness of the environmental impact of food and reluctance to eat less meat as part of a sustainable diet. *Appetite* **96**, 487–493.

Moate, P.J., Williams, S.R.O., Grainger, C., Hannah, M.C., Ponnampalam, E.N., Eckard, R.J. 2011. Influence of cold-pressed canola, brewers grains and hominy meal as dietary supplements suitable for reducing enteric methane emissions from lactating dairy cows. *Animal Feed Science and Technology* **166**, 254–264.

Morton, J.F. 2007. The impact of climate change on smallholder and subsistence agriculture. *Proceedings of the National Academy of Sciences of the United States of America* **104**, 19680–19685.

Myers, S.S., Zanobetti, A., Kloog, I., Huybers, P., Leakey, A.D.B., Bloom, A.J., Carlisle, E., Dietterich, L.H., Fitzgerald, G., Hasegawa, T., Holbrook, N.M., Nelson, R.L., Ottman, M.J., Raboy, V., Sakai, H., Sartor, K.A., Schwartz, J., Seneweera, S., Tausz, M., Usui, Y. 2014. Increasing CO_2 threatens human nutrition. *Nature* **510**, 139–142.

Nelson, D.R., Lemos, M.C., Eakin, H., Lo, Y.-J. 2016. The limits of poverty reduction in support of climate change adaptation. *Environmental Research Letters* **11**, 094011.

Niles MT, Ahuja R, Barker T, Esquivel J, Gutterman S, Heller MC, Mango N, Portner D, Raimond R, Tirado C, Vermeulen S. 2018. Climate change mitigation beyond agriculture: a review of food system opportunities and implications. *Renewable Agriculture and Food Systems* **33**, 1–12.

Niles, M.T., Brown, M., Dynes, R. 2016. Farmer's intended and actual adoption of climate change mitigation and adaptation strategies. *Climate Change* **135**, 277–295.

Niles, M.T., Lubell, M., Brown, M. 2015. How limiting factors drive agricultural adaptation to climate change. *Agriculture, Ecosystems & Environment* **200**, 178–185.

Niles, M.T., Lubell, M., Haden, V.R. 2013. Perceptions and responses to climate policy risks among california farmers. *Global Environmental Change* **23**, 1752–1760.

Payne, C.L., Scarborough, P., Cobiac, L. 2016. Do low-carbon-emission diets lead to higher nutritional quality and positive health outcomes? A systematic review of the literature. *Public Health Nutrition* **19**, 2654–2661.

Pradhan, P., Reusser, D.E., Kropp, J.P. 2013. Embodied greenhouse gas emissions in diets. *PLoS One* **8**, e.0062228

Ray, D.K., Gerber, J.S., MacDonald, G.K., West, P.C. 2015. Climate variation explains a third of global crop yield variability. *Nature Communications* **6**, 5989.

Thornton, P.K., van de Steeg, J., Notenbaert, A., Herrero, M. 2009. The impacts of climate change on livestock and livestock systems in developing countries: a review of what we know and what we need to know. *Agricultural Systems* **101**, 113–127

Tilman, D., Clark, M. 2014. Global diets link environmental sustainability and human health. *Nature* **515**, 518–522.

United States Department of Agriculture (USDA). 2014. *2012 Census of Agriculture*. Available at: https://quickstats.nass.usda.gov/ [accessed 8 November 2018].

United States Environmental Protection Agency (EPA). 2017. *Inventory of U.S. Greenhouse Gas Emissions and Sinks: 1990–2014*. Available at: https://www.epa.gov/ghgemissions/inventory-us-greenhouse-gas-emissions-and-sinks-1990-2014 [accessed 4 November 2018].

van der Werf, G.R., Morton, D.C., DeFries, R.S., Olivier, J.G.J., Kasibhatla, P.S., Jackson, R.B., Collatz, G.J., Randerson, J.T. 2009. CO_2 emissions from forest loss. *Nature Geoscience* **2**, 737–738.

Varanasi, A., Prasad, P.V.V., Jugulam, M. 2016. Impact of climate change factors on weeds and herbicide efficacy. *Advances in Agronomy* **15**, 107–146.

Weber, C.L., Matthews, H.S. 2008. Food-miles and the relative climate impacts of food choices in the United States. *Environmental Science & Technology* **42**, 3508–3513.

Williams, J.W., Jackson, S.T., Kutzbach, J.E. 2007. Projected distributions of novel and disappearing climates by 2100 AD. *Proceedings of the National Academy of Sciences of the United States* **104**, 5738–5742.

Energy

Paul Behrens

Chapter Overview

- Energy is vital in human societies. Food production uses around 15–20% of the total energy produced for human needs.

- Two critical issues in energy use are the improved availability of energy in poorer countries and the implementation of low-carbon technologies in all countries.

- The transition to a zero-carbon energy system is termed the energy transition. We must transition fast to avoid the negative impacts of climate change, as well as other impacts such as air pollution and energy conflicts.

- There are two main ways the energy transition will continue to impact food systems: (1) the reduction and replacement of fossil fuels in food systems, and (2) the potential use of agriculture to enable the energy transition (such as biofuels and biofuel carbon capture and sequestration).

- Biofuels compete with food production, causing direct and indirect land use change, sometimes leading to larger greenhouse gas emissions.

- Bioenergy carbon capture and sequestration are increasingly used in future climate and energy scenarios. It is an extremely speculative approach to compensate for pollution today with untested future technologies.

Introduction

Throughout human history, societies have used increasing amounts of energy to improve the productivity of food systems (see Table 7.1). In the same way as other animals, early hominoids relied on the energy of their own bodies when foraging for food. Early humans discovered the use of clubs and levers which improved the energy efficiency of food collection. For 90% of human history, humans lived in these small groups, but beginning around 10,000 years ago (at the start of the Holocene; see *Chapter 1: Introduction*), humans developed food systems, enabling them to settle in communities. This development began a slow but accelerating process of increasing crop yields with new technologies.

A key technological innovation was the domestication of other animals such as oxen, many times more powerful than humans. With these animals, humans could increase food production tremendously, but

Table 7.1 Energy inputs and harvests for different forms of food collection and production

	Energy input (GJ/ha)	Food harvest (GJ/ha)	Population density (persons/km²)
Foraging	0.001	0.003–0.006	0.01–0.9
Pastoralism	0.01	0.03–0.05	0.8–2.7
Early agriculture	0.04–1.5	10.0–25	10–60
Traditional farming	0.5–2	10–35	100–950
Modern agriculture	5–60	29–100	800–2000

As energy inputs increase, food harvests increase and allow for increasing human population densities. Keep in mind that this is only for the production stage and that the post-harvest food chain (transport, refrigeration, etc.) uses much more energy. For example, the energy for a cold beer is dominated by the electricity needed for the fridge. The units are in gigajoules (GJ) and hectares (ha).

(Simmons, 1993)

farmers were still limited by the power of the animals. The Industrial Revolution changed everything, leading to an explosion of human activity as societies exploited the huge energy potential of fossil fuels, moving quickly from primitive tractors to today's massive combine harvesters, thousands of times more powerful than a single human (see Figure 7.1).

In *Chapter 6: Climate Change*, we introduced the concepts, challenges, and impacts of climate change on the food system. We said that over 90% of emissions arise from either energy or food system activities and discussed the challenge the food system faces in mitigation and adaptation. In this chapter, we

explore the energy system, to investigate the ways in which energy is currently used in the food system and how this use may develop in the future. The global energy transition is one of the most important challenges we face as a species, with unique problems for the food system, as we will see.

First, we outline the physical nature of energy and power, and describe how we convert between different sources of energy. We then explore how energy is used in society and the nature of energy transition, with a focus on food. We investigate how food systems can be decarbonized, and finally we look at the role agricultural systems could play in energy transition itself.

a)

b)

Figure 7.1 (a) Time-intensive manual harvesting in Northern Ethiopia, compared to (b) an industrial combine harvester in New South Wales, Australia. Not only do combine harvesters collect the grain, but they also process it too, combining cutting, threshing, and separating of grain, husk, and straw into a single machine. This harvester does the equivalent amount of work as a small village of approximately 2500 people (assuming a human can do continuous work at a power of around 60 W and a combine harvester 150,000 W).

Source: (a) A. Davey/Flickr (CC BY 2.0); (b) Alan Davey/Flickr (CC BY 2.0).

7.1 What is the difference between power and energy?

As discussed in the introduction, **energy** is helping drive human progress, but the definition of energy is not as straightforward as it might seem. Energy is the **potential** of one system to do work to another system. In physics, doing work is analogous to **doing something**. In most cases, doing something involves providing light, heat, or motion.

It is important to realize that **potential** energy is still a quantity of energy, even when it is not in use. For example, the fuel in a tank still has the potential to do something even if it is not used, and the water captured behind a dam in a reservoir even if it is not released downhill, still has the potential to be released to do work (for example, to bring a turbine into motion). The most common forms of energy are discussed in Box 7.1.

The unit of energy is the joule (with the symbol J). Technically, 1 joule is equivalent to the work done on a system when 1 newton of force is applied in the direction of motion for a distance of 1 metre (1 newton is equivalent to 1 kg m/s^2 in standard units). Put simply, 1 J is approximately the amount of energy required to lift an apple 1 m from the ground. Since

BOX 7.1

Why did it take so long to define different types of energy?

Although energy sounds like a very intuitive concept, it took a surprisingly long time to coin and define. Before the 1700s, the primary way of thinking about movement was through Newton's laws of motion. These laws deal with forces and momentum, with which you can compute many quantities that were important at the time (like cannonball trajectories). In 1725, Johannes Bernoulli defined a very early concept of energy, but it took some time for this to catch on. One of the first energy equations of energy was defined in 1802 by Thomas Young who wrote that 'the product of the mass of a body into the square of its velocity may properly be termed its energy'. This may sound familiar (if a bit archaic); it is actually the equation for kinetic energy, as shown in Table A. As decades went by, more and more types of energy were discovered, with heat and gravitational potential energy quickly following the discovery of kinetic energy.

In Table A, a few key types of energy and their general formulae are listed. Some similarities can be seen across these different energy types. For example, mass is important and proportional in several cases; an object of double the mass has double the kinetic energy when at the same velocity, or double the potential energy when held the same height above a reference point.

Table A Forms of energy and their formulae

Energy type	Formula	Description	Example energy generator
Kinetic	$KE = 0.5mv^2$	m, mass of object (kg) v, velocity (ms^{-1})	Wind turbines
Gravitational potential	$GPE = mgh$	g, acceleration due to Earth (ms^{-2}) h, height above reference point	Hydropower plant
Heat	$Q = mc\Delta T$	c, heat capacity of object (JK^{-1}) ΔT, change in temperature of an object (K^{-1})	Fossil fuel plant
Nuclear	$E = mc^2$	c, speed of light (ms^{-1}) m, mass converted in reaction (kg)	Nuclear plant
Thermodynamic	$U = -p\Delta V$	p, pressure (Nm^{-2}) ΔV, change in volume (m^3)	Fossil fuel plant

this is a tiny amount of energy, we usually use prefixes to describe much larger quantities of energy used in daily life such as 1 kJ (1000 joules), 1 MJ (1 million joules), or 1 GJ (1 billion joules). For reference, a toaster needs about 162 kJ, or 0.16 MJ, of heat energy to toast two slices of bread.

A crucial concept of energy is the **law of energy conservation**, which states that energy is never created or destroyed, just converted into different forms. If you turn on the engine of a car and drive, the chemical energy in the gasoline held in the tank is converted into kinetic energy (motion) and heat energy in the engine of the car. The chemical energy from the gasoline is equal to the kinetic energy of the car in motion and the heat and noise energy produced by the engine.

Energy is converted ubiquitously by both living and non-living systems. A natural example is photosynthesis where plants use light energy to build carbohydrates, a form of stored chemical energy providing energy for growth and reproduction (see the concepts of GPP, NPP, and HANPP in *Chapter 2: Biodiversity*). The flow of this energy to higher trophic levels also describes how food webs are structured. The size of trophic levels within the food pyramid can only be fully explained by the energy within the trophic level, not by the number of individuals or the biomass of the individuals. In the example of food webs, the potential chemical energy within a trophic level is converted up the food pyramid, from lower trophic levels to higher trophic levels (see Figure 7.2).

This conversion is not perfect. In fact, as we saw in *Chapter 2: Biodiversity*, only 10–20% of the energy in one trophic level is transferred to the next. As energy cannot be lost, the remaining 80–90% of the energy is transformed into waste heat and used for respiration by the plant, which is not available for organisms at the next trophic level. The fact that energy conversions are everywhere led the ecologist Charles Elton to suggest that energy is the currency in the economy of nature. In fact, because the human economic system is a sub-system of nature, we can also consider energy as the currency of economics (Daly and Farley, 2010).

Not only are we interested in the amount of energy that a system has, but also in how fast we can convert that energy to useful forms. For example, when we discuss energy generators (that is, wind turbines or coal power stations), we often talk in terms of **power**. Power is the rate of energy conversion. Consider a reservoir of water behind a dam (a source of potential energy), with some water allowed to flow through a set of gates. The potential energy of the water is converted to kinetic energy as the water flows downwards. If this leads to a hydroelectric power plant the flowing water moves through a turbine and converts the kinetic energy of the flowing water into electrical energy. If we let more water through we generate a higher electrical

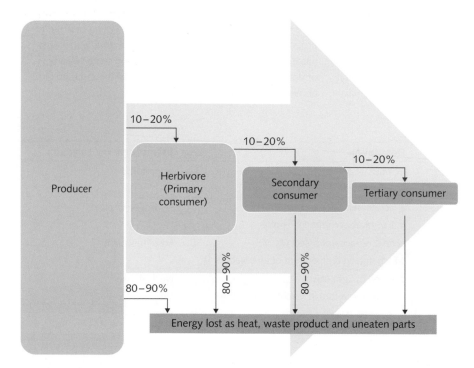

Figure 7.2 The trophic pyramid is based on energy flows, not the mass or number of organisms. In the conversion of energy between trophic levels, not all energy can be converted and much is lost to the environment as waste heat.

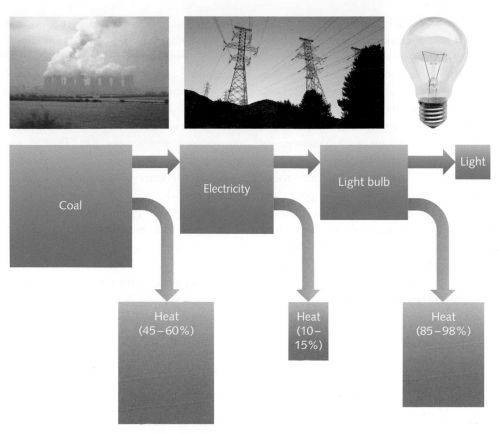

Figure 7.3 An overview of energy losses through the production of light by a coal power station.
Source: Alan Zomerfeld / Wikimedia Commons (CC BY-SA 2.5) (left); Varistor60 – Own work / Wikimedia Commons (CC BY-SA 4.0) (middle); KMJ / Wikimedia Commons (CC BY-SA 3.0) (right).

power. The unit of power is called the watt and has the symbol W. It is a compound unit of the number of joules converted each second, that is W = J/s.

This leads us to the last concept of energy described in this section—**efficiency**. Energy can never be perfectly converted into a useful type of energy; there will always be some amount of waste heat which is lost to the environment. This is as true for electricity generated by coal power plants as it is for chemical energy in food webs. Consider switching on the light on your desk, with the energy following a complicated series of transitions, from a lump of coal to the illumination of this textbook. Each transition along the way results in useful energy lost in the form of waste heat (see Figure 7.3). First, the energy generated at the power station is only 40–55% of the chemical energy in the fossil fuel with the remaining energy lost as heat to water. Water is usually used to cool power plants, which can lead to other environmental issues. Next, about 10–15% of the remaining energy is lost in the electricity transmission system, since because electrons do not move perfectly in conductors they heat up the surroundings. Finally, the electricity is used to generate light, which, depending on the type of bulb, loses even more energy in the form of heat (LEDs lose about 60% of energy as heat, but incandescent bulbs around 98%). These losses in energy conversions are a fundamental property of the natural world and physics (you may see it discussed in other texts as the concept of **entropy**).

7.2 How do we power society?

As a simple premise, energy is a necessity for modern life because it frees time for other tasks. For example, take a day in the life of a farmer, instead of spending hours tilling soil by pulling a plough, collecting cooking wood, and cleaning clothes by hand a farmer could use a tractor, a gas stove, and a washing machine to free time in which she can educate her children, design a new method of harvesting, or just simply relax. The use of energy not only generates increased production, but also frees time for innovation, the spreading of ideas, and further production.

Energy **demand** and **supply** are two common perspectives of the energy system. Energy demands are the services that society requires to function, for example heating a house, cooling food in a fridge, transport by car or train, and lighting in your room. Energy supply is the way in which we supply these demands with different energy types (for example, the use of gas, coal, or wind) and the networks which transmit this useful energy (for example, electric transmission lines and oil pipelines).

7.2.1 How much energy does society need each year?

From tractors and cargo ships (converting fossil fuels to kinetic energy) to cooking stoves (converting gas to heat), energy conversions are the invisible drivers in our society and our food system. Modern society is simply not possible without large-scale conversion and the use of energy. The number of joules harnessed by all humans over the year is staggering and needs almost an entire line of this textbook to even write down:

$$510,000,000,000,000,000,000 \text{ J}$$

We can shorten this by using scientific notation (510×10^{18} J), but this is still too large a number to comprehend. We can get a better understanding by offering comparisons. Let us define a new unit—the *Tour de France* (or TdF), one of the most physically demanding sporting events in the world. One TdF is the amount of energy that a cyclist needs to complete the 3500 km-long TdF. In western Europe, the average person uses around 5 TdF of primary (before conversions) energy per day. That is, each day the average person in western Europe uses the equivalent amount of energy to one professional cyclist completing the whole TdF five times.

Energy use varies widely between countries, with the average Indian citizen using 0.6 TdF per day per person (approximately 12% of citizens in western Europe). In China, having undergone the fastest and largest transition from poverty the world has witnessed, energy use has increased from 0.5 TdF per person per day in 1970 to 2.1 TdF in 2011.

Energy is used in many different ways and comes from many different resources. The oil in the tank of a tractor comes from biological marine deposits which have formed under millions of years of high temperature and pressure, whereas the electricity used in a hairdryer can be generated from the falling water in a hydropower dam or from a wind turbine. Generally, we like to think in terms of the different services that energy supplies such as heating your house, cooling your fridge, providing light, and transport. Figure 7.4 shows the average domestic energy use for different types of things per person per day in the UK, with all values in litres of gasoline.

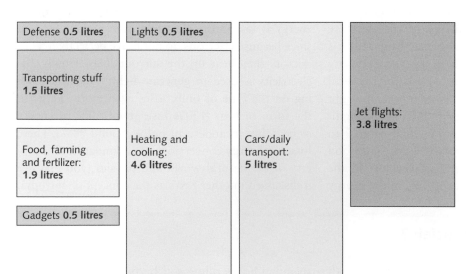

Figure 7.4 The breakdown of energy use in the UK per person per day in litres of petrol equivalent. Energy embodied in imports is not included and accounts for another 50% of energy use.

Source: Adapted from *Sustainable Energy – Without the Hot Air* by David J. C. MacKay. © 2009 David J. C. MacKay.

From Figure 7.4, it is clear that heating and cooling (refrigerators, radiators, air conditioning, etc.), cars, and jet flights are large users of energy in high-income countries. Direct food production (not including processing and distribution) is the fourth largest. Other uses of energy in the food system are embodied in other categories such as 'transportation'. There is one important energy use missing from the figure—it does not show the energy embodied in the imports from other countries. Imports can account for as much as 50% of total domestic energy use (Peters et al., 2012).

Where do we get our energy from?

Figure 7.5 shows the share of global energy use by energy type. The large majority of energy is derived from energy stocks, most notably fossil fuels: gas, coal, and oil (see an explanation of the difference between energy stocks and flows in Box 7.2). The remainder, only 12.9% of total global energy use, comes from renewable energy resources. Of the renewable energies, the majority is bioenergy, at 10.2% of the total. Though this may sound like modern biofuels (for example, ethanol used in gasoline), it is in fact predominantly the combustion of wood with which to cook in lower-income countries. Hydropower is the second largest renewable energy type,

at 2.3%, with the remaining 'modern' renewables, such as solar and wind comprising much less but growing fast.

Figure 7.5 is useful to see the overall types of energy at a global level but energy use varies greatly between countries. In countries with large amounts of renewable resources, such as Norway, Iceland, and New Zealand the amount generated from renewables is much greater. Conversely, for countries with large amounts of cheap fossil fuels, such as those in the Middle East and Venezuela, almost all energy comes from fossil fuels. There are many poorer countries in the world which are in a state of **energy poverty**, limiting the ability of people to do anything other than time-intensive manual labour and threatening human survival (for example, if energy for heating or cooling is unavailable) (see Figure 7.6).

In your experience

Imagine that you have no access to modern energy providers (electricity, fossil fuels, etc.). Consider how different your day might be if you had to wash your clothes by hand, collect wood for heating/cooking, feed horses for transport, etc. Considering the detail in *Chapter 8: Nutrition*, how would this impact your lifestyle in terms of your health?

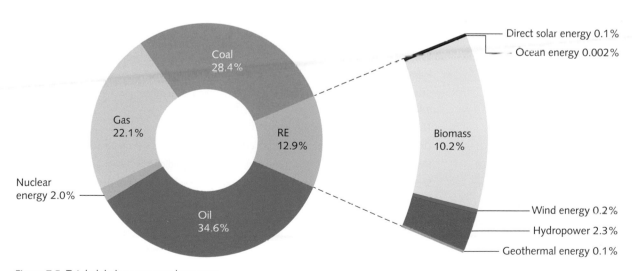

Figure 7.5 Total global energy uses by source.

Source: Intergovernmental Panel on Climate Change. (2014). Summary for Policymakers. In Climate Change 2013 – The Physical Science Basis: Working Group I Contribution to the Fifth Assessment Report of the Intergovernmental Panel on Climate Change (pp. 1-30). Cambridge: Cambridge University Press. (c) Intergovernmental Panel on Climate Change.

BOX
7.2

Energy stocks and flows

Energy can be supplied to users in many different ways and can be roughly divided between energy stocks and energy flows. Energy **stocks** are those that have formed over a long period of time, and once used, they are gone (at least in the time frame of millions of years). For this reason, they are non-renewable resources. The two main types of energy

stock are fossil and nuclear fuels. Energy **flows** are those that flow through nature and are therefore renewable resources, for example solar, wind, wave, and hydropower (see Figure A for some example uses of wind and solar energy on agricultural land). Table A lists the major types of energy available to us and the origin of their generation.

Figure A Left: wind turbines on agricultural land in India. The people pictured may not have access to electricity generated by wind turbines, as often communities are not electrified. Right: solar panels on barns in Switzerland.

Source: (left) Vestas/Flickr (CC BY 2.0); (right) Gabrielle Merk/Wikimedia Commons (CC BY-SA 4.0).

Table A Types of energy and their immediate origin

Energy type		Origin
Energy stocks	Fossil fuels (gas, oil, coal)	Biological matter (after millions of years at high pressure and temperature)
	Nuclear fission	Nuclear bonds in some elements created in supernovae (the death of previous stars)
Energy flows	Solar	Sunlight
	Bioenergy	Plants absorb energy from the sun and converts it into material to grow
	Wind	Second-generation solar energy—the sun heats the Earth, differences in temperature (and rotation) drive winds
	Wave	Third-generation solar energy—winds drive waves
	Tidal	Difference in gravitational forces experienced by Earth due to the moon and sun
	Geothermal	Heat energy left over from the formation of Earth and radioactive heating
	Hydropower	Second-generation solar energy—evaporation by the sun gives water potential energy in clouds which then falls in river systems

However, we know that much of the energy we use is not sustainable. When fossil fuels are burnt, carbon–hydrogen bonds separate, releasing carbon into the atmosphere and producing CO_2. As discussed in

Chapter 6: Climate Change, CO_2 is the largest factor producing climate change. Therefore, rapid and wide-ranging changes to energy systems are critical if we are to avoid irreversible damage to ecosystems (IPCC,

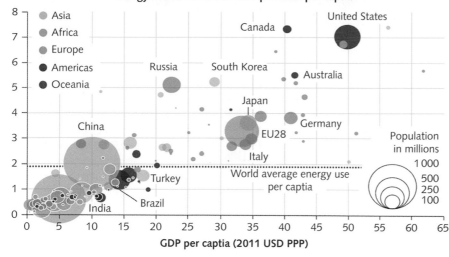

Figure 7.6 The relationship between energy use and gross domestic product (GDP). We can see that they are highly correlated.

Source: Georgetown Public Policy Review Journal (http://gppreview. com/2016/02/26/6091/).

2013). The transition towards low-carbon, renewable energy is commonly called the **energy transition**.

There are two main ways in which the food system will be involved in energy transition: (1) the food system itself will need to transition to low-carbon energy use, and (2) agricultural systems may be used to help facilitate the energy transition more generally, for example the use of biofuels or bioenergy **carbon capture** and **sequestration**. First, we will look at the use of energy in the food system and the ways in which the food system itself can transition to low-carbon energy use.

7.3 How does the energy system feature in the production of food?

Consider the food on your plate at your next meal. An enormous amount of energy has been used to produce your food (not to mention the equipment in which you stored the food and on which you cooked the food, the utensils with which you will eat the food, etc.). If the food came in a packet take a look at the ingredients on the side and where the food was made, and try to imagine all the energy needed to grow, process, package, and transport the food along the way.

Globally, between 15% and 20% of total energy use is in the food system (Beckman et al., 2013). However, energy use varies a lot from country to country. The largest energy users such as the USA and the Netherlands use several times more energy than lower-income countries, such as Ethiopia and the Philippines, due to different techniques and models of production (such as those we discussed in the introduction).

In Figure 7.7, we see the development of energy use in agriculture over the last few decades. Generally, we see that higher-income countries continue to use high amounts of energy (in the USA and the UK, for example). In low-income countries, such as Ethiopia, energy use in agriculture is still very low and correlates with poverty (see Figure 7.6). In some low-income countries, energy use has increased significantly over time, for example in Indonesia and India. The increase in agricultural energy use in India is a reflection of the increasing wealth of the region and the opportunity for increasing energy inputs. The slowly reducing energy use in UK agriculture can be attributed to efficiency improvements and increased imports, as it moved from a manufacturing to services economy through the late decades of the twenty-first century.

Different types of food production have different impacts on energy usage. For example, some countries like the Netherlands use heated greenhouses which need vast amounts of energy (almost double that in average US production). Conversely, food grown organically in the appropriate climate (for

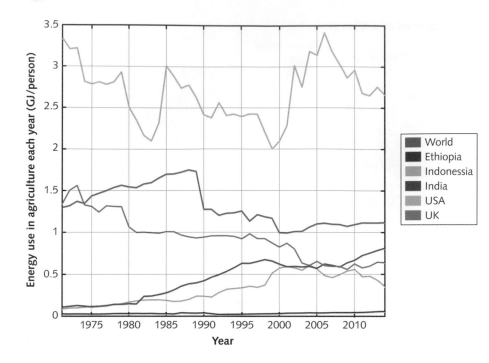

Figure 7.7 There are major differences in energy use per person among countries. This figure shows the average energy use per person in agriculture/forestry for two higher-income (the USA and the UK) and three lower-income (Ethiopia, Indonesia, and India) countries during the period 1961–2011.

Source: Data from IEA, 2016

example, grapes in France) can have very low energy inputs. As a general rule, if more energy is used then food yields are greater.

So far, we have mainly discussed the production of food. However, energy is used along the entire food production chain, from the growth phase to the consumption phase. The LCA approach we introduced in *Chapter 6: Climate Change* becomes useful once again and is the dominant method for investigating energy in food chains (Pelletier et al., 2011). Using it, we can sum the energy used at different points in the food production chain and calculate an estimate of energy use for each portion of food we might buy. Unfortunately, data on energy use in food sectors are scarce for many countries and typically only available for higher-income countries. Therefore, often it is only possible to make general statements about energy use in higher-income countries (as data are sparse for lower-income countries, it is not possible to draw broad conclusions). A country for which a lot of data are available is the USA, which, for example, shows that animal products consumed in the American diet account for 50% of total food-related energy demands (Pimentel et al., 2008).

We will split the next part of this section into two: production and post-production. Production includes the growth and harvesting of food, whereas post-production includes the processing, packaging, transportation, preparation, and cooking of food.

Connect the dots

In *Chapter 8: Nutrition*, we discuss the sustainability issues surrounding nutrition and diets. What types of diets might reduce energy use and why? Can we harness waste products, and for what could we use them?

7.3.1 How much energy do we need in the production of food?

A massive amount of energy is needed to produce food. Energy use in the food production phase includes the energy needed for operating agricultural machinery such as tractors, and the energy needed to produce fertilizers and pesticides (sometimes these are termed pre-production since they are often made industrially off-site; see *Chapter 3: Pollution*). Table 7.2 shows the disparity in energy use in production between the USA and the Philippines in terms of different agricultural activities (this difference is easily seen in Figure 7.8). It is easy to see that productive yields by both mass and energy increase with greater energy inputs (this can also be seen in Table 7.1).

Modern US agricultural methods for rice production need energy inputs of approximately 65,000 MJ/ha (about 65 barrels of oil per hectare), yielding 5.8 tons of rice per hectare. Compare this to

Table 7.2 Energy inputs and productive yields for different types of agricultural methods

	Rice production			Maize production	
	Modern (USA)	Transitional (Philippines)	Traditional (Philippines)	Modern (USA)	Traditional (Mexico)
Energy input (MJ/ha)	64,885	6386	170	30,034	170
Productive yield (kg/ha)	5800	2700	1250	5083	950
Energy input yield (MJ/kg)	11.19	2.37	0.14	5.91	0.18

(FAOSTAT, 2014)

a)

b)

Figure 7.8 (a) Rice production in Bohol, the Philippines. (b) Rice harvesting in Louisiana, USA. They show very different production practices. Note that these are also different varieties of rice.

Source: (a) Øyvind Holmstad/Wikimedia Commons (CC BY-SA 4.0); (b) https://rememberingletters.wordpress.com/2012/07/02/harvesting-rice-in-louisiana/

traditional Philippine methods with 1/500th the energy input and yields of 20% the US amount. So while the US yield is much higher, it is also less energy-efficient. Generally, the monetary value of energy used is much less than the value of agricultural outputs, which means that high-energy input agriculture is ultimately more cost-effective.

One key energy sink in the production phase of food is the production of artificial fertilizers, and specifically nitrogen fertilizers by using the Haber–Bosch process (see Box 7.3). As we saw previously in *Chapter 3: Pollution*, fertilizers are crucial for producing high yields with some estimates suggesting that 30–50% of current yields are attributable to their application (Stewart et al., 2005). Production of artificial fertilizer alone uses 1–2% of total global energy. In Chinese agriculture, 60–70% of all energy inputs are accounted for by chemical fertilizer and pesticide production.

In organic crop production, artificial fertilizers are not used at all, which may offer opportunities for reducing the overall production of artificial fertilizers. However, as discussed in *Chapter 3: Pollution*, organic agriculture currently has difficulties in competing with industrial production for many crucial crops around the world such as cereals, oilseeds, and potatoes (Pelletier et al., 2008). Closed systems, such as greenhouse production or urban farming, may also reduce fertilizer inputs (see Case study 7.1), but the energy costs of heated greenhouses can be so high that it is often preferable to ship food from warmer climates (this also relates to the concept of food miles; see *Chapter 6: Climate Change*). Both options tend to be more expensive at the moment, limiting their consumption to higher-income groups or those who want to be certain they are not ingesting residues of artificial fertilizers and pesticides.

BOX 7.3

Energy use in fertilizer production

We can see from Figure A that energy efficiency in fertilizer production has improved dramatically over the past century, with modern steam reforming needing less than 20% of the energy in the first forms of artificial fertilizers. However, we can also see that reductions in energy use have slowed over time, so that the most promising option we have now is probably to use fewer fertilizers overall (which would also improve nitrogen and phosphorus cycles, as discussed in *Chapter 3: Pollution*). We could do this by growing crops which have better nutrient efficiency. Another option to improve energy use is to harness the heat produced in the production of fertilizers in district heating.

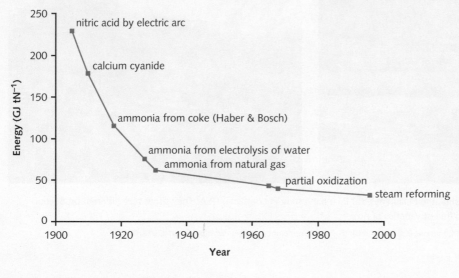

Figure A Historical development in energy requirements in nitrogen fixation for nitrogen fertilizers.

Source: Woods J, Williams A, Hughes JK, Black M, Murphy R (2010) Energy and the food system. *Philos Trans R Soc B Biol Sci* 365(1554):2991–3006.

CASE STUDY 7.1

Urban farming in The Hague, the Netherlands

As the name suggests, urban agriculture is any food production in urban areas such as villages, towns, and cities. In some cases, production can occur on top of buildings themselves. One example of this was *The Urban Farmers*, at one stage the largest urban farm in Europe, located in The Hague, which began production in 2016. On the top floor of the building (Figure A.a) was a greenhouse, in which grew many different types of vegetables, including tomatoes,

aubergines, cucumbers, and peppers (Figure A.e). On the floor below was an aquaculture fish farm (Figure A.d), which provided natural fertilizers for the growth of the vegetables. Plants were grown hydroponically, that is, they were not grown in soil (for more details on hydroponics, see Box 5.5 in *Chapter 5: Soils*). In this case, a coconut fibre substrate was used (see Figure A.c). The Urban Farmers shut down in 2018, saying that it was not profitable due to competition from

a)

b)

c)

d)

e)

Figure A The Den Haag Urban Farm. Greenhouses can be seen at the top of the building (a). Urban farms are also places to bring people living in cities, who may have had little exposure to farming, closer to their food (b). Food is grown in a substrate (base) of coconut fibres, and water with nutrients is run through pipes at the base of the white trays, shown in (c). Fish waste is used for plant nutrients, and they are sold as food when they become too large (d). Several varieties of plants are grown, including tomatoes (e).
Source: (a) Martijn Zegwaard; (b) UrbanFarmers Netherlands; (c)-(e) authors' own photogtaphs.

land-based greenhouses in the west of the Netherlands. As with many new concepts and technologies, there are often difficulties in making pilot programmes fully profitable.

The farm was owned by a parent company operating three other smaller-scale farms in Basel, Berlin, and Zurich. They aim to make urban farming systems modular, in order to bring costs down and offer rooftop farms, mobile farming units in shipping containers, and systems for aquaculture integration with existing hydroponic farms (where plants are grown without soil).

Urban farming of this type has some environmental advantages to extensive rural farming. Generally, urban farms aim to be a closed system, so it is easier to recycle water and nutrients as the plants respire. The Urban Farmers in The Hague were able to produce natural fertilizers using

fish waste, and because there are fewer threats from pests on the top floors of buildings, high above the ground, they used natural predators sparingly (see *Chapter 3: Pollution* for more details). This raises the energy efficiency of the process; there were plans to heat the greenhouse by geothermal heat, instead of fossil fuels. They both grew and sold 45 tons of vegetables and 19 tons of fish a year, from an area roughly one-third the size of a football field.

As the closing of the urban farm shows, the long-term profitability of urban farming is yet to be proved, as it is a relatively new industry. Initial attempts show that it will be hard to compete with other types of production on quantity and so have focused on improving taste and quality as their unique selling point.

In your experience

In *Chapter 3: Pollution* and in this chapter, we discuss the impact of organic crop production with other production methods. Given what you know about the ecotoxicological impacts and energy use of organic food, would you now choose organic over conventional food? How much more would you be willing to spend, and how much would this be driven by concerns over your health, compared to the health of the environment?

Connect the dots

Case study 7.1 describes urban farming in the Netherlands. Do you think urban farming has a future on a large scale? What might be the limitations? How might this impact biodiversity (see *Chapter 2: Biodiversity*), pollution (see *Chapter 3: Pollution*), water use (see *Chapter 4: Water*), soils (see *Chapter 5: Soils*), and climate change (see *Chapter 6: Climate Change*)? Do you think it is possible to go completely circular (see *Chapter 11: Consumption*)?

While we have made progress towards renewable electricity using wind turbines and solar panels, it has been harder to transition to renewable energy for many areas of food production. In Europe, 15% of the overall electricity mix is renewable, but in the food sector, this drops to only 7% (Eurostat, 2014). This is because the transition of energy use

on farms presents some unique difficulties. By their nature, farms usually cover large amounts of land, and energy is used diffusely over the whole area. This is very different to other sectors which can generate electricity centrally and transmit this through electricity lines to the consumer.

Due to the large space required for production in food systems, it is sometimes impractical to electrify processes. Generally, heavy mobile machinery, such as tractors and combine harvesters, are used to reap and process food on the fields. These machines need large, mobile sources of energy to operate, which at the moment forces us to use oil (which has a high energy content per unit volume, usually referred to as a high energy density). A portable electrical alternative in the form of batteries or fuel cells is not yet competitive for heavy machinery, but these technologies are developing quickly.

7.3.2 Post-production

Post-production includes all other processes in the food system, including the processing, packaging, transportation, and preparation of food. Within this, the industrial processing of food is the largest energy user and accounts for around 28% of all energy used in the food sector in the EU.[1] Industrial processing includes any process which converts raw ingredients into edible food for consumption. It includes physical processing (for example, slicing, mincing, liquefaction, emulsification), cooking, pickling, pasteurization, washing, and preservation (see Figure 7.9 for two examples).

a)

b)

Figure 7.9 (a) Industrial cheese production in China. (b) Industrial sausage making.
Source: (a) Matthias Kabel/Wikimedia Commons (CC BY-SA 3.0); (b) Robert Lawton/Wikimedia Commons (CC BY-SA 3.5).

[1] Data for many countries are not available, but this statistic is probably similar in many high-income countries.

Industrial processing is typically undertaken in large, centralized buildings and powered by the electricity grid. This means it is easier to convert industrial processing to renewable energy than in the production phase. In some cases, these processes can be highly integrated within the surrounding economy using only sustainable resources, allowing for a circular flow of energy, nutrients, and food (see an example of this in Case study 7.2).

Once the food has been processed, it is packaged and transported, accounting for about 22% of total energy use in the food sector. Much of this energy use is in the form of oil for large cargo ships and trucks. Similar to the heavy machinery used in the production phase, it is difficult to transition the transport sector to lower carbon emissions and energy use, although there have been recent developments in electric vehicles (see Figure 7.10) (see *Chapter 6: Climate Change* for an in-depth discussion).

Next, food is delivered to vendors such as supermarkets. The energy use in supermarkets is similar to other retail activities except for the large amount of energy needed for refrigeration. Refrigeration is key for the long-term storage of food, not just in supermarkets, but also in transportation. As discussed in *Chapter 9: Food Security*, refrigeration is vitally important for lower-income countries, which

Figure 7.10 An electric truck from Tesla.
Source: Tesla © 2019.

currently lose a lot of food through rotting due to inadequate options for storage. Fortunately, many refrigerators are powered with electricity which is easier to replace with renewable energy. The last step is purchase, storage, and cooking of food by consumers, which is also easier to transition into renewable energy sources. For example, we can use electric induction ovens and stove tops powered by renewable energy, instead of gas (other approaches are covered in *Chapter 6: Climate Change* and *Chapter 11: Consumption*).

CASE STUDY 7.2

Miraka milk-drying plant (New Zealand)

Miraka is a milk-drying operation in North Island of New Zealand (see Figure A). It processes over 250 million litres of milk into powder products each year. The energy used comes from a local geothermal power plant. Electricity is used in the local facilities and households nearby, and the waste heat generated is used to dry milk to make powder for export.

When the plant was first built, the facility found that a lot of organic waste material was produced by the drying of milk. Instead of disposing of it, it is taken to worm farms locally to produce fertilizers for plants in greenhouses on site. Capsicums (bell peppers) and courgettes are grown in the greenhouses using nutrients from the processed milk powder

waste and are heated by waste heat from the geothermal plant. Additional CO_2 for the plants is pumped into the greenhouses from waste streams on site (see Figure B).

While the production of milk still takes place in the form of extensive cattle farming in the surrounding regions, this is an example of a 'circular' approach to post-production processing. A circular approach means that we extract the maximum amount of value from each process, harnessing as many waste streams as possible, to produce goods to sell and to increase sustainability (circular approaches are described in further detail in *Chapter 11: Consumption*). Some of the profits are reinvested socially, into education grants and further innovation on the site.

Figure A The Miraka milk powder processing plant.

Source: © Miraka Ltd.

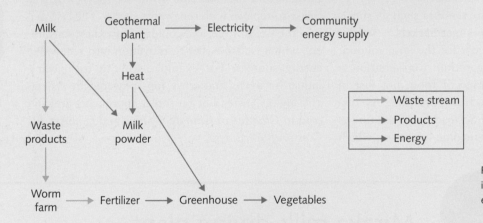

Figure B A schematic of the flows in the production of milk powder, energy, and vegetables.

7.4 **Can we make food into energy?**

In Section 7.3, we considered how energy is used in the food system, exploring both the types and amount of energy used at each stage in food production. However, there is also another important perspective—the use of agricultural products in the energy system itself. Since food contains chemical energy (captured by photosynthesis; see *Chapter 2: Biodiversity*) and it is something that we can grow and renew, could we develop food systems which support our energy system? And can this be used to transition to a low-carbon society?

There are two key avenues for the use of food in the energy system. The first is bioenergy production (including biofuels) which could replace fossil fuels. The second is the combined use of bioenergy and carbon capture technologies—termed **bioenergy carbon capture and sequestration (BECCS)**—to reduce the total amount of CO_2 in the atmosphere (plants use CO_2 during photosynthesis). This might allow us to continue burning fossil fuels in the short term, while making up for it in the long term. BECCS is now a critical part of models that allow us to limit warming to 1.5°C and 2°C. As we will see later, this approach would have huge impacts on the food system and sustainability. Given the importance of these technologies, in this section, we will introduce both of these concepts in more detail.

7.4.1 What is bioenergy?

Bioenergy is any energy that is made available from organic sources. In this sense, bioenergy is nothing new; wood has been a key source of bioenergy since the discovery of fire at least 125,000 years ago. Bioenergy can be thought of as a second-generation solar energy—sunlight provides the energy for plants to grow, which we then burn (see Table A in Box 7.2). Historically, we have used bioenergy released from fire for many things, including warmth, cooking, and smelting metal.

Bioenergy is of interest in energy transition because, in principle, bioenergy is **carbon neutral**. When plants grow they capture CO_2 from the atmosphere, storing carbon in their leaves. When plants are then used in combustion this carbon is released back into the atmosphere, completing a cycle in which no extra carbon has been added. The flow of carbon through the process is supposed to be carbon neutral. In contrast, fossil fuels are a stock of carbon developed over millions of years; once the carbon is released, it cannot be converted quickly back into fuels and will stay in the atmosphere.

With modern technologies, some crops have the potential to produce fuels that can replace fossil fuels. Modern bioenergy is already a prominent alternative in countries like Brazil and the USA, and some see it playing an even bigger role in the future. The production of bioenergy has direct and indirect links to our food system. At present, there are three major forms of bioenergy usage globally:

1. Direct combustion of biomass, for example burning wood for heat or burning high-quality wood pellets for electricity generation in power plants

2. Production of crops which are turned into liquid biofuels such as biodiesel or ethanol, generally for use in transport

3. Collection of biogas from decaying biomass or organisms, which are then burnt in the same way we burn natural gas (see Figure 7.11).

Figure 7.11 A biogas generator on a local farm in Pabal, India.
Source: http://bio-gas-plant.blogspot.nl/2011/08/biogas-generator.html.

In Section 7.4.2, we will focus specifically on liquid biofuels because many policy-makers in different countries have high hopes for their use in energy transition.

7.4.2 Liquid biofuels

A primary goal of biofuels is to replace different types of oil in the combustion engines of vehicles such as cars, trucks, planes, and ships. Without any modification, cars can already run on low mixes of biofuels (that is, 10–15% of gasoline is replaced by biofuels), and with alteration, they can run on levels above 85%. This higher mix is very popular in Brazil where car owners can choose to fill up with either biofuels or fossil fuels—consumers can choose between whichever is cheapest at the time (see Figure 7.12). The major global biofuel crops are palm

Figure 7.12 Choices for consumers at a petrol station in Brazil. The ethanol pump is seen on the left, and the gasoline pumps on the right.
Source: HenriqueWestin / Shutterstock.com.

oil or soy bean for biodiesel, and corn or sugarcane for bioethanol. These are processed in different ways, but both result in a liquid fuel with a high-energy density (J/L) that is easily stored and transported.

This sounds promising so far; with some small modifications, we can completely replace gasoline with biological substitutes and eliminate approximately 35% of total carbon emissions (see *Chapter 6: Climate Change*). However, there is a fundamental issue—plants convert only 0.5–1.5% of incoming light into chemical energy (yet more energy is lost converting plants into a liquid fuel and in combustion in vehicle engines). Compare that to solar panels with an efficiency of 15–21% (which includes conversion production and transportation). Because plants are so inefficient, we need to grow large amounts in massive areas of land to produce enough or perhaps increase photosynthetic efficiency further using genetically modified crops (see *Chapter 3: Pollution* for a discussion on genetic modification). The amount of land we

need is enough that energy crops can influence the availability and price of food (explore this complex issue below in Food controversy 7.1).

How might we get a handle on how much land we need for biofuels? And how can we compare this to other energy sources? We need a statistic with which we can compare them. We call this statistic **power density**—this is the amount of power [W] we can produce by an energy system on 1 square metre of land. For example, if we build a large solar power facility, we can produce between 4 and 10 W/m². With gas combustion, we can produce much more, around 2000 W/m². In comparison, biofuels struggle to reach 0.5 W/m² (Smil, 2010; van Zalk and Behrens, 2018). With our current biofuel technology, around 20 times more land is needed than other renewable options, and 4000 times that of fossil fuel options. The low power density of biofuels has important consequences when we consider replacing fossil fuels with them (see Box 7.4).

BOX 7.4

Example calculation of arable land needed for liquid biofuels in the Netherlands

---

The Netherlands consumed a total of 34 million tons of petroleum for transportation in 2014 (Eurostat, 2014). As 1 ton equals 42 GJ, we can calculate the total energy use in the Netherlands in 2014:

$$33,946,000 \text{ [tons]} \times 42,000,000,000 \text{ [J]} = 1.4 \times 10^{18} \text{ [J]}$$

Would it be an option to produce that amount of energy using biofuels? In order to answer that question, we need to compare this to the power density (W/m²) of biofuels. To achieve this, first we need to convert energy use (which is expressed in J) to the average power usage (in W, or J/s). The first step is to calculate an equivalent average power of petroleum used over the year.

As calculated above, the Netherlands uses 1.4×10^{18} J of petroleum per year. To determine the average power usage, we need to divide this number by the number of seconds in a year:

$$Power = \frac{Energy}{Time} = \frac{1.4 \times 10^{18} \text{ [J]}}{60 \times 60 \times 24 \times 365 \text{ [S]}} = 4.4 \times 10^{10} \text{ [W]}$$

If we now want to calculate the surface area of biofuels needed if we want to power the Netherlands with biofuels, we have to equate this to the power density of biofuels of 0.5 W/m²:

$$\text{Area needed} = \frac{4.4 \times 10^{10} \text{ [W]}}{0.5 \text{ [W/m}^2\text{]}} = 8.8 \times 10^{10} \text{ m}^2 = 88,000 \text{ km}^2$$

If the Netherlands were to replace all liquid fossil fuels with biofuels, an area of around 88,000 km² would be needed. But the total arable land in the Netherlands is 22,825 km², and the total surface area is 41,542 km². The Dutch would require an area of land twice the size of the country to become self-sufficient in liquid fuels from biological sources and would have no space left to grow food (the Dutch are a major producer and exporter of food). This also leads to the conclusion that if biofuels are to play a major role in replacing petroleum, many countries will have to import these from overseas, leading to more pressures elsewhere (see Food controversy 7.1).

As we have just discussed, modern agricultural systems require large amounts of energy inputs during production and post-production, which is also true when producing biofuels. In order to see if the biofuel process is worthwhile we have to look at the net energy output in an LCA (see *Chapter 6: Climate Change*). Studies estimate that US corn-based ethanol has an energy return of around 1.6 (Wang et al., 2012). This means that we only produce 60% more energy from the process of growing the crops (that is, for every litre of oil used in production and post-production, biofuels generate an equivalent of 1.6 L as a finished product). However, once uncertainties in the calculations are taken into account, some studies find that the benefit is statistically indistinguishable from 1. This means that according to some studies we cannot be certain we get any energy benefits from biofuels (Murphy et al., 2011).

Since the benefits of biofuels in the USA are generally marginal and they are costly to produce, US corn-based ethanol is heavily supported by subsidies from the government. US corn-based ethanol is widely considered as the poorest performing biofuel worldwide. The best performing biofuel is Brazilian sugarcane, managing an energy return of 4.3. For every 1 L of energy used in (post-) production, over four times the energy is harvested in the final product (Wang et al., 2012).

Even though bioenergy has a low energy density, there are still several reasons to pursue biofuels as a replacement to fossil fuels. In some cases the production of biofuels aligns with national targets for energy independence (that is, increasing the domestic supply of energy sources). For example, as a result of the oil shocks in 1973, Brazil undertook a long process to increase the local bioenergy output, so they could provide security in the event of further oil shocks. Now Brazil produces around 50% of the total liquid transport fuel market from biofuels (Yergin, 2012). Energy security is also one of the stated targets for biofuels in the USA and Europe.

However, if the rationale for biofuel development is the mitigation of climate change, we must investigate the GHGs released during the production of biofuels (these investigations have caused much concern to policymakers; see Food controversy 7.1). This needs

---

FOOD CONTROVERSY 7.1

Biofuels in the European Union

In 2003, the EU issued a target to replace at least 5.75% of all transport fuels with biofuels by 2010, later increased to 10% by 2020. There was a lot of excitement about reducing carbon emissions using existing vehicles and infrastructure (petrol stations, pipelines, etc.) (see Figure A). However, evidence of poor environmental outcomes of some biofuels soon started to pile up (increasing land use change leading to more GHG emissions, loss of biodiversity, etc.). Many people started to become concerned that the targets were actually doing more harm than good. Of particular concern was the use of palm oil (the impacts of palm oil plantations on biodiversity are described in detail in *Chapter 2: Biodiversity*) (see Figure B). Recent work estimates that carbon emissions from biofuel-mixed fuels are actually higher than pure gasoline due to indirect land use changes in locations such as Indonesia. Valin et al. (2015) found emissions of 79 g CO_2e per MJ of biofuel-mixed fuel, compared to 71 g CO_2e per MJ for gasoline.

However, early concerns did not translate into cancellation or reduction of the biofuel targets, but an ongoing policy discussion on how to ensure that biofuels in the European energy system were produced more sustainably. The decision-making process was complicated by the interaction of many different interest groups, with competing national and international views. It is useful to realize that, as Oxfam reported, the 400 lobbyists in Brussels employed by biofuel producers outnumber the European Commission's entire energy directorate (Oxfam, 2016).

From 2008, even those who promoted the targets started to have serious concerns, including the politicians who recommended the policies in the first place (Harrabin, 2008), but it took 7 years for action to be formalized, with new regulations only coming into force in 2015 (for an overview, see EC, 2016). This is a good example of a legal framework, by its time-consuming nature, taking a long time to respond to specific scientific concerns (and associated policy debate).

Figure A Fuel pumps in the UK offering different grades of biofuel mixes.

Source: Pete Birkinshaw/Flickr (CC BY 2.0).

Figure B A palm oil plantation in Malaysia.

Even with these new policies, and with the 2020 target almost met, there are still deep concerns regarding the overall sustainability of the policy. In 2017, the European Parliament voted to ban the use of palm oil in biofuels by 2020 (Nelson, 2017). In the words of one reporter, this ' . . . *stirred a diplomatic hornet's nest, with seven countries—including* *Indonesia, Malaysia, Costa Rica and Ecuador—warning of a trade dispute if the ban is acted upon'*. The vote was non-binding, needing further approval, but it shows how some decisions quickly become important internationally, as national economics become dependent on biofuel production.

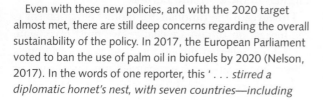

to include direct emissions (for example, during production and processing), but also other effects like land use change. We discussed the contribution of direct and indirect land use change to climate change in *Chapter 6: Climate Change*, and both could play an important role in increasing GHG emission from the biofuel sector.

Given that large amounts of carbon emissions are released during land use change, how can we make sure that biofuels are beneficial? We can use a concept called the carbon payback time. This is the time taken for a crop to 'earn back' the carbon released in the initial land use change (which can be thought of as a carbon debt). Studies estimate this payback time to be between 17 and 423 years, depending on the location and crop type (see Figure 7.13). That is, we will only start benefitting from biofuels after 17 and 423 years of that land being in production. We see this large variation because, as we discussed previously, different crops, such as corn or sugarcane, have different energy yields and different types of land, such as forests or grasslands, store different amounts of carbon in their natural vegetation before land use change.

Similar numbers have been found in many different studies and appear to be robust to various methodological approaches (Lapola et al., 2010; Searchinger et al., 2008). The only scenario where the payback time is less than 1 year is if the original land was marginal (not productive enough to farm) or abandoned cropland. The options for both these types of land are limited because, by definition, marginal land is hard to make a profit from and abandoned cropland is extremely rare.

Given we have an urgent, short-term need to reduce emissions and a longer-term requirement to live at zero or near-zero carbon emissions we need to be very careful when pursuing biofuels for the aim of mitigating climate change.

So far, the types of biofuel we have discussed are called '**first-generation biofuels**'. These are fuels which are made from the primary products of crops (and so compete with food production). An option which may hold more promise are **second-generation biofuels** that are made from waste, and third-generation biofuels which are made from aquatic

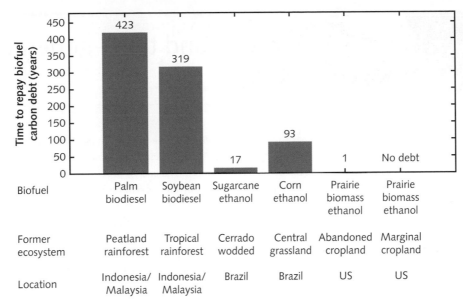

Figure 7.13 Carbon payback period for different land use changes, different biofuels, and different countries.

Source: adapted from Fargione et al., 2008.

plants such as seaweed (these types are discussed further in Box 7.5).

7.4.3 A future relying on bioenergy carbon capture and sequestration?

Finally, in this section, we look at an increasingly important future technology in the energy transition—**bioenergy carbon capture and sequestration (BECCS)**. If BECCS is used, it will directly compete with the food system on a massive scale. Traditional carbon capture and sequestration (CCS) is to remove CO_2 from the atmosphere by capturing it and burying it underground (see Figure 7.14). This has been discussed as an option for fossil fuel generation whereby CO_2 produced by power plants is captured and buried. There are now several demonstration CCS power plants around the world, but none are commercially competitive yet.

Years ago, it was probably possible that we could have reduced GHG emissions to meet the most popular 1.5° or 2° temperature targets using conventional renewable technologies or alternatives like CCS. However, progress has been so slow that most climate change models now require negative emission technologies in the second half of this century. A recent, hypothetical extension of this is to burn bioenergy

in CCS power plants, making them BECCS power plants. Plants would extract CO_2 from the atmosphere in growth, which would then be harvested and burnt to generate electricity; CO_2 would be captured and stored underground. The net result is CO_2 being removed from the atmosphere, and so bioenergy carbon capture is called a 'negative emission technology'.

The majority of influential climate change models used by the United Nations Intergovernmental Panel on Climate Change (IPCC) project the need for large amounts of BECCS power plants under almost all scenarios. In fact, the need is so great that, in the average scenario, high-income countries will have to start large-scale BECCS power plants by the end of 2030 and have to capture all global emissions by around 2055 (see Figure 7.15).

While bioenergy carbon capture may play a role in the future, we must be sceptical about this and consider the impacts critically. The natural and social impacts of the large-scale use of this technology are poorly understood. For example, we do not know what the large-scale use of this technology would do to food prices or land use change emissions. Additionally, these models assume that the cost of bioenergy carbon capture in future decades is less than the cost of deep mitigation, such as renewable energy on a huge scale, today.

BOX 7.5

Second- and third-generation biofuels

First-generation biofuels are produced from the easy-to-access primary products of crops. For example, fermenting corn kernels to produce ethanol and squeezing palm oil seeds to produce biodiesel. These parts of the plant are also used for food, and so first-generation biofuels compete with food production.

However, plants have many other parts that contain sugars (chemical energy) from which we can make fuel, for example the stalk, husks, and leaves. The problem is that these sugars are held in complex structures which are difficult to unlock. If future biotechnologies could speed up the unlocking process cheaply, we might be in a position where biofuel production does not directly compete with food production. The corn ears can be taken to the food market, and the remaining material can be fermented into fuels.

When we use the wastes from crops for biofuels, we term this 'second-generation biofuels' or 'cellulosic fuels' (see Figure A). The energy productivity is still limited because it is limited by the power density we have described previously, but it sidesteps many of the issues of land use change and market competition. That said, some wastes are used as animal feed, and so second-generation biofuels may still compete indirectly with food production. Technologies are under development to make second-generation biofuels more cost-effective, including new enzymes and genetically modified plants.

A third type of biofuel is from aquatic plants such as seaweeds and algae, and are referred to as third-generation biofuels. These technologies are at a very early stage but have the potential to avoid the problems associated with first- and second-generation biofuels, as they do not compete for land or food.

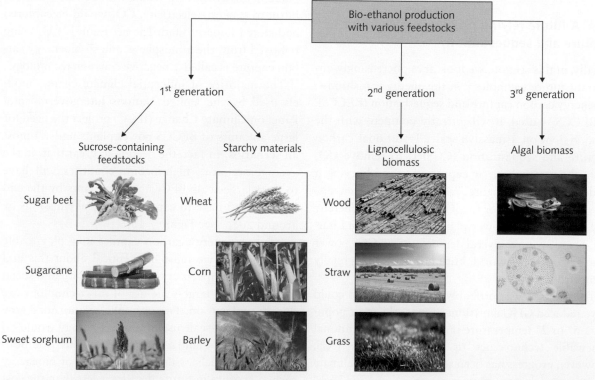

Figure A The difference between first-, second-, and third-generation biofuels.

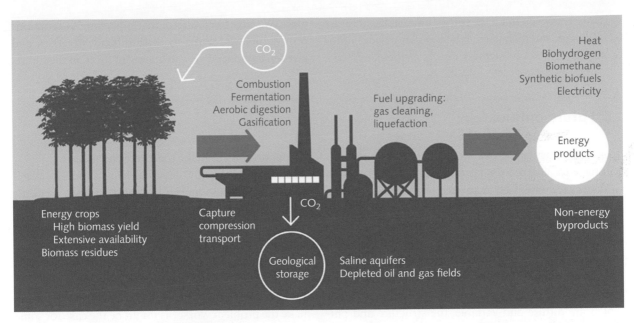

Figure 7.14 Bioenergy capture and sequestration.

Source: Sanchez, Daniel & H Nelson, James & Johnston, Josiah & Mileva, Ana & Kammen, Daniel. (2015). Biomass enables the transition to a carbon-negative power system across western North America. *Nature Climate Change*. 5, 230-234. Copyright © 2015, Springer Nature.

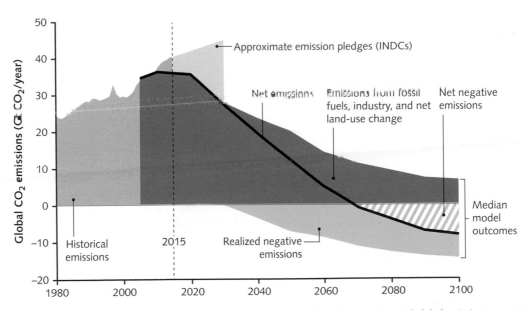

Figure 7.15 Projections for carbon emissions and BECCS through to 2100. On the y-axis are the total global emissions per year, and on the x-axis the year. The grey areas show the CO_2 already emitted, and the light grey areas show how much we plan to emit under the Paris climate accords. The net emissions show the total emissions over time, taking into account BECCS, which draws down carbon into the Earth (note that the y-axis extends into negative numbers).

Source: Anderson and Peters, 2016.

However, we do not have perfect knowledge of future technologies and their costs, much less knowledge of the economic and social settings in which these will take place. We have to be careful that modelling assumptions provide political cover to defer difficult decisions and give the impression we can 'make up for it' in the future.

7.5 **Future outlook**

To avoid catastrophic climate change, energy systems will have to undergo the fastest and deepest transition we have ever witnessed in any sector of human society. Based on historical data, such as the transition from coal to oil, it takes, on average, 50 years to complete an energy transition (see, for example, Smil, 2003). Given the urgency of climate change, we must now make a deeper and more fundamental transition in 15–30 years. As we will be relying on renewable energies which are diffuse and variable, the structure and philosophy of our energy systems need to change. In some areas, such as the electricity sector, there are promising signs of decarbonization in the form of solar and wind energy; in other sectors, such as transport and industry, the transition is more complex and will take longer.

The food sector, as a major user of energy worldwide, will also need to undergo this transition. This brings unique challenges due to the reliance on liquid fuels for heavy machinery. However, there are parts of the food production chain that are easier to decarbonize—the low-hanging fruit. Wherever there is a stationary processing point, such as washing or packaging, there is the opportunity to make the electricity used completely renewable. As this low-hanging fruit is picked, it will become harder and harder to remove carbon from the system. Further research on battery technology or perhaps fuel cells will need to be undertaken for the large machinery the food industry uses every day.

The food sector may provide part of the solution to combat climate change in the form of biofuels or BECCS. But we must be very careful that we do not make the problem worse, as might be the case for some forms of first-generation biofuels. Second- and third-generation biofuels hold great promise in smoothing the transition from fossil fuels by providing high-quality liquid fuel that can be used without much modification, that does not require additional land use change, and that does not need large amounts of energy inputs.

It is important to remember that many of the things discussed in this chapter are technological in nature. However, as we will explore in the next chapters of this textbook, it is clear that social changes will be as important in order to transition into a more sustainable food and energy system.

● QUESTIONS

7.1 Describe the way in which energy flows through trophic levels, and relate this to efficiency.

7.2 Americans use an equivalent of 6.7 tons of oil per year per person. What is this as average power through the year in watts (J/s)?

7.3 What are the three major services of energy in personal consumption?

7.4 Describe the concept of power density when referring to energy generation. Relate this to the production of biofuels, and explain why it is hard to rely solely on biofuels when transitioning towards renewable energy.

7.5 Why is land use change important when considering biofuels?

● FURTHER READING

McKay, D. 2008. *Sustainable Energy: Without the Hot Air*. Cambridge: UIT Cambridge Ltd. (**An excellent book with a new and fresh way of calculating energy issues. It has a focus on the UK but is also very relevant to other countries. It is available free online at: www.withouthotair.com**)

Smil, V. 2010. *Energy Myths and Realities: Bringing Science to the Policy Debate*. Washington, DC: American Enterprise Institute for Public Research. (**A book about the facts and fictions of the energy debate—will we run out of oil? Are biofuels the future? The prolific scientist and author Valclav Smil answers these questions and more.**)

Yergin, D. 2012. *The Quest, Energy, Security, and the Remaking of the Modern World*. New York, NY: Penguin. (**Pulitzer**

Prize-winning author Daniel Yergin tells the history of energy through the ages, with a special focus on modern times and

recent developments. Includes a lot of discussion about the geopolitics of energy.)

● REFERENCES

Anderson, K., Peters, G. 2016. The trouble with negative emissions. *Science* **354**, 182–183.

Beckman, J., Borchers, A., Jones, C.A. 2013. *Agriculture's Supply and Demand for Energy and Energy Products*. United States Department of Agriculture. Economic Research Service. Available at: https://www.ers.usda.gov/webdocs/publications/43756/37427_eib112.pdf?v=0 [accessed 4 November 2018].

Daly, H.E., Farley, J. 2010. *Ecological Economics: Principles and Applications*. Washington, DC: Island Press.

European Commission (EC). 2016. *Sustainability Criteria. Energy*. Available at: https://ec.europa.eu/energy/en/topics/renewable-energy/biofuels/sustainability-criteria [accessed 2 September 2018].

Eurostat. *Electricity Generation Statistics: First Results*. Available at: https://ec.europa.eu/eurostat/statistics-explained/index.php?title=Electricity_production_and_supply_statistics&oldid=59327 [accessed 8 November 2018].

FAOSTAT. 2014. Available at: http://www.fao.org/faostat/en/#home [accessed 4 November 2018].

Fargione, J., Hill, J., Tilman, D., Polasky, S., Hawthorne, P., 2008. Land clearing and the biofuel carbon debt. *Science* **319**, 1235–1238.

Harrabin, R. 2008. EU rethinks biofuels guidelines. *BBC News*. Available at: https://www.bbc.com/news [accessed 4 November 2018].

Intergovernmental Panel on Climate Change (IPCC). 2013. Summary for Policymakers 2013. In: Stocker, T.F., D. Qin, G.-K. Plattner, M. Tignor, S.K. Allen, J. Boschung, A. Nauels, Y. Xia, V. Bex and P.M. Midgley (eds). *Climate Change 2013: The Physical Science Basis. Contribution of Working Group I to the Fifth Assessment Report of the Intergovernmental Panel on Climate Change*. Cambridge and New York, NY: Cambridge University Press.

International Energy Agency (IEA). 2016. *Key World Energy Statistics 2016*. Paris: International Energy Agency.

Lapola, D.M., Schaldach, R., Alcamo, J., Bondeau, A., Koch, J., Koelking, C., Priess, J.A. 2010. Indirect land-use changes can overcome carbon savings from biofuels in Brazil. *Proceedings of the National Academy of Sciences of the United States of America* **107**, 3388–3393.

Mackay, D. 2009. *Sustainable Energy—Without the Hot Air*, 1st ed. Cambridge: UIT Cambridge Ltd.

Murphy, D.J., Hall, C.A.S., Powers, B. 2011. New perspectives on the energy return on (energy) investment (EROI) of corn ethanol. *Environment Development and Sustainability* **13**, 179–202.

Nelson, A. 2017. MEPs vote to ban the use of palm oil in biofuels. *The Guardian*. Available at: https://www.theguardian.com/sustainable-business/2017/apr/04/palm-oil-biofuels-meps-eu-transport-deforestation-zsl-greenpeace-golden-agri-resources-oxfam [accessed 4 November 2018].

Oxfam. 2016. *Burning Land, Burning the Climate*. Oxfam Briefing Paper. Available at: https://d1tn3vj7xz9fdh.cloudfront.net/s3fs-public/bp-burning-land-climate-eu-bioenergy-261016-en_0.pdf [accessed 4 November 2018].

Pelletier, N., Arsenault, N., Tyedmers, P. 2008. Scenario modeling potential eco-efficiency gains from a transition to organic agriculture: Life cycle perspectives on Canadian canola, corn, soy, and wheat production. *Journal of Environment Management* **42**, 989–1001.

Pelletier, N., Audsley, E., Brodt, S., Garnett, T., Henriksson, P., Kendall, A., Kramer, K.J., Murphy, D., Nemecek, T., Troell, M. 2011. Energy intensity of agriculture and food systems. *Annual Review of Environment and Resources* **36**, 223–46.

Peters, G.P., Davis, S.J., Andrew, R. 2012. A synthesis of carbon in international trade. *Biogeosciences* **9**, 3247–3276.

Pimentel, D., Williamson, S., Alexander, C.E., Gonzalez-Pagan, O., Kontak, C., Mulkey, S.E. 2008. Reducing energy inputs in the US food system. *Human Ecology* **36**, 459–471.

Searchinger, T., Heimlich, R., Houghton, R.A., Dong, F., Elobeid, A., Fabiosa, J., Tokgoz, S., Hayes, D., Yu, T.-H. 2008. Use of U.S. croplands for biofuels increases greenhouse gases through emissions from land-use change. *Science* **319**, 1238–40.

Simmons, I.G. 1993. *Environmental History: A Concise Introduction*. Oxford: Oxford University Press.

Smil, V. 2003. *Energy at the Crossroads: Global Perspectives and Uncertainties*. Cambridge, MA: The MIT Press.

Smil, V. 2010. *Power Density Primer: Understanding the Spatial Dimension of the Unfolding Transition to Renewable Electricity Generation (Part I—Definitions)*. Available at: http://vaclavsmil.com/wp-content/uploads/docs/smil-article-power-density-primer.pdf [accessed 4 November 2018].

Stewart, W.M., Dibb, D.W., Johnston, A.E., Smyth, T.J. 2005. The contribution of commercial fertilizer nutrients to food production. *Agronomy Journal* **97**, 1–6.

Valin, H (IIASA). 2016. *The Land Use Change Impact of Biofuels Consumed in the EU: Quantification of Area and Greenhouse Gas Impacts*. Report for the European Commission, Energy Directorate by the International Institute for Applied Systems Analysis (IIASA). Available at: https://ec.europa.eu/energy/sites/ener/files/documents/globiom_complimentary_2016_published.pdf [accessed 9 November 2018].

van Zalk, J., Behrens, P. 2018. The spatial extent of renewable and non-renewable power generation: a review and meta-analysis of power densities and their application in the U.S. *Energy Policy* **123**, 83–91.

Wang, M., Han, J., Dunn, J.B., Cai, H., Elgowainy, A. 2012. Well-to-wheels energy use and greenhouse gas emissions of ethanol from corn, sugarcane and cellulosic biomass for US use. *Environmental Research Letters* **7**, 45905.

Woods, J., Williams, A., Hughes, J.K., Black, M., Murphy, R. 2010. Energy and the food system. *Philosophical Transactions of the Royal Society B: Biological Sciences* **365**, 2991–3006.

Yergin, D. 2012. *The Quest*. New York, NY: Penguin.

PART 2

Food and Society

What food do people actually need to eat to survive? How do diets change as incomes increase? In light of the global obesity epidemic, should we tax sugar? What do mobile phones have to do with getting enough to eat? Do rich or poor countries waste more food? What is the public perception of organic food versus reality? And why do people say one thing, then do another?

In *Part 1*: *Food and Environment*, we explored the various ways in which food systems impact resources and the environment. Here in *Part 2: Food and Society*, we look at the way human societies themselves interact with food. We investigate the social and environmental sustainability of many of these interactions.

In *Chapter 8: Nutrition*, we lay out what people actually need to eat to survive and what happens when people consume too little of different parts of the diet and if they consume too much. We describe the intersection between nutrition and disease. Next, the way in which diets change with increasing affluence is described. These changes are commonly called the global nutrition transition, a key concept when considering future dietary changes. The environmental impacts of different types of food are outlined, and we discuss whether it is possible to eat both healthy and environmentally sustainable diets.

Chapter 9: Food Security describes the availability, accessibility, utilization, and stability of food supply. If there are problems in any of these components, food insecurity can happen, possibly resulting in the poor nutritional outcomes highlighted in the previous chapter. We describe major factors of food insecurity, from being able to afford food in the first place to making sure people can cook healthily. We describe the ways in which new technologies can help improve food security worldwide.

Later, in *Chapter 10: Food Aid*, we describe food aid as a key approach to improving food security. We describe the main ways in which aid is provided and the efficacy of these different methods. The major international efforts are discussed, and the different supply chains for food aid are outlined. Finally, we investigate the perverse outcomes food aid can have, from encouraging local violence to smothering local food production, prolonging food insecurity.

Chapter 11: Consumption focuses on the consumers of food—the people purchasing food or growing their own. It first defines sustainable consumption and provides an insight into how people actually make food choices. The chapter investigates the interaction of price sensitivity and society, providing a case study of the 2006–2008 food riots. The gap between what people say they want to do and what they actually end up doing is highlighted and explored. Methods for influencing food choices are outlined, from product labelling and advertising to policies developing a circular economy in consumption.

Image credit: Mike Goldwater/Alamy Stock Photo

8

Nutrition

How are diets linked to environmental impacts?

Jessica Kiefte-de Jong and
Paul Behrens

Chapter Overview

- Humans need two main types of nutrients to survive: macronutrients to supply bulk energy, and micronutrients for specific functions. Any difference between what is needed and what is eaten is called malnutrition. Malnutrition includes undernutrition, micronutrient deficiencies, and overnutrition (including obesity).

- Poor nutrition can result in communicable and non-communicable diseases, and, in reverse, these diseases can influence nutrient use and absorption.

- The global nutrition transition describes how diets evolve with developing lifestyles, typically as regions become richer and move out of poverty. This transition is characterized by an increased consumption of animal and processed foods, leading to a higher prevalence of non-communicable diseases (often related to obesity).

- Countries in the middle of the nutritional transition can experience a double burden of malnutrition, in which both obesity and undernutrition occur at the same time.

- As different food types place different stresses on the environment, the development of diets is interconnected with changes in environmental impacts.

- Alternative diets, such as vegetarianism and veganism, reduce environmental impacts substantially. Sometimes they offer health improvements, when compared to average diets.

- Food-based dietary guidelines are able to give advice to the public on how to improve dietary patterns but are not yet balanced against environmental impacts.

Introduction

A healthy lifestyle is something that most human beings desire (whether we manage to live one or not), and you are probably aware that nutrition plays a key role in preventing different diseases. In addition, the previous chapters have highlighted how food production impacts the environment. Is it possible

a)

b)

Figure 8.1 Are these two photos things you associate with a healthy lifestyle? Or are they stereotypical? What do you consider to be a healthy lifestyle, and are there multiple ways to be healthy?

Source: (left) "Mike" Michael L. Baird/Flickr (CC BY 2.0); (right) © United States Department of Agriculture.

to eat healthily, while minimizing your impacts on the environment? If so, how can this be promoted? These are some of the key questions addressed in this chapter.

Humans need a set of different nutrients to live healthily, which can only come from a varied diet with sufficient amounts of food (see Figure 8.1). Generally, as countries get richer, undernutrition and famine are replaced with urban-industrial lifestyles, characterized by increased food consumption associated with obesity-related diseases. This transition is associated with significant environmental impacts which are set to increase with increased global affluence. For this reason, there has been more emphasis on diets which balance social, environmental, health, and economic priorities.

In this chapter, we first outline the key components of diets, including **macronutrients** and micronutrients (Section 8.1). We then examine malnutrition, which happens when the consumption of nutrients does not match the body's requirements, and link malnutrition to health impacts and the prevalence of major diseases around the world (Section 8.2). Then we describe the way diets have changed around the world in response to the alleviation of poverty and the global increase in incomes (Section 8.3). We then investigate the impact of different food types on the environment, and how these may change in the future (Section 8.4). Finally, we take a special look at the environmental impact of different diets and outline a case study on nationally recommended diets, an important path for advising the public on healthy diets (Section 8.5).

8.1 You are what you eat—what are the key components of diets?

Humans need a diverse, balanced diet, with two main types of nutrient: (1) **macronutrients**, which are chemical compounds which humans require in high quantities, and (2) **micronutrients** that provide vitamins, minerals, and trace elements, which are needed in smaller amounts for specific functions. Humans do need other vital substances to be healthy, such as water and dietary fibre, but these do not have specific nutritional value. The key components of diets appear on our food labels, which can act as a guide to healthy eating (see Figure 8.2). Given the importance of macro- and micronutrients, we will discuss these in detail below.

Nutrition Facts

Serving Size 2/3 cup (55g)
Servings Per Container About 8

Amount Per Serving

Calories 230	Calories from Fat 40

	% Daily Value*
Total Fat 8g	**12%**
Saturated Fat 1g	**5%**
Trans Fat 0g	
Cholesterol 0mg	**0%**
Sodium 160mg	**7%**
Total Carbohydrate 37g	**12%**
Dietary Fiber 4g	**16%**
Sugars 1g	
Protein 3g	

Vitamin A	10%
Vitamin C	8%
Calcium	20%
Iron	45%

* Percent Daily Values are based on a 2,000 calorie diet. Your daily value may be higher or lower depending on your calorie needs.

	Calories:	2,000	2,500
Total Fat	Less than	65g	80g
Sat Fat	Less than	20g	25g
Cholesterol	Less than	300mg	300mg
Sodium	Less than	2,400mg	2,400mg
Total Carbohydrate		300g	375g
Dietary Fiber		25g	30g

Figure 8.2 An example food label from the United States Department of Agriculture. First, the label lists different types of macronutrients, then underneath the micronutrients such as vitamins. Food labels also provide advice on recommended daily consumption (at the bottom of the label)

8.1.1 Macronutrients

The three main classes of macronutrients are proteins, carbohydrates, and fats, and they each play different roles in the diet.

Proteins: protein in our diet is broken down during digestion into basic building blocks—amino acids. We use the amino acids from our diet to make new proteins, which are the work horses in our body. Proteins function as hormones (for example, insulin) and enzymes and have a role in our immune system. In addition, they are important building blocks for many different parts of our body, including muscles, bones, teeth, and hair.

Animal products (for example, dairy, cheese, eggs, meat, fish, and seafood) are often high in protein content. Some plants also produce relatively high levels of protein, for example legumes (for example, beans, peas, lentils) and can be good alternatives for meat (see Figure 8.3). There are a total of 20 different amino acids; some of these can be made by our body, but we receive others from our diet (these are called **essential amino acids**). Generally, plant-based protein sources have a lower content of essential amino acids. Therefore, individuals on a plant-based diet need to eat a higher quantity of vegetable protein, compared to individuals who have animal products included in their diet (Millward et al., 2008).

Carbohydrates: carbohydrates are the most abundant source of energy in the human diet, contributing 40–80% of all energy requirements. The unit of energy used in nutrition is the kilocalorie (kcal) (this is different to other fields of study; see *Chapter 7: Energy*). The required average daily intake for humans is between 2000 and 2200 kcal/day (for reference, an apple contains around 52 kcal, and a hamburger 563 kcal). Carbohydrates supply around 4 kcal/g, whereas fat supplies 9 kcal/g (Gibney, 2009). Carbohydrates are made from sugar molecules, with different types of carbohydrate such as simple carbohydrates (for example, monosaccharides such as glucose and fructose) and complex carbohydrates (for example, polysaccharides such as starch and glycogen), defined by the number of different sugar molecules in a chain. In our body, nearly all carbohydrates are transformed into monosaccharides during digestion.

The source of carbohydrates matters from a public health perspective, but this is linked to the additional nutritional value provided by **micronutrients**. For example, it is better to eat fruit and vegetables for carbohydrates, rather than candy and sweets, because the latter have no micronutrient value. Food sources rich in complex carbohydrates, such as grains and legumes, often also include dietary fibre.

Fats: fats are important for energy storage and cell membranes. As with carbohydrates, there are

a)

b)

Figure 8.3 Proteins can be obtained from plant and animal sources. (a) A selection of legumes, including kidney beans, chick peas, and green peas. (b) Animal proteins are found in meats, cheese, eggs, and fish.

different types of fat, grouped into saturated and unsaturated fats. Important sources of saturated fats are animal products (for example, meat and dairy), as well as processed foods. Unsaturated fats are divided into trans-fatty acids (found in margarine and processed food), monounsaturated fats (found in red meat, whole milk products, and some oils), and polyunsaturated fats (found in some oils, fish, nuts, and seeds).

Each of these have different dietary and health impacts. The health impacts of saturated fats are mixed, with early studies finding a relationship with cardiovascular disease, but recent studies failing to confirm this (de Souza et al., 2015), because it depends on how saturated fat is substituted in the diet. Trans-fatty acids have been linked to an increase in the risk of **cardiovascular disease**, and several countries have placed legal limits on them (de Souza et al., 2015). Mono- and polyunsaturated fats are thought to have beneficial health effects, for example by lowering the risk of cardiovascular disease when saturated fats are replaced by polyunsaturated fats.

8.1.2 **Micronutrients**

There are many different types of micronutrients, but those most important from a public health perspective are vitamins A and D, zinc, iron, and iodine.

Table 8.1 gives an overview of these major micronutrients, their function, and the public health and environmental challenges of their supply.

In most cases, our diets provide sufficient amounts of these micronutrients, but when they do not, there is the potential for developing diseases. For example, iodine is essential for thyroid functioning and brain development (Gibney, 2009). Iodine is naturally present in some marine fish and plants. Some smaller quantities are found in the air and soil, but the amount depends on how much iodine is deposited from the oceans to the atmosphere and terrestrial systems (Smyth et al., 2011). As a result, iodine deficiency can occur in diets low on fish and seafood.

Iodine deficiency was fairly common around the world, so many high-income countries began **fortification** programmes, whereby important nutrients (in which people might be deficient) are added to staples (common foods), in this case adding iodine to staple foods such as bread and salt. These programmes have been hugely successful, eradicating iodine deficiency in most high-income countries (Charlton and Skeaff, 2011). In countries without these programmes, adverse health effects related to iodine deficiency are still frequently observed, including goitre, swelling of the thyroid gland (see Figure 8.4).

Table 8.1 Important micronutrients linked to public health and environmental challenges

Micronutrient, and source	Description and function	Challenges
Vitamin A Retinol: liver, fish, dairy, and eggs Carotenoids: spinach, carrots, cabbage, and egg yolk	Two compounds: (1) retinoids, needed for pigment in eyes and crucial for vision (2) carotenoids, used in the immune system and as antioxidants	Public health: vitamin A deficiency is the leading cause of blindness and mortality due to infectious diseases in children and pregnant women in low-income countries Environment: higher levels in animal foods, which are limited in low-income countries which rely instead on plant-based food sources (which have lower bioavailability). The availability and variation of these plant-based food sources are threatened by reduced biodiversity
Vitamin D Fish and mushrooms Sometimes artificially fortified in milk and margarine	Two forms: (1) formed in the skin under sunlight (vitamin D2); (2) formed in some plants (vitamin D3). Both have similar functions in bone health	Public health: deficiency is highly prevalent in Western and higher-latitude countries. Deficiency is also associated with mortality from non-communicable diseases (Chowdhury et al., 2014) Environmental: sun-induced vitamin D synthesis is influenced by season, latitude, altitude, as well as air pollution, which is one reason that deficiency is common in Western countries (Voipio et al., 2015)
Zinc Grains, meat, liver, cheese, eggs, and fish	Used in the skeleton, liver, skin, and muscles. Important role in enzymes, protein, growth and tissue regeneration, and in the immune system	Public health: deficiency is prevalent in low- and middle-income countries and characterized by diarrhoea, skin problems, impaired cognitive development, as well as impaired growth of children Environmental: low-income countries rely on zinc from grains which have limited bioavailability. Zinc content of foods also depends on soil quality which can be poor in several countries, as well as preparation methods
Iron Heme iron: animal products, meat, eggs, and fish Non-heme iron: green vegetables	Mainly located in red blood cells. Important for oxygen transport in the body. Heme iron has higher bioavailability than non-heme iron	Public health: deficiency leads to anaemia, associated with impaired cognitive development and growth in children. Can result in complications in pregnancy. Iron deficiency is further exacerbated by gastrointestinal diseases such as diarrhoea and malaria Environmental: heme iron is only found in animal foods and is limited in several low-income countries, resulting in a reliance on plant-based iron with lower bioavailability. The availability and variation of these sources are challenged by reduced biodiversity
Iodine Fish and seafood Note: middle- and high-income countries fortify salt and bread with iodine	Used in thyroid hormone, and deficiency can cause thyroid malfunction. Thyroid hormones are important for the growth and development of children	Public health: iodine deficiency is one of the most common causes of intellectual disability in children. Iodine deficiency is mainly common in countries without fortification policies Environmental: iodine concentration depends on the place where the food is grown. Food grown near the coast has higher iodine content due to water and fog from the sea, whereas lower contents are found inland (Smyth et al., 2011)

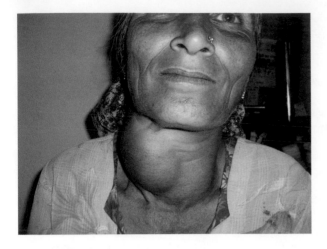

Figure 8.4 Goitre, as seen here in a woman in India, is a medical condition directly linked to iodine deficiency.

Source: Dr. J.S.Bhandari, India (CC BY-SA 3.0).

8.2 How is malnutrition defined, and how does it impact our health?

Any imbalance between dietary intake and **physiological** demands (demands of the body and its systems) is called malnutrition. Given the different types of nutrients described in Section 8.1, it can be quite complex to exactly define and measure malnutrition since it takes many different forms. Officially, the World Health Organization defines malnutrition as:

'The cellular imbalance between the supply of nutrients and energy and the body's demand for them to ensure growth, maintenance, and specific functions.'

(De Onis et al., 1993)

This definition encompasses malnutrition in the obvious case of macro- and micronutrient deficiencies, but it also includes overnutrition and obesity (Box 8.1 describes the main ways we measure and quantify malnutrition at the individual and population levels). The most extreme effect of macronutrient deficiencies is famine and death due to starvation. However, malnutrition can also cause a variety of other effects, some of which can result in an increased risk of diseases and reduced life expectancy.

Diseases are categorized as either communicable or non-communicable. **Communicable diseases** are passed from person to person through a specific infectious agent, for example a virus or bacteria. Some diseases can impact the health of infected individuals very rapidly (for example, the Ebola virus), while others take a much longer time to show adverse effects [for example, the human immunodeficiency virus (HIV)]. In contrast, **non-communicable diseases** do not spread from human to human and often have a long duration and slow progression.

8.2.1 Nutrition and communicable diseases

The relationship between malnutrition and communicable (infectious) diseases is often bidirectional. On the one hand, malnutrition can increase the risk of developing infectious diseases by lowering the body's immune function, while on the other, infectious diseases can also increase the risk of some types of malnutrition. For example, research has shown that zinc deficiency can increase the risk of diarrhoea. In turn, diarrhoea can result in the loss of important nutrients, including zinc, due to impaired intestinal absorption (Schaible and Kaufmann, 2007). This, in turn, can impact further vulnerability for other infectious diseases, leading to a vicious cycle (see Figure 8.5). Globally, there are five important infectious diseases for which a complex interrelationship with malnutrition is well described: HIV, malaria, tuberculosis (TB), measles, and diarrhoea. These diseases are outlined in depth in Table 8.2.

BOX 8.1

How can you measure nutritional status?

There are two main ways to evaluate **nutritional status**: on the individual level or the population level. On the individual level, we generally investigate body composition (for example, height and weight, fat-free mass), dietary intake, or biomarkers (for example, blood tests for nutrients). The best-known test for body composition is the body mass index (BMI), given by the equation:

$$BMI = weight\ [kg]/height^2[m^2]$$

BMI is closely related to body fat percentage and is used widely in adult populations. A threshold of above 30 kg/m² is defined as adult obesity (WHO, 1995). There is an ongoing debate whether these BMI thresholds are applicable to all populations, because different ethnic groups have different body fat distribution. For example, Asian people have a higher risk of specific diseases with a BMI of above 22 kg/m² (for example, type 2 diabetes and cardiovascular disease). For this reason, more stringent thresholds are sometimes used (Barba et al., 2004).

For children, alternative measures on the basis of weight and height are used to define malnutrition. This includes their weight relative to their height, and their height relative to their age. For example, the presence of **stunting** (low height for age) and **wasting** (low weight for height) can be detected using simple height and weight measures (see Figure A). Other ways to measure individual levels of malnutrition are skinfold thickness and mid-arm circumference (see Figure B).

On a population level, direct measurements on individuals are not practical. As an alternative, average food intake is used to estimate malnutrition. A common method in national analyses is the use of food balance sheets. Food balance sheets give an approximation of food intake on the basis of the balance of food production, import, and export (Shim et al., 2014). However, these data do not take into account consumer waste, and, as a result, they can sometimes overestimate the actual nutritional intake (Del Gobbo et al., 2015). Other measurements of diets can be made using food records, 24-hour recall food frequency questionnaires (in this case, individuals are asked to list all the food they used for the last 24 hours or what they eat on a regular basis in the past month) (Shim et al., 2014). As with food balance sheets, results need to be interpreted with caution, as there can be issues with poor recollection, recording, or reporting by individuals.

Child stunting
Low height for age

Child wasting
Low weight for height

Child overweight
High weight for height

Adult overweight
Carrying excess body fat with a body mass index ≥ 25

Figure A An overview of child stunting and wasting, as well as children and adults classified as overweight.

Source: Global Nutrition Report 2016.

Figure B Measuring skinfold thickness to assess body composition.

Source: Authors' own photo.

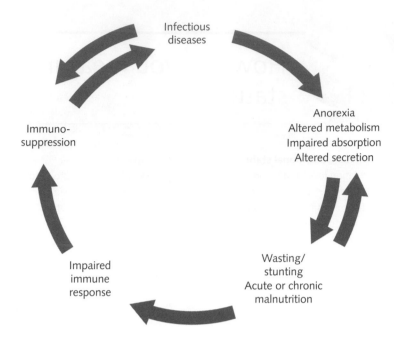

Figure 8.5 The vicious cycle of malnutrition and infectious diseases.
Source: Authors' own work.

8.2.2 Nutrition and non-communicable diseases

Compared to communicable diseases, non-communicable diseases are chronic (persisting and long-lasting). They often have a long period before symptoms show and so are often seen in ageing populations (WHO, 2014). The most common non-communicable diseases globally are cardiovascular disease (for example, heart attacks and stroke), cancer, type 2 diabetes, and chronic respiratory diseases (Lozano et al., 2012). Although risk factors for these diseases vary from country to country, the most important are poor diets, smoking, high blood pressure, and being over- or underweight (Forouzanfar and Zhu, 2016; Lim et al., 2012).

8.3 How do diets develop with increased affluence? The global nutrition transition

As affluence has increased over the last decades, health challenges related to diets have shifted. Diets with insufficient nutritional value are still an important problem in poor regions, leading to diet-related infectious diseases, but as incomes increase, there has been a marked shift towards diet-related non-communicable diseases such as cancer and cardiovascular disease (through obesity). As a result, by some estimations, there are now more people suffering from overnutrition than undernutrition (Lozano et al., 2012).

 Connect the dots

Think about the social factors behind malnutrition on a local scale. How might advertising, food availability, and gender impact what you eat? We will discuss these factors in *Chapter 9: Food Security* and *Chapter 11: Consumption*.

Historically, food policies have focused on supplying sufficient food amounts and food variety to

Table 8.2 Important nutrition-dependent communicable diseases

Disease	Characteristics	Interaction with nutrition
HIV	A severe viral infection compromising the immune system, leading to the development of infections and acquired immune deficiency syndrome (AIDS). HIV/AIDS is a pandemic in sub-Saharan Africa where almost two-thirds of patients live	Nutrient deficiencies are often seen in patients and can be due to poor availability of foods. The disease itself can affect appetite and metabolism (Piwoz and Prebble, 2000). The benefit of food supplementation is debated, but vitamin A and zinc are linked to reduced mortality
Malaria	A parasitic infection transmitted to humans through mosquito bites. Children and pregnant women are particularly vulnerable, since children do not have immunity and women lose acquired immunity during pregnancy. Mainly found in Sub-Saharan Africa and countries near the equator (WHO, 2015)	A common symptom of malaria is red blood cell (RBC) deficiency (anaemia/low iron levels). Iron supplementation is often provided in some areas. But iron also seems to be required for malaria parasite reproduction, so iron deficiency may actually offer some protection against the disease. Iron supplements in children with normal RBC levels may increase the risk (Neuberger et al., 2016). Hence, wide-scale iron supplementation in risk areas is recommended, only when regular malaria surveillance and treatment services are in place
TB	A bacterial infectious disease that mainly affects the lungs. It often occurs in people with an impaired immune system such as HIV or in malnourished individuals	Immune response after vaccination can be impaired due to malnutrition but can be reversed with protein supplements (Hoang et al., 2015). Vitamin D deficiency is extremely common, with around 90% of TB patients having vitamin D deficiency, due to reduced sun exposure, as well as skin pigmentation and use of HIV/TB therapies (Keflie et al., 2015). Vitamin D supplements may correct deficiencies but do not seem to influence the progression or treatment effect
Measles	A highly contagious airborne viral infection. Although it has been markedly reduced by immunization programmes, it remains one of the leading causes of deaths among young children in low- and middle-income countries	Likely to occur in individuals with vitamin A deficiency. Vitamin A supplementation in children with measles can reduce mortality by approximately 80% and reduce the risk of other infections (Jensen et al., 2015). For this reason, many vaccination programmes are often combined with vitamin A supplementation (WHO, 2014)
Diarrhoea	Results from disturbed absorption and secretion in the gut due to bacteria, viruses, toxins, or parasites. A main cause of child mortality in low-income countries and often related to poor hygiene and sanitation. The WHO estimates that the burden of food-borne illnesses is comparable to HIV/AIDS, malaria, and TB combined. Common in Africa, South East Asia, and the Eastern Mediterranean region (Havelaar et al., 2015)	Caused by consumption of unsafe/contaminated foods or drinking water with pathogens. Food-borne illnesses have a higher impact on people in poor health, infants, and pregnant women (Havelaar et al., 2015). Deaths due to diarrhoea are mainly due to extreme fluid loss. The main treatment of diarrhoea is rehydration with oral salts. Zinc supplementation can also be beneficial in preventing diarrhoea (Liberato et al., 2015)

prevent undernutrition and micronutrient deficiencies (see *Chapter 9: Food Security* for more details). As a result, the availability and consumption of food have increased globally. Unfortunately, the consumption of 'unhealthy' foods (for example, processed food and sugary drinks) has increased to a greater extent than the consumption of 'healthy' foods (for example, fruits and vegetables), particularly in low- and middle-income countries (Del Gobbo et al., 2015). We can see this by looking at the trends in consumption in countries over time. Figure 8.6 shows that the intake of meat protein, empty calories

(for example, confectionary, sugary drinks), and dietary calories per person has increased in all regions from 1961 to 2009. This trend has a major impact on public health and the environment and is termed the **nutrition transition**. The nutrition transition is characterized by increased food availability, but not necessarily **healthier foods**. It also heavily depends on the demographic and **epidemiological** transition of the past few decades, as well as the process of the agricultural revolution (see Figure 8.7).

Although food intake has increased, minimizing the risk of undernutrition, there are still major concerns on

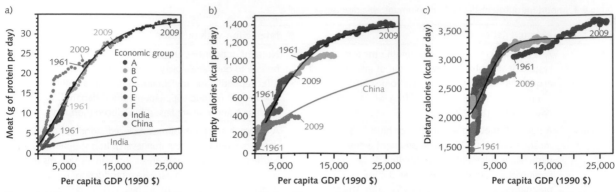

Figure 8.6 Development of diets in different groups over time with respect to income. Economic groups are ranked from 'A' (highest income) to 'F' (lowest income), with two additional categories for India and China. The figure shows the relationship of (a) meat intake, (b) empty intake, and (c) energy intake with income.

Source: Tilman, D. and M. Clark, Global diets link environmental sustainability and human health., *Nature,* vol. 515, no. 7528, pp. 518–522. Copyright © 2014, Springer Nature.

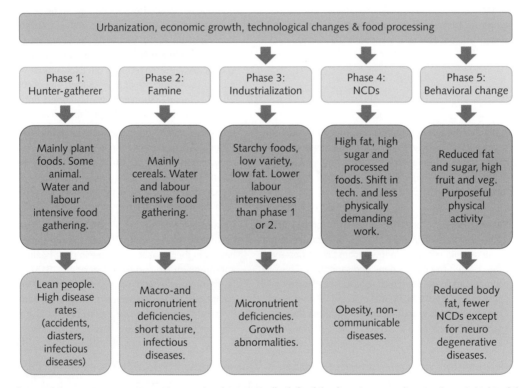

Figure 8.7 Phases of the nutrition transition. Phase 1 develops vertically (left of the figure), proceeding to phase 5 (right of the figure). NCD, non-communicable diseases.

Source: Barry M Popkin; Global nutrition dynamics: the world is shifting rapidly toward a diet linked with noncommunicable diseases, The American Journal of Clinical Nutrition, Volume 84, Issue 2, 1 August 2006, Pages 289–298. Copyright © 2006, Oxford University Press.

a country level about the presence of socio-economic inequalities in dietary quality. In many countries, the overall dietary patterns of low-income groups tend to be less healthy, with a lower consumption of whole grains, fruits, and vegetables, and a high consumption of fats, soft drinks, processed meats, and refined sugar. Refined sugars are one of the main components of **empty calories,** products which supply little or no nutritional value, and excess consumption is closely related to the obesity epidemic. This has led government efforts to encourage a lower consumption of unhealthy foods, using taxes on fat- and sugar-rich foods that are currently being implemented in several countries (see Case study 8.1) (Thow et al., 2016).

CASE STUDY 8.1

Should we tax sugar?

Taxes can be designed to incentivize people to reduce the consumption of certain goods, for example cigarettes and alcohol. Often these types of taxes are applied to goods which have 'external' costs, that is, the costs of the things we consume are borne by society, not by the individual—think of the national health care needed for a smoker who has lung cancer (see *Chapter 14: Collective Action* for more on external costs). With rising obesity resulting in increased health care costs, many governments have turned to sugar taxes as a way to reduce the economic pressures.

Sugary drinks can significantly contribute to high-caloric intake. To reduce intake, sugar taxes are already prevalent in many countries and often target sugary drinks, as they are a relatively straightforward product to tax. In Mexico, the estimated impact of the tax is a 12% drop in purchases. In 2018, the UK government began a sugar tax on soft drinks whereby manufacturers have to pay a levy, based on the amount of sugar added. The UK has a severe obesity problem, and the average person consumes over twice the recommended amount of sugar across all age ranges (see Figure A). In response, several manufacturers have already reduced the sugar content in their drinks.

However, it has been argued that the effects of taxes on fats or sugar on health outcomes may be small. In addition, these taxes may even penalize poorer families, since they typically consume more sugar than higher-income families. For this reason, some see this as a **regressive**, rather than **progressive**, step (one that increases inequality). As an alternative, a fiscally neutral (net zero change in money going to the government) policy which has a fat/sugar tax on one side but subsidizes for fruit and vegetables may be more progressive (Thow et al., 2014; Tiffin and Salois, 2012). In the UK, the sugar tax was made more palatable by using the revenue for early childhood dietary education.

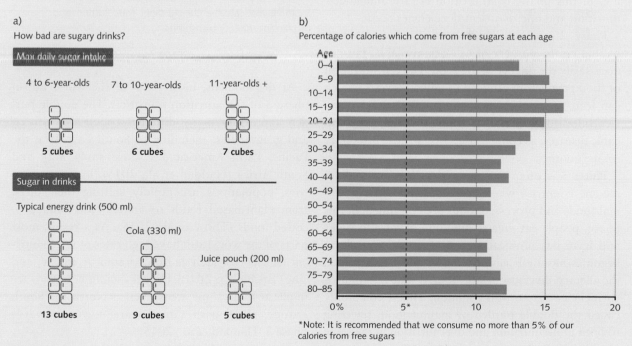

Figure A (a) Sugary drinks contain large amounts of sugar. Just one bottle of a typical energy drink contains up to 13 cubes of sugar, and a cola bottle 9 cubes. (b) People in the UK consume far more sugar than advised, across all ages.

Source: (a) Public Health England: Sugar Reduction: The evidence for action; (b) Source: Public Health England: National Diet and Nutrition Survey, 2008-2014.

In your experience

Ask someone of an older generation what the most common food they used to eat was as they grew up, and the size of the portions they consumed (for example, what was a normal-sized steak for them?). How is this different to what you eat now? What might have been possible malnutrition effects from what they ate? How did this change as they got older? Did they continue with the same dietary habits, or did they change?

Focusing further on Figure 8.7, each phase is an important step in the nutrition transition. In any categorization of diets globally, there will always be large generalizations, but this transition is a useful description of the overall trends. Going through from phase 1 to phase 5, we have:

Phase 1 is characterized by consumption of plant-based food sources and some animal foods. This requires high levels of physical activity for hunting, when compared to planting, harvesting, and processing. In this phase, overnutrition is rare, but undernutrition and micronutrient deficiencies occur.

Phase 2 societies begin centralized agriculture, increasing the number of calories available, but not necessarily the quality of the food. The centralization of food production can lead to failed harvests, which can lead to severe famine and nutrient deficiencies. Centralization increases the prevalence of communicable diseases, as communities live in close proximity to one another.

Phase 3 is often encountered in low- to middle-income countries where increased amounts of animal foods and processed foods are consumed. In this phase, people eat greater amounts of fat and sugar and have less physical activity, because agriculture becomes more efficient. At the same time, however, the dietary diversity is still limited and micronutrient deficiencies are still common. This is commonly called the **double burden of malnutrition**, referring to both undernutrition and overnutrition in the same

country (Popkin, 2006; Popkin et al., 2012). Different forms of malnutrition can even occur in the same person, for example someone who is obese and yet still has an iron deficiency.

Phase 4 shows further reduced physical activity due to technological changes. On top of this, diets are characterized by even higher intake of fats and sugars. Countries in this phase have a high prevalence of obesity and obesity-related non-communicable diseases.

Phase 5 is characterized by an increased consumption of healthier foods such as whole grains, fruit, and vegetables, while consumption of saturated fats reduces. In this final phase, physical activity is increased for recreation and populations move to a healthier balance of food intake and activity (Popkin et al., 2012).

Pause and think

Do you think it is possible to 'skip' the nutrition transition and move from phase 3 to phase 5 directly? How would culture influence this? Choose one country in phase 3 (for example, Guatemala, China, or India), and investigate the current diets, food culture, and plausible dietary transitions.

At the moment, most parts of the world are in phase 3 of the nutrition transition. The double burden appears in many different countries, cultures, and geographies, including in South East Asia, the Pacific, Latin America (see Case study 8.2), and South Africa (Haddad et al., 2015). In these countries, populations are rapidly shifting their diets, from plant-based foods to animal-based and processed foods (Thow et al., 2016). As a result, most parts of the world still have high rates of micronutrient deficiencies such as iron, vitamin A, iodine, and zinc (Tulchinsky, 2010). Higher-income countries are in phase 4, but some areas, such as North and South Europe, along with Canada, are moving towards phase 5 (Haddad et al., 2015).

CASE STUDY 8.2

The double burden of malnutrition: an example in Guatemala

The effects of both undernutrition and overnutrition are not only seen across countries as a whole, but also on a local, household scale. In an example from Latin America, families can have stunted (micronutrient-deficient) children and an overweight/obese mother living under the same roof (see Figure A) (Oddo et al., 2012). In South East Asia and the Pacific, nearly half of the population sees stunting in children under the age of 5 years, while half of the adult women are overweight (Haddad et al., 2015).

Urbanization, reduced food prices, and slow improvements in water and sanitation systems seem to explain part of the phenomenon (Haddad et al., 2015). Urbanization and reduced food prices make animal-based and high-fat foods more available and accessible, but infectious diseases and diarrhoea due to poor sanitation lead to micronutrient deficiencies and undernutrition, predominantly in children.

In Guatemala, the food transition has resulted in a sequence of adverse health effects on the population. Up to the mid-1970s, VAD used to be one of the main forms of malnutrition in Guatemala. Vitamin A is very important for a proper immune response (Jensen et al., 2015). As a result of VAD, the country had high child mortality rates due to infectious diseases. In response, an effective fortification programme was implemented in the mid-1970s. In this programme, table sugar was fortified with vitamin A (Pineda, 1998).

During the past decade, Guatemala experienced a major nutrition transition, with newer diets characterized by high intakes of sugar and fat-rich foods, while low in micronutrients (Makkes et al., 2011).

A survey among households showed that more than a third of infants have stunting (low growth), while almost half of mothers are overweight or obese. In 17% of families, both stunting and obesity were found (Doak et al., 2016). Guatemala now has one of the highest rates of obesity in Latin America, and non-communicable diseases, such as type 2 diabetes and cardiovascular failure, contribute more to mortality rates than infectious diseases (WHO, 2014). The increase in obesity can be directly linked to the changed diet.

The direct causes of childhood undernutrition have been linked to high microbial exposure and environmental contamination, as well as the early introduction of sweetened herbal drinks (Agüatas) in the first year of life. This combination increases the chance of diarrhoea and contribute to high levels of communicable diseases found in Guatemalan children. The Government of Guatemala and the World Food Programme have embraced policies to improve nutrition which focuses on maternal and child nutrition, and in particular appropriate breastfeeding practices and fortified foods for young children, to combat stunting. There are also efforts to improve childhood education in schools (see Figure B). Nonetheless, limited attention is currently paid to the double burden (Ramirez-Zea et al., 2014).

Figure A A Guatemalan child is measured, as part of a national campaign against malnutrition.

Source: Rodrigo Abd/AP/Shutterstock.

Figure B A field worker discusses the benefits of good nutrition at an urban primary school in Guatemala.

Source: (Photo: CIIPEC) https://www.iaea.org/newscenter/news/iaea-impact-guatemala-works-control-double-burden-malnutrition.

8.4 Can we eat our way to environmental sustainability?

As discussed in the first half of this book, the global food system has many different impacts on the environment, including on water (for example, pollution and eutrophication; see *Chapter 3: Pollution*), on land use (for example, erosion; see *Chapter 5: Soils*), and on the atmosphere (see *Chapter 6: Climate Change*). Importantly, not all food types have the same environmental impact. For example, animal-based products generally impact the environment more, compared to plant-based products.

Consequently, the diets we consume and the food choices we make influence the environmental impacts of food systems. In this section, we will first look at the quantitative environmental impacts of different types of food; then we will discuss the environmental impacts of the global nutrition transition. Finally, we will examine how nutritional and environmental outcomes could be balanced in the future.

8.4.1 Which foods have the highest impact?

How do you assess the impact of your diet on the environment? Do you focus on biodiversity loss? Soil erosion? The use of freshwater resources? Or the emission of GHGs? Importantly, although there

are differences in the impact of diets among these environmental parameters, they generally tend to have similar patterns across food groups. Here we first focus on the impact of food on the emission of GHGs, one of the biggest risks facing the environment at the moment. Using an LCA approach, we can estimate the relative impact of different types of food on GHG emissions (see *Chapter 6: Climate Change* for details). The GHGs emitted by the production of different food types (for example, per serving or per calorie) vary by almost 100-fold, from maize to ruminant meat such as cattle (see Figure 8.8).

As we saw in *Chapter 2: Biodiversity*, with every step in a food chain, energy is lost. Eating animal-based products is generally a more inefficient way to produce energy for human consumption, compared to plant-based products. Among these, cattle have the highest impacts, as they are slaughtered later than other animals and release large amounts of CH_4 during the process of digestion, a potent GHG (further dynamics of this are described in *Chapter 6: Climate Change*). Fish are also associated with large GHG impacts due to high energy needs, for example for operating trawlers, recirculation systems on fish farms, and intensive refrigeration (see Figure 8.9).

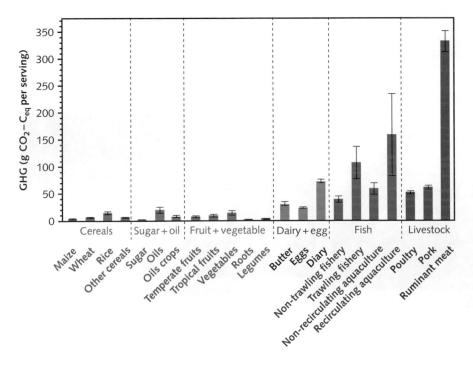

Figure 8.8 The different impact of food types on GHG emissions (in grams of CO_2 equivalent per serving). Plant-based foods are on the left of the figure, and animal-based foods on the right.

Source: Adapted from Tilman, D, and M. Clark. 'Global diets link environmental sustainability and human health', *Nature*, vol. 515, no. 7528, pp. 518–522, 2014.© 2014, Springer Nature.

a)

b)

Figure 8.9 (a) A fishing trawler. (b) Fish refrigerated on a supermarket shelf. All of these activities need large amounts of energy, either in the form of oil (in ships) or electricity (pumps and refrigeration).

Source: (a) John Nuttall / Flickr (CC BY 2.0); (b) Ania Mendrek/Flickr (CC BY-ND 2.0)

There are some specific, currently small-scale approaches for rearing animals which would result in much lower impacts, for example rearing pigs using waste products (which would not require the production of additional animal feed). Importantly, there are examples of fruits and vegetables having very large GHG impacts if they are air-freighted, rather than transported by ship. Fortunately, the amount of food flown is still reasonably low, making up only 1% of food in the UK; however, this has grown in recent years (food miles are covered in full in *Chapter 7: Energy*).

8.4.2 How does the global nutrition transition impact the environment?

At the moment, the majority of countries and people in the world are undergoing a nutritional transition from phase 3 to phase 4 (see Box 8.2). This transition is typified by increased animal-based proteins and higher calorie intake. Since increases in consumption and an increase in the relative intake of animal-based products both increase environmental pressures, we can safely assume environmental impacts will increase in the near future (assuming no fundamental changes in the diets of individuals worldwide). If diets do transition along historically observed paths, then we will see an approximately 25% increase in diet-related GHG emissions, along with increases in land use for food production (see

Figure 8.11). However, alternatives such as **Mediterranean, pescatarian**, and **vegetarian** diets may reduce impacts substantially, with the highest reductions among these options associated with vegetarianism (some alternative diets are outlined below in Box 8.3).

The different phases of the food transition give important information about the environmental impact of diets. Diets in high-income countries, such as the USA (in phase 4 of the transition), have impacts three times higher than lower-income countries such as Indonesia (in phase 3 of the transition) (see Figure 8.11). Meat, fish, and dairy products contribute the most to GHG emissions in high-income countries (>60%). In low-income countries, these food groups contribute to around a third of GHG emissions. Other adverse environmental effects, such as eutrophication and agricultural land use, show similar trends across the income status of countries, with high-income countries impacting the environment more, compared to low- and middle-income countries (see Figure 8.10).

Vegans all round then?

The message is clear then—we should all go **vegan**, and as fast as possible! And if we cannot go vegan, we should at least all go vegetarian. Well not so fast. While the general, highly stylized picture is that many animal products are harmful to the environment,

BOX 8.2

The nutrition transition in Brazil, China, and India

The impact of the nutrition transition on environmental impacts is clear to see in countries such as Brazil, China, and India, which have undergone large socio-economic changes in recent decades. Since Brazil, China, and India represent around 38% of the world population, the food transitions in these countries alone have a huge global impact, and they have very different culinary cultures, as all countries do (see Figure A). Here we focus on the environmental impact of two key components of their diets which have changed due to the food transition: meat consumption and the use of cereals and vegetable oils.

Brazil and China have seen a marked increase in the consumption of meat, growing at a faster rate than in European countries (Tilman and Clark, 2014). This is not seen in India, as a large portion of the population are vegetarian (see Figure 8.6). In addition, in all three countries, cereal and vegetable oil consumption has increased enormously. This increase in vegetable oil consumption mainly drives land use changes (the conversion of natural habitat to agricultural lands), whereas the increase in meat production in Brazil and China causes significant increases in water resources and GHG emissions (Gill et al., 2015).

Increased demand for plant-based foods in countries with climate extremes can also impact the environment, since these foods are often threatened by increased temperatures and precipitation, and are thus subject to increased losses (Reichstein et al., 2013). As a result, the second food group with the highest water footprint (following livestock) are fruit and vegetables in Brazil, China, and India (Gill et al., 2015).

a)   b)   c)

Figure A (a) Traditional Chinese (sweet and sour pork), (b) Indian (dry dal), and (c) Brazilian (churrasco) meals. The existing food cultures will influence the way the nutrition transition develops in different countries.

Source: (a) Alpha/Flickr (CC BY-SA 2.0); (b) Shabnam Shamsuddin/Wikimedia Commons (CC BY-SA; 4.0) (c) L.Miguel Bugallo Sánchez/Wikimedia Commons (CC BY-SA 3.0)

there are at least three major issues with simplistically applying this:

1. Specific product types: the detailed picture is far more complex than the generalized issues mentioned so far. The specific impacts of a diet depend on the specific food items bought. We have to understand the different environmental impacts of the production, transportation, and consumption modes for each individual food product. For example,

Californian almonds made into almond milk may have lower GHG emissions than dairy but require huge amounts of freshwater. In this case, other alternatives, such as oat milk, may be better to minimize environmental impact, but further assessments would have to be done for each specific product (see Figure 8.12).

2. Nutritional outcomes for everyone: even if alternative food types are available (next time you

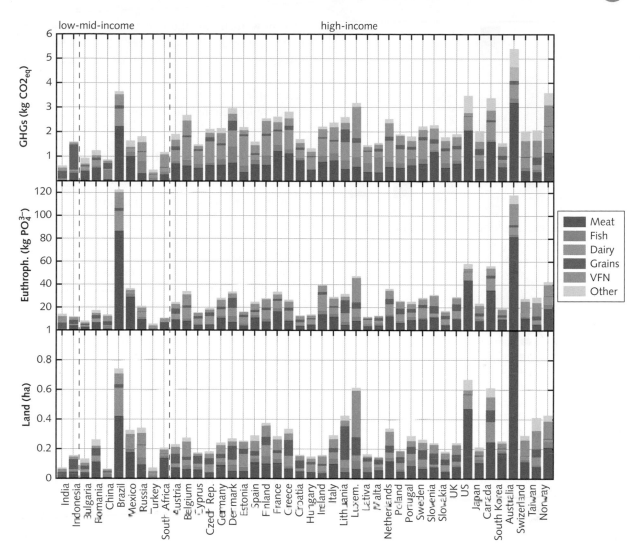

Figure 8.10 Impacts of average diets worldwide on three aspects of the environment: climate change, eutrophication, and land use. Low-, middle-, and high-income countries are shown from left to right. The contribution of different food types to impacts are shown in different colours. These amounts include effects from all consumption, including embodied impacts in trade.

Source: Behrens et al., 2017. Evaluating the environmental impacts of dietary recommendations. *Proceedings of the National Academy of Sciences of the United States of America* 114, 13412–13417.

go shopping, check whether almond milk or oat milk are even available in your local shop), these decisions have very different nutritional outcomes. Environmentally friendly diets may have negative nutritional impacts in some regions. For example, in low-income countries without access to a variety of high-protein vegetable substitutes, foregoing animal protein on the basis of the environment could result in serious health problems.

3. The value of food cultures: finally, and importantly, diets are not just some technical dial that

we can turn and change on a whim. Even if specific products were beneficial for the environment and health, if they are available at a reasonable price locally, and if people had all the nutritional information needed to stay healthy while making a shift, the way we eat is deeply connected with our culture, identity, and taste. These factors are crucial. We make personal and social trade-offs with environmental impacts all the time; consider flights taken to spend time with family, while increasing carbon footprints. Food is no different to this and should not be seen differently.

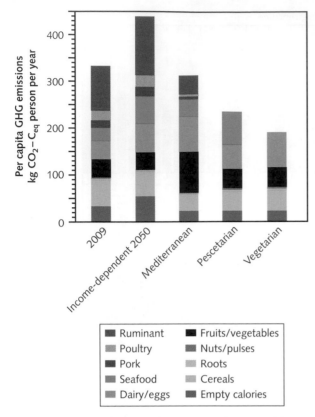

Ruminant | Fruits/vegetables
Poultry | Nuts/pulses
Pork | Roots
Seafood | Cereals
Dairy/eggs | Empty calories

Figure 8.11 The development of environmental impacts of diets by 2050 with the assumption that people will eat in the future much like they do today, compared to alternative diets. Average global per capita emissions per year are shown per food type and diet type.

Source: Tilman, D, and M. Clark, 'Global diets link environmental sustainability and human health, *Nature*, vol. 515, no. 7528, pp. 518–522. Copyright © 2014, Springer Nature.

We now focus on the first two points (the third point is covered fully in *Chapter 11: Consumption*). How could we balance nutritional and sustainability outcomes for a so-called sustainable diet? And if we are able to formulate such a balanced diet, how would we let people know about it?

8.4.3 Can we balance nutrition and sustainability?

Balancing nutritional and sustainability requirements is a new frontier in public health, bridging two quite different disciplines to provide solutions which optimize both factors.

Solutions will often have to be specific to countries and local problems and may vary widely from country to country. For example, high-income countries may benefit from reduced environmental and health impacts by eating fewer calories and animal foods. As discussed in Box 8.3, an important option is to switch to a plant-based diet or to increase the amount of plant-based meals (Perignon et al., 2016). However, these options are often not appropriate in lower-income countries. Meat and animal products provide essential nutrients, such as protein, iron, and vitamin A (see Table 8.1), and need to be replaced with plant- or insect-based alternatives with similar nutritional value. In low-income countries, these alternatives are often not available and switching to a vegetarian diet would result in an increase in cereal consumption, resulting

a) b) c)

Figure 8.12 (a) Austrian cow's milk, (b) US almond milk, and (c) Swedish oat milk. Which milk would you choose, and why? Have you tried any of these before? What are the differences in flavour?

Source: (a) Tiia Monto/Wikimedia Commons (CC BY-SA 3.0); (b) Veganbaking.net from USA [CC BY-SA 2.0]; (c) counterculturecoffee/Flickr (CC BY-NC-ND 2.0).

BOX 8.3

Alternative diets

There are several well-established alternative diets. Here are a selection of a few popular diets.

Vegetarianism

One environmentally and health-friendly option is vegetarianism, which has, for a long time, been associated with ethical and religious choices, rather than environmental or health concerns. Since vegetables contain different types and amounts of nutrients than meats, a small number of micronutrients need to be supplemented in the vegetarian diet with a good variety of foods. The three main micronutrients in a vegetarian diet which need attention are fatty acids, vitamin B, and calcium. These needs can be met with an increase in the consumption of pulses such as soy, nuts such as walnuts, or other animal products such as eggs or dairy.

Veganism

Veganism extends the vegetarian philosophy to all animal-derived products (see Figure A), including eggs and dairy products. As well as making the changes needed for a healthy vegetarian diet, there is one nutritional requirement which is not made by plants at all—vitamin B12. This is usually supplemented in cereals and other food products or is available from certain types of mushroom.

Entomophagy (insects as food)

Not a diet in itself, but the practice of consuming insects as food is widespread in many areas of the world, with 80% of countries eating 1000 different types of insect. Insects can provide high levels of protein, without the environmental impacts of animal proteins. This is because insects convert energy from primary producers more efficiently than most other animals. There are several companies which are attempting to increase the consumption of insects in western markets; they generally focus on novel foods, such as mealworm lollipops, in order to provoke interest (see Figure B).

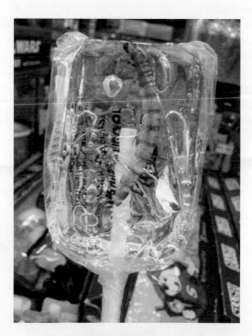

Figure A A vegan meal of asparagus, rice, tomatoes, avocado, and salad.

Ula Zarosa/Flickr (CC BY 2.0).

Figure B A mealworm in a lollipop.

Source: Wikimedia Commons.

in malnutrition (cereals lack certain essential nutrients) and a backward step in the nutrition transition.

In your experience

Would you consider changing your diet? If you would, why would you? What sort of factors would you consider, for example health, environmental, or ethical? What would be the major reasons for not changing?

In general, healthier diets (compared to the average diet) also have environmental and economic co-benefits. For example, a global transition to plant-based diets could reduce mortality by 6–10% and food-related GHG emissions by 29–70%, and provide economic benefits of 0.4–13% of the GDP (Springmann et al., 2016). However, these benefits assume large changes in consumption around the world. One way these benefits could be harnessed is by better communication about healthy diets to the public (see Food controversy 8.1).

FOOD CONTROVERSY 8.1

How can nationally recommended diets be a political question?

One key avenue for informing dietary choices are national recommended diets (NRDs). NRDs are often issued by national health or nutrition organizations (see Figure A). Due to different health challenges, cultures, and institutional settings around the world, countries have different dietary recommendations, which vary in environmental impact (Behrens et al., 2017). Presently, NRDs focus generally on health impacts and do not take environmental impacts into account. To date, only Sweden makes any serious attempt to include environmental sustainability in their NRD. This is

beginning to change with more mentions of sustainability in NRDs over the past few years, including in the Chinese, UK, and Dutch guidelines, although it should be noted that mentions of sustainability are fairly limited in the case of the Chinese and UK guidelines.

The idea of including environmental and sustainability factors in dietary guidelines is nothing new and goes back to at least 1986 (Gussow and Clancy, 1986). However, efforts to include these impacts in some countries have been met with political resistance. In the USA, nutrition experts responsible for updating the NRD decided to collect information on the environmental impacts of diets. Shortly after this was announced, the US Congress issued a directive (non-binding instruction) expressing concern that the advisory committee '*is showing an interest in incorporating agriculture production practises and environmental factors*' (Charles, 2014). In the directive, Congress asked the Obama administration to disregard the environment in the guidelines.

Now, why did the US Congress ask for the Obama administration to disregard the environment? As with other jurisdictions, many stakeholders are involved in decisions regarding food policy, leading to fierce political debates. Part of the reason Congress directed the advisory group to avoid accounting for environmental impacts may have been the fact that many congressmen and women represent regions in which many jobs are dependent on cattle and other animals (Hamblin, 2015). On its own, the beef industry is worth $95 billion to the USA, and in 2014, the meat industry spent $10.8 million on political campaigns (Shanker, 2015).

Figure A Government advice like this aims to simplify the message by showing people the proportion of foods recommended according to nutritional advice. This example is from the US Department of Agriculture.

Source: https://www.choosemyplate.gov.

As a result, progress on dietary guidelines has been intermittent for a long time. For example, in 1991, *The Washington Post* reported that pressure from meat and dairy industries forced the Agriculture Department to overturn a decision to de-emphasize meat and dairy products in their food pyramid (see Figure B) (Sugarman and Gladwell, 1991). Progress has been made; however, the new guidelines shown in Figure A have vegetables and fruits making up half of the plate and now include only dairy as an animal product, with the source of proteins left up to the consumer.

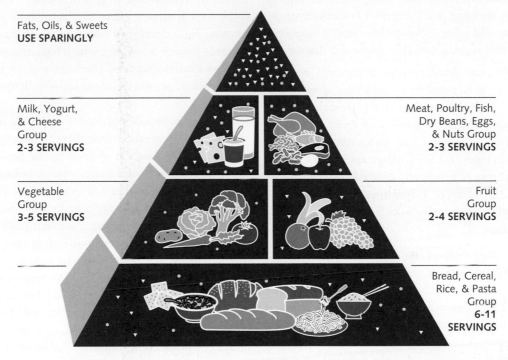

Fats, Oils, & Sweets
USE SPARINGLY

Milk, Yogurt, & Cheese Group
2-3 SERVINGS

Meat, Poultry, Fish, Dry Beans, Eggs, & Nuts Group
2-3 SERVINGS

Vegetable Group
3-5 SERVINGS

Fruit Group
2-4 SERVINGS

Bread, Cereal, Rice, & Pasta Group
6-11 SERVINGS

Figure B Suggested 1991 US dietary pyramid which de-emphasized animal products.
Source: US Department of Agriculture.

8.5 Future outlook

Like so many natural systems, the interaction between diets and the environment will evolve dynamically in very unpredictable ways. Aside from the existing pressures we have covered at length in this book, we also know that low- and middle-income countries following the dietary path trodden by high-income countries (phase 3 to phase 4 of the nutrition transition) would put tremendous pressure on the environment in the future. In the meantime, current high-income countries will likely continue to be responsible for the majority of food-related environmental pressures in the form of high-impact diets.

There are two main ways in which these issues may be abated:

1. Diet developments (social): there is the possibility that countries in phase 4 of the nutrition transition will move swiftly onto phase 5, reducing impacts and improving health. Meanwhile, lower-income countries could 'skip' phase 4 of the transition, moving directly from phase 3 to phase 5. This may come down to personal and social choices and would reflect a cultural change in the attitudes to food across the world, as well

as the availability and affordability of sustainable foods, as described in *Chapter 9: Food Security*. However, there is also an opportunity for governments and other institutions to actively steer in this direction, for example by optimizing NRDs, using taxes and subsidies and using other techniques such as nudging (see *Chapter 11: Consumption*). This would also allay public health concerns about diet-related non-communicable diseases.

2. Technological developments (technical): we may find that it is possible to continue eating traditionally high-impact diets if we are able to technologically substitute foods with more sustainable alternatives; consider the cultured/printed meat shown in Figure 6.13 in *Chapter 6: Climate Change*, certain types of genetically modified organisms (see Box 3.7 in *Chapter 3: Pollution*), insects as a protein replacement, and growing algae-based food in the oceans.

Given the magnitude of our sustainability problems, we would need both of these developments to happen relatively quickly. There have been indications of reasonably rapid growth in the uptake of alternative diets in some countries; at the same time, the costs of food alternatives continue to plummet. Many people could pay the lower price for the cultured meat if it were indistinguishable from farmed meat.

These are also not completely independent of one another, as society and technology interact in complex and unpredictable ways. Simplistically, and hypothetically, as global environmental awareness is raised, perhaps through lobbying or crisis, it may become anti-social to eat farmed meat, resulting in further enhancement of social changes and the stimulation of technological substitutions. This would result in a feedback effect similar to how views on smoking have developed over time (think of attitudes towards passive smoking), which has, in part, driven technological substitutes (such as e-cigarettes) and further social pressure (indoor smoking bans in many countries).

With these broad statements in mind, it is also important to remember that there is no single solution to the combination of good public health nutrition and environmental sustainability. Solutions will have to be developed in context and specific to the regions in which problems occur. In low- and middle-income countries, increasing the nutritional value of food, instead of simply recommending energy-dense foods, will be key. This might be achieved by promoting the production of a wide variety of nutrient-rich plants or by the industrial enrichment of food. Solutions in high-income countries will be tied to the capability of countries to reduce animal product intake without compromising nutrition. The good news is that in cases where food variety and quantity are available, it is perfectly possible to consume alternative diets with substantially lower environmental impacts, while also improving health outcomes.

QUESTIONS

8.1 How would you define and assess malnutrition?

8.2 Which micronutrients are most important globally? Are these the same for low- and high-income countries?

8.3 How can the environment affect the development of malnutrition? And vice versa?

8.4 Describe the concept of the double burden of malnutrition, and find another example of this in a country other than Guatemala.

8.5 In many Western countries, nutrition policies focus on the prevention of obesity. Do you agree with this approach?

FURTHER READING

Birn, A. 2017. *Textbook of Global Health*, 4th ed. Oxford: Oxford University Press. **(This book provides a basic understanding of the most critical issues in the field of global public health. It includes an interdisciplinary discussion on health inequality and the relationship with development, human rights, nutrition, and the environment.)**

Mason, P., Lang, T. 2017. *Sustainable Diets: How Ecological Nutrition can Transform Consumption and the Food System*, 1st ed. New York, NY: Routledge. **(This book describes the mismatch of humans and the planet and gives a definition of sustainable diets and what needs to be done to implement these by giving equal weight to nutrition,** public health, the environment, socio-cultural issues, and economics.)

Ronald, C., Adamchak, R. 2018. *Tomorrow's Table*, 2nd ed. Oxford: Oxford University Press **(This book alternates views from a geneticist and an organic farmer. They investigate what the future of agriculture and nutrition may look like.)**

Willet, W. 2012. *Nutritional Epidemiology*, 3rd ed. Oxford: Oxford University Press. **(This provides insight into how to conduct and interpret studies of diet and health. It covers the methodology involved in assessing the nutritional status of diets.)**

● REFERENCES

Barba, C., Cavalli-Sforza, T., Cutter, J., Darnton-Hill, I., Deurenberg, P., Deurenberg-Yap, M., Gill, T., James, P., Ko, G., Nishida, C. 2004. Appropriate body-mass index for Asian populations and its implications for policy and intervention strategies. *The Lancet* **363**, 157–163.

Behrens, P., Jong, J.K., Bosker, T., Koning, A. De, Rodrigues, J.F.D., Tukker, A. 2017. Evaluating the environmental impacts of dietary recommendations. *Proceedings of the National Academy of Sciences of the United States of America* **114**, 13412–13417.

Charles, D. 2014. Congress to nutritionists: don't talk about the environment. *National Public Radio*. Available at: https://www.npr.org/sections/thesalt/2014/12/15/370427441/congress-to-nutritionists-dont-talk-about-the-environment [accessed 4 November 2018].

Charlton, K., Skeaff, S. 2011. Iodine fortification: why, when, what, how, and who? *Current Opinion in Clinical Nutrition and Metabolic Care* **14**, 618–624.

Chowdhury, R., Kunutsor, S., Vitezova, A., Oliver-Williams, C., Chowdhury, S., Kiefte-De-Jong, J.C., Khan, H., Baena, C.P., Prabhakaran, D., Hoshen, M.B., Feldman, B.S., Pan, A., Johnson, L., Crowe, F., Hu, F.B., Franco, O.H. 2014. Vitamin D and risk of cause specific death: systematic review and meta-analysis of observational cohort and randomised intervention studies. *BMJ* **348**, g1903.

De Onis, M., Monteiro, C., Akre, J., Clugston, G. 1993. The worldwide magnitude of protein-energy malnutrition: an overview from the WHO global database on child growth. *Bulletin of the World Health Organization* **71**, 703–712.

Del Gobbo, L.C., Khatibzadeh, S., Imamura, F., Micha, R., Shi, P., Smith, M., Myers, S.S., Mozaffarian, D. 2015. Assessing global dietary habits: a comparison of national estimates from the FAO and the Global Dietary Database. *American Journal of Clinical Nutrition* **101**, 1038–1046.

de Souza, R.J., Mente, A., Maroleanu, A., Cozma, A.I., Ha, V., Kishibe, T., Uleryk, E., Budylowski, P., Schünemann, H., Beyene, J., Anand, S.S. 2015. Intake of saturated and trans unsaturated fatty acids and risk of all cause mortality, cardiovascular disease, and type 2 diabetes: systematic review and meta-analysis of observational studies. *BMJ* **351**, h3978.

Doak, C.M., Campos Ponce, M., Vossenaar, M., Solomons, N.W. 2016. The stunted child with an overweight mother as a growing public health concern in resource-poor environments: A case study from Guatemala. *Annals of Human Biology* **43**, 122–130.

Forouzanfar, M., Alexander, L., Anderson, H.R., et al. (2015). Global, regional, and national comparative risk assessment of 79 behavioural, environmental and occupational, and metabolic risks or clusters of risks in 188 countries, 1990-2013: a systematic analysis for the Global Burden of Disease Study 2013. *The Lancet* **386**, 2287–2323.

Forouzanfar, M.H, Zhu, J. 2016. Global, regional, and national comparative risk assessment of 79 behavioural, environmental and occupational, and metabolic risks or clusters of risks, 1990–2015: a systematic analysis for the Global Burden of Disease Study 2015. *The Lancet* **388**, 1659–1724.

Gibney, M.J. 2009. *Introduction to Human Nutrition*. Oxford: Wiley-Blackwell.

Gill, M., Feliciano, D., Macdiarmid, J., Smith, P. 2015. The environmental impact of nutrition transition in three case study countries. *Food Security* **7**, 493–504.

Gussow, J.D., Clancy, K.L. 1986. Dietary guidelines for sustainability. *Journal of Nutrition Education* **18**, 1–5.

Haddad, L., Cameron, L., Barnett, I. 2015. The double burden of malnutrition in SE Asia and the Pacific: priorities, policies and politics. *Health Policy Plan* **30**, 1193–1206.

Hamblin, J. 2015. How agriculture controls nutrition guidelines. *The Atlantic*. Available at: https://www.theatlantic.com/health/archive/2015/10/ag-v-nutrition/409390/ [accessed 9 November 2018].

Havelaar, A.H., Kirk, M.D., Torgerson, P.R., et al. 2015. World Health Organization global estimates and regional comparisons of the burden of foodborne disease in 2010. *PLoS Medicine* **12**, e1001923.

Hoang, T., Agger, E.M., Cassidy, J.P., Christensen, J.P., Andersen, P. 2015. Protein energy malnutrition during vaccination has limited influence on vaccine efficacy but abolishes immunity if administered during *Mycobacterium tuberculosis* infection. *Infection & Immunity* **83**, 2118–2126.

Jensen, K.J., Fisker, A.B., Andersen, A., Sartono, E., Yazdanbakhsh, M., Aaby, P., Erikstrup, C., Benn, C.S. 2015. The effects of Vitamin A supplementation with measles vaccine on leucocyte counts and *in vitro* cytokine production. *British Journal of Nutrition* **115**, 619–628.

Keflie, T.S., Nölle, N., Lambert, C., Nohr, D., Biesalski, H.K. 2015. Vitamin D deficiencies among tuberculosis patients in Africa: a systematic review. *Nutrition* **31**, 1204–1212.

Liberato, S.C., Singh, G., Mulholland, K. 2015. Zinc supplementation in young children: a review of the literature focusing on diarrhoea prevention and treatment. *Clinical Nutrition* **34**, 181–188.

Lim, S.S., Vos, T., Flaxman, A.D., et al. 2012. A comparative risk assessment of burden of disease and injury attributable to 67 risk factors and risk factor clusters in 21 regions, 1990–2010: a systematic analysis for the Global Burden of Disease Study 2010. *The Lancet* **380**, 2224–2260.

Lozano, R., Naghavi, M., Foreman, K., et al. 2012. Global and regional mortality from 235 causes of death for 20 age groups in 1990 and 2010: a systematic analysis for the Global Burden of Disease Study 2010. *The Lancet* **380**, 2095–2128.

Makkes, S., Montenegro-Bethancourt, G., Groeneveld, I.F., Doak, C.M., Solomons, N.W. 2011. Beverage consumption and anthropometric outcomes among schoolchildren in Guatemala. *Maternal & Child Nutrition* **7**, 410–420.

Millward, D.J., Jackson, A.A. 2004. Protein/energy ratios of current diets in developed and developing countries compared with a safe protein/energy ratio: implications for recommended protein and amino acid intakes. *Public Health Nutrition* **7**, 387–405.

Neuberger, A., Okebe, J., Yahav, D., Paul, M. 2016. Oral iron supplements for children in malaria-endemic areas. *Cochrane Database of Systematic Reviews* **2**, CD006589.

Oddo, V.M., Rah, J.H., Semba, R.D., Sun, K., Akhter, N., Sari, M., De Pee, S., Moench-Pfanner, R., Bloem, M., Kraemer, K. 2012. Predictors of maternal and child double burden of malnutrition in rural Indonesia and Bangladesh. *American Journal of Clinical Nutrition* **95**, 951–958.

Perignon, M., Masset, G., Ferrari, G., Barré, T., Vieux, F., Maillot, M., Amiot, M.-J., Darmon, N. 2016. How low can dietary greenhouse gas emissions be reduced without impairing nutritional adequacy, affordability and acceptability of the diet? A modelling study to guide sustainable food choices. *Public Health Nutrition* **19**, 2662–2674.

Pineda, O. 1998. Fortification of sugar with vitamin A. *Food and Nutrition Bulletin* **19**, 131–136.

Piwoz, E.G., Prebble, E.A. 2000. *HIV/AIDS and Nutrition: A Review of the Literature and Recommendations for Nutritional Care and Support in Sub-Saharan Africa*. Washington, DC: United States Agency for International Development.

Popkin, B.M. 2006. Global nutrition dynamics: the world is shifting rapidly toward a diet linked with noncommunicable diseases. *American Journal of Clinical Nutrition* **84**, 289–298.

Popkin, B.M., Adair, L.S., Ng, S.W. 2012. Global nutrition transition and the pandemic of obesity in developing countries. *Nutrition Reviews* **70**, 3–21.

Ramirez-Zea, M., Kroker-Lobos, M.F., Close-Fernandez, R., Kanter, R. 2014. The double burden of malnutrition in indigenous and nonindigenous Guatemalan populations. *American Journal of Clinical Nutrition* **100**, 1644S–1651S.

Reichstein, M., Bahn, M., Ciais, P., Frank, D., Mahecha, M.D., Seneviratne, S.I., Zscheischler, J., Beer, C., Buchmann, N., Frank, D.C., Papale, D., Rammig, A., Smith, P., Thonicke, K., Van Der Velde, M., Vicca, S., Walz, A., Wattenbach, M. 2013. Climate extremes and the carbon cycle. *Nature* **500**, 287–295.

Schaible, U.E., Kaufmann, S.H.E. 2007. Malnutrition and infection: complex mechanisms and global impacts. *PLoS Medicine* **4**, e115.

Shanker, D. 2015. The US meat industry's wildly successful 40-year crusade to keep its hold on the American diet. *Quartz*. Available at: https://qz.com/523255/the-us-meat-industrys-wildly-successful-40-year-crusade-to-keep-its-hold-on-the-american-diet/ [accessed 4 November 2018].

Shim, J. S., Oh, K., Kim, H.C. 2014. Dietary assessment methods in epidemiologic studies. *Epidemiology and Health* **36**, e2014009.

Smyth, P.P., Burns, R., Huang, R.J., Hoffman, T., Mullan, K., Graham, U., Seitz, K., Platt, U., O'Dowd, C. 2011. Does iodine gas released from seaweed contribute to dietary iodine intake? *Environmental Geochemistry and Health* **33**, 389–397.

Springmann, M., Godfray, H.C.J., Rayner, M., Scarborough, P. 2016. Analysis and valuation of the health and climate change cobenefits of dietary change. *Proceedings of the National Academy of Sciences of the United States* **113**, 1–6.

Sugarman, C., Gladwell, M. 1991. U.S. drops new food chart. *Washington Post*, 27 April 1991: A1, A10.

Thow, A.M., Downs, S., Jan, S. 2014. A systematic review of the effectiveness of food taxes and subsidies to improve diets: understanding the recent evidence. *Nutrition Reviews* **72**, 551–565.

Thow, A.M., Fanzo, J., Negin, J. 2016. A systematic review of the effect of remittances on diet and nutrition. *Food and Nutrition Bulletin* **37**, 42–64.

Tiffin, R., Salois, M. 2012. Inequalities in diet and nutrition. *Proceedings of the Nutrition Society* **71**, 105–111.

Tilman, D., Clark, M. 2014. Global diets link environmental sustainability and human health. *Nature* **515**, 518–522.

Tulchinsky, T.H. 2010. Micronutrient deficiency conditions: global health issues. *Public Health Reviews* **32**, 12.

Voipio, A.J.W., Pahkala, K.A., Viikari, J.S.A., Mikkilä, V., Magnussen, C.G., Hutri-Kähönen, N., Kähönen, M., Lehtimäki, T., Männistö, S., Loo, B.-M., Jula, A., Marniemi, J., Juonala, M., Raitakari, O.T. 2015. Determinants of serum 25(OH)D concentration in young and middle-aged adults. The Cardiovascular Risk in Young Finns Study. *Annals in Medicine* **47**, 253–262.

World Health Organization (WHO). 1995. *Physical status: the use and interpretation of anthropometry*. Report of a WHO Expert Committee. World Health Organization Technical Report Series No. 854. Available at: http://www.who.int/childgrowth/publications/physical_status/en/ [accessed 4 November 2018].

World Health Organization (WHO). 2004. Appropriate body-mass index for Asian populations and its implications for policy and intervention strategies. *The Lancet* **363**, 157–163.

World Health Organization. 2014. *Global Status Report on Noncommunicable Diseases 2014*. Geneva: World Health Organization.

World Health Organization. 2015. *World Malaria Report 2015*. Available at: http://www.who.int/malaria/publications/world-malaria-report-2015/report/en/ [accessed 4 November 2018].

9

Food Security

Meredith T. Niles and Molly E. Brown

Chapter Overview

- There are four pillars of food security: availability, access, utilization, and stability.

- We currently produce enough food to feed the global population, but this will change in the future, depending on diet, technologies, and climate change.

- Despite the global availability of food, many countries and regions do not produce enough food to feed their populations and use trade to help meet their food needs.

- Many people are food-insecure because they lack economic access to food or even the resources to buy food.

- Economic and climatic shocks can affect the stability of food availability and reduce economic access to food for vulnerable populations.

- Food security also includes utilization, which encompasses the preparation, cooking, and how we utilize food to provide adequate nutrition, which is critical for healthy, active lives, especially important in the first 1000 days of life.

Introduction

In the twenty-first century, for the first time in history, getting enough food is not something many people in high-income countries have to worry about. In fact, a study in 2016 estimated that more people are currently obese than underweight (NCD, 2016). However, while many people in high-income countries can quickly and easily obtain food, often from a grocery store (see Figure 9.1a), this is not universal around the world. In some regions, obtaining food can be much more complex and challenging, especially in low- and middle-income countries (see Figure 9.1b).

This chapter will discuss concepts of food security and its connection to sustainability. We look at where food is produced, by whom it is produced, and how we then move and use this food around the planet. First, we outline the definition of food security, as this is the core concept to whether a group of people are able to eat or not. We then introduce the four pillars of food security: availability, access, utilization, and stability.

a)

b)

Figure 9.1 Contrasting levels of food security. (a) A grocery store in a high-income country carrying a wide variety of fresh produce. (b) Protesters in Venezuela in 2017 protesting against food insecurity (the sign reads: *I am protesting against [food] scarcity. Where can I get these items?*)

Source (b) María Alejandra Mora (SoyMAM)/Wikimedia Commons (CC BY-SA 3.0)

9.1 **What do we mean by food security?**

The concept of **food security** is a relatively recent one and has been built upon, and expanded through, several iterations (see Box 9.1). It is more than simply having enough food, although that is part of it. The Food and Agricultural Organization of the United Nations (FAO) gives the most widely recognized definition of food security as: '*When all people, at all times, have physical, social, and economic access to sufficient, safe, and nutritious food to meet dietary needs for a productive and healthy life*' (FAO, 2001). This combines earlier economic and physical food security with newer concepts, including social factors. More recently, there have been discussions to also include concepts of **culturally appropriate food** and having sufficient micronutrients to support cognitive and physical development (see *Chapter 8: Nutrition* for a discussion on micronutrient needs). The evolution of this definition reflects both the growing scientific understanding of nutritional needs, as well as an increased recognition that cultural values and **food cultures** (the culture surrounding food) are important for many societies.

Currently, there are significant levels of food insecurity worldwide. The most recent *Hunger Report* from the United Nations FAO indicated that 815 million people are undernourished and food-insecure (FAO, 2017). These people are most likely to reside in low-income countries. Importantly, while rates of undernourishment and food insecurity have been dropping in recent years, there was an increase recently (in 2017), which was largely attributed to political unrest in regions such as the Middle East and Africa. Nevertheless, it is possible to be both overweight and undernourished (both are considered malnourishment), as your body may not be consuming all of the vital nutrients that it needs, even if you are achieving adequate caloric intake (see *Chapter 8: Nutrition* for more information).

To give further structure to the 2001 definition above, the FAO has defined four pillars that encompass food security: availability, access, utilization, and stability. All of these pillars are necessary to achieve food security; they all form the 'support' of food security (see Figure 9.2). Briefly, **food availability** addresses the supply-side of food and is determined by food production, food stocks, and food trade. **Food access** relates to how households are able to obtain food and usually focuses on both physical access and economic access. **Food utilization** involves how the body digests food, how food

BOX
9.1

The evolution of food security

The definition of food security has expanded over time, beginning with narrower concerns of providing enough food and, more recently, encompassing social and cultural issues. The first definitions were developed in response to widespread food shortages and famines at the start of the 1970s, with the initial focus on the physical availability of 'food supplies'.

While this concept did address the issue of the time, by the early 1980s, researchers and policymakers realized that the concept of food security should incorporate more general, chronic issues such as the lack of economic access to food (people too poor to buy food). A further shift in thinking led, in 1986, to an expanded definition that included the concept that food should contribute to an active and healthy life and, in 1996, that food should also be safe and nutritious. The chronological development of these definitions is listed in Table A.

Table A The developing definitions of food security (key concepts for each iteration are in boldface)

Date	Source	Definition
1974	UN World Food Summit, *Report of the World Food Conference*	'availability at all times of adequate world **food supplies** of basic foodstuffs to sustain a steady expansion of food consumption and to offset fluctuations in production and prices'
1983	FAO, *World Food Security: a reappraisal of the concepts and approaches*	'ensuring that all people at all times have both physical and economic **access** to the basic food that they need'
1986	World Bank Report, *Poverty and Hunger: options for food security in developing countries*	'access of all people at all times to **enough food** for an **active, healthy life**'
1996	World Food Summit, *Declaration of World Food Security*	'Food security, at the individual, household, national, regional and global levels [is achieved] when all people, at all times, have physical and economic access to sufficient, **safe and nutritious** food to meet their dietary needs and **food preferences** for an active and healthy life'
2001	FAO, *The State of Food Insecurity*	'Food security [is] a situation that exists when all people, at all times, have physical, **social** and economic access to sufficient, safe and nutritious food that meets their dietary needs and food preferences for an active and healthy life'

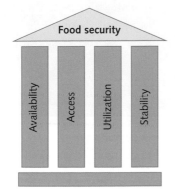

Figure 9.2 Food security can only be achieved when all the four pillars of food security (availability, access, utilization, and stability) are realized.

is prepared, and whether food is safe to eat, which ultimately influences the health of people. Finally, **food stability** refers to the foundations of the three other pillars. All four pillars are needed to achieve food security; the absence of any single one suggests food insecurity.

We discuss the first three pillars separately in Section 9.3. Because the final pillar stability is so important to all the other pillars, we include stability as a subsection in each of the other three pillars. As we will see, environmental issues and sustainability feature heavily in this stability concept.

9.2 Pillar 1: Availability: what constitutes food availability?

Availability means ensuring that food is actually available in a given place and time. Here we discuss three steps in food availability: from production (see Section 9.2.1) to processing and packaging (see Section 9.2.2), and finally transportation (see Section 9.2.3). The final subsection (see Section 9.2.4) describes the stability of these steps and how environmental change can influence overall availability.

9.2.1 Production

As discussed in *Chapter 1: Introduction*, we have been able to continue to increase both yields and the total amount of agricultural area during the Green Revolution. This has allowed global food production to keep up with population growth (see Figure 9.3).

High-income countries typically produce a significant surplus of food ('**food surplus countries**') through using technologies such as: high-yielding seeds, mechanization, and inputs such as fertilizers, pesticides, and irrigation (see Figure 9.4). However, the shifts in diets that include more calories consumed (see Figure 9.5), as well as an increased demand for animal products and increased food waste globally, may mean that greater food supply is necessary in the future (Kearney, 2010).

In low-income countries, populations have continued to increase, and much of the projected increase in global population is concentrated in Africa (Gerland et al., 2014). Africa already has high rates of food insecurity and malnutrition, compared with other continents, along with the lowest agricultural yields and food availability (van Ittersum et al., 2016). Low per capita incomes, coupled with poor investment in rural infrastructure and agricultural technology, are likely to reduce African food availability even further. However, improvements are being made in the incidence of **malnutrition** and reductions in the resulting impaired growth in children across Africa, though, in some cases, at slower rates than other low-income countries in the past decade (FAO, 2017).

As discussed in *Chapter 8: Nutrition,* the dietary choices in low- and middle-income countries are shifting towards more westernized diets with growing affluence, resulting in an increased concern over whether current agricultural production can meet future food demands (as the Western diet can be

Figure 9.4 Many high-income countries produce an exportable surplus of food due to the use of technologies such as heavy machinery. Here we see multiple combines and tractors with grain carts harvesting a large field in the USA.

Source: Corn production Source: University of Kentucky College of Agriculture, Food and Environment (CC BY-NC-ND 2.0).

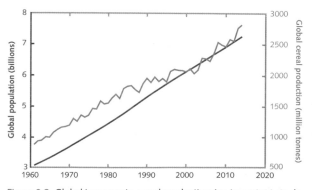

Figure 9.3 Global increases in cereal production (an important staple food in large parts of the world) and population from 1961 to 2014.

© FAO 2018 Food Balance Sheets [http://www.fao.org/faostat/en/#data/FBS] [10/07/2018].

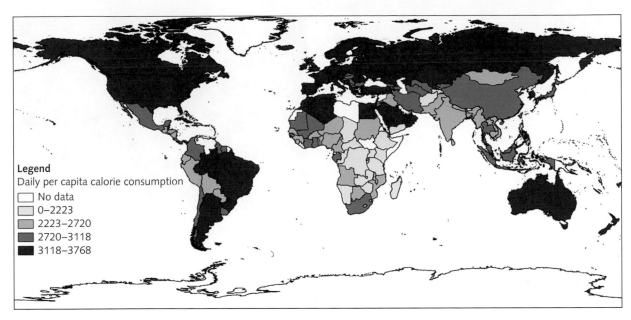

Figure 9.5 Distribution of average daily per capita calorie consumption by country in 2014.

Source: © FAO 2018 Food Balance Sheets [http://www.fao.org/faostat/en/#data/FBS] [10/07/2018].

more resource-intensive). An additional stress on food supplies is the diversion of crops to produce biofuels (Rosegrant et al., 2006) (a trend which is set to increase; see *Chapter 7: Energy*).

In your experience

Reflect on your current diet and caloric intake. Given the growing demand for food, what changes in your diet could you make that would enhance global food security? Would you be willing to make these changes? If not, what would stop you from doing so?

Around the world, dependence on locally produced food is very high. Only a small proportion of land produces enough surplus food to sell to global markets. In Africa and Asia, most food is grown by small family farms operated by **smallholders** (who generally have less than 2 ha of land) who sell their produce in local markets or to other neighbouring households (Bosc et al., 2013). A good example is rice (see Figure 9.6), with only 9% of total production sold on international markets (Headey, 2011). In total, smallholders are responsible for about 75% of

Figure 9.6 Most of the global rice production, as seen here in Bangladesh, is produced by smallholder farms.

Source: Bread for the World / Flickr (CC BY-ND 2.0).

the world's agriculture and produce the majority of most crops (Lowder et al., 2016).

Given that a large proportion of food is produced by smallholders and that smallholders are concentrated in low-income regions, you may think these regions produce enough food to support the local population. In fact, these areas are heavily dependent on food imports. For example, Mali has to import rice, fish, milk products, and cooking oils, even though three-quarters of its population rely on agriculture for their income (UN, 2017). In many cases, farmers within these communities have switched to growing **cash crops** (that is, tobacco, cotton, sugarcane) that they then export (see Figure 9.7), rather than growing food crops, which are less stable in price. Like Mali, food deficit countries are generally geographically isolated from global trade networks, with poor roads and few exports, and are concentrated in tropical regions. Of course, situations can vary, depending on the economic and political stability in a region.

When agricultural productivity and opportunities face challenges, many people in low- and middle-income countries decide to emigrate to the city, either within their own country or in other regions. In some cases, this may help alleviate food insecurity, as there are more opportunities for earning a living in urban areas. In Africa, these urban opportunities have generally failed to materialize; furthermore, poor rural infrastructure often prevents food produced in rural regions from reaching urban centres (see Section 9.2.3) (Andersson Djurfeldt, 2015). Many urban areas continue to source food from outside the region and country, because

Figure 9.7 These smallholder farmers in Mali are growing cotton, an important cash crop. Many smallholder famers have switched to producing cash crops (such as sugarcane, tobacco, and cotton), which can be exported, rather than food crops, which can adversely impact food security.
Source: Olivier EPRON Olivierkeita/Wikimedia Commons (CC BY-SA 3.0).

neighbouring rural regions do not produce a surplus. Urban agriculture, which is becoming more common, may be one way of relieving the food pressures of urban populations by growing food *in situ* (see Box 9.2).

BOX 9.2      Urban agriculture

Urban agriculture is the production of food in cities and urban areas (see Figure A). Urban agriculture is an important part of both income (when it is sold locally) and food provision in many rapidly growing cities (Brown and McCarty, 2017). Fresh fruit and vegetables are most economically produced within urban centres, because they can be transported quickly to the consumer. In addition, urban agricultural activities contribute to food availability, providing employment and income, and can be an important part of food security and nutrition of urban dwellers, particularly in low-income countries (FAO, 2012).

An excellent example of the promise of urban agriculture to reinvigorate cities can be found in Detroit (in the USA). After years of being a manufacturing powerhouse (building cars

and other machinery), Detroit suffered a steep decline due to increased overseas competition and economic pressures. These forces, coupled with the global financial crisis, led to a struggling city with high unemployment and low economic opportunities. With people leaving the city, houses and gardens became vacant, causing public hazards and crime. However, in recent years, there has been a programme of urban renewal, taking vacant lots and turning them into urban farming operations (see Figure B). There are now over 1000 urban farms and gardens in the city, along with many businesses and entrepreneurs. These developments have driven the growth of new jobs, engagement in the public community, and new restaurants. The *New York Times* recently called Detroit a culinary oasis, something that would have seemed far-fetched only a decade ago (Conlin, 2014).

a)

b)

Figure A Urban agriculture examples in (a) San Francisco (USA) and (b) the Philippines.

Source: (a) Arnoud Joris Maaswinkel/Wikimedia Commons (CC BY-SA 4.0); (b) Judgefloro/Wikimedia Commons (CC BY-SA 4.0).

Figure B An urban farm in Detroit, locally called the 'agrihood'.

Source: © Copyright 2018, Michigan Urban Farming Initiative.

Another promising, effective, and affordable way to increase food security in the future to deal with these challenges is to collect and use agricultural information to better effect, which would also improve overall profitability (see Case study 2.1).

There are multiple challenges to our ability to produce sufficient food in the future, and we have covered many of these challenges in this book so far.

 Pause and think

How might cities need to change to promote additional urban agriculture? What might urban agriculture do to city planning? Could land prices hamper efforts in the future as cities expand? In fact, where could we grow food in cities?

CASE STUDY 9.1

Mobile phones in Kenya: opportunities for improving food security

Historically, the main focus on increasing agricultural yields has been on using more inputs such as synthetic fertilizers, pesticides (see *Chapter 3: Pollution*), and energy (see *Chapter 7: Energy*). More recently the Internet, mobile phones, and technologies that collect, store, analyse, and share data are changing the lives of many farmers around the world. Even in low-income countries, more people have access to mobile phones than to proper sanitation or electricity (66% of the global population had access to mobiles in 2017) (GSMA, 2017). New agricultural techniques exploit satellite information and Internet-connected devices to improve farm management, using observations and measurements to respond to inter- and intra-field variability in crops (McBratney et al., 2005). These techniques, developed in the USA and Europe (see Figure A), are now spreading across low- and middle-income countries.

However, information alone may not enable farmers to produce enough nutritious food for everyone. Markets work by price signals; if there is less food available in town A, compared to town B, then the price of food may be higher in town A. The idea of markets is to align producers and consumers. A farmer would get better prices in town A, so could head there to sell their food for a higher price. This would provide more food in town A and eventually reduce overall prices, as more food becomes available. Both the farmer and consumer win. The farmer also now has more money to invest in improving yields. However, in many cases, farmers do not have information on the best places to sell their food. This is where mobile phones can step in again, connecting farmers to food markets.

Figure A Three false colour images showing the application of remote sensing in agriculture. The pictures were taken using aircraft flying over an agricultural centre in Arizona in the USA. The top image shows the vegetation density in different plots—the dark blues and greens show lush vegetation, and the red bare soil. The middle image shows where water deficits are located, with greens and blues showing wet soil and red dry soil. The bottom image shows fields where crops are under serious stress (you can see fields 120 and 119 have many red and yellow pixels); these fields were irrigated the next day.

Source: Susan Moran, Landsat 7 Science Team and USDA Agricultural Research Service.

One good example has been the use of mobile phones in Kenya. While mobile phones have become commonplace in most high-income countries, they are also rapidly expanding in low-income countries as well. In particular, Africa has seen a dramatic increase in mobile phone use, and Kenya is leading the way, with an estimated 82% of people owning cell phones in 2014, compared with 89% in the USA (Poushter et al., 2015). The expansion of mobile phones in Kenya has enabled the simultaneous expansion of mobile

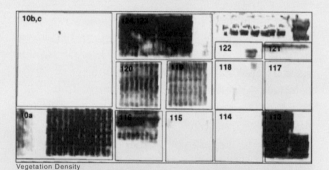

Vegetation Density

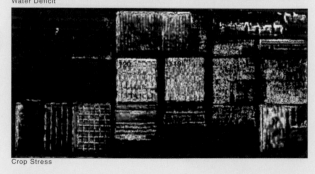

Water Deficit

Crop Stress

banking, including M-PESA, the largest mobile banking system in the world. Estimates from researchers suggest that people who were registered for M-PESA were 32% more likely to have savings (Demombynes and Thegeya, 2012). Access to mobile phones is also used by farmers to optimize their profits. The combination of mobile phone banking and the Kenya Agricultural Commodity Exchange (KACE) helps farmers get information on agricultural produce and market prices. This allows them to identify favourable markets for their products and cut out the middle men.

In addition, farmers also have better access to weather reports and technical assistance to improve productivity (see Figure B). Increasing both savings and farm income may have significant impacts on a farmer's food security, enabling them to either purchase more food in the marketplace or grow more food for themselves.

Figure B A farmer checks his mobile phone in Wote, Kenya.
Source: C Schubert (CCAFS).

Some of the key challenges, among others, are: climate change (see *Chapter 6: Climate Change*) (Rosenzweig et al., 2014), an existing and continuing reduction in soil fertility (see *Chapter 5: Soils*) (Doran, 2002), increasingly damaging animal and plant diseases (Sundström et al., 2014), and a lack of affordable phosphorus fertilizers (see *Chapter 3: Pollution*) (Cordell and White, 2011).

9.2.2 Processing and packaging

In an increasingly globalized world, we rely more on food distribution to move products around the world. However, even in local and regional food systems, we need distribution networks to take food from farms to shoppers. Before food can be distributed, it often needs processing and packaging to prevent rotting or spoiling during transport. For this reason, the amount of processed and packaged food has been on the rise for decades (Kearney, 2010; Thow, 2009). On the one hand, some processed foods can also be associated with poor nutritional value. In the USA, processed foods, including ready-to-eat meals, make up 75% of all calorie intake and are significantly higher in fat and sodium (Poti et al., 2015), causing significant health impacts (see *Chapter 8: Nutrition*). However, food processing can also be beneficial in ensuring the longevity of food (that is, frozen vegetables or canned fruit) and, in some cases, can provide

excellent nutrition as long as extra sugar and salt are not added. In both developed[1] and developing countries, a significant amount of food is wasted before it reaches the consumer (see Figure 9.8). In developing countries, food is more commonly wasted in production, handling, and storage. A general lack of infrastructure and transportation can often prevent food from reaching a market (see Figure 9.8) (FAO, 2011). The increased availability of equipment to process or package fruits, vegetables, grains, or root crops has therefore been hugely beneficial for the farmers and for food security (FAO, 2014).

Processing and packaging can have a significant impact on GHG emissions and other environmental impacts—food packaging accounts for half of all packaging materials in all sectors. This also contributes significantly to waste disposal challenges (Boye and Arcand, 2013). Some possible solutions include new packaging materials such as bioplastics made from plant residues (see Figure 9.9). Many new packaging technologies, such as active packaging which includes antimicrobial properties, have the ability to keep food fresher for longer, providing both

---

[1] This chapter uses the terms low-, middle-, and high-income countries to describe the varying economic situations in countries around the world, rather than 'developing' or 'developed'. However, because the original data from the FAO uses the terms 'developing' and 'developed', we maintain the original terms here. We use low-, middle-, and high-income throughout the rest of the chapter.

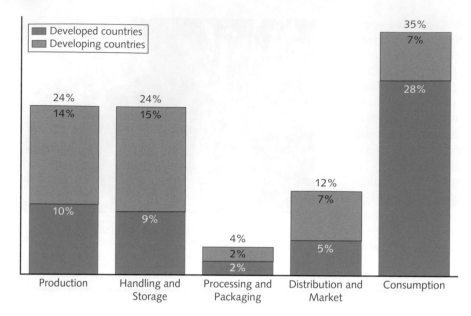

Figure 9.8 Loss of food at different points in the supply chain. Food waste in developed countries is most common during consumption, while 29% of all food waste globally is in production, handling, and storage in developing countries.

Source: Data from Food and Agriculture Organization (FAO) of the United Nations.

Figure 9.9 Bioplastics are a new material for the processing and packaging of food products. Bioplastics are made of plant residues and are compostable, providing a sustainable alternative to regular plastic, which is made of oil and does not decompose.

Source: Christian Gahle, nova-Institut GmbH/Wikimedia Commons (CC BY-SA 3.0).

retailers and consumers a greater opportunity to use food without having it go to waste. Given that the majority of food waste in high-income and developed countries is from retail sectors and consumers themselves (Conrad et al., 2018) (see Figure 9.8), packaging technologies will play a large part in future sustainable solutions (FAO, 2011).

9.2.3 Transportation

The average US grocery store carried around 39,500 different items in 2015 (FMI, 2017). Such astonishing selection highlights the recent trends towards a globalized, connected world, moving from a local and regional system to a global one. It is now possible to buy avocadoes, grown in the tropics, at a store in Alaska, and Alaskan snow crab in a supermarket in Spain. Adequate infrastructure (roads, rail, law enforcement, etc.) is needed to ensure reasonable transportation costs for low-value, high-bulk goods, such as grain, in areas far from ports or other trade networks (Brown et al., 2013).

You might think that the environmental impacts of transportation would depend on how far the food travels and would be increasing with globalization. In fact, as discussed in *Chapter 6: Climate Change*, food miles are not a significant portion of the total environmental impact of most products. Instead, the transportation impact is not from the distance travelled, but the way food travels. A product transported by plane, instead of via rail, ship, or truck (see Figure 9.10), will result in far greater GHG emissions.

In some places, such as in parts of Africa, it is a lack of transportation altogether that influences food security. In many low-income countries, economic isolation and stagnation in manufacturing have caused the degradation of existing road and rail networks (Shively, 2017; UNCTAD, 2012). Since transportation networks, by definition, cover a lot of distance, they can often intersect with areas of natural and social importance, which can drive conflict between communities, governments, businesses, and other stakeholders. A good example of this for railway development in Kenya is described in Food controversy 9.1.

9.2.4 The stability of food availability

The stability of food availability can change over time. Once stable, regions with a high availability of food can become unstable, for example due to climatic changes. Many regions will be increasingly affected by extreme weather events such as heat waves, tropical cyclones,

Figure. 9.10 Transporting tomatoes in a truck in California. Increasing transportation of many kinds of fruits and vegetables may mean some fruits and vegetables are bred for new characteristics.
Source: Steven Damron/Flickr (CC BY 2.0).

FOOD CONTROVERSY 9.1 Food security over biodiversity?

Kenya has struggled with a degraded rail and road network for a very long time (see Figure A). It has been extremely challenging to get food from rural areas and coastal ports, such as Mombasa, to the capital Nairobi. In fact, freight transported from the port of Mombasa to Nairobi fell from 4.8 million tonnes in the 1980s to just 1.5 million tonnes in 2012. In 2017, a 4 billion US dollar new railway network was opened, connecting Mombasa with Nairobi. Interestingly, this railway was financed and built by the Chinese government and is one of the many projects financed by China (both to be able to export products to Africa, as well as import food and products to China).

The goal is to make a regional railway network, which connects the port of Mombasa to other East African countries, including Uganda, the Democratic Republic of Congo, Rwanda, and Burundi, and northward into South Sudan and Ethiopia (see Figure B).

This improved infrastructure can greatly cut down transportation costs and can result in improved food security in East Africa, especially in urban areas, which are projected to grow rapidly (for example, Nairobi is expected to see an increase of 80% in population between 2010 and 2025).

However, to complete the railway, the Chinese-owned construction company (China Road and Bridge Corporation)

Figure A Kibera slum in Nairobi, Kenya. Access to good infrastructure is needed to transport foods into cities and to increase food security. In many low-income countries, existing infrastructure, such as rail networks, have been seriously degraded due to poor maintenance.
Source: Trocaire/Flickr (CC BY 2.0).

wants to extend the railway through the Nairobi National Park, a wildlife reserve at the edge of Nairobi and home to a wide variety of wildlife, including lions, giraffes

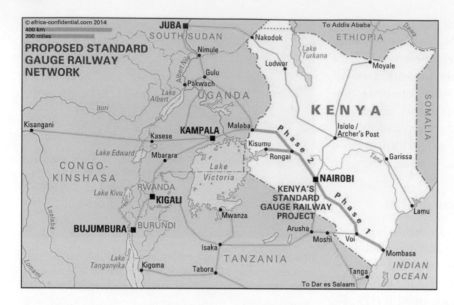

Figure B The proposed rail network from the port at Mombasa to the African interior.

Source: Reproduced with permission from Vol 55 No 4 www.africa-confidential.com.

(see Figure C), and the endangered black and white rhinos. Conservationists resisted the decision to route the railway through the park, and in 2016, the Kenyan High Court halted the project, so that all sides of the argument could be heard. However, the government and company have continued construction in the interim, prompting protests (see Figure D).

With the expected expansion of African economies over the next century, increasing tensions such as these are likely. The power dynamics between different groups will be increasingly apparent, as developments are either stopped or rerouted, or continue regardless.

Figure C A giraffe in the Nairobi National Park, which is located at the edge of Nairobi and hosts a wide variety of wildlife.

Source: Mkimemia/Wikimedia Commons (CC BY-SA 3.0).

Figure D Protestors opposing the Kenyan government's decision to build a railway line across the Nairobi National Park.

Tony Karumba/AFP/Getty Images.

and intense rainfall that will damage or destroy agricultural crops. Even during the initial stages of climatic changes, during the 1964–2007 period, extreme heat, droughts, and floods reduced cereal production on average by 9–10% (Lesk et al., 2016). As the climate warms further, natural disasters pose an increasing risk to the food security of the local population. Serious problems would then arise if no government assistance is available to compensate for damages or help communities recover (Linnerooth-Bayer and Hochrainer-Stigler, 2015). Instability of food availability could then ensue, much like the famines of the 1970s and 1980s where, in many cases, food was not available due to long-term droughts.

9.3 Pillar 2: Food access

An adequate supply of food in a community, region, or country does not guarantee that an individual or household has access to it, even within that same region. Lack of income and transportation, or marginalization such as social stigma, can restrict the ability of an individual or household to access food. Lack of access is the most common reason for food insecurity in both low- and high-income countries. There are two types of access—physical (see Section 9.3.1) and economic (see Section 9.3.2).

9.3.1 Physical access

Almost everyone accesses food through markets. Even smallholders who may grow most of their own food would likely sell any surplus for money, in order to purchase other products (for example, shoes, clothes, school supplies, and building materials). Population expansion during the past few decades has resulted in large increases in demand, which has been met largely from globally marketed cereals because of their relatively low cost, ease of use, and high number of calories. Countries that are not able to produce all of their own food and must import food are termed as being at a **food deficit**.

Trading food is a key component of ensuring food access, especially in places where insufficient food is grown domestically. However, that food has to be able to move freely and movement can be hindered by political, economic, or physical barriers that isolate the deficit region. Isolation further inhibits agricultural production through restriction of new technologies, such as improved plant varieties, agricultural inputs (such as synthetic fertilizers), or the flow of agricultural knowledge from and to farmers (Brown et al., 2013).

9.3.2 Economic access

Although there is an increasing demand for food, many farmers are finding that their incomes and productivity are either stagnating or even falling over time, particularly in low-income countries with poor trading ties to other regions (Anderson et al., 2014). Many farming families around the world have diversified their income sources, so that they can provide for their family. Farmers often find other work in rural markets, livestock production, crafts, and other jobs (Abdulai and CroleRees, 2001). This is common not only in low-income countries, but also in high-income countries. In the USA, nearly 80% of total farm household income came from off-farm income sources (USDA, 2016). Although this has increased income and, to some extent, has improved the standard of living, it also means that most farmers are net purchasers of food from the market (Bryceson, 2002).

Food affordability is a primary cause of food insecurity for low-income people who rely on the market to purchase food across all regions of the world. Affordability depends on the amount of disposable income an individual or family has, relative to the cost of food (Ploeg et al., 2012). Low-income households who spend a large portion of their incomes on food are more vulnerable to rapid changes in food prices than middle- and high-income households (Grosh et al., 2008). Government safety nets, such as food banks and the United States' Supplemental Nutrition Assistance Program (SNAP), or the Ethiopian Productive Safety Net Program (PSNP), provide financial assistance to low-income people to ensure that everyone has access to sufficient food, even when local food prices are high. However, many governments fail to provide sufficient resources for its citizens in such programmes when conditions change rapidly (Alderman and Haque, 2006).

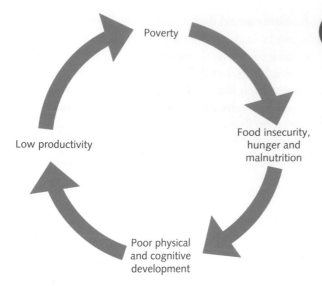

Figure 9.11 Poverty, malnutrition, and food insecurity are deeply interrelated.

Source: Based on Glewwe and King, 2001.

Having enough money requires participation in the local economy, through producing agricultural or other goods for sale or working for a wage. People need good health to work and family support for dependents, so that parents are free to work. Hunger due to poverty is both a cause and an effect of food insecurity where poor physical and cognitive development can lead to low productivity, which leads to further poverty in a vicious cycle (see Figure 9.11).

Pause and think

What are the trade-offs of having low-cost food? How does cheaper food affect consumers and producers? Can you think of the nutritional content of low-cost food? What nutrients are easier to provide cheaply? What is the role, if any, of government policies to influence food prices?

9.3.3 The stability of food access

Political upheaval, social unrest, or terrorism cause significant disruptions that can prevent people and food from getting to markets (Akresh et al., 2009). In 2011, a combination of a persistent drought, conflict, economic breakdown, and rising food crises caused a famine in Somalia (Maxwell and Fitzpatrick, 2012). Even when food is abundant in other areas of a country, food may not be available in these sorts of crisis zones. The main factors are the theft of livestock and crops, the forced displacement of people from their homes, leading to more pressure on food supplies, and increased exposure to diseases (see Figure 9.12) (Akresh et al., 2009). These effects can be long-lasting; civil wars have been shown to lead to significant increases in mortality and disability for years after fighting ends, with a particularly

Figure 9.12 A refugee camp close to the Rwandan border. Rwanda has been intermittently hit by war and drought, leading to recurrent food insecurity, famine, and, in the 1980s, genocide.

Source: Julien Harneis/Flickr (CC BY-SA 2.0).

strong impact on women and children (Ghobarah et al., 2003).

Rapid changes in local food prices due to global economic crises can reduce food security in a region that has previously adequate food availability (Brown and Kshirsagar, 2015). During the 2008 food price crisis, export restrictions imposed by major rice-exporting countries were responsible for tripling the cost of rice over 4 months during a time of record production. Restricting trade (on either imports or exports) may protect the domestic population from the impacts of regional and global economic shocks in the short term (Carr, 2011; Do et al., 2013), but over the long term, producers cannot properly respond with production changes, prices are higher, technology uptake is lower, adaptation is more difficult, and climate effects on food security are worse (Brown et al., 2015).

9.4 Pillar 3: Utilization

Up until now, we have looked at food security in terms of how food is produced and accessed. However, there is another important pillar—utilization—food needs to be nutritious to contribute the nutrients we need to lead productive lives, and hygienic so that it does not make people sick. Here we outline three key components of utilization: nutrition (see Section 9.4.1), cooking (see Section 9.4.2), and food safety (see Section 9.4.3).

9.4.1 Nutrition

People do not just need available food (see Section 9.2), with the means of access to markets or shops (see Section 9.3); they also need food with the required nutrients, particularly in combinations with other foods, to ensure maximum benefit. This is why certain meals (rice with beans, for example) are considered complete—because they offer the basic nutritional building blocks that your body needs to be healthy and productive (in this case, rice with beans contain carbohydrates, proteins, iron, etc.). For people who are not able to obtain these complete meals, nutrition and health can be compromised, which is especially concerning for young children.

The first 1000 days, just under 3 years, of a child's life, including the pregnancy of the mother, are absolutely critical to overall development and health trajectory (Wrottesley et al., 2016). In many high-income countries, specific programmes exist to help care for this demographic. For example, in the 1970s, the US Women, Infants, and Children (WIC) programme was established to provide supplementary nutrition to low-income and high-nutritional risk women and their children under the age of 5 (NWA, 2017). Today the WIC programme serves 53% of all infants born (see Figure 9.13) (USDA, 2015).

Changes in micronutrients, protein, or other essential components of food through environmental stresses and changes could have significant impacts on the health of the global population (see *Chapter 6: Climate Change* for more information). Young children are, again, most vulnerable to these changes.

Figure 9.13 A woman pictured with her children, recovering from malnutrition at a health clinic in southern Ethiopia. Nutrient utilization in their first 3 years of life is particularly important for the long-term health of her children.

Source. Tanya Axisa/Department for International Development (CC BY 2.0).

Pause and think

Are there other ways that new technologies in genetic engineering could be used to ensure crops maintain or increase their nutritional capacity? What level of research should be done on both people and the environment to ensure the safety of new genetically engineered crops? What kinds of policies and regulations might be necessary (or not) to ensure the safety of these crops?

9.4.2 Cooking

The nutrients in some foods are not fully available without some form of cooking and processing. In many cases, cooking depends on the knowledge and facilities that people have available. Being able to cook may not be what first comes to mind when thinking about food security; you may be more likely to think about food on shelves or having enough money to eat what you need for a healthy lifestyle. But because cooking makes some foods nutritionally available and kills bacteria, preventing illness, it is actually inextricably linked to food security.

Cooking in the home is less and less frequent, especially in high-income countries. In the USA, the share of money spent on prepared food outside the home, for example at restaurants or on ready-made food, rose from 26% of all food expenditures in 1970 to 43% in 2012 (see Figure 9.14). There are several reasons for this, including a greater number of women employed outside the home, higher incomes, the rise in more affordable and convenient restaurant and fast-food locations, and increased advertising by food service chains and restaurants (USDA, 2016). In Britain, it is estimated that there was a 43% increase in fast-food restaurants between 1990 and 2008, an increase which was seen mostly in the lower socio-economic regions of the country (Maguire et al., 2015).

In your experience

How do you make the decision of whether to cook at home or eat out? Do you feel you are able to prepare and cook all the food you would need to feel healthy? Would some of this food already be pre-processed?

While growth in food eaten outside the home in high-income countries is associated with many positive social outcomes (greater number of women in the workforce, higher incomes, etc.), it has also had significant impacts on nutrition and sustainability. Ready-to-eat-meals have been estimated to have a much higher total impact on GHG emissions. Compared to using only fresh ingredients for dinners, ready-to-eat dinners result in the same GHG emissions as driving 900 km per person or around 10% of emissions from driving in high-income countries (Hanssen et al., 2017). These shifts also affect nutritional quality, because meals and snacks prepared away from home generally contain more calories and more fat and have fewer nutrients, including calcium, fibre, and iron (USDA, 2016).

In low-income countries, challenges for cooking are significantly different, revolving around fuel availability, adverse human health impacts from cooking types, and environmental impacts. In 2013, it was estimated that 40% of the world's population still relied on traditional biomass (wood, crop residues, animal dung, etc.) for cooking (Lee et al., 2013) (see Case study 9.2).

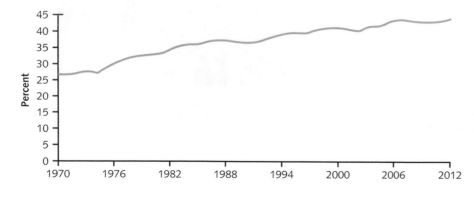

Figure 9.14 In the USA, food expenditures out of home have been rising over the past several decades, which has implications for food security and nutrition.

Source: © United States Department of Agriculture/Economic Research Service.

CASE STUDY 9.2

Clean cooking stoves for sustainable cooking

In many low-income countries, traditional methods of cooking are still common, including the use of biomass (for example, wood, dung, and crop residues). People often cook on fires inside a home, which exposes families to black carbon, a small particulate released from the combustion of biomass. When these particulates are inhaled, they can cause cardiopulmonary diseases such as asthma, lung cancer, and heart attacks.

One study in Ghana found that particulate and black carbon exposures from cooking were significantly higher for women, as they traditionally do most of the cooking (Van Vliet et al., 2013). As a result, many new efforts focus on providing cleaner, more efficient cooking stoves to families that still use these methods (see Figure A).

In addition, there is an interesting additional benefit of cleaner stoves. GHG emissions from wood fuels result in more than 2% of all global GHG emissions. A successful switch to 100 million improved cooking stoves globally could reduce these emissions up to 17%, while also improving black carbon emissions and health (Bailis et al., 2015). New efforts by the WHO include a Clean Households Energy Solutions Toolkit, which aims to help policymakers to build household energy policies and programmes at both national and local levels.

An additional benefit of cleaner cooking stoves is that they burn the fuel more efficiently, reducing the demand for fuel. It is estimated that up to around a third of wood used for cooking is not managed sustainably. One key issue is the unsustainable production of charcoal in many tropical regions of the world (Chidumayo and Gumbo, 2013). Charcoal is often produced using traditional earth and pit kilns, which are oven(-like) structures, in which layers of wood are covered with leaves and soil.

Converting wood to charcoal has an efficiency rate of only 20%, and therefore, a large amount of wood is needed to produce charcoal. As a result, trees are logged, causing a reduction in biodiversity (see Figure B). With the removal of trees, the soil is susceptible to erosion (see *Chapter 5: Soils*) and the water-holding capacity of the soil is also reduced, potentially reducing primary production in the system (Chidumayo and Gumbo, 2013). With cleaner cooking stoves, the charcoal is used more efficiently, reducing the overall demand per family.

Figure A A woman in Tanzania showing a new clean cooking stove.

Source: Russell Watkins/Department for International Development (CC BY 2.0).

Figure B Wood cut and trimmed for conversion into charcoal in Kenya.

Source: © The authors.

9.4.3 Food safety

With a growing and globalized food system, the supply chain of many products has become longer and longer (see *Chapter 12: Food Systems* and *Chapter 10: Food Aid*). As more actors have the potential to handle a food product in the supply chain, this presents new potential challenges for food safety—food could be available and accessible but may not be edible. A recent WHO report cited globalization, trade, travel, and migration among key aspects of increasing food safety concerns (Brickman and Kruse, 2015). Food safety issues can occur at any point in the food chain, from raising backyard chickens to food preparation inside the home.

One of the most significant food safety concerns are **aflatoxins**, naturally produced by many fungi. These toxins occur when key staple crops (including corn, rice, and cassava) are not properly dried after harvest. Consumption of food contaminated with aflatoxins at high levels can result in poisoning, and even at low levels, they can contribute to long-term liver and immune diseases (PACA, 2015). In high-income countries, this is not generally a problem, as crops are able to dry quickly under shelter. But in some low-income regions like Africa, aflatoxin levels are significantly higher than what is allowed in the EU and USA. With increased rainfall events predicted by climate change, the control of aflatoxins may get harder since it will be more difficult to dry crops.

Beyond the farm, there has been an expansion of the **cold-chain**, which refers to the ability to keep food cold throughout its lifespan (critical for products like dairy, meat, and frozen goods), for example by using refrigerated shipping containers (see Figure 9.15). This has enabled higher standards of food safety because colder temperatures prevent the growth of bacteria. Although globally 40% of all food requires refrigeration (James and James, 2010), in practice, only 10% of perishable foods are actually refrigerated. This suggests that the cold-chain is likely to expand across low- and middle-income countries in the coming decades. This will be particularly important in a changing climate, as it is estimated that bacterial growth rates roughly double every 10°C above 10°C (James and James, 2010). Importantly, it is estimated that if cold-chain installation were accompanied by efficiency measures, expansion of

Figure 9.15 A shipping container with a refrigerator unit in blue. These shipping containers are vital for preserving food from where it is produced to consumption.

Source: Gazouya-japan/Wikimedia Commons (CC BY-SA 4.0).

the cold-chain could occur without increasing GHG emissions (James and James, 2010).

9.4.4 Food utilization and stability

Food utilization requires stability across the supply chain. Nutritionally, climate change can impact the types of foods that we grow, which could compromise the nutritional value (discussed in greater detail in *Chapter 6: Climate Change*). Stable transportation systems are also critical to ensuring that cold storage is possible as foods travel around the globe and food safety is not compromised. Reliable sources of electricity or fuel are also critical for many of the components of utilization—without adequate access to fuel or electricity, many foods cannot be prepared in a way that is nutritionally optimal. Further, lack of access to refrigeration can compromise food safety. As global populations increase and demands for fuel and electricity also increase (both because of a growing population and a warming climate), these factors will be critical to ensuring continued utilization of safe and nutritious food.

Connect the dots

How could changes in diet (see *Chapter 8: Nutrition*) influence the availability of food in the world? Under future climate change (see *Chapter 6: Climate Change*) conditions, what diet changes might we see or might we need to adopt?

9.5 Future outlook

Food security encompasses four pillars: food availability, food access, food utilization, and food stability. The sustainability of our food system influences food security, and how food security is achieved also has implications for the sustainability of food systems. Globally, we produce enough food to feed all people on Earth, though there are estimates that we may need to increase global food supplies for the future. Further, many places may not meet their food availability needs and often employ trade to ensure domestic food security.

Most people who are food-insecure lack access to food, either because their region does not produce enough food or they do not have economic, physical, or social access to it. Nutrition and utilization from food are changing as diets shift; cooking is changing in both low- and high-income countries, and food safety concerns are significant with both a lack of access to technologies in low-income countries and increasing global food supply chains. The stabilization necessary for food security is a critical underpinning, which could be fundamentally threatened with climate change.

There are many future potential pathways for food security and its implications for sustainability. New technologies and means of communication are driving fast change in the food system, which has enabled greater access to markets and improved agricultural production. Simultaneously, dietary shifts towards more westernized diets may present new challenges to meet these demands with fewer resources. What the future looks like depends on how society continues to change its eating patterns, the amount of food we waste, and the expansion of technologies and practices that can enable higher agricultural production, particularly in regions with lower-yielding agriculture.

● QUESTIONS

9.1 What are the four pillars of food security, and what do they encompass?

9.2 Do we currently produce enough food to feed our global population? How will this change in the future? What factors might influence that?

9.3 What is a food deficit country? A food surplus country?

9.4 How does food waste differ between low-income and high-income countries?

9.5 What are some negative and positive aspects of processing foods?

● FURTHER READING

Brown, M.E., Antle, J.M., Backlund, P., Carr, E.R., Easterling, W.E., Walsh, M.K., Ammann, C., Attavanich, W., Barrett, C.B., Bellemare, M.F., Dancheck, V., Funk, C., Grace, K., Ingram, J.S.I., Jiang, H., Maletta, H., Mata, T., Murray, A., Ngugi, M., Ojima, D., O'Neill, B., Tebaldi, C. 2015. *Climate Change, Global Food Security, and the U.S. Food System*. Available at: http://www.usda.gov/oce/climate_change/FoodSecurity2015Assessment/FullAssessment.pdf [accessed 4 November 2018]. **(This report explores how climate change will affect food security both in low-income countries—where it is commonly discussed more frequently—and in the USA. It examines the entire food system and the multiple pillars for food security.)**

High Level Panel of Experts on Food Security and Nutrition (HLPE). 2017. *Nutrition and food systems*. September 2017.

Rome: United Nations Food and Agriculture Organization. Available at: http://www.fao.org/3/a-i7846e.pdf [accessed 4 November 2018]. **(This report analyses how food systems influence diets and nutrition by exploring the role of diets as a link between food systems, health, and nutrition outcomes. It also explores the impacts of agriculture and food systems on economic, social, and environmental sustainability.)**

West, P.C., Gerber, J.S., Engstrom, P.M., Mueller, N.D., Brauman, K.A., Carlson, K.M., Cassidy, E.S., Johnston, M., MacDonald, G.K., Ray, D.K., Siebert, S. 2014. Leverage points for improving global food security and the environment. *Science* **345**, 325–328. **(This article looks at the many ways that we can improve food security and environmental outcomes throughout the food system.)**

● REFERENCES

Abdulai, A., CroleRees, A. 2001. Determinants of income diversification amongst rural households in Southern Mali. *Food Policy* **26**, 437–452.

Akresh, R., Verwimp, P., Bundervoet, T. 2009. Crop failure, civil war and child stunting in Rwanda. *Economic Development and Cultural Change* **59**, 777–810.

Alderman, H., Haque, T. 2006. Countercyclical safety nets for the poor and vulnerable. *Food Policy* **31**, 372–383.

Anderson, K., Ivanic, M., Martin, W.J. 2014. Food price spikes, price insulation, and poverty. In: Chavas, J.-P., Hummels, D., Wright, B.D. (eds). *The Economics of Food Price Volatility*. Chicago, IL: University of Chicago Press, pp. 311–339.

Andersson Djurfeldt, A. 2015. Urbanization and linkages to smallholder farming in sub-Saharan Africa: Implications for food security. *Global Food* Security **4**, 1–7.

Bailis, R., Drigo, R., Ghilardi, A., Masera, O. 2015. The carbon footprint of traditional woodfuels. *Nature Climate Change* **5**, 266–272.

Boye, J.I., Arcand, Y. 2013. Current trends in green technologies in food production and processing. *Food Engineering Reviews* **5**, 1–17.

Brickman, S., Kruse, H. 2015. *Complex Food Chain Increases Food Safety Risks*. Geneva: World Health Organization.

Brown, M.E., Antle, J.M., Backlund, P., Carr, E.G., Easterling, W.E., Walsh, M.K., Ammann, C., Attavanich, W., Barrett, C.B., Bellemare, M.F., Dancheck, V., Funk, C., Grace, K., Ingram, J.S.I., Jiang, H., Maletta, H., Mata, T., Murray, A., Ngugi, M., Ojima, D., O'Neill, B., Tebaldi, C. 2015. *Climate Change, Global Food Security, and the U.S. Food System (Report)*. USDA Technical Document. Available at: https://www.usda.gov/oce/climate_change/FoodSecurity2015Assessment/FullAssessment.pdf [accessed 4 November 2018].

Brown, M.E., Kshirsagar, V. 2015. Weather and international price shocks on food prices in the developing world. *Global Environmental Change* **35**, 31–40.

Brown, M.E., McCarty, J.L. 2017. Is remote sensing useful for finding and monitoring urban farms? *Applied Geography* **80**, 23–33.

Brown, M.E., Silver, K.C., Rajagopalan, K. 2013. A city and national metric measuring isolation from the global market for food security assessment. *Applied Geography* **38**, 119–128.

Bryceson, D.F. 2002. The scramble in Africa: reorienting rural livelihoods. *World Development* **30**, 725–739.

Carr, E.R. 2011. *Delivering Development: Globalization's Shoreline and the Road to a Sustainable Future*. New York, NY: Palgrave Macmillan.

Chidumayo, E.N., Gumbo, D.J. 2013. The environmental impacts of charcoal production in tropical ecosystems of the world: a synthesis. *Energy for Sustainable Development* **17**, 86–94.

Conlin, J. 2014. In Detroit, Revitalizing Taste by Taste. *New York Times*. Available at: https://www.nytimes.com/2014/10/12/travel/in-detroit-revitalizing-taste-by-taste.html [accessed 4 November 2018].

Conrad, Z., Niles, M.T., Neher, D.A., Roy, E.D., Tichenor, N.E., Jahns, L. 2018. Relationship between food waste, diet quality, and environmental sustainability. *PLoS One* **13**, e0195405.

Cordell, D., White, S. 2011. Peak phosphorus: clarifying the key issues of a vigorous debate about long-term phosphorus security. *Sustainability* **3**, 2027–2049.

Demombynes, G., Thegeya, A. 2012. *Kenya's mobile revolution and the promise of mobile savings*. World Bank Policy Research Working Paper 5988. Available at: http://documents.worldbank.org/curated/en/900911468047101453/pdf/WPS5988.pdf [accessed 4 November 2018].

Do, Q.-T., Levchenko, A., Martin, R. 2013. *Trade insulation as social protection. World Bank Policy*. Available at: https://elibrary.worldbank.org/doi/pdf/10.1596/1813-9450-6448 [accessed 4 November 2018].

Doran, J.W. 2002. Soil health and global sustainability: translating science into practice. *Agriculture, Ecosystems & Environment* **88**, 119–127.

Food and Agricultural Organization of the United Nations (FAO). 2001. Human energy requirements: Report of a Joint FAO/WHO/UNU Expert Consultation. Available at: http://www.fao.org/3/a-y5686e.pdf [accessed 4 November 2018].

Food and Agricultural Organization of the United Nations (FAO). 2011. *Global Food Losses and Food Waste: Extent, Causes and Prevention*. Available at: http://www.fao.org/docrep/014/mb060e/mb060e00.pdf [accessed 4 November 2018].

Food and Agricultural Organization of the United Nations (FAO). 2012. *Food, Agriculture and Cities: The Challenges of Food and Nutrition Security, Agriculture and Ecosystem Management in an Urbanizing World*. Rome: FAO.

Food and Agricultural Organization of the United Nations (FAO). 2014. *Appropriate Food Packaging Solutions for Developing Countries*. Available at: http://www.fao.org/docrep/015/mb061e/mb061e00.pdf [accessed 4 November 2018].

Food and Agricultural Organization of the United Nations (FAO). 2016. *FAOSTAT. Food and Agriculture Data*. Available at: http://www.fao.org/faostat/en/ [accessed 12 November 2018].

Food and Agricultural Organization of the United Nations (FAO), International Fund for Agricultural Development (IFAD), UNICEF, World Food Programme (WFP), World Health Organization (WHO). 2017. *The State of Food Security and Nutrition in the World*. Available at: http://www.fao.org/3/I9553EN/i9553en.pdf [accessed 4 November 2018].

Food Market Institute (FMI). 2017. *Supermarket Facts*. Available at: https://www.fmi.org/our-research/supermarket-facts [accessed 4 November 2018].

Gerland, P., Raftery, A.E., Ševčíková, H., Li, N., Gu, D., Spoorenberg, T., Alkema, L., Fosdick, B.K., Chunn, J., Lalic, N., Bay, G., Buettner, T., Heilig, G.K., Wilmoth, J. 2014. World population stabilization unlikely this century. *Science* **346**, 234–237.

Ghobarah, H.A., Huth, P., Russett, B. 2003. Civil wars kill and maim people—long after the shooting stops. *American Political Science Review* **97**, 189–202.

Glewwe, P., King, E.M. 2001. The impact of early childhood nutritional status on cognitive development: does the timing of malnutrition matter? *The World Bank Economic Review* **15**, 81–113.

Grosh, M., del Ninno, C., Tesliuc, E., Ouerghi, A. 2008. *For Protection and Promotion: The Design and Implementation of Effective Safety Nets*. Washington, DC: World Bank.

GSMA, 2018. *The Mobile Economy 2018*. Available at: https://www.gsma.com/mobileeconomy/ [accessed 12 November 2018].

Hanssen, O.J., Vold, M., Schakenda, V., Tufte, P.A., Møller, H., Olsen, N.V., Skaret, J. 2017. Environmental profile, packaging intensity and food waste generation for three types of dinner meals. *Journal of Cleaner Production* **142**, 395–402.

Headey, D. 2011. Rethinking the global food crisis: the role of trade shocks. *Food Policy* **36**, 136–146.

High Level Panel of Experts on Food Security and Nutrition (HLPE). 2013. *Investing in Smallholder Agriculture for Food Security: A Report by the High Level Panel of Experts on Food Security and Nutrition*. Available at: http://www.fao.org/fileadmin/user_upload/hlpe/hlpe_documents/HLPE_Reports/HLPE-Report-6_Investing_in_smallholder_agriculture.pdf [accessed 4 November 2018].

James, S.J., James, C. 2010. The food cold-chain and climate change. *Food Research International* **43**, 1944–1956.

Kearney, J. 2010. Food consumption trends and drivers. *Philosophical Transactions of the Royal Society of London B: Biological Sciences* **365**, 2793–2807.

Lee, C.M., Chandler, C., Lazarus, M., Johnson, F.X. 2013. *Assessing the Climate Impacts of Cookstove Projects: Issues in Emissions Accounting*. Stockholm Environment Institute, Working Paper No. 2013-01. Available at: https://www.sei.org/mediamanager/documents/Publications/Climate/sei-wp-2013-01-cookstoves-carbon-markets.pdf [accessed 6 November 2018].

Lesk, C., Rowhani, P., Ramankutty, N. 2016. Influence of extreme weather disasters on global crop production. *Nature* **529**, 84–87.

Linnerooth-Bayer, J., Hochrainer-Stigler, S. 2015. Financial instruments for disaster risk management and climate change adaptation. *Climate Change* **133**, 85–100.

Lowder, S.K., Skoet, J., Raney, T. 2016. The number, size, and distribution of farms, smallholder farms, and family farms worldwide. *World Development* **87**, 16–29.

Maguire, E.R., Burgoine, T., Monsivais, P. 2015. Area deprivation and the food environment over time: A repeated cross-sectional study on takeaway outlet density and supermarket presence in Norfolk, UK, 1990–2008. *Health & Place* **33**, 142–147.

Maxwell, D., Fitzpatrick, M. 2012. The 2011 Somalia famine: context, causes, and complications. *Global Food Security* **1**, 5–12.

McBratney, A.B., Whelan, B., Ancev, T., Bouma, J. 2005. Future directions of precision agriculture. *Precision Agriculture* **6**, 7–23.

National WIC Association (NWA). 2017. *WIC Program Overview and History*. Available at: https://www.nwica.org/overview-and-history [accessed 4 November 2018].

NCD Risk Factor Collaboration (NCD-RisC), Di Cesare, M., Bentham, J., Stevens, G.A., Zhou, B., Danaei, G., Lu, Y., Bixby, H., Cowan, M.J., Riley, L.M., Hajifathalian, K., Fortunato, L.,

Taddei, C., Bennett, J.E., Ikeda, N., Khang, Y.-H., Kyobutungi, C., Laxmaiah, A., Li, Y., Lin, H.-H., Miranda, J.J., Mostafa, A., Turley, M.L., Paciorek, C.J., Gunter, M., Ezzati, M. 2016. Trends in adult body-mass index in 200 countries from 1975 to 2014: a pooled analysis of 1698 population-based measurement studies with 19·2 million participants. *The Lancet* **387**, 1377–96.

Partnership for Aflatoxin Control in Africa (PACA). *Aflatoxin Impacts and Potential Solutions in Agriculture, Trade, and Health*. Available at: http://www.un.org/esa/ffd/wp-content/uploads/sites/2/2015/10/PACA_aflatoxin-impacts-paper1.pdf [accessed 6 November 2018].

Ploeg, M. Ver, Breneman, V., Farrigan, T., Hamrick, K., Hopkins, D., Kaufman, P., Lin, B.-H., Nord, M., Smith, T.A., Williams, R., Kinnison, K., Olander, C., Singh, A., Tuckermanty, E. 2012. *Access to Affordable and Nutritious Food: Measuring and Understanding Food Deserts and Their Consequences*. Available at: https://www.ers.usda.gov/webdocs/publications/42711/12716_ap036_1_.pdf [accessed 4 November 2018].

Poti, J.M., Mendez, M.A., Ng, S.W., Popkin, B.M., 2015. Is the degree of food processing and convenience linked with the nutritional quality of foods purchased by US households? *American Journal of Clinical Nutrition* **101**, 1251–1262.

Poushter, J., Bell, J., Carle, J., Cuddington, D., Deane, C., Devlin, K., Drake, B., Keegan, M., Kent, D., Parker, B., Poushter, J., Schwarzer, S., Simmons, K., Smith, B., Stokes, B., Wike, R., Zainulbhai, H. 2015. Cell phones in Africa: communication lifeline. *Pew Research Center* 1–16. Available at: http://www.pewglobal.org/2015/04/15/cell-phones-in-africa-communication-lifeline/ [accessed 4 November 2018].

Rosegrant, M.W., Msangi, S., Sulser, T., Valmonte-Santos, R., 2006. *Biofuels and the Global Food Balance. 2020 Vision for Food, Agriculture and the Environment*. Available at: http://www.globalbioenergy.org/uploads/media/0612_IFPRI_-_Biofuels_and_the_Global_Food_Balance_01.pdf [accessed 4 November 2018].

Rosenzweig, C., Elliott, J., Deryng, D., Ruane, A.C., Müller, C., Arneth, A., Boote, K.J., Folberth, C., Glotter, M., Khabarov, N., Neumann, K., Piontek, F., Pugh, T.A., Schmid, E., Stehfest, E., Yang, H., Jones, J.W. 2014. Assessing agricultural risks of climate change in the 21st century in a global gridded crop model intercomparison. *Proceedings of the National Academy of Sciences of the United States of America* **111**, 3268–3273.

Shively, G.E. 2017. Infrastructure mitigates the sensitivity of child growth to local agriculture and rainfall in Nepal and Uganda. *Proceedings of the National Academy of Sciences of the United States of America* **114**, 903–908.

Sundström, J.F., Albihn, A., Boqvist, S., Ljungvall, K., Marstorp, H., Martiin, C., Nyberg, K., Vågsholm, I., Yuen, J., Magnusson, U. 2014. Future threats to agricultural food production posed by environmental degradation, climate change, and animal and plant diseases—a risk analysis in three economic and climate settings. *Food Security* **6**, 201–215.

Thow, A.M. 2009. Trade liberalisation and the nutrition transition: Mapping the pathways for public health nutritionists. *Public Health Nutrition* **12**, 2150–2158.

United Nations. *United Nations Comtrade Database*. Available at: https://comtrade.un.org [accessed 4 November 2018].

United Nations Conference on Trade and Development (UNCTAD). 2012. *Economic Development in Africa Report 2012: Structural transformation and sustainable development in Africa*. Geneva: United Nations Conference on Trade and Development. Available at: https://unctad.org/en/PublicationsLibrary/aldcafrica2012_embargo_en.pdf [accessed 4 November 2018].

United States Department of Agriculture (USDA). 2015. *Women, Infants and Children (WIC). WIC Funding and Program Data*. Available at: https://www.fns.usda.gov/wic/wic-funding-and-program-data [accessed 12 November 2018].

United States Department of Agriculture (USDA). 2016. *Food-Away-from-Home*. Available at: https://www.ers.usda.gov/topics/food-choices-health/food-consumption-demand/food-away-from-home.aspx [accessed 4 November 2018].

van Ittersum, M.K., van Bussel, L.G.J., Wolf, J., Grassini, P., van Wart, J., Guilpart, N., Claessens, L., de Groot, H., Wiebe, K., Mason-D'Croz, D., Yang, H., Boogaard, H., van Oort, P.A.J., van Loon, M.P., Saito, K., Adimo, O., Adjei-Nsiah, S., Agali, A., Bala, A., Chikowo, R., Kaizzi, K., Kouressy, M., Makoi, J.H.J.R., Ouattara, K., Tesfaye, K., Cassman, K.G. 2016. Can sub-Saharan Africa feed itself? *Proceedings of the National Academy of Sciences of the United States of America* **113**, 14964–14969.

Van Vliet, E.D.S., Asante, K., Jack, D.W., Kinney, P.L., Whyatt, R.M., Chillrud, S.N., Abokyi, L., Zandoh, C., Owusu-Agyei, S. 2013. Personal exposures to fine particulate matter and black carbon in households cooking with biomass fuels in rural Ghana. *Environmental Research* **127**, 40–48.

Wrottesley, S. V, Lamper, C., Pisa, P.T. 2016. Review of the importance of nutrition during the first 1000 days: Maternal nutritional status and its associations with fetal growth and birth, neonatal and infant outcomes among African women. *Journal of Developmental Origins of Health and Disease* **7**, 144–162.

Food Aid

How can food aid effectively reduce food insecurity?

Caroline Archambault and David Ehrhardt

Chapter Overview

- Food aid is defined as any voluntary transfer of money or food items that is aimed to increase food security of a specific population.

- There are many different ways in which food aid is delivered, which can be differentiated by the donor, the distance between donors and recipients, the time frame for which the aid is provided, and the type of aid that is given (in kind, vouchers, or cash).

- Five key ways of providing food aid are supplementary feeding (providing in-kind food to specific populations), food stamps (providing vouchers for food to eligible populations), food-for-work (FFW) (exchanging in-kind food for labour), food banks (food distribution points run by civil society organizations), and food sharing (sharing of food or money within social networks).

- While food aid is often effective at relieving food insecurities of recipients, there are several challenges to food aid efficacy, including: targeting (that is, how efficiently the aid achieves its goals); conditionalities that serve donors' interests, sometimes at the expense of recipients; spoiler behaviour by middlemen; and undesirable effects on recipient food systems (for example, lower agricultural productivity and aid dependency).

Introduction

As *Chapter 9: Food Security* showed, food insecurity can be driven by a range of causes, including droughts, poverty, and civil wars (see Figure 10.1). Food aid can help increase food security in these situations. Who should provide such aid? What kinds of food aid can they provide? How can we make sure that food aid actually reduces food insecurity for the

people who need it the most? And what are the wider impacts of food aid on those who give and receive it?

These questions are meant to make you think about a specific part of the global food system: food aid, or all voluntary transfers aimed directly at reducing food insecurity of a particular population. *Chapter 8: Nutrition* has explained what people need to consume

a)

b)

Figure 10.1 The Yemeni civil war (from 2015 to date) has devastated agricultural livelihoods, replacing (a) markets with (b) food aid handouts. More than 17 million Yemenis, almost two-thirds of the total population, are now food-insecure and a reported 7 million are on the brink of famine.

Source: (a) Rod Waddington/Flickr (CC BY-SA 2.0); (b) AP/Shutterstock.

to survive, and *Chapter 11: Consumption* will take you through the ways in which most people in the world go about choosing, buying, growing, and consuming that food. But *Chapter 9: Food Security* has shown that, despite producing more food than we can consume, more than 800 million people do not have such choices—they face food insecurity and hunger (WFP, 2017).

In this chapter, we explore how food aid has been provided to address this problem and sketch out key challenges to its efficacy. Why does this matter to food sustainability? First, it matters because many people around the world rely on food aid to survive and flourish. Second, food aid has supply chains, defined as the sequences of processes and organizations involved in the production, processing, and distribution of food items (see Section 12.1 in *Chapter 12: Food Systems*). These chains impact the environment, either negatively (for example, pollution caused by transporting food aid) or positively (for example, reducing food waste by donating excess or unwanted food).

In this chapter, we first define food aid and explain key concepts that help to understand food aid supply chains (Section 10.1). Subsequently, we explore different donors of food aid (Section 10.2) and the different ways in which food aid can be delivered to recipients (Section 10.3). Finally, we explore key challenges to effective food aid provision, both in terms of their immediate goal of increasing food security and their longer-term impacts on donor and recipient food systems (Section 10.4).

10.1 What is food aid?

Food aid is a response to acute or chronic food insecurity, which can be caused by anything, from poverty and violent conflict to seasonal weather patterns or the environmental impacts of climate change (see Figure 10.2). It is defined as the voluntary transfer of resources for the purpose of enhancing food security in a specific population (Barrett and Maxwell,

2007). In *Chapter 9: Food Security*, you have learnt that food security encompasses food availability, access, utilization, and stability. Food aid can focus on enhancing any or all of these pillars. Based on these four pillars, and by including sustainability concerns, we can specify the requirements for food aid to be effective:

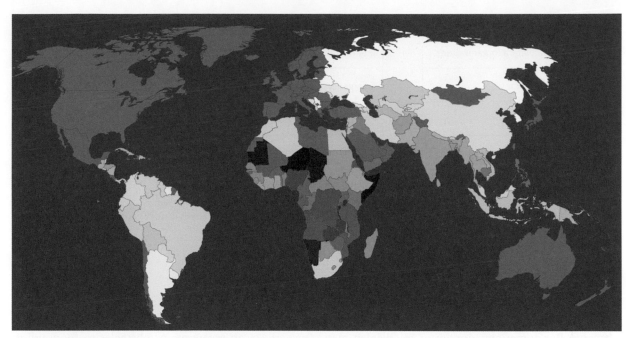

Figure 10.2 Map displaying the vulnerability of food systems to climate-related hazards, from low risk (light yellow) to high (dark red). Organization for Economic Cooperation and Development (OECD) and EU countries are excluded, because it is assumed that they can access global food markets to compensate for the negative impact of weather events.

Source: Food Insecurity and Climate Change Vulnerability map. © UN World Food Programme (WFP) and the Met Office Hadley Centre.

1. It should make food available, accessible, and utilizable to those in need

2. It should be sufficiently stable and provided for as long as the insecurity persists, and

3. It should not compromise the longer term sustainability of food systems connected to the aid supply chain.

Food aid is a food supply chain running from a donor to a recipient (or beneficiary). Donors are organizations or individuals who finance the aid or provide aid in cash or in kind (that is, provide food products directly). Examples include governments, food retailers like supermarkets, farmers, or neighbours. Many aid supply chains include middlemen, or agents who play a part in transferring food from donor to recipient. Examples include handling procurement, storage, or transportation; customs officers; or organizations tasked with the distribution of the food aid. Recipients are those who receive and utilize food aid.

As with any supply chain, stakeholders in the chain (donors, middlemen, and recipients) may have different interests and motivations to participate in these transfers. Interests may be aligned, for example when someone volunteers at a soup kitchen to provide free meals to the homeless (see Figure 10.3). In this case, the interest of the food donors is to help the

Figure 10.3 Volunteer soup kitchen. The interests of donors, middlemen, and recipients may be aligned (if all are interested in providing and receiving soup). But there may also be tensions, for example if the donors are providing poor-quality soup ingredients, the middlemen take ingredients home for their own families, or if some of the recipients are not actually poor but do not want to cook.

Source: Pixel-Shot/Shutterstock.com.

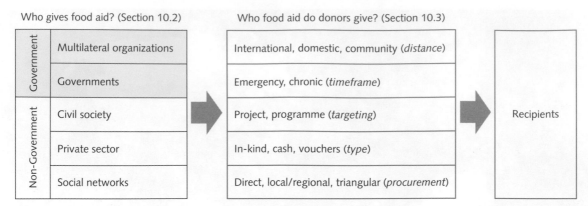

Figure 10.4 Flow chart showing the main food aid donors (discussed in Section 10.2) and different ways in which food aid can be given (see Section 10.3).

homeless; the volunteer (middleman) wants to provide free labour, and the homeless recipients want to eat soup. But sometimes donors or middlemen may have interests that are at odds with each other or with the needs of supposed beneficiaries. Such unaligned incentives can create major obstacles to food aid efficacy, as you will see in Section 10.4.

Food aid chains can vary considerably. Figure 10.4 maps out the main characteristics of food aid supply chains that give rise to this variation by donor, donor–recipient distance, time frame, targeting, type of aid, and procurement. These will be introduced in the next two sections; Section 10.2 discusses the donors and distance, and Section 10.3 explores the kinds of food aid that can be given (time frame, targeting, kind, and procurement) and analyses five real-life examples of important food aid efforts.

In your experience

What are some of the kinds of food aid that are given in your community, neighbourhood, or city? For one of these situations, try and draw out the supply chain as best you can, from the donor to the recipient. Think about whether it was a long or short supply chain and whether there were potential differences in interests among those who participated in the transfer.

10.2 Who gives food aid?

There are five main types of food aid donors (see Figure 10.4): multilateral organizations (international organizations with country governments as members), governments, civil society organizations, private sector actors, and social networks. When governments voluntarily transfer resources to enhance food security of recipients across national borders, we call that **international food aid** (Barrett and Maxwell, 2007) (see Box 10.1). Governments can do this bilaterally, in the form of a transfer from one government to another. Alternatively, they can pool their contributions with other governments and donate their assistance collectively through multilateral organizations [like the World Food Programme (WFP)] (see Box 10.1).

Pause and think

Do people living in high-income, food-secure countries have a moral obligation to feed those who are hungry? Why or why not? Should they feed the hungry in their own countries before assisting those elsewhere?

BOX
10.1

International food aid flows

There is a long history of international food aid during periods of crisis. Take, for example, the international relief efforts, most notably by the USA, during the Great Famine in Ireland from 1845 to 1849. Or the support of the Red Cross (one of the first international organizations providing food aid) to victims of World War II (see Figure A.a).

A more structured approach to international food aid emerged in the 1950s, when the USA began developing programmes to transfer their agricultural surpluses to low-income countries. This was a precursor for **multilateral food aid**, which was institutionalized in the early 1960s under the leadership of the United Nations WFP. Established in 1961, the WFP aimed to de-politicize food aid through the creation of a multilateral, intergovernmental agency for eradicating hunger and promoting development. Headquartered in Rome and governed by an Executive Board with representatives from 36 member states, it is the largest humanitarian agency

providing food aid worldwide (see Figure A.b), working in more than 80 countries.

Despite the media attention it receives, international food aid has been comparatively modest in terms of its impact on global food security. It is a small share (less than 3% since 2000) of global **official development assistance (ODA)** flows, which the OECD defines as all government aid designed to promote the economic development and welfare of low- and middle-income countries (for example, for education, health, or infrastructure) (OECD, 2018). Similarly, as a source of food, food aid is dwarfed by local food production and commercial trade (Barrett and Maxwell, 2007). In absolute terms, international food aid has decreased steadily over time, from 11 billion kilograms transferred in 2001 to 5 billion kilograms transferred in 2012 (WFP, 2013) (see Figure B).

The USA is the largest contributor to international food aid. In 2012, it contributed 44% of total food

a)

b)

Figure A (a) Canadian Red Cross advertisement asking for donations for food aid during World War II. (b) A boy in Raymah (Yemen) collecting WFP wheat in 2016.

Source: (a) © Canadian Red Cross; (b) Julien Harneis/Flickr (CC BY-SA 2.0).

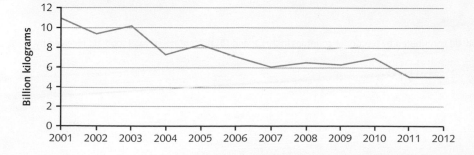

Figure B Global food aid deliveries in billion kilograms from 2001 to 2012.

Source: 2012 Food Aid Flows (2013). © World Food Programme.

aid deliveries, followed by Japan (8%), Brazil (7%), Canada (6%), China (5%), and the rest of the 58 donors contributing the remaining 30% (WFP, 2013) (see Figure C.a). Eight countries received 49% of the food aid deliveries: Ethiopia (16%), the Democratic People's Republic of Korea (8%), Yemen (5%), and Bangladesh, Kenya, Pakistan, Somalia, and Sudan (4% each) (WFP, 2013) (see Figure C.b).

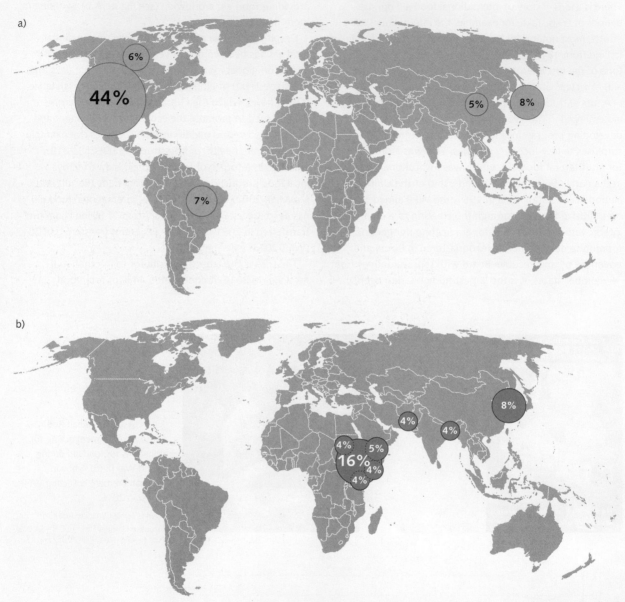

Figure C (a) Percentage of total food aid contributed by top donor countries in 2012; and (b) percentage of total food aid received by the largest recipient countries in 2012.

Source: 2012 Food Aid Flows (2013). © World Food Programme.

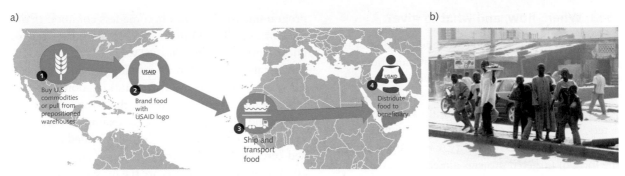

Figure 10.5 (a) Basic schematic of the United States Agency for International Development (USAID) food aid supply chain; and (b) young Quranic students (*almajirai*) in northern Nigeria begging for money to buy food.

Source: (a) United States Agency for International Development; (b) David Ehrhardt.

Governments can also transfer resources to enhance food security of their own citizens (or residents); this is called **domestic food assistance**. Domestic food assistance is organized through officially sanctioned channels involving public bureaucracies. Of course, private businesses, such as farmers or transportation companies, can also be contracted to participate in the supply chain, as can civil society organizations.

Civil society is a term used to describe organizations that are neither the state nor private businesses; it represents specific groups or interests. It includes, for example, NGOs, community-based organizations (CBOs), faith-based organizations (FBOs), labour unions, charitable foundations, professional associations, or sports clubs. Some of these organizations provide food aid to their own members, while others assist food-insecure populations in neighbourhoods or cities, across entire countries, and even internationally. The private sector can also be a donor of food aid. Supermarkets, for example, may donate their surplus or non-marketable food as part of their corporate social responsibility or to avoid the costs of food waste disposal.

Social networks finally are defined as the connections you have with other people, for example through extended family (kinship) ties, friendship, religious affiliations, schooling, or work. Food aid can be transferred between members of the same network, but also can be provided across networks. Networks can be small and localized, for example in a single neighbourhood, but they can also be geographically spread out, as, for example, cross-national migrant networks. The flow of international remittances, that is the money sent by migrants to their home communities, can be an example of cross-border food aid within social networks.

Real-life aid chains can include multiple donors, middlemen, and recipients. Moreover, they can vary considerably in the physical distance from donor to recipient. Take, for example, an international shipment of wheat from the USA destined to help displaced persons in East Africa (see Figure 10.5a). Contrast this with the cash you could give to young boys begging on the street in northern Nigeria (see Figure 10.5b), and you get a sense of the variation involved. This variation is important to the nature and efficacy of aid, as Section 10.4 and Section 10.5 will show in more detail.

Connect the dots

Think back to the concept of food miles introduced in *Chapter 6: Climate Change*. Can it be applied to aid supply chains? Does it always imply that shorter aid chains are more sustainable than longer ones? Why (not)?

10.3 What food aid do donors give?

This section introduces two key distinctions between food aid efforts (time frame and type; see Section 10.3.1) and provides illustrations of five different types of efforts (supplementary feeding, food stamps, FFW, food banks, and food sharing; see Section 10.3.2). Combined, the two sections describe the intricacies and variety of approaches to food aid. They also illustrate the extent to which food aid affects lives around the world.

10.3.1 When, how, and what to give?

Once donors have decided to give food aid, they must decide on the time frame of aid provision, how to target the aid, and what to actually give. Donors can give short-term **emergency aid** in response to urgent needs, for example to victims of natural or man-made disasters. They can also provide **chronic aid** to respond to long-lasting food insecurity. Chronic aid can take the form of targeted projects aimed at specific populations (see Section 10.3.2). It can also be given in untargeted ways, for example when donors give recipient governments budget support in the hope that they will use it to improve food security (this is referred to as **programme aid**). In recent years, international programme food aid has become less common (from 58% of overall ODA in 1988 to 3% in 2012), due to uncertainty about its efficacy (WFP, 2013).

In terms of what donors actually provide, they can either give food (called in-kind aid) or money to buy food. In international food aid, the large majority of in-kind aid (86%) are cereals (for example, wheat and wheat flour, coarse grains, rice, and blended and fortified cereals), due to their high calorie content (WFP, 2013). Pulses, oils, and fats are the most common non-cereal foods and are provided because of their high protein content. In-kind food aid can also vary in the ways in which it is distributed, utilized by recipients, and where it is procured by the donors (see Box 10.2).

BOX 10.2 Procuring in-kind food aid

If donors want to give in-kind food aid, they must decide where to get it (procurement)—a decision that can have important implications for aid efficacy and efficiency (see Section 10.4.3). Generally, the choice is between procurement in the recipient community or elsewhere. In international food aid, there are three procurement options:

1. Direct transfers, where the food is grown or purchased in the donor country and sent to the recipient country (for example, wheat grown in Europe, bought by European donors, and sent to South Sudan)

2. Local and regional purchases, where food is purchased within the recipient country (for example, wheat grown in South Sudan, purchased by European donors, and delivered to recipients in South Sudan), and

3. Triangular purchases, where the purchase is made by the donor in a third-party country (often nearby the recipient country) and delivered to the recipient (for example, wheat grown in Kenya, bought by European donors, and delivered to recipients in South Sudan).

Within international food aid, direct transfers have made up the majority of food aid delivery (see Figure A), but the percentage declined in recent years (from over 80% in 2001 to 60% in 2012) (WFP, 2013). The remaining proportion is roughly split evenly between local and triangular purchases.

Figure A Emergency food aid of wheat provided by USAID for refugees at Dolo Kobe camp in Ethiopia—an unmonetized direct transfer (from the USA) delivered in a central point in the refugee camp.

Source: United States Agency for International Development.

Connect the dots

Try to imagine you live in circumstances where you have hardly any access to good-quality food for your family and you receive emergency aid. What would you like to receive in a WFP food basket? Thinking back to *Chapter 8: Nutrition*, which food products would have your priority, and why? What do you think is actually in the box? For the answer to the last question, check: https://www.wfp.org/food-assistance/kind-food-assistance/wfp-food-basket

Instead of in-kind aid, donors can also provide money for recipients to purchase their own food. This can come in the form of cash (physical bills and coins), paper vouchers that can be exchanged for food in specific locations, or as digital payment cards that can be used as debit or credit cards. Vouchers and digital cards can be restricted in their use (for example, by designating the shops where they can be used or the types of food that they can be used to purchase); digital cards allow donors to monitor where and how the money is spent (see Box 13.4 in *Chapter 13: Governance* for a WFP example of this approach). There are many generic reasons why being paid in cash or vouchers might be preferable to being paid in food—it gives beneficiaries autonomy over

their household management; it is more likely to allow them to diversify their diet, and cash or vouchers are generally easier to deliver than food (Metz et al., 2012). Yet it requires food markets to be available and accessible to recipients—in other words, people must be able to use the cash or vouchers to buy food. This may not be the case in situations of drought or civil war, or if a food-insecure community lives far away from the local market (or the roads to the market are poor). Moreover, it may be costly or unsafe to deliver cash or vouchers in sufficient quantities at the right times, and it is more difficult to ensure and monitor that recipients actually spend the cash on food (Mercy Corps, 2007).

10.3.2 Examples of food aid

Preceding sections have highlighted six variables along which food aid can vary—by donor, donor–recipient distance, time frame of aid provision, targeting strategy, the type of aid that is provided, and the way it is procured. This section aims to give a sense of how food aid really works, by showcasing five real-life examples: supplementary feeding, food stamps, FFW, food banks, and food sharing (see Figure 10.6).

While these five examples are not comprehensive of all food aid efforts, they cover a wide range. This is highlighted in Table 10.1. Note that, for simplicity,

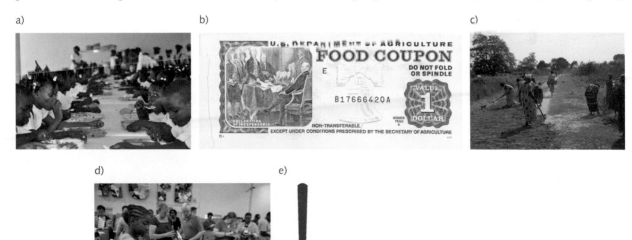

Figure 10.6 Examples of food aid efforts. (a) Supplementary school feeding programme in Nigeria. (b) Food stamps (vouchers) in USA. (c) FFW project in Sri Lanka. (d) Food bank in Atlanta, USA. (e) Online food sharing initiative in Germany.

Source: (a): https://tvcnews.tv/2018/03/osinbajo-lauds-home-grown-school-feeding-programme; (b) United States Department of Agriculture; (c) Food for Work, Sri Lanka (via Wikimedia Commons); (d) © Atlanta Community Food Bank; (e) Foodsharing.de.

Table 10.1 Overview of the five food aid examples discussed in Section 10.3.2 and their features in terms of donors, donor–recipient distance, time frame, and type

	Food aid examples				
	Supplementary feeding	Food stamps	Food-for-work	Food banks	Sharing
Donors	International, domestic, civil society	International, domestic, civil society	International, domestic, civil society	Civil society	Networks
Donor–recipient distance	International, domestic	International, domestic	International, domestic	International, domestic, community	International, domestic, community
Time frame	Chronic, emergency	Chronic	Chronic, emergency	Chronic	Chronic, emergency
Type of aid	In-kind	Vouchers	In-kind	In-kind	In-kind, cash

we organize distance into three categories: **international** (donor and recipient are in different countries), **domestic** (donor and recipient are in the same country, but not in the same immediate community), and **community** (donor and recipient are in the same geographically bounded community). We discuss each of the examples below, but the general message of Table 10.1 is that food aid efforts are flexible and can be given by different donors, over varying distances, and for either chronic or emergency purposes.

Example 1: supplementary feeding

Supplementary feeding directly provides food to people who are food-insecure, rather than giving cash or vouchers (Barrett, 2002a; FAO, 2001). Included in these are all emergency food donations (for example, by the WFP; see Box 10.1), as well as longer-term projects that focus on vulnerable groups like school lunches (see Case study 10.1), pregnant women, or the elderly.

In your experience

Did you bring your own lunch to school, or did your school serve school meals? If your school provided food, what was on offer? Were these healthy lunches? Did the school pay for your lunch, or did your parents have to? Do you know anything about the sustainability of the supply chains of this food? And do you think all schools should provide food to students?

Example 2: food stamps

Food stamps are vouchers given to eligible individuals, which can be used in exchange for food items in stores (Barrett, 2002a). The US SNAP (see Case study 10.2) is one of the largest and best-known examples of a food stamp programme, but other countries have (had) similar schemes, funded either domestically or internationally.

Pause and think

In February 2018, US President Trump proposed to reform SNAP. Instead of offering all the food aid through the EBT digital credit card, half would be given in kind. The in-kind half would take the form of a food package with '*shelf-stable milk, ready to eat cereals, pasta, peanut butter, beans and canned fruit and vegetables*' (Hunzinger et al., 2018). All products would be grown and produced in the USA. What do you think are the possible benefits and risks of this reform?

Example 3: food-for-work

Food-for-work (FFW) schemes make the provision of food aid conditional on providing labour. Such efforts offer food-insecure people temporary jobs on projects (often building public infrastructure like roads, dams, or irrigation systems) in exchange for food. The amounts and types of food are determined on the basis of local needs and market conditions. FFW is often used in post-emergency situations that are stable enough to allow for construction, but in which people

CASE STUDY 10.1

Sustainable school meals

From the USA to Japan and from Nigeria to Finland, one of the most common ways in which multilateral organizations, governments, and NGOs provide food aid is through school meals (Drake et al., 2016; Harper et al., 2008). The details of the programmes vary widely in content, as well as in the methods of provision, the funding structure, and the supply chains. While in some countries (for example, Sweden and Japan), virtually all students eat school-provided lunches, in others (for example, the USA or Hong Kong), the meals are targeted to vulnerable groups within the school. Similarly, the degree of subsidization varies greatly; in Sweden and Brazil, all food in schools is free, while in Australia, virtually all food has to be paid for by the pupils (which means these school meals are not food aid).

This variation reflects the different goals and priorities that donors aim to achieve with school feeding, including promoting healthy diets, enhancing food security, incentivizing children to come to school, and making food systems more environmentally sustainable.

The school meal system in Brazil (*Programa Nacional de Alimentação Escolar*, PNAE) is a good example of an attempt to address all these goals at once (see Figure A). Although school meals have existed in Brazil since the 1950s, the PNAE was expanded and reformed as part of the larger Brazilian Zero Hunger policy programme implemented in 2003 (Otsuki, 2011). The current PNAE provides around 45 million school lunches to Brazilian children every day (Sidaner et al., 2013).

Figure A Children eating school lunch in Paragominas, Brazil.

Source: Hilario Junior/Shutterstock.com.

It targets all students enrolled in basic education from the age of 6 months, providing up to 70% of the nutritional needs of children enrolled in full-time education. The content of the meals must respect local eating cultures (as Brazil has a lot of regional variation), minimize the intake of sugar and fat, and ensure the consumption of at least 200 g of fruit per week. Nutrition education is also part of the policy, teaching children about healthy consumption and lifestyle. Finally, the government specifies that 30% of the subsidies provided for school food must be spent on food procured from local farms (ideally organic and ecological), with the twin goal of supporting family agriculture and enhancing the sustainability of the supply chains for school food (Sidaner et al., 2012).

There is considerable evidence that PNAE, together with the other parts of the Zero Hunger policy approach, has reduced food insecurity and hunger in Brazil (de Mattos and Bagolin, 2017). As such, Brazil's policy has come to be an example of how school meals can effectively reduce food insecurity. But in addition, it highlights that careful procurement strategies (in this case, focused on strengthening local agriculture) can contribute to enhancing sustainable food systems.

CASE STUDY 10.2

The Supplemental Nutrition Assistance Program (SNAP)

In 2017, the US government spent around $70 billion on SNAP, a food stamp programme that supports over 42 million Americans (13% of the population) in their food purchasing (Center on Budget and Policy Priorities, 2018). How does it work? American households that live on or below the poverty line are eligible to apply for SNAP benefits, and if their application is accepted, they receive an Electronic Benefit Transfer (EBT) (see Figure A) that functions like a debit card and is charged with the appropriate funds every month. SNAP recipients receive dollar-equivalent credit on their EBTs, depending on their income and assets.

Beneficiaries can use this EBT to pay for food purchases in over 250,000 stores around the USA, including large retail chains such as Walmart. SNAP credit can be used to buy all staple foods and other processed food items, including candy or soft drinks. 'Luxury items', such as alcohol, tobacco, and non-food items, are ineligible. Hot food is also not included, except in some cases where qualified recipients can use their SNAP credit to buy ready-made meals at certain restaurants (USDA, 2017).

How effective has SNAP been in reducing US food insecurity? Surveys suggest that SNAP reduces the risk of being food-insecure by 20–30% (Ratcliffe et al., 2011). But in terms of diet quality and nutrition levels, the evidence of SNAP's impacts is inconclusive (Gregory et al., 2013). The main obstacles to SNAP improving recipient diets include the marketing of unhealthy food to recipient communities, the high cost of healthy foods, and lifestyle determinants of low-income diets (Blumenthal et al., 2014).

Figure A EBT cards are distributed by state governments in the USA and can be used to pay for food in SNAP.

Source: United States Department of Agriculture.

a)   b)   c)

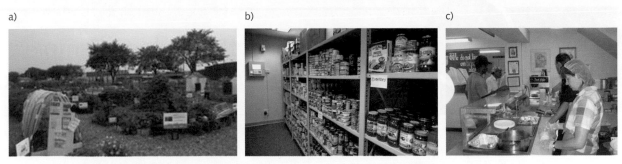

Figure 10.7 Examples of: (a) a neighbourhood garden; (b) a food pantry; and (c) a soup kitchen.
Sources: (a) Wikimedia Commons (CC BY 2.0); (b) U.S. Air Force photo by Airman 1st Class Jeffrey Parkinson; (c) Nancy Heise via Wikimedia Commons.

cannot easily access food markets. Alternatively, donors can pay the workers in cash, allowing them to purchase the food themselves (often referred to as cash-for-work, or CFW).

Example 4: food banks

Civil society organizations are involved in producing, collecting, and redistributing food in a variety of ways—for example, some run neighbourhood gardens in which food is grown and distributed to those who are food-insecure; food pantries that look like grocery stores but are actually places where people can pick up food free of charge; or soup kitchens where they can be served a free meal (see Figure 10.7).

Many of these efforts rely on food banking, that is the collecting, sorting, and preparation of food donations to be redistributed to the hungry (Bazerghi et al., 2016). These **food banks** can either provide their food aid directly to recipients (as in the Langar in Case study 10.3) or they can distribute it to other organizations, like soup kitchens or pantries, that will then directly give to recipients. Most food banks rely on voluntary donations of food, and much of this food comes from excess production along the food supply chain. Think, for example, of leftovers from supermarkets past the 'sell-by' date or products that do not meet aesthetic standards (for example, oddly shaped fruit). Food is also collected through public solicitation, like food drives, and individual donations. Food banks thus not only provide food aid, but also reduce waste and make food systems more sustainable. In 2010, an estimated 7% of people in high-income countries (60 million people) used food banks (Gentilini, 2013); the model is also spreading to many lower-income countries (see https://www.foodbanking.org/what-we-do/our -global-reach/ for a list).

Example 5: food sharing

Food sharing through social networks happens all around the world and for many different reasons. In many cases, it may only serve to build relationships (for example, sharing cake with your neighbours) or to diversify diets (for example, exchanges of different food between farmers) and, as such, thus does not constitute food aid (see Figure 10.8). But in other cases, sharing is also aimed at enhancing food security. Although systematic research is scarce, sharing as food aid occurs in high- and low-income countries. It appears to be particularly common in situations where other forms of food aid are not available or accessible, for example because governments do not provide well-targeted food aid (see Case study 10.4) or where donors cannot reach those in need (for example, in civil wars). Case study 10.4 provides one example from rural USA, highlighting both the importance of food sharing (for several reasons, including food insecurity) and some of the constraints on official forms of food aid, particularly for vulnerable populations like the elderly.

Figure 10.8 Sharing bar food with friends—under what conditions might this be an example of food aid?

The Langar of the Golden Temple: the largest soup kitchen in the world

Donations (alms) from faith-based organizations can function in similar ways to food banks. For example, many Muslims practice *zakat*, which requires them to give a percentage of their savings and annual business revenue (2–3%) or of their harvest (5–10%) to the poor. They can do this individually, in which case it constitutes food sharing, but they can also donate through their mosque, which then distributes it as a food bank. Similarly, Sikh temples often offer *langar* (Punjabi for kitchen), a free meal to all visitors, regardless of faith and socio-economic circumstance (see Figure A).

In the Indian city of Amritsar stands Sri Harmandir Sahib, informally called the Golden Temple, considered the Sikh's holiest shrine and the world's largest soup kitchen. Every day, around 50,000 people are fed a free, simple, hot meal. Numbers can double on weekends and special occasions. To feed these numbers, more than 450 staff and hundreds of volunteers prepare thousands of *rotis* (Indian flatbreads) and litres of *dal* (lentil soup). One estimate of the daily ingredient requirements include 5000 kg of wheat flour, 1400 kg of rice, 1800 kg of lentils, and 700 L of milk (Malhotra, 2016). The meal is cooked on firewood and gas stoves, and in electronic breadmakers (see Figure A).

a)

b)

c)

d)

Figure A (a) Dinner time at Sri Harmandir Sahib, informally referred to as (b) the Golden Temple in Amritsar, India; it features (c) automatic chapati making and (d) many, many plates for *langar*.

Source (a) Ravneetn13/Wikimedia Commons; (b) Manu (via Pexels.com); (c) Ajay Goyal/Flickr; (d) anijdam/Flickr (CC BY 2.0).

CASE STUDY 10.4

Food sharing and the elderly in North Carolina

Older persons tend to be vulnerable to food insecurity for a variety of reasons. For example, ageing and declining health lead to changing nutritional needs. Older persons also find it more difficult to access food, as their income declines, their capacity to produce their own food diminishes, or they find it increasingly difficult to shop for food. Such is the case for the elderly population living in two rural counties in North Carolina (USA), a case study that highlights the benefits and limitations of food sharing as a form of food aid (Quandt et al., 2001).

Like many other communities and neighbourhoods around the world, rural elders in North Carolina participate actively in food sharing with their networks of extended family, neighbours, and churchmates. Following 145 multi-ethnic senior residents (aged 70+) over a 1-year period showed that food sharing was the norm, rather than the exception (Quandt et al., 2001). Their food exchanges were mostly in person, although occasionally food was left anonymously on someone's doorstep.

Why do people turn to informal sharing? First, food sharing is a way to build community and norms of

Figure A Shared meals are one form of food sharing for elderly people.

generalized reciprocity, that is, giving without the immediate expectation that it will be returned (as opposed to balanced reciprocity where immediate returns are expected) (Quandt et al., 2001). Sharing is also driven by conservation where community members would rather give leftovers to other people than see them go to waste (Falcone and Imbert, 2017). Neither of these types of sharing are directly food aid. Food sharing is food aid for these older members of the community when it is given as a response to food shortages, or unbalanced diets, or an inability to utilize the foods they have access to (Quandt et al., 2001).

Interestingly, many people in these two rural communities are eligible for state-funded nutritional programmes for the elderly. They are also entitled to make use of national food stamp schemes and other domestic programmes. But much of this help is inaccessible to them. For example, the bureaucracy involved is complex and requires a lot of form filling. Also, distribution points take a long time to get to (or cost a lot of money in transport). The long distances mean that meal delivery services are not available, as the food cannot stay fresh enough to meet health safety standards. And some formal programmes, like food stamps, are considered stigmatizing, leading some people to reject them (Quandt et al., 2001).

There is evidence to suggest that food sharing is a common, often effective, coping strategy for food-insecure populations, even if official food aid is also available (see, for example, Cohen, 2018; Garasky et al., 2006; Tam et al., 2014). But more research remains to be done. And there are limitations to the strategy as well. For example, in the North Carolina case, some individuals could not participate—wealthy individuals or individuals with a weak social network were less likely to receive food. More importantly, the amounts of food received varied, for example from a fridge full to a head of lettuce, and donations were sometimes unpredictable (Quandt et al., 2001).

Connect the dots

Chapter 11: Consumption describes how much food is wasted every year: up to 180 kg by each EU citizen. Food banks and food sharing can be used to reduce waste. Case study 10.4 provides an example of waste reduction through food sharing in a rural North Carolina community, while in Germany, an online social networking platform helps members to reduce their waste by organizing food exchanges using their website (Ganglbauer et al., 2014). If you search the Internet, can you find other innovative efforts to reduce waste through food aid? How effective can such projects be? And what do you think is their potential contribution to waste reduction globally?

Connect the dots

Consider the different ways of providing food aid in all five examples that have been provided in Section 10.3.2. Thinking back to *Chapter 6: Climate Change*, in what ways might the different types and examples of food aid delivery mode impact our environment? Are some more sustainable than others? Why (not)?

10.4 What are the challenges to providing effective food aid?

Since World War II, food aid has helped millions of people to overcome food insecurity and hunger (Barrett, 2006). Although large-scale, comparative evaluations are only available for international aid, the case studies described in Section 10.3.2 have already shown that many forms of food aid can be effective—from American food stamps to Brazilian school meals, and from WFP food drops over famine-hit regions to food sharing through networks. But they also have limitations and side effects. We will use this section to highlight some of these issues, which can be used to develop approaches to further enhance food aid efficacy.

Section 10.4.1 introduces targeting, arguably the key condition for effective food aid. Subsequent sections deal with incentive problems in supply chains: donors (Section 10.4.2), middlemen (Section 10.4.3), and recipients (Section 10.4.4). Together, they highlight areas of attention to predict and prevent negative side effects. Many of the examples are based on international aid (both multilateral and bilateral), mostly because the problems are best studied in these aid efforts. However, the principles behind the challenges can be applied to all forms of aid.

10.4.1 Project design: targeting

Targeting assesses whether food aid reaches the people who actually need it (without flowing to people who do not), at the moment they need it, in the appropriate form, and at the lowest possible cost (Barrett, 2002b). Better targeting means more effective projects, both improving food security and preventing unwanted side effects. As such, it may be the key requirement for effective food aid (Barrett, 2002a)—yet aid efforts vary considerably in their level of targeting. For example, programme aid is poorly targeted, as it is given to governments without further requirements. In contrast, FFW schemes are designed to be self-targeting in the sense that only those who really need the food aid will apply to do the labour (see Figure 10.9).

Successful targeting of food aid is challenging as it requires (i) high quality information, (ii) good timing, and (iii) balancing trade-offs between speed and precision (Barrett 2006, 2010).

Regarding information quality, targeting requires knowledge about recipient communities (Barrett, 2002b). Project designers should know who needs the food aid and what their specific needs are, as well as the reasons why they are food-insecure. They also need to know the context in which the aid is delivered. For example, are local food markets functioning? How high are transportation costs? What is the security situation? Are there other stakeholders that can either facilitate or spoil delivery? And what will be the impact of the food aid on local food systems? Answers to these questions are necessary but often difficult to get before the start of a project, particularly if there are large distances between donors, middlemen, and recipients.

a)

b)

c)

d)

Figure 10.9 Different food aid types with different targeting strategies: (a) US Airforce drops pallets of water and food to Mirebalais, Haiti, 2010; (b) recipients lining up to collect their rations from the back of a WFP helicopter in South Sudan; (c) food basket delivered to someone's front door; (d) cash transfers given out to eligible recipients. Can you say anything about the targeting efficacy of each of these examples?

Source: (a) U.S. Air Force photo by Tech. Sgt. James Harper Jr; (b) © Alexandr Podolian; (d) teegardin/Flickr (CC BY-SA 2.0).

Targeting also depends on timing—food aid must be delivered at the right time in order to be effective. In emergencies, this simply means as quickly as possible, but evidence on international aid suggests that this is difficult—the median response time of US emergency food aid is a lengthy 5 months (Barrett, 2010). Timing also matters for seasonal forms of food insecurity, as food aid during harvesting time has been shown to decrease food prices, risking the livelihoods of local farmers and undermining local food production efforts (see also Section 10.4.4).

Third, there is tension between the pressure to respond quickly and the need to design well-targeted projects. The former requires speed and decisive action. The latter, in contrast, demands time-consuming deliberation and careful implementation, for example by testing a delivery mechanism through a well-monitored pilot project. If the pilot runs well, then the programme can be expanded (or scaled up). This is a costly and long-term strategy—and an emergency may be over by the time the project is ready for implementation. In some cases, therefore, targeting can be traded for speed; but knowing when to do this and how to then avoid the pitfalls of in-kind food aid remains challenging.

Pause and think

Find at least three examples of different types of food aid in this chapter, and compare their strengths and weaknesses in terms of targeting. Are some better targeted than others? Why (not)? And what do you think is the impact of this variation in targeting?

10.4.2 Donor motivations

Reducing food insecurity is not the only donor motivation to give food aid. Aid can be a (geo-)political tool, by providing support to 'friendly' governments. Conversely, it can be withheld to make demands from recipients (Clapp, 2015; Wallerstein, 1980). Aid deliveries give opportunities to enter new markets, while setting up the institutional apparatus and networks necessary for commercial trade (Barrett, 2002c). Food aid can also be used to stimulate the agricultural exports of donors or to give away surplus food as in-kind aid (Barrett, 2006). Aid that is offered on the condition that it be used to procure goods or services from the donor is called **tied aid**.

When are donor motivations problematic? In principle, aid given for any reason has the potential to be effective. The important factor is whether or not donor and recipient interests are aligned. Take the controversial practice of **surplus dumping**—situations in which donors use surplus food (for example, from agricultural overproduction) as in-kind aid. In international aid, surplus dumping is highly controversial because it can have a doubly negative effect—first, facilitating donor overproduction and second, reducing the productive capacity of the recipients (see Figure 10.10 and Food controversy 10.1).

Yet surplus dumping also has the potential to serve both recipient and donor interests at once. Take North Carolina's informal food providers (see Case study 10.4). They sometimes give away what they do not need and, much like surplus-dumping governments, thus give aid on the basis of their own interests, rather than those of the recipients (who may or may not like the food they get). Yet if their interest is to minimize waste, and the recipients actually like the food they get, interests are aligned and everybody wins.

The impact of donor motivations on food aid efficacy thus depends on the extent to which they are aligned with recipient interests. Of course, identifying these motivations and interests is complex, as stakeholders can have multiple interests at once and they can shift over time. For example, the early US food stamp programme in the 1930s was driven both by food insecurity of Americans hit by the Great Depression and the simultaneous overproduction of certain agricultural products by American farmers. It subsidized products that US farms were producing in unsellable quantities (for example, eggs, butter, and beans), which served both the interests of the poor and the farmers. But historians argue that, over time, the food stamp policy became increasingly motivated by the political need to raise agricultural incomes at the expense of reducing food insecurity for the poor (Poppendieck, 2014).

Pause and think

Tied aid is often less efficient than in-kind food aid procured locally, costing at least 30% and up to 50% more (Clay et al., 2006). As a result, international food aid donors have been trying to reduce tied aid, but some (including the USA) continue to use it (Clay et al., 2009). What might be reasons for donors to give tied aid? And besides efficiency, what might be other benefits of untied aid? In what situations do you think donors should give tied or untied aid?

a)

b)

Figure 10.10 (a) In 2012, the USA provided in-kind emergency food aid to the Horn of Africa through Kenya. Under what conditions might this reduce the incentive for (b) local Kenyan farmers to produce the same product?

Source: (a) United States Agency for International Development; (b) CIAT/Flickr.

10.4.3 Middlemen and thieves: spoilers

As you have seen throughout this chapter, food aid often travels considerable distances from donor to recipient. In these supply chains, there are many opportunities for **spoilers**—stakeholders or third parties working in their own interests and against aid efficacy. For example, there are many instances in which food aid is stolen before it reaches recipients (Polman, 2010) (see Box 10.4). Some, but not all, spoiling is done by third-party thieves; it also involves the middlemen in the aid supply chains.

Spoiler problems are a risk to all forms of food aid but are more common as supply lines increase in length or complexity. The opportunities for spoiling increase, as monitoring becomes more difficult and donors and recipients have too little information about the aid delivery process to keep middlemen in check. In economics, this problem is referred to as the **principal agent problem**, in which a principal hires an agent for a specific task. If there is some tension in the interests between the principal and agent (**split incentives**) and the principal does not have enough information about the agent's capabilities or actions (**asymmetric information**), the agent is likely to cheat the principal (and, in this case, reduce the impact of food aid) (Mankiw, 2014).

In international aid, the distance between donors and recipients increases the likelihood of asymmetric information, allowing agents to siphon off aid money or food (the Yemeni example in Box 10.4). Asymmetric information can be particularly problematic in emergency situations, when things are happening quickly and monitoring is difficult or dangerous (the Boko Haram example in Box 10.4). Domestic assistance by governments, food aid by civil society organizations, and even food sharing can also suffer from principal agent problems, depending on the extent to which donor–agent–recipient incentives are aligned and donors are informed about the efficacy of delivery (the Zimbabwean and Boko Haram examples in Box 10.4).

BOX 10.4   Stealing food from the poor

Who steals food from a hungry person? It turns out that many people do. It is a risk in virtually all places where food aid is provided and involves a range of spoiler middlemen, from insurgents and criminal looters to soldiers, local leaders, and even formal political parties. The sample of stories below are illustrations of this, collected over the period of writing this textbook.

June 2017: Perhaps half of all the food aid sent for Boko Haram victims in Nigeria is 'diverted' (a euphemism for theft), including donations from multilateral organizations and NGOs (see Figure A). A week earlier, 200 tonnes of dates donated to Nigeria by Saudi Arabia for Ramadan were found on sale at local markets (BBC, 2017).

October 2016: A report is released that details a huge amount of aid, perhaps worth millions of dollars and including basic food, stolen by Yemeni leaders (tribal chiefs, political and rebel leaders) and sold to merchants for resale (below market price) or for distribution in their networks (Culbertson, 2016).

September 2016: The Zimbabwean Human Rights Commission has accused the ruling Zanu-PF party of food aid theft and political discrimination in its distribution,

Figure A Nigerian army demonstrating their tactics against Boko Haram, an insurgency operating in the north east of the country. But is it enough to prevent the movement from stealing food aid to sustain itself?

Source: VOA/Nicolas Pinault.

denying opponents of the ruling party access to the aid that was largely government-provided (The Guardian, 2016).

10.4.4 **Recipients: food system distortions**

One of the most complex challenges of food aid is the impact it has on recipient food systems. Next to the direct positive effects of food aid, there are potential negative side effects of aid provision, and there are many examples where such impacts are clear. However, systematic evaluations of the theory are scarce (Awokuse, 2011). As a result, it is unclear how pervasive the problems are and when they are most likely to appear (Barrett, 2006).

How can food aid have a negative impact? First, it can lead to **aid dependency**—a situation in which recipients come to rely on aid flows for some of their basic needs and activities (see Figure 10.11). This may be driven by labour disincentives (that is, recipients feel less pressure to work because they receive food for free) or production disincentives due to price decreases (if free food floods a market, locally produced food prices will drop dramatically). Food aid can also lead to shifts in local consumption away from locally produced goods (thus further reducing local food prices), as food aid products become more appealing. There is proof that this has happened in the West African Sahel in the 1970s and 1980s, as food aid flooded the region with wheat (not indigenous to the region), but systematic evidence of this pattern is hard to find (Barrett, 2006).

Second, international food aid, civil society aid, or aid through social networks can lead to more food insecurity if it reduces pressure on governments to solve the underlying causes of the insecurity. Food aid can be a disincentive for policy reform or provide an incentive for governments to reform policies to facilitate food aid (see Food controversy 10.1). Even more dramatically, there is evidence that emergency food aid can be used as a weapon. For example, governments have cleared populations from a certain area by restricting the availability of food within that area and enhancing food aid outside it. They have also used food aid to feed their soldiers (Barrett and Maxwell, 2007; Keen, 1994). As such, there is evidence that food aid may even sustain civil wars (Nunn and Qian, 2014).

Of course, given the positive contribution food aid has made to food security over the years, these negative side effects are not an argument to abandon food aid. And positive side effects can also occur; for example, if food prices decrease at a time when vulnerable populations face financial hardship, (somewhat) lower food prices can enhance food security without necessarily destroying demand for local products (Barrett, 2006).

In your experience

If you give money for food to a homeless person on the street, are you making them aid-dependent or supporting them to become self-sufficient? How can you know?

Figure 10.11 Cartoon illustrating the risk of aid dependency.

Source: © Popa Matumula.

FOOD
CONTROVERSY
10.1

Helping Haiti or adding insult to injury?

On 12 January 2010, a catastrophic earthquake with a magnitude of 7.0 shook Haiti, killing an estimated 160,000 people, demolishing buildings, blocking roads, and destroying communication networks (Kolbe et al., 2010) (see Figure A). About 4 million people became food-insecure overnight, and emergency relief efforts were set in motion. Within 24 hours, American military air-dropped biscuits and ready-to-eat meals as an untargeted food aid strategy. A few days later, distribution points were set up with more precise targeting. According to officials of USAID, the earthquake damaged local food supply chains and markets so drastically that in-kind food assistance was a critical early response (Kushner, 2012).

Following up on the early food drops, USAID responded to the crisis with a 1-year 140-million-dollar food aid programme, including voucher and cash distribution to facilitate local food purchases and a direct transfer of in-kind aid. Among the various crops delivered was a large shipment of rice—this shipment, in particular, triggered a controversy about the sincerity and efficacy of American food aid to Haiti. There has been a long history of subsidized rice imports from the USA that, over time, many feel disrupted local Haitian rice production. Shipping rice as food aid became highly controversial—to some, it was genuine support for earthquake victims; to others, it was the continuation of self-interested and disruptive agricultural trade policies.

There is considerable evidence for the negative impact of US trade policies on Haiti's agricultural sector. In the 1970s, Haiti was self-sufficient in rice, sugar, poultry, and pork, and only 19% of the country's food needs were imported. Today this has increased to 51% (Dupuy, 2010). For rice, the country's staple food, the numbers are even more extreme—80% of rice consumed in Haiti is imported, the large majority from the USA.

Part of the explanation for this shift is local. Haiti's agricultural sector faces many challenges that push down rice yields, which have remained at half of the international average yields for 25 years (Furche, 2013). Production is small-scale, with farmers generally operating on less than 1 ha of land (Furche, 2013; McGuigan, 2006). Farmers also face environmental obstacles, for example soil erosion and flooding (Churches et al., 2014). In addition, Haiti's poor-quality irrigation infrastructure makes agriculture largely rain-fed and susceptible to drought. Markets are difficult to access due to poor infrastructure, and a lack of storage facilities prevent farmers from accessing markets at profitable times. There is also limited access to credit and government support for production, preventing farmers from investing in improved technologies (see Figure B for an example of a project aimed at enhancing production efficiency).

So why blame US rice imports? The key issue is that all of the structural constraints on rice production could

a)

b)

c)

Figure A (a) Earthquake damage in Haiti, January 2010. (b) A women carrying a 55-pound bag of rice distributed by WHO at a food distribution site in Port-au-Prince. (c) A local Haitian market.

Source: (a) Photo Marco Dormino/The United Nations Development Programme; (b) U.S. Navy photo by Chief Mass Communications Specialist Robert J. Fluegel; (c) Photo by Fred W. Baker III/Department of Defense.

Figure B An aid-funded initiative in Haiti, aimed at improving rice productivity.

Source Photo by Erika Styger, Cornell University/Climate-Resilient Farming Systems.

have been overcome with effective food governance. But instead of supporting and reforming domestic production, macro-economic policies in Haiti and the USA consistently promoted cheap rice imports. Throughout the twentieth century, Haiti's politics have been turbulent and violent, with external stakeholders playing a large role—in particular, the US government (Girard, 2004). As a result of this influence, economic policy in the 1990s comprised a dual strategy: development of an urban, industrial garment sector and reducing protections for agriculture. Haiti was pressured by the USA to reduce subsidies for local production, slash tariffs (taxes) on imported food, and lift import quotas (maximum amounts allowed to be imported). For example, tariffs on rice were reduced from 50% to 3% in 1995, the lowest tariff on rice across the Caribbean.

All these measures increased the price of Haiti's domestically produced rice, while it had to compete with cheaply produced international rice. Ironically, US rice production has long been heavily subsidized, up to an estimated 72% of the production costs (Raworth and Green, 2005). This makes US rice much cheaper than locally produced rice; and at the time of the earthquake, US rice was almost half of the price of local rice (Dodds, 2010). As a result, it has become very difficult to make a living for Haitian rice farmers; as an Oxfam study in 2001 indicated, farming families in the rice regions of Haiti have the highest concentration of malnutrition and poverty in the country (Chery, 2001). Many are left with no other choice than to abandon their farms and look for a livelihood elsewhere.

Former US President Bill Clinton, who served as co-chair on Haiti's earthquake recovery commission, publicly apologized for championing policies that ultimately destroyed Haiti's rice production. *'It may have been good for some of my farmers in Arkansas, but it has not worked. It was a mistake'* (quoted in Katz, 2010). He then attended an international donor's conference to spearhead fundraising for Haiti in which he told reporters that what should have been done was to *'help them be self-sufficient in agriculture'* (Dupuy, 2010).

The controversy about American food aid after the earthquake should be understood in the context of this longer history of US involvement. But what kind of food aid should have been given? Grassroots activists in Haiti do not reject emergency relief but demand agricultural and trade reform alongside it. Real aid, they argue, would be to repeal trade liberalization, reinstate subsidies for local production and export, rebuild the necessary infrastructure (for example, transport, communications, irrigation) to access markets, and invest in knowledge, services, and finance for farmers to adapt, innovate, and grow.

These Haitian activists are not alone. Many activists and scholars around the world argue that reducing the extent to which high-income countries protect and promote their own agricultural sectors (including the EU through its Common Agricultural Policy; see Box 13.1 in *Chapter 13: Governance*) could be far better at reducing food insecurity than food aid (Borrell and Hubbard, 2000; Elinder, 2005).

Pause and think

The Haitian controversy could be taken to suggest that food aid functions as a Band-Aid, rather than a cure—it fixes the symptoms of food insecurity, rather than its causes. Given all the examples of food aid you have seen in this chapter, do you support this argument? Why (not)?

10.5 **Future outlook**

What should be the future of food aid? The trend away from untargeted and tied aid will likely continue, but targeted forms of emergency and chronic food aid will remain vital. There are some obvious avenues for reform and to enhance efficacy, for example by ensuring that food aid is only provided when it is the most effective means of addressing food needs, that it is integrated into wider approaches to reduce poverty and stimulate development, and that it is effectively targeted (Barrett and Maxwell, 2007). In May 2016, 18 major donor countries and 16 multilateral organizations and international NGOs agreed on a Grand Bargain to improve general emergency assistance along these lines (The World Humanitarian Summit, 2016) (see Figure 10.12). Agreements included:

- Increasing systematic evaluation and greater transparency of aid impacts, as well as improving needs assessments, all of which can lead to better information to guide efforts

- Increasing coordination between different donors and middlemen, which can lead to reduced inefficiencies and fewer risks of spoiler behaviour through better monitoring

- Providing more facilitation for local responders and participation of recipients, all of whom have better information than international donors and can target projects more effectively.

It is too early to tell whether these agreements will be implemented and if they will help. But they do highlight that aid donors are trying to find ways to adjust their aid practices to address the efficacy challenges raised in this chapter, particularly those around targeting. Yet this chapter has also raised issues for food aid efficacy that are not visible in the Bargain agreements. First, it has underlined the connections between food aid and other international or domestic policies such as trade restrictions and agricultural subsidies. These interactions, painfully illustrated in Food controversy 10.1, signal the need to put food aid firmly in the context of wider food systems and food governance, not just other development policies (see *Chapters 12: Food systems* and *Chapter 13: Governance*).

Second, this chapter has shown several ways in which food aid can be integrated with efforts to make food systems more sustainable. The Brazilian school feeding programme (see Case study 10.1) illustrated how procurement for food aid can stimulate

Figure 10.12 Participants of the World Humanitarian Summit in May 2016 in Istanbul where major development donors agreed on a Grand Bargain to improve development assistance.

Source: Oktay Çilesiz / Anadolu Agency.

sustainable food production. Food banks and sharing practices showed how food aid can be combined with efforts to reduce waste, by using surplus food as in-kind aid. The focus on donor–recipient distance highlighted the need to think of food miles in food aid supply chains to prevent needless carbon emissions—a goal that aligns well with the need for reducing direct transfers in order to minimize the negative impacts of in-kind aid on recipient food systems. In all these examples, the means to enhance aid efficacy can also increase its environmental sustainability. Future food aid approaches can, and should, explore such synergies more effectively.

Connect the dots

As you read the next *Chapter 11: Food Systems*, try to keep the different food aid types in mind. Where do they fit into the complex models used to describe food systems? What impacts do they have? And does the food system approach help you see ways in which food aid flows might be better integrated to enhance their sustainable impact?

QUESTIONS

10.1 Can you define and give examples of international food aid, domestic food assistance, and community food sharing?

10.2 What are the possible negative impacts of food aid on recipient societies? Using the food supply chains you have mapped out, can you think of ways to assess and ultimately rank their environmental sustainability?

10.3 How can social/geographical distance between donors and recipients compromise the efficacy of food aid?

10.4 Can you think of a food aid strategy that would combine the three different types of food aid effectively?

10.5 Knowledge of local politico-economic conditions and information about the behaviour of middlemen and the impacts of food aid are crucial to prevent negative impacts on recipient incentives. Domestic and informal forms of food aid are therefore less likely to negatively impact recipients. Can you explain this connection?

FURTHER READING

Awokuse, T.O. 2011. Food aid impacts on recipient developing countries: a review of empirical methods and evidence. *Journal of International Development* 23, 493–510. **(Surveys the economic literature on the impacts of food aid on recipient countries, highlighting the conceptual and methodological challenges that make such evaluations difficult and understudied.)**

Barrett, C.B., Maxwell, D. 2007. *Food Aid After Fifty Years: Recasting its Role*. Abingdon: Routledge. **(Traces the** development of food aid practices and policies over the past 50 years, examining the current status, future prospects, and proposing strategies for a more recipient-oriented food aid system.)

Clapp, J. 2015. *Hunger in The Balance: The New Politics of International Food Aid*. Ithaca, NY: Cornell University Press. **(Explores major international food aid controversies, including the tying of food aid, subsidies, and surplus dumping, and the use of genetically modified crops.)**

REFERENCES

Awokuse, T.O. 2011. Food aid impacts on recipient developing countries: a review of empirical methods and evidence. *Journal of International Development* 23, 493–510.

Barrett, C.B. 2002a. Food security and food assistance programs. *Handbook of Agricultural Economics* 2, 2103–2190.

Barrett, C.B. 2002b. *Food Aid Effectiveness: It's the Targeting, Stupid!* Cornell University Applied Economics and Management Working Paper No. 2002-43. Available at SSRN: https://ssrn.com/abstract=431261 or http://dx.doi.org/10.2139/ssrn.431261.

Barrett, C.B. 2002c. *Food Aid and Commercial International Food Trade*. Trade and Markets Division Background Paper. Paris: Organisation of Economic Cooperation and Development (OECD).

Barrett, C.B. 2006. *Food Aid's Intended and Unintended Consequences*. Available at SSRN: https://doi.org/10.2139/ssrn.1142286

Barrett, C.B. 2010. Measuring food insecurity. *Science* 327, 825828.

Barrett, C.B., Maxwell, D. 2007. *Food Aid After Fifty Years: Recasting its role*. Abingdon: Routledge.

Bazerghi, C., McKay, F.H., Dunn, M. 2016. The role of food banks in addressing food insecurity: a systematic review. *Journal of Community Health* **41**, 732–740.

BBC News. 2017. *'Half' Nigeria Food Aid for Boko Haram Victims Not Delivered*. Available at: http://www.bbc.com/news/world-africa-40325043 [accessed 9 November 2018].

Blumenthal, S.J., Hoffnagle, E.E., Leung, C.W., Lofink, H., Jensen, H.H., Foerster, S.B., Cheung, L.W., Nestle, M., Willett, W.C. 2014. Strategies to improve the dietary quality of Supplemental Nutrition Assistance Program (SNAP) beneficiaries: an assessment of stakeholder opinions. *Public Health Nutrition* **17**, 2824–2833.

Borrell, B., Hubbard, L. 2000. Global economic effects of the EU common agricultural policy. *Economic Affairs* **20**, 18–26.

Center on Budget and Policy Priorities. 2018. *Policy Basics: The Supplemental Nutrition Assistance Program (SNAP)*. Available at: https://www.cbpp.org/research/policy-basics-the-supplemental-nutrition-assistance-program-snap [accessed 1 June 2018].

Chery, J.M.R. 2001. *Etude de L'Impact de la Libéralisation Commerciale dans le Secteur Rizicole*. Oxfam GB.

Churches, C.E., Wampler, P.J., Sun, W., Smith, A.J. 2014. Evaluation of forest cover estimates for Haiti using supervised classification of Landsat data. *International Journal of Applied Earth Observation and Geoinformation* **30**, 203–216.

Clapp, J. 2015. *Hunger in The Balance: The New Politics of International Food Aid*. Ithaca, NJ: Cornell University Press.

Clay, E., Riley, B., Urey, I. Clay, E. 2006. *The Development Effectiveness of Food Aid: Does Tying Matter?* Paris: OECD Publishing.

Clay, E.J., Geddes, M., Natali, L. 2009. *Untying Aid: Is it Working? An Evaluation of the Implementation of the Paris Declaration and of the 2001 DAC Recommendation of Untying ODA to the LDCs*, Copenhagen.

Cohen, A.L. 2018. The motivations for food exchanges in the lives of rural older adults. *Rural Sociology*. Available at: https://doi.org/10.1111/ruso.12217 [accessed 9 November 2018].

Culbertson, A. 2016. *Foreign Aid Farce: Millions Siphoned Off by Yemen Government from Desperate Civilians*. Daily Express. Available at: https://www.express.co.uk/news/world/718760/Yemen-British-UN-United-Nations-aid-stolen [accessed 9 November 2018].

de Mattos, E.J., Bagolin, I.P. 2017. Reducing poverty and food insecurity in rural Brazil: the impact of the zero hunger program. *EuroChoices* **16**, 43–49.

Dodds, P. 2010. *Food Imports Hurt Struggling Haitian Farmers*. NBC News. Available at: http://www.nbcnews.com/id/35608836/ns/world_news-americas/t/food-imports-hurt-struggling-haitian-farmers/ [accessed 9 November 2018].

Drake, L., Woolnough, A., Burbano, C., Bundy, D. 2016. *Global School Feeding Sourcebook: Lessons from 14 Countries*. London: Imperial College Press.

Dupuy, A. 2010. Disaster capitalism to the rescue: the international community and Haiti after the earthquake. *NACLA Report of the Americas* **43**, 14–19.

Elinder, L.S. 2005. Obesity, hunger, and agriculture: the damaging role of subsidies. *BMJ* **331**, 1333.

Falcone, P.M., Imbert, E. 2017. Bringing a sharing economy approach into the food sector: the potential of food sharing for reducing food waste. In: Morone, P., Papendiek, F., Tartiu, V.E. (eds). *Food Waste Reduction and Valorisation*. Cham: Springer, pp. 197–214.

Food and Agricultural Organization of the United Nations (FAO). 2001. *Targeting for Nutrition Improvement: Resources for Advancing Nutritional Well-being*. Rome: FAO.

Food and Agricultural Organization of the United Nations (FAO). 2017. *Yemen Crisis: FAO in Emergencies*. Available at: http://www.fao.org/emergencies/crisis/yemen/en/ [accessed 9 November 2018].

Furche, C. 2013. *The Rice Value Chain in Haiti: Policy Proposal*. Oxfam America.

Ganglbauer, E., Fitzpatrick, G., Subasi, O., Guldenpfennig, F. 2014. *Think Globally, Act Locally: A Case Study of a Free Food Sharing Community and Social Networking*. Proceedings of the 17th ACM Conference on Computer Supported Cooperative Work & Social Computing, of CSCW 14. New York, NY: ACM, pp. 911–921.

Garasky, S., Wright Morton, L., Greder, K.A. 2006. The effects of the local food environment and social support on rural food insecurity. *Journal of Hunger and Environmental Nutrition* **1**, 83–103.

Gentilini, U. 2013. *Banking on Food: The State of Food Banks in High-Income Countries*. IDS Working Papers 415. Brighton: IDS, pp. 1–18.

Girard, P. 2004. *Clinton in Haiti: The 1994 US Invasion of Haiti*. New York, NY: Palgrave Macmillan.

Gregory C, Ver Ploeg M, Andrews M, Coleman-Jensen A. 2013. *Supplemental Nutrition Assistance Program (SNAP) Participation Leads to Modest Changes in Dietary Quality*. Economic Research Report No. 147. Washington, DC: United States Department of Agriculture, Economic Research Service.

Harper, C., Wood, L., Mitchell, C. 2008. *The Provision of School Food in 18 Countries*. School Food Trust.

Hunzinger, E., Charles, D., Godoy, M., Aubrey, A. 2018. *Trump Administration Wants To Decide What Food SNAP Recipients Will Get*. NPR. Available at: https://www.npr.org/sections/thesalt/2018/02/12/585130274/trump-administration-wants-to-decide-what-food-snap-recipients-will-get [accessed 31 May 2018].

Katz, J.M. 2010. *With Cheap Food Imports, Haiti Can't Feed Itself*. Available at: http://archive.boston.com/news/world/latinamerica/articles/2010/03/20/with_cheap_food_imports_haiti_cant_feed_itself/ [accessed 9 November 2018].

Keen, D. 1994. The functions of famine in Southwestern Sudan: implications for relief. In: Macrae, J., Zwi, A.B. (eds). *War and Hunger: Rethinking International Responses to Complex Emergencies*. London: Zed Books in association with Save the Children Fund (UK).

Kolbe, A.R., Hutson, R.A., Shannon, H., Trzcinski, E., Miles, B., Levitz, N., Puccio, M., James, L., Noel, J.R., Muggah, R. 2010. Mortality, crime and access to basic needs before and after the Haiti earthquake: a random survey of Port-au-Prince households. *Medicine, Conflict and Survival* **26**, 281–297.

Kushner, J. 2012. *Haitian Farmers Undermined by Food Aid*. The Center for Public Integrity. Available at: https://www.publicintegrity.org/2012/01/11/7844/haitian-farmers-undermined-food-aid [accessed 9 November 2018].

Malhotra, N. 2016. *50,000 Free Hot Meals a Day and 9 Other Amazing Facts About the Langar at the Golden Temple*. The Better India. Available at: https://www.thebetterindia.com/53531/golden-kitchen-10-things-didnt-know-langar-golden-temple-amritsar/ [accessed 31 May 2018].

Mankiw, N.G. 2014. *Principles of Macroeconomics*, 7th ed. Stamford, CT: Cengage Learning.

McGuigan, C. 2006. *Agricultural Liberalisation in Haiti*. London: Christian Aid.

Mercy Corps. 2007. *Guide to Cash-for-Work Programming*. Portland, OR: Mercy Corps International.

Metz, M., Biel, M., Kenyi, H.A. 2012. *Comparing the Efficiency, Effectiveness and Impact of Food and Cash for Work Interventions*. Berlin: Deutsche Gesellschaft für Internationale Zusammenarbeit (GIZ) GmbH.

Nunn, N., Qian, N. 2014. US food aid and civil conflict. *American Economic Review* **104**, 1630–1666.

Organisation for Economic Cooperation and Development (OECD). 2018. *Net ODA*. Available at: https://data.oecd.org/oda/net-oda.htm [accessed 31 May 2018].

Otsuki, K. 2011. Sustainable partnerships for a green economy: a case study of public procurement for home-grown school feeding. *Natural Resources Forum* **35**, 213–222.

Polman, L. 2010. *The Crisis Caravan: What's Wrong With Humanitarian Aid?* New York, NY: Metropolitan Books.

Poppendieck, J. 2014. *Breadlines Knee-Deep in Wheat: Food Assistance in the Great Depression*. Berkeley, CA: University of California Press.

Quandt, S.A., Arcury, T.A., Bell, R.A., McDonald, J., Vitolins, M.Z. 2001. The social and nutritional meaning of food sharing among older rural adults. *Journal of Aging Studies* **15**,145–162.

Ratcliffe, C., McKernan, S.-M., Zhang, S. 2011. How much does the Supplemental Nutrition Assistance Program reduce food insecurity? *American Journal of Agricultural Economics* **93**, 1082–1098.

Raworth, K., Green, D. 2005. Kicking down the door: how upcoming WTO talks threaten farmers in poor countries. *Oxfam Policy and Practice: Agriculture, Food and Land* **5**, 43–111.

Sidaner, E., Balaban, D., Burlandy, L. 2013. The Brazilian school feeding programme: an example of an integrated programme in support of food and nutrition security. *Public Health Nutrition* **16**, 989–994.

Tam, B.Y., Findlay, L., Kohen, D. 2014. Social networks as a coping strategy for food insecurity and hunger for young Aboriginal and Canadian children. *Societies*, **4**, 463–476.

The Guardian. 2016. *Zimbabwe Urged to Put Compassion Before Politics in Distribution of Food Aid*. Available at: https://www.theguardian.com/global-development/2016/sep/10/zimbabwe-compassion-before-politics-food-aid-distribution [accessed 9 November 2018].

The World Humanitarian Summit. 2016. *The Grand Bargain: A Shared Commitment to Better Serve People in Need*. Istanbul, Turkey, 23 May 2016. Available at: https://reliefweb.int/sites/reliefweb.int/files/resources/Grand_Bargain_final_22_May_FINAL-2.pdf [accessed 25 November 2017]

United Nations. 2017. *Yemen Facing Largest Famine the World Has Seen for Decades, Warns UN Aid Chief*. UN News Centre. Available at: http://www.un.org/apps/news/story.asp?NewsID=58058#.Whq-W0qnGUm [accessed 25 November 2017].

United States Department for Agriculture (USDA). 2017. *Supplemental Nutrition Assistance Program (SNAP)*. Available at: https://www.fns.usda.gov/snap/supplemental-nutrition-assistance-program-snap [accessed 31 May 2018].

Wallerstein, M.B. 1980. *Food for War/Food for Peace. United States Food Aid in a Global Context*. Cambridge, MA: The MIT Press.

World Food Programme (WFP). 2013. *Food Aid Flows 2012 Report*. Rome: WFP.

World Food Programme. 2017. *Zero Hunger*. Available at: http://www1.wfp.org/zero-hunger [accessed 4 December 2017].

Consumption

How can we promote sustainable food consumption?

Bríd Walsh, Daniela Vicherat-Mattar, and David Ehrhardt

Chapter Overview

- Sustainable food consumption promotes consumer consumption of goods and services to allow us to meet our needs and preferences, while minimizing the use of natural resources and pollutants and reducing waste and emissions throughout the entire product life cycle.

- Despite global societal awareness about the environmental impacts of food consumption increasing, consumer choices are often not influenced by these concerns.

- Instead, food choices of consumers are influenced a range of determinants, such as social and cultural influences (for example, heritage attachments, status, and fashion), cost and price, and the attitude–behaviour gap.

- Strategies to enhance sustainable food consumption include pricing, food labelling, nudging, green advertising, nudging, and system-wide reforms (for example, building a circular economy).

Introduction

How do people choose the food they eat? Are they aware of how the food they consume impacts the environment? What are the factors that motivate individuals to switch to a more sustainable diet in the first place? How does culture influence the food choices people make? And what strategies might help people consume more sustainably? As we have seen in *Chapter 8: Nutrition*, if people shift towards a more environmentally friendly diet, it could be a major driver to make our food systems more sustainable. Yet, to date, food production has been associated with large-scale environmental degradation.

Increased public awareness about the environmental impacts of food has resulted in several global initiatives focused on addressing them. For example, in December 2015, in the lead-up to the climate negotiations in Paris, millions of people took to the streets in marches across 175 countries. In April 2017, Science Marches took place all over the world to show public support for evidence-based policy-making on environmental issues and to celebrate the value of science to society (see Figure 11.1). And in 2018, Extinction Rebellion emerged as a non-violent,

Figure 11.1 Science marches took place in over 600 cities on Earth Day 2017. The picture shows Museumplein, Amsterdam.

Source: Courtesy of Tammy Sheldon Photography.

global resistance movement against climate change, biodiversity loss, and ultimately the risk of human extinction.

These types of initiatives give a clear signal that many are concerned about the environmental impacts of human activities. But does this general concern translate into pro-environmental behaviour at a local level— in the home, in the supermarket, and in other parts of everyday life? Many consumers around the world, in high-, middle-, and low-income countries alike, are faced with an increasing range of choices when it comes to food products they consume. In this chapter, we explore the factors that shape consumer choice and discuss potential strategies to increase sustainable consumption at an individual and societal level.

The chapter is split into three sections. First, we explore the concept of sustainable consumption, that is, the consumer use of goods and services that have fewer environmental impacts throughout the life cycle of a product (Section 11.1). After this, we discuss the determinants of consumption patterns (Section 11.2). We do this by discussing the social and cultural factors that inform the food choices we make, the price of food, and the influences that create the so-called attitude–behaviour gap. Lastly, we examine the advantages and disadvantages of existing efforts to increase sustainable consumption, focused on changing prices; shifting social and cultural factors through product labelling, nudging, and green advertising; and implementing more structural reforms such as policies strengthening the circular economy (Section 11.3).

11.1 **What is sustainable consumption?**

Increased awareness of the environmental impacts of consumption in high-income countries has brought about a sustainable food movement where eating healthily and being 'green' is increasingly popular. Today, consumers have a wide range of choice of sustainable foods, including products that are labelled Rainforest Alliance Certified™, Fairtrade, sustainable, locally grown, or organic. These various components of the sustainable food movement can be understood in the context of what is known as sustainable consumption.

The concept of **sustainable consumption** was introduced at the Johannesburg World Summit on Sustainable Development in 2002. It promotes the consumer use of goods and services that allows us to meet our needs, while minimizing the use of natural resource stocks, pollutants, and contaminants, and reducing waste and emissions throughout the entire product life cycle (Vermeir and Verbeke, 2006). It emphasizes the environmental, social, and economic impacts of our daily consumer choices (see *Chapter 8: Nutrition*) and aims to develop a link between the use of resources (such as energy resources and water) and the levels of environmental pollution and productivity (Johannesburg Implementation Plan, 2002).

Being a sustainable consumer does not necessarily mean consuming less, but it does mean consuming differently, mitigating the negative impacts of your choices on a daily basis. Despite the apparent rise in awareness of environmental issues and the emergence of the concept of sustainable consumption, only a minority of consumers actually consider environmental aspects when making decisions (Vermeir and Verbeke, 2006). Consumers are increasingly detached from food production, as fewer people are employed in the agricultural sector. In addition, the food sector has also become less localized, and food production is increasingly high-tech and difficult to understand (Evans and Miele, 2017).

But what can we do to help people consume more sustainably? A first step would be to clarify what food choices are sustainable (see *Chapter 6: Climate Change* and *Chapter 8: Nutrition*). To illustrate how complex this question can be, we explore the consumer perceptions of organic food in Box 11.1.

BOX 11.1

Perception versus reality: the case of organic food

Demand for organic food is growing rapidly, with the highest demand in Europe and North America. Global sales increased fivefold to 72 billion USD between 1999 and 2013 (Reganold and Wachter, 2016). Consumers see organic food as a symbol of good ethics (for example, animal welfare concerns), environmentally friendly behaviour (for example, addressing soil and water conservation concerns), healthy eating (for example, food produced without pesticides/chemicals), and social status (for example, fear for one's reputation among peers). Consumers are therefore willing to pay more for food that is produced organically (Bauer et al., 2013).

Table A presents three common consumer perceptions of organic farming, and we compare these to the scientific literature. As you can see, there is a mismatch between the commonly held perceptions and the scientific evidence, particularly in regard to three common ideas:

- All organic food is locally sourced from small farms
- All organic food is more nutritious than conventionally produced food, and
- All organic food is more environmentally sustainable than conventionally produced food.

Table A Common consumer perceptions of organic farming (left column) versus the reality of organic farming (right column)

Perception	Reality
Organic farms are small scale and locally owned	The relationship between farm size and organic production is complex, and a farmer's choice to use (partial) organic production methods is often separate from the decision to seek organic certification (Veldstra et al., 2014). Overall, this means that there are both small-scale and large-scale organic food producers (Earthbound Farms in California is an example of the latter)
Organic food is better for you	There is a lack of strong evidence to support this statement (Smith-Spangler et al., 2012). Some studies find organic food to be more nutritious [for example, higher quantities of vitamin C and iron (Worthington, 2001)], but there is debate about whether these findings are nutritionally important (see Reganold and Wachter, 2016). Conversely, some researchers have found no difference at all (see, for example, Dangour et al., 2009). There is evidence to suggest that consuming organic food reduces exposure to both antibiotic bacteria and pesticide residues (Mie et al., 2017; Smith-Spangler et al., 2012). Sometimes research on the health benefits of organic food are complicated by the fact that those who consume organic products tend to live a healthier lifestyle (Mie et al., 2017)
Organic food is better for the environment	Organic food is environmentally friendly in some ways, with better soil quality and less soil erosion (Gomiero et al., 2011), but more damaging in other ways with high land requirements, though this is highly variable per crop (Tuomisto et al., 2012). An analysis of 71 peer-reviewed papers found that organic farms are: (i) more environmentally friendly per unit area, with higher species and organism richness, and (ii) less environmentally friendly per production unit due to generally lower yields (Tuomisto et al., 2012), though this is difficult to assess, as there are high levels of variation within both organic and conventional systems (Tuomisto et al., 2012). For example, in most cases, organic milk has higher GHGs, as compared to non-organic milk, due to the lower production per animal and higher CH_4 emissions (Thomassen et al., 2008) (see *Chapter 6: Climate Change* for more details). In contrast, organic beef has been found to have lower GHGs due to lower use of industrial inputs produced using fossil fuels (Casey and Holden, 2006).

(Adapted from Bauer et al. 2013, 2015, and Hughner et al. 2007)

11.2 Determinants of consumption: how do people make food choices?

If we disregard, for a minute, any economic availability or social barriers we might face in accessing food (see *Chapter 9: Food Security*), there are probably foods we like and foods we dislike. These preferences vary from person to person and from culture to culture; moreover, they change over time (Monteleone et al., 2017). In many cultures, for example, snails, ants or grasshoppers (see Figure 11.2) are viewed as delicacies and a good source of protein, while in others, they may be seen as unappetising (see *Chapter 8: Nutrition* for an overview of entomophagy).

In addition, we also have preferences based on health benefits, price, and ease of access (Monteleone et al., 2017). These can be weighted differently across cultures; for example, while French attitudes are affected by the sensory aspects of food such as taste, American attitudes are more influenced by the perceived health impacts of food (Cervellon and Dubé, 2005). These differences in preferences and food choices are influenced by many factors, ranging from cultural influences to the diets of our parents, which is reported to have an 'anchoring function' in the formation of children's dietary preferences (Guidetti and Cavazza, 2008; Monteleone et al., 2017).

We engage in a complex game of calculations and trade-offs, as we come to a decision on what food to buy (see *Chapter 14: Collective Action* for more discussion of this kind of cost–benefit decision-making). The **determinants** of food consumption choices are the factors that affect what foods we choose to eat and drink. We may be aware of some of these determinants (for example, cost), while others may have a subconscious influence (for example, how we have been socialized or the social expectations of our friends) (see Figure 11.3). In the next subsections, we discuss the impact of three factors in depth: social and cultural influences, cost and price, and the attitude–behaviour gap.

In your experience

What foods do you like and dislike? If you reflect on your food choices, can you figure out what factors determine them (also considering Figure 11.3)? Do you think these are good reasons for choosing your food, or can you think of better ones?

11.2.1 What are the social and cultural influences on consumption choices?

As food is a fundamental aspect of our social identity, food choices can help build and maintain connections to our cultural heritage and family history. Consequently, both cultural and social influences are powerful determinants of food choices and have the potential to promote sustainable consumption (Chekima et al., 2016). Cultural factors that influence

Figure 11.2 Grasshopper on a stick—delicacy or disgusting?

Determinants of food choice	
Hunger and appetite	Health impacts
Taste	Environmental impacts
Convenience	Brand preferences
Country of origin	Food quality
Cost	Emotional determinants
Availability and access	Cultural influences
Nutritional information	Social influences

Figure 11.3 The main biological food choice determinants in blue, economic and physical determinants in green, attitudinal determinants in orange, psychological determinants in pink, and social and cultural determinants in purple.

Source: Adapted from EUFIC, 2006.

food choice include religious norms (see Box 11.2) and cultural constraints. For example, immigrants may cook traditional food in order to retain a small bit of 'home' in their new country of residence. In a similar vein, food can also prompt memories and emotions. Maybe you have a favourite 'comfort food' when you are feeling down, or maybe your family has eaten the same meal every year on Christmas Day or during Eid al-Fitr, which marks the end of Ramadan (see Figure 11.4). Here we explore the importance of socio-cultural influences on food choice by using two examples; the first is focused on meat consumption, and the second on eco-consumerism.

Meat is one of the biggest causes of environmental pollution, as discussed in detail in Part 1 of this book. So what makes people choose to eat meat, or to reduce their consumption? The '4 N' framework explains why people continue to choose to eat meat—to many meat-eaters, it is Natural, Normal, Necessary, and Nice (Piazza et al., 2015). Natural refers to the belief that '*a proper meal has to include meat*' and that it is '*part of our staple diet*' (MacDiarmid et al., 2012). Normal links to the idea that many people eat meat out of habit (Saba and Di Natale, 1998). Necessary refers to the belief that meat is needed in our diet because of its nutritional

a)

b)

Figure 11.4 Food can prompt emotions and memories, for example (a) to a traditional Christmas meal (b) or to celebrate the end of Ramadan with friends and family.

Source: (a) Dianne Rosete/Flickr (CC BY-ND 2.0); (b) Benreis/Wikimedia Commons (CC BY-SA 4.0).

BOX
11.2

Why are pigs not kosher?

The religious restriction on eating pork, found in Islam, Judaism, and some Christian denominations, is an example of a long-standing cultural determinant of food consumption. Given its global prominence and persistence, an elaborate scholarly debate has evolved trying to explain its historical origins. For some, it was the fact that the pig did not fit neatly in the standard biological taxonomies of the time that explains why taboos were constructed around it (Douglas, 1966). In contrast, it may also be the case that the restrictions are caused by the genetic similarities between pigs and human beings, and humans trying to distinguish themselves from pigs through dietary restrictions (Hitchens, 2008).

Yet others consider the ecological and economic requirements for herding pigs. For example, the need of pigs for mud to cool off has been linked to their lack of popularity in the Middle Eastern religions (see Figure A). Keeping pigs can also simply be more expensive than relying on other animal sources of food (Harris, 1998). Finally, evidence has recently been presented that the chicken outcompeted the pig as the main source of protein in the Middle East, because it is smaller, produces eggs, and is a more efficient source of protein than pigs (Redding, 2015). Regardless of the true reason for the pig's exclusion from consumption in many religions, the taboo remains a prime example of the power of religious norms on food choices and consumption patterns.

Figure A On the basis of their religions, Muslims and Jews around the world consider pigs unfit for human consumption. There is debate on the origin of this, and one reason put forward is the need for pigs to cool off in mud, which may be considered impure and also requires (often scarce) water resources.

Source: Mark Peters/Flickr (CC BY 2.0).

value (MacDiarmid et al., 2012). Nice refers to the consumption of meat simply because people like the taste (Saba and Di Natale, 1998).

There can be additional reasons. For some, meat consumption signifies a particular socio-economic status in their culture, that is they have the financial ability to purchase meat. For others, meat consumption is used to express masculinity (Schösler et al., 2015), and some men are worried about being perceived as 'wimpy' if they do not eat meat (Lea and Worsley, 2008; MacDiarmid et al., 2012).

Conversely, there are several key reasons why people reduce or stop their meat consumption. People may choose to reduce their consumption of meat due to financial constraints, environmental concerns, religious rules, potential health benefits, or because

their partner is vegetarian (MacDiarmid et al., 2012). In addition, the influence of personal values and social norms—that is, 'what most people do or approve of' (Zandstra et al., 2017)—is influential in shaping attitudes towards meat consumption. Such norms are powerful social factors and can influence behaviour.

In your experience

Do you eat meat? If so, reflect on the '4 N' concept, and use it to explain what the main driver is for your meat consumption. If you do not eat meat, how do you explain that decision?

A second example of socio-cultural factors that can influence our consumption is the popularity of environmentally friendly or 'green' shopping (often called **eco-consumerism**). In its early years, eco-consumerism was perceived as the domain of hippie environmental activists; but today it is a sign of status, a clear marker of distinction for the upper classes.

Eco-chic is the term used to describe this pattern of behaviour—a consumer lifestyle that emerged as a result of our modern, fast-paced lives, in combination with environmental awareness and socio-spiritual trends that aim to reconnect human beings with nature (Barendregt and Jaffe, 2014). Eco-chic consumers are considered to have '*a nostalgic longing for times when things were less complex*' (Barendregt and Jaffe, 2014). The popularity of this new form of consumption is visible through the proliferation of green and organic alternatives in supermarkets and slow food movements. Eco-chic consumption is visible in daily consumer choices such as Fairtrade coffee, organic vegetables, and organic clothing (see Figure 11.5). Whether it promotes sustainable consumption, however, is still to be determined.

Figure 11.5 'Green is the new black': eco-chic purchasing is a sign of status, a clear marker of distinction for the upper classes.

Source:: http://obiter-dicta.ca/2014/03/10/a-little-sheep-told-me-green-is-the-new-black/.

Connect the dots

Chapter 8: Nutrition focused on the types of food the human body needs and the environmental impacts of this food. There is a contemporary fashion of 'dieting cultures' where many people are focused on the latest and greatest diets. Health concerns appear to be an important determinant of these food choices, but sustainability concerns are also often highlighted. To what extent are these concerns really about promoting healthy diets? How do they relate to socio-cultural factors and food prices? And to what extent are 'healthy diets' more sustainable?

11.2.2 What is the impact of food prices on consumption choices?

In *Chapter 9: Food Security*, we described how food must be economically available to attain food security. In this section, we discuss the impact of income on food consumption from a consumer perspective, addressing the role of food prices in consumer decision-making, and consumer sensitivity to price changes. The cost of food is an important determinant of food choice, particularly in the interaction with the level of income (or wealth) of an individual or household. **Price sensitivity** is key to understanding the relationship between income, food price, and consumer choice, and it is defined as the extent to which prices affect a consumer's valuation of a product's attractiveness (or utility) (Erdem et al., 2002). Case study 11.1 provides an example of the extreme consequences of high price sensitivity and price shocks in low- and middle-income countries.

Outside of crisis situations, such as those described in Case study 11.1, price sensitivity is dynamic and varies greatly between and within countries and regions. As a relatively simple illustration, think about the differences in price sensitivity between low- and high-income groups within countries. Generally, people choose healthier foods as income rises or the price of healthier food decreases (Waterlander et al., 2010). Low-income groups, as a result, tend to choose food that is high in calories to ensure a high-calorie/dollar-spend ratio (Drewnowski and Spectre, 2004). For example, low-income US households consume fewer fruits and vegetables and higher amounts of fatty foods, compared to high-income households (Drewnowski and Spectre,

The food riots of 2006–2008

Between 2006 and 2008, global cereal prices rose rapidly due to factors such as high oil prices, expansion of the biofuel sector, trade shocks associated with drought, panic buying of staple cereals, and export restrictions (Headey and Fan, 2010). Generally, low- and middle-income countries cannot endure long periods of high-food prices without experiencing significant social tensions, as citizens do not have a safety net of high wages or savings and price sensitivity is correspondingly high (Berazneva and Lee, 2013). As a result of the high food prices, riots broke out across the world, from Bangladesh to Burkina Faso. In Africa alone, food riots took place in 14 countries across the continent between 2006 and 2008 (see Figure A). For example, in Côte d'Ivoire, 1500 protestors took to the streets in the capital Abidjan, and following clashes with police, 11 protesters were wounded and one was killed (Berazneva and Lee, 2013).

A similar dynamic characterized the Mexican tortilla riots in 2006–2007. Tortillas are made from corn flour and a staple of the Mexican diet, economy, and national identity. In the period 2000–2010, the annual global production of bioethanol and biodiesel increased from 16 million litres to over 100 million litres, placing considerable pressure on the supply of corn, sugarcane, and vegetable oil (Fairley, 2011). In 2006, the price of tortillas in Mexico rose by up to 400%. This was linked to the sharp increase in the demand for corn to be used in biofuel production, although there is debate about the size of this effect (de Ita, 2008; Fairley, 2011; Kleiner, 2008; Wise, 2011). Whatever the cause, tortillas suddenly became unaffordable for much of the population, threatening to increase undernourishment of low-income groups and undo Mexico's earlier successes in enhancing food security (see Figure B).

Figure A A boy standing by a burning car in the 2008 food riots in Mozambique.

Source: Sergio Costa/AFP via Getty Images.

a)

b)

Figure B In 2006, tortilla riots erupted in Mexico. At the heart of this was the increased cost of corn flour used in tortillas (a), a staple of the Mexican diet. This was caused, at least in part, by the increased global demand for corn to produce biofuels (b).

Source: (a) ProtoplasmaKid/Wikimedia Commons (CC BY-SA 3.0); (b) Diaper/Flickr (CC BY 2.0).

2004). Generally, low-income households also spend significantly more on fast food than their high-income counterparts (French et al., 2010).

These are simple illustrations of the way in which income and prices have strong, and often predictable, impacts on food consumption in high-income countries, particularly in terms of the nutritional quality of food that is consumed. But there are many other factors that also influence purchasing decisions (Chekima et al., 2016; Gatersleben et al., 2014). In fact, there is evidence that in many high-income countries, other factors than food prices are more influential in explaining sustainable consumption, including education level, cultural values, environmental advertising, and gender (women are more likely to have pro-environmental attitudes and high sustainable purchasing intentions) (Chekima et al., 2016).

Pause and think

Think about the different socio-economic groups in low- and middle-income countries, from informal market salespersons and subsistence farmers to the formal-sector middle class and wealthy state employees. Remember also the 'double burden of malnutrition' (see *Chapter 8: Nutrition*) that plagues many of these countries. How do you think income, price, and food choices are related in these societies? Are the relationships different than in high-income countries? And what are they like in your society?

11.2.3 Attitude–behaviour gap: is environmental awareness enough to motivate sustainable behaviour?

Even if many consumers want to behave sustainably, this motivation does not transfer seamlessly to the purchase of sustainable food products. In fact, even when information is available and understandable, it often does not lead to a change in consumption patterns (Grunert et al., 2014). This is captured by what is known as the **attitude–behaviour gap**—the disconnect between a consumer's knowledge and their behavioural choices.

Such gaps can be seen across many different sectors, from food and nutrition to waste management. For example, a recent UK study (from a sample of

251 consumers) found that 19.3% of consumers interviewed on the street indicated that they would avoid products with high 'food miles' (their attitude); however, in a supermarket environment, just 3.6% of the consumers purchased local produce (actual behaviour) (Kemp et al., 2010).

Social scientists across disciplines have sought to explain the gap between awareness and behaviour (see, for example, Auger and Devinney 2007; Carrington et al., 2010; Kollmuss and Agyeman 2002; Vermeir and Verbeke, 2006). In general, there is consensus that the choices involved in sustainable behaviour do not follow a simple progression from environmental knowledge (for example, knowing the impact of meat consumption on the environment) to sustainable attitudes (for example, feeling responsible for reducing this impact) and sustainable behaviour (for example, choosing to reduce consumption of meat).

The debate on what factors obstruct this process is still ongoing. They likely include consumer perceptions (for example, low perceived availability or efficacy of sustainable products), psychological factors (for example, strength of motivation or self-control), and social factors (for example, social peer pressure). Moreover, wider political, economic, and other institutional factors can also push people towards unsustainable food choices, even if they have pro-environmental attitudes (Kollmuss and Agyeman, 2002; Vermeir and Verbeke, 2006), for example if debates about particular food choices become polarized—that is, they become cast in binary terms that push people to take up extreme, rather than nuanced, viewpoints. Food controversy 11.1 provides an example of such a debate around the use of genetically modified foods (see also Box 3.7 in *Chapter 3: Pollution*).

In your experience

What is your position in the GMO debate? Which arguments in favour of, or against, the use of GM in food production matter to you? What are some other types of food that are debated in a polarized way? And what are the impacts of these debates on your own food choices?

FOOD
CONTROVERSY
11.1

'Say no to GMO'?

GM food is the subject of intense public and political debate and a source of globalized public protests, recurrent destruction of GM crops and research facilities, and even threats of personal violence (see Figure A). The intensity of the debate, as well as its binary 'you are either for or against' nature, hide the fact that it actually covers several different, highly complex issues at once—debates about environmental impacts and health risks, as well as ethical concerns about experimentation with nature and the acceptability of the global capitalist economy.

For proponents, genetic modification is a safe and efficient way to enhance crops in ways that are beneficial for both consumers and the environment. They highlight that scientific efforts to test the safety of GM foods so far have found no significant hazards of the use of available GM crops per se (see, for example, Nicolia et al., 2014). And, proponents say, GM has many benefits.

For example, as explained in Box 3.7 of *Chapter 3: Pollution*, one of the contributions of GM crop development has been to help plants develop a crystal protein that functions as a built-in insecticide (in so-called Bt-crops). This reduces the need to apply insecticides, reducing the effects of insecticide pollution on the environment. Another example is the modification of crops to enhance their nutrient content such as vitamin A in 'Golden Rice'. Given that rice is an important source of food across the world, and vitamin A deficiency can cause major adverse health effects (including blindness in children), Golden Rice has great potential for improving public health.

The opposing camp has concerns about the environmental and health impacts of GM crops, for example suggesting that viral resistance in GM plants might lead to the formation of more resistant viruses, that new GM foods pose immunological, allergenic, or toxic risks, or that pest- or herbicide-resistant plants can lead to the evolution of 'superbugs' and 'superweeds' (for example, Bawa and Anilakumar, 2013).

But in addition, many opponents have principled, ethical grounds to reject GM research and development, irrespective of the actual risks of genetic modification (Reiss and Straughan, 2001), for example opposing GM on the basis that experimentation with God's creation is unethical (for example, SGP, 2010). Finally, a strand of GM criticism focuses on what they see as the overly large influence of multinational companies in the development and production of GM foods coming (see Food controversy 12.1 in *Chapter 12: Food Systems* which discusses this issue in more detail).

While all three issues—environmental and health risks, ethical concerns, and the role of capitalist multinationals—are genuine scientific debates with nuanced evidence and ethical arguments, public discussions have often combined and recast them in the simplistic controversy of 'say yes or no to GMO'. While this has put the issue in the public eye, it has risked obscuring the complexity of the issue (Mayer and Stirling, 2004).

Figure A Opponents of GM food, such as those walking in this anti-genetically modified organism (GMO) protest in Vancouver in 2013, criticizing GM foods for many different reasons, which complicates the public debate.

Source: Rosalee Yagihara/Flickr (CC BY 2.0).

11.3 How to make consumers choose more sustainable products?

In this section, we explore ways to make food consumption more sustainable. The concept of sustainable consumption has brought together a huge range of local to global stakeholders from all levels of society. But what strategies are available to these stakeholders? We discuss three important examples (Reisch et al., 2013). First, we start by considering pricing strategies, followed by approaches that rely on disseminating information, raising awareness of the environmental impacts of different food choices, and 'nudging' consumers towards more sustainable choices. Finally, we consider more structural reforms, in particular the potential of promoting a circular economy to change consumer behaviours and reduce the negative environmental impacts of food systems.

11.3.1 What is the effect of pricing strategies on food choices?

Pricing strategies rely on the fact that prices are a determinant of food choice, as demonstrated in Section 11.2. Food consumption can be made more sustainable by artificially reducing the price of 'sustainable food' or raising the price of 'unsustainable food' (Kim and Kawachi, 2006; Ní Mhurchu et al., 2009). These strategies are used extensively in the promotion of healthy food, for example through discounts on healthy food options, taxing unhealthy products such as food and drinks with an artificially high sugar content, or reducing multi-buy offers (see Figure 11.6 and Case study 8.1 on sugar tax in *Chapter 8: Nutrition*) (Brownell and Frieden, 2009; Jacobson and Brownell, 2000). These strategies can also be used to promote sustainable food consumption (Reisch et al., 2013).

Of course, pricing strategies also have weaknesses. For example, the ecological footprint of food consumption is not evenly distributed across income groups, with greater responsibility falling on middle- to high-income households (Csutora and Mozner, 2014). Unfortunately, these groups are also less likely to be sensitive to pricing strategies than lower-income groups. Moreover, pricing strategies often involve some form of government intervention in food markets, either by taxing unsustainable food or subsidizing sustainable food (see *Chapter 13: Governance*). Such interventions may be effective if they are carefully targeted, monitored, and evaluated, but often they can create severe market inefficiencies, unintended side effects, and even corruption, particularly if they are large-scale and ambitious projects in countries where patronage is pervasive.

Figure 11.6 Multi-buy offers in supermarkets as a source of unsustainable overconsumption, because they motivate consumers to buy more than they need.

Pause and think

If you were the manager of your local supermarket, how could you use pricing strategies to motivate consumers to buy more sustainable food? And if you were the Minister of Finance in charge of taxation policies, how could you use taxes to motivate citizens to buy more sustainably?

11.3.2 What are the effects of food labelling, environmental advertising, and nudging on food choices?

Next to food prices, there are many other ways in which consumers can be incentivized to buy more sustainable food. Here we discuss three key methods: food labelling, environmental advertising, and nudging. **Food labels** provide product information to consumers, for example nutritional information or information on parts of the production process (see Figure 11.7). The logic of food labelling is that providing consumers with accurate information about the sustainability of food products can make consumers more likely to buy sustainably. In

pursuit of this goal, there are an enormous number of different labelling schemes available to indicate sustainable production techniques. In the EU alone, for example, there are approximately 130 public and private labelling schemes with a focus on sustainability (EC, 2012).

In your experience

Do you read the nutritional labels of your food products? If not, why not? And do you use labels such as the Fairtrade or organic labels that provide information on the production processes? If so, what information do you look for?

Unfortunately, this proliferation of eco-labels in itself can lead to consumer confusion about their meaning and significance (Brécard, 2014). Moreover, the use of labels is quite limited by the fact that many consumers have difficulty understanding them (Grunert et al., 2014; Harper et al., 2007). This situation is further complicated by the increasing emergence of misleading labels in product marketing such

a)

b)

c)

Figure 11.7 Labels are used to provide information about (a) nutritional value, but also about specific production processes, for example (b) Fairtrade products with a focus on ethical production conditions for local farmers or (c) products that have been produced without the use of artificial fertilizers and pesticides.

Source: (a) Conor Alexander/Wikimedia Commons (CC BY-SA 4.0);
(b) © Fairtrade Foundation 2018;
(c) © European Union, 1995-2018.

as 'farmhouse' and 'natural', which have no specific meaning and are not regulated by any group (that is, the opposite of labels such as Fairtrade). Such labels are seen as representing healthy food, which is not necessarily the case (Harris et al., 2011). These labels are an example of **greenwashing**, a situation in which companies promote a superficial perception of the 'environmental friendliness' of their products, when, in reality, of the production/distribution processes are (partially) unsustainable.

More broadly, **green advertising** promotes products, services, and companies that promise to mitigate negative environmental impacts (Kim et al., 2016). It is defined as a type of advertisement that highlights the relationship between a product/service and the natural environment and promotes an environmentally friendly lifestyle (Banerjee et al., 1995; Sheehan and Atkinson, 2012). Generally, it seems that the stronger a consumer's pro-environmental attitude is, the more likely she is to respond positively to green advertising (Kim et al., 2016). But the details of green advertising also matter; positive responses are more likely if advertisers use a high level of numerical precision in their adverts, for example by advertising a '19.56% decrease in carbon emissions', rather than a 20% one (Xie and Kronrod, 2012).

Specificity in messaging is thus one way to make green advertising more effective and to promote more eco-friendly consumption (Fowler and Close, 2012). However, many consumers remain sceptical of green advertising, fuelled by reports of companies greenwashing their products to appeal to consumers (see Figure 11.8). Moreover, consumers also often rely on other (and often more misleading) aspects of product packaging than informative food labels or green advertisements such as front-of-package visuals (this is even true for parents making decisions about the food for their children; see Box 11.3). Addressing these issues can involve improving label design, for example by reducing the amount of information on product labels, or regulating the information that packaging can contain (Grunert et al., 2014; Harper et al., 2007; Sørensen et al., 2012).

As an example, research has found that parents rely strongly on front-of-package labels (Abrams et al., 2015). The front packaging of products tends to be flashy, combining health claims, pictures, and colours to attract the attention of consumers (van Herpen and

Figure 11.8 Green advertising or greenwashing?
Source: Kristoffer Tripplaar / Alamy Stock Photo.

van Trijp, 2011) (see Figure A). Unfortunately, they are also often misleading; there is little regulation of descriptions such as 'naturally flavoured', which has no specific meaning and are often used on unhealthy and unsustainable products (Wirtz et al., 2013).

Parents also believe that products with images of fruit and those displaying health claims (for example, '100% vitamin C') are healthier, while cartoons and other such visuals are associated with unhealthy options (Abrams et al., 2015). But in reality, there often is no direct correlation between 'healthy claims' made by labels and the actual nutritional content of food or their sustainability. Yet these graphics and claims, in combination with the price, marketing, and requests from children, are powerful determinants of parents' food purchasing behaviours (Maubach et al., 2009).

Finally, in addition to food labelling and green advertising, **nudging** can also be used to promote more sustainable consumption. Nudging involves the use of people's cognitive boundaries, biases, and other weaknesses, to subtly incentivize particular forms of behaviour (Hansen, 2016; Reisch et al., 2013). For example, supermarkets can promote vegetable consumption by designating one part of their shopping carts as vegetables-only (see Figure 11.9a), or housing estates can promote waste reduction via a colour-coded recycling station (see Figure 11.9b) (Thompson, 2013).

To further illustrate, a study was recently conducted to measure fruit consumption in two schools in the USA (Lehner et al., 2016). In the first, cafeteria

BOX 11.3

A label too far? What informs parents' food choices in a supermarket environment?

The 'child consumer' is an important market group (Schor, 2014). Today's parents are bombarded with advice on every aspect of parenting, from breastfeeding versus bottle-feeding to healthy eating habits. In supermarkets, parents are faced with an array of choices per product, with each brand seeking to be eye-catching and appealing (see Figure A). This has led to information overload in supermarkets, an environment that is already highly pressured (Sørensen et al., 2012).

Figure A The health claims for three cereal brands. The vitamin D content of Honey Smacks™ cereal is being promoted, while the use of whole grain rice in Rice Krispies™ is being advertised, and for Frosted Flakes™, nine essential vitamins are highlighted.

Source: Robert Couse-Baker/Flickr (CC BY 2.0).

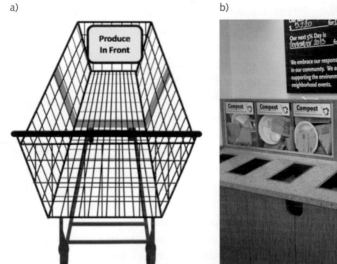

Figure 11.9 Examples of (a) a shopping cart nudging customers to buy more fruit and vegetables (produce), and of (b) a recycling station that nudges consumers towards composting and recycling goods.

Source: (a) Wansink, B., 2017. Healthy Profits: An Interdisciplinary Retail Framework that Increases the Sales of Healthy Foods. Journal of Retailing, The Future of Retailing 93, 65–78. © 2017 New York University. Published by Elsevier Inc. All rights reserved.

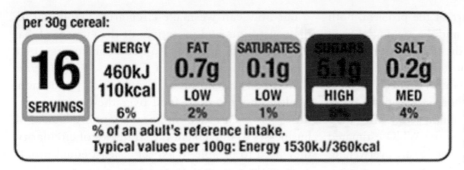

per 30g cereal:

16 SERVINGS

| ENERGY 460kJ 110kcal 6% | FAT 0.7g LOW 2% | SATURATES 0.1g LOW 1% | SUGARS 5.1g HIGH | SALT 0.2g MED 4% |

% of an adult's reference intake.
Typical values per 100g: Energy 1530kJ/360kcal

Figure 11.10 A colour-coded nutritional label, based on daily dietary guidelines for calories, sugar, fat, saturated fat, and salt. Source: NHS.

workers gave a verbal prompt to children by asking whether they would like to add fruit or juice to their lunch. No verbal prompt was provided in the second. Compared to no verbal prompt, 70% of the children consumed a piece of fruit in the first school, compared to less than 40% in the second.

There is increasing evidence that nudges such as these are highly effective, and they are relatively easy and cheap to implement. For example, the UK's Behavioural Insights Team has suggested that applying a colour-coded system to food to indicate the amount of salt, fat, sugar, and calories in a product, as a percentage of your total recommended daily intake, may promote healthier consumption (see Figure 11.10). Given its reliance on subconscious responses, rather than conscious consideration, however, nudging has raised an ethical debate about how appropriate it is in different situations (Hansen, 2016; Hausman and Welch, 2010).

11.3.3 How can structural reforms promote sustainable consumption?

In addition to the approaches described in Section 11.3.2, policymakers can also enhance sustainable food consumption through policies aimed at restructuring food systems (see also *Chapter 12: Food Systems*). One particularly ambitious effort is the goal of developing a **circular economy**. It is important to remember that an economy is a social creation and we can change the way it is organized. In other words, economies are not governed by the same natural laws discussed in many chapters in Part 1 of this book. Rather, it is largely structured by human-made rules that can be changed through food governance (see *Chapter 13: Governance*).

A circular economy is an economy in which economic productivity, and perhaps even growth, does not come at the expense of resource depletion

(Ghisellini et al., 2016) (see Figure 11.11). Our current food system relies on high resource inputs (for example, energy and raw materials) and capital (for example, infrastructure and finance) to produce goods for consumers. The current model is problematic in at least two fundamental ways (Andersen, 2007):

1. Many inputs, such as fossil fuels, soils, and phosphorus, are not renewable and cause environmental pollution, and

2. There is a lot of waste associated with the current production and consumption of products.

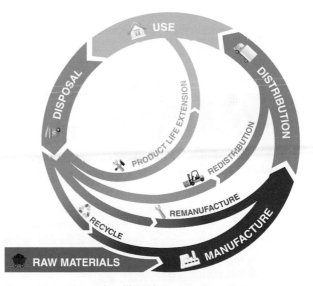

Figure 11.11 The product life cycle in a circular economy, from raw material extraction to the manufacture of products, distribution, consumer use, and disposal. Here we can see that, at the point of disposal in a circular economy, consumers have several options—they can recycle the product or send the product for remanufacturing, redistribution, or life extension to keep waste materials circulating in the economy and displacing the need for new inputs.

Source: Knowledge Transfer Network.

Consumers are a large part of the problem, both in their consumption choices and the amount of food they waste. For example, in the EU, the average person wastes at least 179 kg of food each year (Mirabella et al., 2014). Globally, Gustavsson et al. (2011) suggest that one-third of all food produced is wasted, this at a time when over 800 million people are undernourished. Moreover, in 2016, the WHO estimated that close to 2 billion people (over 18) are overweight, and 650 million obese. These numbers suggest staggering inefficiencies that, if reduced, would make considerable contributions to sustainability. A circular economy in food would do this by constraining wasteful consumer behaviour. But what can be done to enhance the circularity of our food economy?

Several strategies can be used (Ghisellini et al., 2016):

1. **Design:** designing production technologies and products in such a way that they require fewer nonrenewable resources and more effectively use waste, for example by emphasizing disassembly and reuse

2. **Reduction:** producing and consuming less

3. **Reuse:** increasing the reuse rate of non-waste products and technologies

4. **Recycle:** more efficiently recycling food waste and packaging.

But how can we make circularity viable within the constraints and incentives of the current global capitalist economy (see also Food controversy 14.1 about global capitalism in *Chapter 14: Collective Action*)? The enormous growth in population, wealth, and access to food over the last two centuries has been based on linear-economy logic. For businesses to implement the strategies of circularity, they need to be able to expect at least equal profits as from their normal, resource-wasting ones. This is only likely when consumers start demanding more sustainable products and businesses work together more closely with governments, scientists, social enterprises, and other non-profit organizations to supply them (see Case study 11.2 on China's circular economy policies).

Pause and think

The Chinese policies in Case study 11.2 are labelled as efforts in building a circular economy, but to what extent is this accurate? Are renewable energy and energy efficiency a necessary part of a truly circular economy? And if so, how important are they?

CASE STUDY 11.2

The circular economy in China

Since the death of Mao in 1976, China has moved from a tightly state-controlled, planned economy to a more market-oriented one. As China industrialized rapidly in the late 1990s and 2000s, Chinese policymakers became increasingly aware of the immense environmental challenges resulting from this consumption—from intense urban air pollution to rural land degradation (see Figure A). By virtue of its population, China is the largest emitter of GHGs, ahead of the USA (it is worth noting that in terms of per capita emissions, China is far behind the USA, but it is catching up). In response to these challenges, in 2002, the central government accepted the circular economy as a key development strategy (Yuan et al., 2006).

Policies have been implemented at the micro-, meso-, and macro-levels of Chinese society to enhance what the central government calls its circular economy. At the micro-level, firms are pushed by provinces (large administrative regions in China) to make production technologies more environmentally friendly. For example, firms are encouraged to increase manufacturing efficiencies and renewable technology is subsidized (Yuan et al., 2006). Yet consumer-waste recycling, another important pillar of a circular economy, has received much less attention (Miao and Tang, 2016).

At the meso-level, groups of firms are given incentives to increase the resource efficiency in the shared parts of their

a)

b)

Figure A (a) Air pollution in Zhengzhou, China and (b) factories along the Yangtze river are two examples of environmental pollution in China.

Source: (a) Sammy Corfield http://sammysgenericblog.blogspot.ca/2012/05/serious-travel-article-10-from.html; (b) High Contrast/Wikimedia Commons (CC BY 2.0).

supply chains, often in eco-industrial parks (Mathews and Tan, 2011). For example, the manufacturers of printed circuit boards in Suzhou New District near Shanghai use copper that is recovered from waste from elsewhere in the park, rather than newly mined copper (Mathews and Tan, 2016). Finally, at the macro-level, municipalities and cities are targeted by policies to push both consumers and producers to be more efficient (Mathews and Tan, 2011, 2016; Yuan et al., 2006).

It is too early to evaluate the effects of all these policies at the national level, and there are serious methodological challenges in measuring their impact. For example, do we measure raw materials recycled? The energy needed in a system? Or the land required to produce something for consumption? Moreover, given that the circular

economy must be thought of as including all aspects of production and consumption, it is not clear that the policies actually constitute an effort to implement a circular economy as defined above, or is simply a push to reduce the environmental impacts of Chinese economic activity. In some cases, efforts can be politically self-serving, by improving efficiencies in production (increasing economic competitiveness).

Even so, it is likely that circular economy policies have helped Chinese cities to take steps towards a more sustainable economy, with one estimate suggesting that resource use and waste reduction between 2005 and 2013 had improved between 35% and 47% (Geng et al., 2013; Mathews and Tan, 2016; Su et al., 2013).

11.4 **Future outlook**

Consumer choices for sustainable food options have the potential to transform food systems, making them more sustainable (and possibly even circular). After all, if consumers collectively and consistently demand more sustainable products, they can force the rest of the food system to comply with their wishes. Unfortunately, this chapter has highlighted

the many obstacles on that path. Consumer choices are determined by many things, including personal taste, social and cultural norms, food price, availability, and many other variables. Sustainability concerns are not universally shared; information on food's impacts on sustainability is not always easy to acquire or understand, and even for those consumers who want to behave sustainably and have sufficient information, there is a considerable gap between their environmentalist attitudes and knowledge and their actual consumption behaviour.

Furthermore, while individual consumer motivations matter, the situation also has many of the features of a classic collective action problem (see *Chapter 14: Collective Action*). For while sustainable behaviour of an individual consumer can contribute to the overall sustainability of the system, more fundamental changes in the food system likely require a critical mass of people across national borders demanding more sustainable food and being willing to change their lifestyles fundamentally. For example, for consumer demand to drive a transformation of our food systems into circular food economies, it must convince policymakers, NGOs, and businesses to radically change the way in which they engage with food. Such processes generally involve thresholds in terms of the numbers of consumers supporting it (even if it is unclear where this threshold might be) (Pierson, 2004).

What are the chances that consumers will collectively begin to take sustainability seriously in their food choices? Without drastic shifts in the circumstances under which they make those choices, not great. But there are scenarios under which such circumstances can shift quickly and dramatically. First, food choices are partly a function of informal religious or cultural norms, as well as fashion trends. Both norms and fashion trends have the capacity to mobilize and change the behaviour of large groups of people across national borders. Getting sustainable consumption taken up by either is thus an opportunity to facilitate change. Civil society organizations (such as churches or cultural centres), opinion leaders, celebrities, and other actors with influence on social norms and fashion trends can have important roles to play here.

Second, governments can implement and enforce policies that constrain wasteful consumption and promote sustainability. The chapter has highlighted many examples, from the minimal efforts involved in nudging to the full-scale reform of building a circular economy. In all these cases, government policies have the advantage of enforcing collective action on those who are subject to the policies. International cooperation may be required, highlighting a role of international organizations. But they all face information problems, as well as the chicken-and-egg dilemma—what if government action depends on demand from the public, and public attitudes and behaviour towards sustainability depend on government action?

Finally, then, there is the role of crises or other game-changing developments. Theoretically, such exogenous shocks to the food system have the capacity to shift public opinion or government policies quickly and kickstart the process of making food consumption more sustainable. But what would such game-changers look like? The list of possibilities is endless, but one likely candidate is **food scares**—moments of escalating public anxiety (and media attention) over the safety of food products. There is evidence that food scares can result in long-run decreases in demand of the 'scary' products (Böcker and Hanf, 2000; Freidberg, 2004). A combination of a game-changing event, such as a food scares around unsustainable food, with government and civil society efforts in promoting sustainable norms may therefore constitute one scenario where relatively quick change in consumption behaviour is possible.

● QUESTIONS

11.1 What are three of the most important determinants of food choices?

11.2 How does price affect food consumption? Are the effects of food prices the same for everyone? If not, why?

11.3 Can you give examples of how nudging can be used to promote more sustainable food consumption? What are the ethical dilemmas involved with this technique?

11.4 How can food labelling be improved to make it more effective at promoting sustainable food consumption?

11.5 What is circular economy? Why is it important for sustainable food consumption?

● FURTHER READING

Carrington, M.J., Neville, B.A., Whitehall, G.J. 2010. Why ethical consumers don't walk their talk: towards a framework for understanding the gap between the ethical purchase intentions and actual buying behaviour of ethically minded consumers. *Journal of Business Ethics* **97**, 139–158. **(Develops a conceptual framework for the attitude–behaviour gap, drawing on psychological and economic ideas.)**

Cervellon, M.-C., Dubé, L. 2005. Cultural influences in the origins of food likings and dislikes. *Food Quality and Preference* **16**, 455–460. **(Insightful empirical analysis of cross-national differences in food preferences.)**

Chung, C., Myers Jr, S.L. 1999. Do the poor pay more for food? An analysis of grocery store availability and food price

disparities. *Journal of Consumer Affairs* **33**, 276–296. **(Classic analysis highlighting how food spending is determined by the availability of cheap chain stores.)**

Hausman, D.M., Welch, B. 2010. Debate: to nudge or not to nudge. *Journal of Political Philosophy* **18**, 123–136. **(Philosophical evaluation of nudging as a strategy to overcome the perennial tension between freedom and paternalism.)**

Reisch, L., Eberle, U., Lorek, S. 2013. Sustainable food consumption: an overview of contemporary issues and policies. *Sustainability: Science, Practice and Policy* **9**, 7–25. **(Useful synthesis article highlighting the main issues in sustainable consumption, its links to other aspects of food sustainability, and several policy options to enhance it.)**

● REFERENCES

Abrams, K.M., Evans, C., Duff, B.R. 2015. Ignorance is bliss. How parents of preschool children make sense of front-of-package visuals and claims on food. *Appetite* **87**, 20–29.

Andersen, M.S. 2007. An introductory note on the environmental economics of the circular economy. *Sustainability Science* **2**, 133–140.

Auger, P., Devinney, T.M. 2007. Do what consumers say matter? The misalignment of preferences with unconstrained ethical intentions. *Journal of Business Ethics* **76**, 361–383.

Banerjee, S., Gulas, C.S., Iyer, F. 1995. Shades of green: a multidimensional analysis of environmental advertising. *Journal of Advertising* **24**, 21–31.

Barendregt, B., Jaffe, R. 2014. *Green Consumption: The Global Rise of Eco-Chic*. London: Bloomsbury Publishing.

Bauer, H. H., Heinrich, D., Schäfer, D.B. 2013. The effects of organic labels on global, local, and private brands: more hype than substance? *Journal of Business Research, Recent Advances in Globalization, Culture and Marketing Strategy* **66**, 1035–1043.

Bawa, A. S., Anilakumar, K.R. 2013. Genetically modified foods: safety, risks and public concerns—a review. *Journal of food science and technology* **50**, 1035–1046.

Berazneva, J., Lee, D.R. 2013. Explaining the African food riots of 2007–2008: an empirical analysis. *Food Policy* **39**, 28–39.

Böcker, A., Hanf, C.H. 2000. Confidence lost and—partially—regained: consumer response to food scares. *Journal of Economic Behavior & Organization* **43**, 471–485.

Brécard, D. 2014. Consumer confusion over the profusion of eco-labels: lessons from a double differentiation model. *Resource and Energy Economics* **37**, 64–84.

Brownell, K.D., Frieden, T.R. 2009. Ounces of prevention—the public policy case for taxes on sugared beverages. *New England Journal of Medicine* **360**, 1805–1808.

Carrington, M.J., Neville, B.A., Whitwell, G.J. 2010. Why ethical consumers don't walk their talk: Towards a framework for understanding the gap between the ethical purchase

intentions and actual buying behaviour of ethically minded consumers. *Journal of Business Ethics* **97**, 139–158.

Casey, J.W., Holden, N.M. 2006. Greenhouse gas emissions from conventional, agri-environmental scheme, and organic Irish suckler-beef units. *Journal of Environmental Quality* **35**, 231–239.

Cervellon, M.-C., Dubé, L. 2005. Cultural influences in the origins of food likings and dislikes. *Food Quality and Preference* **16**, 455–460.

Chekima, B., Chekima, S., Syed Khalid Wafa, S.A.W., Igau, O.A., Sondoh Jr, S.L. 2016. Sustainable consumption: the effects of knowledge, cultural values, environmental advertising, and demographics. *International Journal of Sustainable Development & World Ecology* **23**, 210–220.

Csutora, M., Vetőné Mózner, Z. 2014. Consumer income and its relation to sustainable food consumption–obstacle or opportunity? *International Journal of Sustainable Development & World Ecology* **21**, 512–518.

Dangour, A.D., Dodhia, S.K., Hayter, A., Allen, E., Lock, K., Uauy, R. 2009. Nutritional quality of organic foods: a systematic review. *American Journal of Clinical Nutrition* **90**, 680–685.

de Ita, A. 2008. *Fourteen Years of NAFTA and the Tortilla Crisis*. Americas Program Special Report. Washington, DC: Americas Program, Center for International Policy (CIP).

Douglas, M. 1966. *Purity and Danger: An Analysis of Concepts of Pollution and Taboo*. London: Routledge.

Drewnowski, A., Specter, S.E. 2004. Poverty and obesity: the role of energy density and energy costs. *American Journal of Clinical Nutrition* **79**, 6–16.

Erdem, T., Swait, J., Louviere, J. 2002. The impact of brand credibility on consumer price sensitivity. *International Journal of Research in Marketing* **19**, 1–19.

European Commission (EC). 2012. *Food Information Schemes, Labelling and Logos*. Internal document DG SANCO.

European Food Information Council (EUFIC) (2006). *The Determinants of Food Choice*. Available at:

http://www.eufic.org/en/healthy-living/article/the-determinants-of-food-choice [accessed 12 April 2017].

Evans, A., Miele, M. 2017. Food labelling as a response to political consumption: effects and contradictions. In: Keller, M., Halkier, B., Wilska, T.-A., Truninger, M. (eds). *Routledge Handbook on Consumption*. New York, NY: Routledge, pp. 233–247.

Fairley, P. 2011. Introduction: next generation biofuels. *Nature* **474**, S2–S5.

Fowler III, A.R., Close, A.G. 2012. It ain't easy being green: Macro, meso, and micro green advertising agendas. *Journal of Advertising* **41**, 119–132.

Freidberg, S. 2004. *French Beans and Food Scares: Culture and Commerce in an Anxious Age*. New York, NY: Oxford University Press.

French, S.A., Wall, M., Mitchell, N.R. 2010. Household income differences in food sources and food items purchased. *International Journal of Behavioral Nutrition and Physical Activity* **7**, 77.

Gatersleben, B., Murtagh, N., Abrahamse, W. 2014. Values, identity and pro-environmental behaviour. *Contemporary Social Science* **9**, 374–392.

Geng, Y., Sarkis, J., Ulgiati, S., Zhang, P. 2013. Measuring China's circular economy. *Science* **339**, 1526–1527.

Ghisellini, P., Cialani, C., Ulgiati, S. 2016. A review on circular economy: the expected transition to a balanded interplay of environmental and economic systems. *Journal of Cleaner Production* **114**, 11–32.

Gomiero, T., David Pimentel, D., Paoletti, M.G. 2011. Environmental impact of different agricultural management practices: conventional vs. organic agriculture. *Critical Reviews in Plant Sciences* **30**, 95–124.

Grunert, K.G., Hieke, S., Wills, J. 2014. Sustainability labels on food products: consumer motivation, understanding and use. *Food Policy* **44**, 177–189.

Guidetti, M., Cavazza, N. 2008. Structure of the relationship between parents' and children's food preferences and avoidances: an explorative study. *Appetite* **50**, 83–90.

Gustavsson, J., Cederberg, C., Sonesson, U., Van Otterdijk, R., Meybeck, A. 2011. *Global Food Losses and Food Waste*. Rome: Food and Agricultural Organization of the United Nations.

Hansen, P.G. 2016. The definition of nudge and libertarian paternalism: does the hand fit the glove? *European Journal of Risk Regulation* **7**, 155–174.

Harper, L., Souta, P., Ince, J., McKenzie, J. 2007. *Food Labelling Consumer Research: What Consumers Want. A Literature Review*. London: Food Standards Agency.

Harris, J.L., Thompson, J.M., Schwartz, M.B., Brownell, K.D. 2011. Nutrition-related claims on children's cereals: what do they mean to parents and do they influence willingness to buy? *Public Health Nutrition* **14**, 2207–2212.

Harris, M. 1998. *Good to Eat: Riddles of Food and Culture*. Long Grove, IL: Waveland Press.

Hausman, D.M., Welch, B. 2010. Debate: to nudge or not to nudge. *Journal of Political Philosophy* **18**, 123–136.

Headey, D., Fan, S. 2010. *Reflections on the Global Food Crisis: How Did it Happen? How Has it Hurt? And How Can We Prevent the Next One?* Washington, DC: International Food Policy Research Institute.

Hitchens, C. 2008. *God is not Great: How Religion Poisons Everything*. London: Atlantic Books.

Hughner, R.S., McDonagh, P., Prothero, A., Shultz, C.J., Stanton, J. 2007. Who are organic food consumers? A compilation and review of why people purchase organic food. *Journal of Consumer Behaviour* **6**, 94–110.

Jacobson, M.F., Brownell, K.D. 2000. Small taxes on soft drinks and snack foods to promote health. *American Journal of Public Health* **90**, 854.

Johannesburg Implementation Plan. 2002. *Plan of Implementation of the World Summit on Sustainable Development*. A/CONF.199/20, Johannesburg, September 2002.

Kemp, K., Insch, A., Holdsworth, D.K., Knight, J.G. 2010. Food miles: do UK consumers actually care? *Food Policy* **35**, 504–513.

Kim, D., Kawachi, I. 2006. Food taxation and pricing strategies to 'thin out' the obesity epidemic. *American Journal of Preventive Medicine* **30**, 430–437.

Kim, Y., Oh, S., Yoon, S., Shin, H.H. 2016. Closing the green gap: the impact of environmental commitment and advertising believability. *Social Behavior and Personality* **44**, 339–351.

Kleiner, K. 2008. The backlash against biofuels. *Nature Reports Climate Change* **2**, 9–11.

Kollmuss, A., Agyeman, J. 2002. Mind the gap: why do people act environmentally and what are the barriers to pro-environmental behavior? *Environmental Education Research* **8**, 239–260.

Lea, E., Worsley, A. 2008. Australian consumers' food-related environmental beliefs and behaviours. *Appetite* **50**, 207–214.

Lee, H-J., Yun, Z-S. 2015. Consumers' perceptions of organic food attributes and cognitive and affective attitudes as determinants of their purchase intentions toward organic food. *Food Quality and Preference* **39**, 259–267.

Lehner, M., Mont, O., Heiskanen, E. 2016. Nudging—a promising tool for sustainable consumption behaviour? *Journal of Cleaner Production*, Special Volume: Transitions to Sustainable Consumption and Production in Cities **134**, 166–77.

MacDiarmid, J.I., Kyle, J., Horgan, G.W., Loe, J., Fyfe, C., Johnstone, A., McNeill, G. 2012. Sustainable diets for the future: can we contribute to reducing greenhouse gas emissions by eating a healthy diet? *American Journal of Clinical Nutrition* **96**, 632–639.

Mathews, J.A., Tan, H. 2011. Progress toward a circular economy in China. *Journal of Industrial Ecology* **15**, 435–457.

Mathews, J.A., Tan, H. 2016. Circular economy: lessons from China. *Nature News* **531**, 440.

Maubach, N., Hoek, J., McCreanor, T. 2009. An exploration of parents' food purchasing behaviours. *Appetite* **53**, 297–302.

Mayer, S., Stirling, A. 2004. GM crops: good or bad? *EMBO Reports* **5**, 1021–1024.

Miao, X., Tang, Y. 2016. China: industry parks limit circular economy. *Nature* **534**, 37.

Mie, A., Andersen, H.R., Gunnarsson, S., Kahl, S., Kesse-Guyot, E., Rembiałkowska, E., Quaglio, G., Grandjean, P. 2017. Human health implications of organic food and organic agriculture: a comprehensive review. *Environmental Health* **16**, 111.

Mirabella, N., Castellani, V., Sala, S. 2014. Current options for the valorization of food manufacturing waste: a review. *Journal of Cleaner Production* **65**, 28–41.

Monteleone, E., Spinelli, S., Dinnella, C., Endrizzi, I., Laureati, M., Pagliarini, E., Sinesio, F., Gasperi, F., Torri, L., Aprea, E. 2017. Exploring influences on food choice in a large population sample: The Italian Taste project. *Food Quality and Preference* **59**, 123–140.

Nicolia, A., Manzo, A., Veronesi, F., Rosellini, D. 2014. An overview of the last 10 years of genetically engineered crop safety research. *Critical Reviews in Biotechnology* **34**, 77–88.

Ní Mhurchu, C., Blakely, T., Jiang, Y., Eyles, H.C., Rodgers, A. 2009. Effects of price discounts and tailored nutrition education on supermarket purchases: a randomized controlled trial. *American Journal of Clinical Nutrition* **91**, 736–747.

Piazza, J., Ruby, M.B., Loughnan, S., Luong, M., Kulik, J., Watkins, H.M., Seigerman, M. 2015. Rationalizing meat consumption. The 4Ns. *Appetite* **91**, 114–128.

Pierson, P. 2004. *Politics in Time: History, Institutions, and Social Analysis*. Princeton, NJ: Princeton University Press.

Redding, R.W. 2015. The pig and the chicken in the Middle East: modeling human subsistence behavior in the archaeological record using historical and animal husbandry data. *Journal of Archaeological Research* **23**, 325–368.

Reganold, J.P., Wachter, J.M. 2016. Organic agriculture in the twenty-first century. *Nature Plants* **2**, 15221.

Reisch, L., Eberle, U., Lorek, S. 2013. Sustainable food consumption: an overview of contemporary issues and policies. *Sustainability: Science, Practice and Policy* **9**, 7–25.

Reiss, M.J., Straughan, R. 2001. *Improving Nature? The Science and Ethics of Genetic Engineering*. Cambridge: Cambridge University Press.

Saba, A., Di Natale, R. 1998. A study on the mediating role of intention in the impact of habit and attitude on meat consumption. *Food Quality and Preference* **10**, 69–77.

Schor, J.B. 2014. *Born to Buy: The Commercialized Child and the New Consumer Cult*. New York, NY: Simon and Schuster.

Schösler, H., de Boer, J., Boersema, J.J., Aiking, H. 2015. Meat and masculinity among young Chinese, Turkish and Dutch adults in the Netherlands. *Appetite* **89**, 152–159.

SGP. 2010. *Verkiezingsprogramma SGP 2010–2014: Daad bij het Woord*. Available at: https://www.parlement.com/9291000/d/2010_sgp_verkiezingsprogramma.pdf [accessed 5 May 2018].

Sheehan, K., Atkinson, L. 2012. Special issue on green advertising: revisiting green advertising and the reluctant consumer. *Journal of Advertising* **41**, 5–7

Smith-Spangler, C., Brandeau, M.L., Hunter, G.E., Bavinger, J.C., Pearson, M., Eschbach, P.J., Sundaram, V., Liu, H., Schirmer, P., Stave, C., Olkin, I., Bravata, D.M. 2012. Are organic foods safer or healthier than conventional alternatives?: A systematic review. *Annals of Internal Medicine* **157**, 348.

Sørensen, H.S., Clement, J., Gabrielsen, G. 2012. Food labels: an exploratory study into label information and what consumers see and understand. *International Review of Retail, Distribution and Consumer Research* **22**, 101–114.

Su, B., Heshmati, A., Geng, Y., Yu, X. 2013. A review of the circular economy in China: moving from rhetoric to implementation. *Journal of Cleaner Production* **42**, 215–227.

Thomassen, M.A., van Calker, K.J., Smits, M.C., Iepema, G.L., de Boer, I.J. 2008. Life cycle assessment of conventional and organic milk production in the Netherlands. *Agricultural Ssystems* **96**, 95–107.

Thompson, J.F. 2013. *'Nudge Marketing' Most Effective Strategy to Push Produce Sales*. Available at: https://www.adweek.com/digital/nudge-marketing-most-effective-strategy-to-push-produce-sales/ [accessed 4 December 2017].

Tuomisto, H.L., Hodge, I.D., Riordan, P., Macdonald, D.W. 2012. Does organic farming reduce environmental impacts? A meta-analysis of European research. *Journal of Environmental Management* **112**, 309–320.

van Herpen, F., Van Trijp, H.C. 2011. Front-of-pack nutrition labels. Their effect on attention and choices when consumers have varying goals and time constraints. *Appetite* **57**, 148–160.

Veldstra, M.D., Alexander, C.E., Marshall, M.I. 2014. To certify or not to certify? Separating the organic production and certification decisions. *Food Policy* **49**, 429–436.

Vermeir, I., Verbeke, W. 2006. Sustainable food consumption: exploring the consumer 'attitude–behavioral intention' gap. *Journal of Agricultural and Environmental Ethics* **19**, 169–194.

Wansink, B. 2017. Healthy profits: an interdisciplinary retail framework that increases the sales of healthy foods. *Journal of Retailing* **93**, 65–78.

Waterlander, W.E., de Haas, W.E., van Amstel, I., Schuit, A.J., Twisk, J.W., Visser, M., Seidell, J.C., Steenhuis, I.H. 2010. Energy density, energy costs and income: how are they related? *Public Health Nutrition* **13**, 1599–1608.

Wirtz, J.G., Ahn, R., Song, R., Wang, Z. 2013. *Selling or Selling Out? A Content Analysis of Children's Snack Packages and Implications for Advertising Practitioners and Educators*. Washington, DC: Association for Education in Journalism and Mass Communication National Conference; August 2013.

Wise, T.A. 2011. Mexico: the cost of U.S. dumping. *NACLA Report on the Americas* **44**, 47–48, 40.

Worthington V. 2001. Nutritional quality of organic versus conventional fruits, vegetables, and grains. *Journal of Alternative & Complementary Medicine* **7**, 161–173.

Xie, G.X. and Kronrod, A. 2012. Is the devil in the details? The signaling effect of numerical precision in environmental advertising claims. *Journal of Advertising* **41**, 103–117.

Yuan, Z., Bi, J., Moriguichi, Y. 2006. The circular economy: a new development strategy in China. *Journal of Industrial Ecology* **10**, 4–8.

Zandstra, E.H., Carvalho, Á.H., Van Herpen, E. 2017. Effects of front-of-pack social norm messages on food choice and liking. *Food Quality and Preference* **58**, 85–93.

Food and Governance

How do different food systems such as organic farming and agribusiness produce, process, distribute, and consume food? Are there ways to make supply chains and food systems more sustainable? How do power and politics shape the way in which you put food on your table? Why is scale a key challenge to sustainable food governance? How does free-riding help to explain unsustainable food systems? And how might privatization make them more sustainable?

Part 1: Food and Environment explored the ways in which food systems impact the environment, while *Part 2: Food and Society* looked at the way human societies themselves interact with food. In *Part 3: Food and Governance*, we explore the ways in which governance can be used to address collective action problems and make food systems more sustainable.

In *Chapter 12: Food Systems*, we define supply chains and food systems. We differentiate between mainstream and alternative food systems and highlight the importance of scale in analysing them. Next, we define what sustainable food systems look like and identify ways to measure food system sustainability. Finally, we investigate methods for enhancing the sustainability of food systems

Chapter 13: Governance then explores the politics and policymaking processes involved in the 'wicked' problem of food governance. We outline the policy cycle in which stakeholders negotiate and compete to promote their interests. We outline the goals these stakeholders have in influencing government policy and other institutions and how these shape food systems. We then explain the key problems that characterize this process, with the aim of illustrating the difficulties in governing food systems to make them more sustainable.

Finally, *Chapter 14: Collective Action* builds on this by zooming in on the fundamental problems at the heart of our collective inability to make food systems more sustainable. Here we introduce the tragedy of the commons, which describes the general human tendency to abuse non-excludable goods, and present different social-scientific models to explain it. Finally, we outline a range of institutionalist solutions to collective action problems.

Image credit: vicspacewalker/Shutterstock.com

Part contents

Food Systems

How are food systems organized in a globalized economy?

Peter Oosterveer and Anke Brons

Chapter Overview

- Food systems encompass the people, institutions, activities, processes, and infrastructures involved in producing and consuming food. They include supply chains, which are the sequences of organizations and processes involved in the production, processing, and distribution of food items.

- Mainstream food systems aim for economic efficiency and rely on the intensive use of technology, including machinery, chemical fertilizers, and pesticides. They focus on large-scale production for a global market and use supermarkets as their main consumer outlet.

- Alternative food systems prioritize quality over efficiency in the food supply chain, for example through organic production processes, ethics of production, and the relationships between producer and consumer.

- Sustainable food systems aim to provide economic, social, and environmental sustainability, while still providing food security and appropriate nutrition.

- We can measure the sustainability of food systems by measuring the level of sustainable diets, waste, and the concept of foodsheds.

- Strategies to enhance the sustainability of food systems include: (1) building on 'what works', for example alternative food systems, local food systems, sustainable intensification, and sustainable diets, and (2) reforming food governance, for example by increasing food sovereignty or making governance systems more adaptive.

Introduction

How many ingredients, from how many continents, go into making your favourite granola bar (see Figure 12.1a)? Can you buy locally produced fruit and vegetables near you (see Figure 12.1b)? Have you ever bought your food directly from a farmer or at a farmer's market? And what percentage of your food do you grow yourself, maybe on your balcony (see Figure 12.1c)?

The questions above are intended to make you think about the systems that put food on your table. How

a)

b)

c)

Figure 12.1 (a) A granola bar. (b) Locally grown products in a supermarket. (c) An 'edible wall' that can fit on your balcony. Source: (b) walmartcorporate/Flickr (CC BY 2.0).

can we define and categorize these food systems? What does a food system need to do to be sustainable? And how can we realize sustainability in food systems? To answer these questions, we introduce you to food system thinking (Section 12.1)—what are the key elements of food supply chains and food systems? Then, we explore different types of food systems (Section 12.2). Finally, we think about ways in which sustainable food systems can be defined, measured, and promoted (Section 12.3).

12.1 What are food systems?

If you think about where your food comes from, an intuitive model is the **food supply chain**—the sequence of processes and organizations involved in the production, processing, and distribution of food items (often referred to as 'from farm to fork'). Figure 12.2 presents the basic elements of food supply chains. First, the agricultural sector produces the raw materials for most foods, using agricultural inputs such as fertilizers and seeds. Agricultural produce is then either sold directly (that is, to consumers, craft producers, such as bakeries, or non-food sectors) or sold to the processing industry that then processes the food to sell to the consumer through wholesalers and retailers.

Food supply chains can be short and simple. Very short chains include subsistence farming, farmers' markets, visiting a farm to buy food there, or purchasing food on an online farm shop. They can also be complicated networks of agricultural suppliers, food processing industries, wholesale companies, and retailers. A Big Mac at McDonalds not only contains over 60 different ingredients (see Figure 12.4 for a full list), sourced from agricultural and other producers, but it also needs to be processed to very specific quality and taste standards all around the world, after which they are distributed to 30,000+ restaurants quickly enough to remain fresh.

In your experience

When you next buy some food at the supermarket, look at the full list of ingredients and think about the supply chain that delivered it to you. How many ingredients does it contain? What agricultural producers would have been needed? Where do you think they come from? What elements of the food are processed? And how complex was the wholesale/retail network involved in distributing your food?

The concept of supply chains helps food companies map out their organizational processes and identify measures to improve their efficiency (see Figure 12.3). This can sometimes improve food quality, lower costs, reduce environmental impacts (for example, resource use and waste), and reduce social costs (for example, labour exploitation) (Barrientos

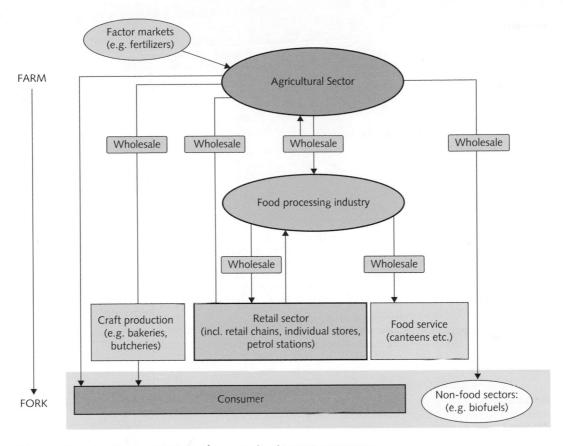

Figure 12.2 Basic elements of food supply chains, from agricultural inputs to consumers.

Source: Bukeviciute, L., Dierx, A. and Ilzkovitz, F. 2009. The functioning of the food supply chain and its effect on food prices in the European Union (Occasional Papers No. 47). Office for infrastructures and logistics of the European Communities. (c) European Union, 1995–2018.

Figure 12.3 From farm to fork used as an example of a Brazilian beef steak consumed in a UK restaurant from production (a) to processing (b) to distribution (c) and consumption (d). If you were managing this supply chain, what could you do to reduce its economic, environmental, or social costs?

Source: (a) Zeloneto/Wikimedia Commons (CC BY-SA 3.0); (d) loustejskal/Flickr (CC BY 2.0).

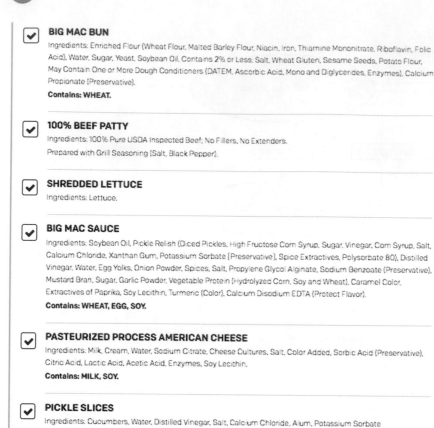

BIG MAC BUN
Ingredients: Enriched Flour (Wheat Flour, Malted Barley Flour, Niacin, Iron, Thiamine Mononitrate, Riboflavin, Folic Acid), Water, Sugar, Yeast, Soybean Oil, Contains 2% or Less: Salt, Wheat Gluten, Sesame Seeds, Potato Flour, May Contain One or More Dough Conditioners (DATEM, Ascorbic Acid, Mono and Diglycerides, Enzymes), Calcium Propionate (Preservative).
Contains: WHEAT.

100% BEEF PATTY
Ingredients: 100% Pure USDA Inspected Beef; No Fillers, No Extenders.
Prepared with Grill Seasoning (Salt, Black Pepper).

SHREDDED LETTUCE
Ingredients: Lettuce.

BIG MAC SAUCE
Ingredients: Soybean Oil, Pickle Relish (Diced Pickles, High Fructose Corn Syrup, Sugar, Vinegar, Corn Syrup, Salt, Calcium Chloride, Xanthan Gum, Potassium Sorbate (Preservative), Spice Extractives, Polysorbate 80), Distilled Vinegar, Water, Egg Yolks, Onion Powder, Spices, Salt, Propylene Glycol Alginate, Sodium Benzoate (Preservative), Mustard Bran, Sugar, Garlic Powder, Vegetable Protein (Hydrolyzed Corn, Soy and Wheat), Caramel Color, Extractives of Paprika, Soy Lecithin, Turmeric (Color), Calcium Disodium EDTA (Protect Flavor).
Contains: WHEAT, EGG, SOY.

PASTEURIZED PROCESS AMERICAN CHEESE
Ingredients: Milk, Cream, Water, Sodium Citrate, Cheese Cultures, Salt, Color Added, Sorbic Acid (Preservative), Citric Acid, Lactic Acid, Acetic Acid, Enzymes, Soy Lecithin.
Contains: MILK, SOY.

PICKLE SLICES
Ingredients: Cucumbers, Water, Distilled Vinegar, Salt, Calcium Chloride, Alum, Potassium Sorbate (Preservative), Natural Flavors, Polysorbate 80, Extractives of Turmeric (Color).

ONIONS
Ingredients: Onions.

Figure 12.4
A screenshot of all the ingredients of a Big Mac.
Source: © 2018 McDonald's.

and Dolan, 2012). This makes food cheaper, as well as more sustainable in environmental and social terms. Businesses will often prioritize economic efficiency, but some measures can help to achieve all these goals simultaneously. For example, shortening supply chains can reduce transportation costs and carbon emissions (see *Chapter 6: Climate Change*), as well as the likelihood of labour exploitation by facilitating supply chain transparency (Trienekens et al., 2012).

Of course, food systems have impacts on the natural environment and human societies (as discussed in depth in Parts 1 and 2), but the environment and societal factors also impact food supply chains. Moreover, food sustainability is about food security, which requires efficient and high-quality food production and distribution, as well as available and affordable food for the world's poor (see *Chapter 9: Food Security*). Supply chain analysis on its own tells us little about these issues; as a result, the concept of

a **food system** has become increasingly popular. Food systems attempt to address the future of food in a more holistic way (see Box 1.1 in *Chapter 1: Introduction* about the concept of holism). This approach can be seen in the analysis presented on the 2007–2008 food crisis (see Box 12.1) (Ingram, 2011).

Although definitions of food systems remain debated, we define them as '*the complete set of people, institutions, activities, processes, and infrastructures involved in producing and consuming food for a given population*' (UNEP, 2016). This includes food supply chains but positions these in their social, economic, political, and environmental contexts (Delaney et al., 2016). Thus, the food systems approach is an example of an interdisciplinary, or even transdisciplinary, and holistic way of analysing food, including as many relevant variables and processes as possible, as well as their interactions.

But what does this look like, concretely? Two examples are presented in Figure 12.5. Figure 12.5a

a)

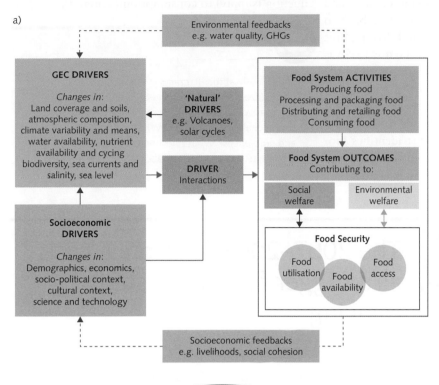

b)

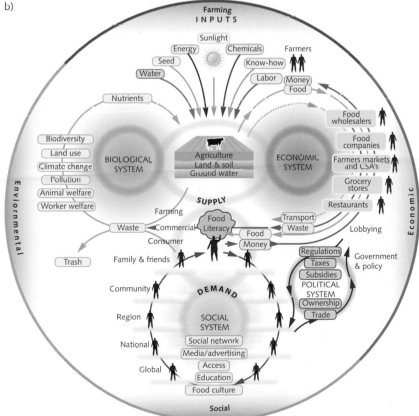

Figure 12.5 Compared to the farm-to-fork thinking shown in Figure 12.2, the food systems approach is a more comprehensive approach to mapping food production, processing, distribution, and consumption.

Source: (a) Ericksen, P. J., Ingram, J. S. I., Liverman, D. M. (2009). Food security and global environmental change: emerging challenges. Environmental Science & Policy 12(4), 373–377. Copyright © 2009 Elsevier Ltd. All rights reserved; (b) Food System Map graphic courtesy of Nourish Initiative, www.nourishlife.org Copyright WorldLink, all rights reserved.

highlights how socio-economic and environmental drivers interact to shape food system activities and outcomes. The food outcomes, in turn, feed back to the socio-economic and environmental drivers, thus creating a circular, dynamic system. Figure 12.5b is similarly circular, but more complex in terms of the numbers of variables and the extent to which scale is incorporated in the model (scale refers to different societal levels from family and friends to the global level in the social system). The chart identifies biological, economic, political, and social systems as key elements of the food system and highlights how they interact with the supply chain from inputs and production (supply) to consumption (demand).

Pause and think

Look closely at the diagrams in Figure 12.5, and summarize the similarities and differences between them, as well as between these models and the food supply chain in Figure 12.2. What are the key similarities and differences? What questions does each of the models help you to address about food sustainability?

BOX
12.1

The 2007–2008 food price crisis as a trigger for food systems thinking

Between 2006 and 2008, global food and other commodity prices rose rapidly. The overall price increase may have been as much as 64% (over 2002–2008), with some foods doubling in price in just a few months (Berazneva and Lee, 2013; Cohen and Garrett, 2010) (see Figure A). It became clear that the price spikes had severe consequences for many of the world's poor, particularly in urban areas, as well as rural areas that were not self-sufficient. According to one estimate, up to 100 million people ended up in poverty because of the price increase (Ivanic and Martin, 2008).

Analysts have tried to estimate the consequences of the price spike and attempted to identify its causes. It turned out that neither task was straightforward; in fact, it forced

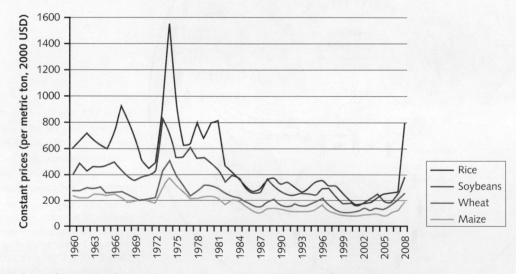

Figure A Trends in real international prices of cereals, rice, soybeans, wheat, and maize (in USD).

Source: Headey, D., Fan, S., 2008. Anatomy of a crisis: the causes and consequences of surging food prices. Agricultural Economics 39, 375–391. Copyright © 2008, John Wiley and Sons.

scholars to turn to informal detective work, rather than running the usual statistical models (Headey and Fan, 2008).

Why the complexity? To start with, the consequences of food prices on poverty and food security are complicated, even in ordinary times. For example, food price increases are only a burden for some of the world's poor—many poor people produce, as well as purchase, food items, so increasing food prices should, in theory, result in larger incomes and a reduction in poverty (Headey and Martin, 2016). But in addition to these differentiated effects of food prices, the 2006–2008 price rises impacted the wider economy and political systems in many low- and middle-income countries. Price increases fuelled widespread anger and, in some cases, led to violent riots (see Case study 11.1 in *Chapter 11: Consumption*). These system-wide impacts have made it difficult to estimate precisely what the full impact of the crisis has been on food security.

Explaining the crisis turned out to be equally challenging. In economics, a price of a good increases when there is a decrease in supply or an increase in demand. If we consider an isolated food, say maize, we might expect price increases during a drought and a decrease during good weather. But what happens when there are multiple pressures all acting at the same time and in complex ways? To take just one example, if

the price of oil increases, it costs more to produce the maize (for example, we need oil for tractors), raising prices. At the same time, increasing oil prices also incentivizes farmers to produce biofuels from the maize, which can be sold near or at the price of oil. Following this incentive, more maize is grown for biofuels, rather than for food, raising the food price further. This is only the tip of the iceberg; dynamics such as these happen in food systems all around the world, and food prices can be influenced by larger trends (for example, currency devaluation can reduce the amount of food people can import).

As an illustration, Figure B outlines the different factors driving the food price crisis (Abbott and Borot de Battisti, 2011). It differentiates demand factors from supply factors and provides the timing and duration for the causes. The figure shows a 'perfect storm'—long-term trends and short-term shocks coming together to rapidly increase food prices.

The approach in Figure B is useful for explaining events after they have occurred, but it is difficult to use for prediction (Abbott and Borot de Battisti, 2011; Pierson, 2004). As an alternative approach, researchers map out the key factors and the way in which they interact. This is commonly called a systems approach—or, when applied to food, a food systems approach. Figure C provides an example of a systems approach to the 2006–2008 crisis,

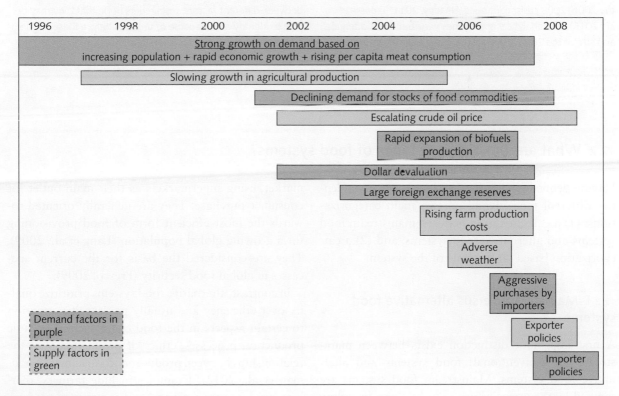

Figure B Selected factors contributing to food price increases in 2006–2008.

Source: Abbott, P., Borot de Battisti, A., 2011. Recent Global Food Price Shocks: Causes, Consequences and Lessons for African Governments and Donors. J Afr Econ 20, i12–i62. Copyright © 2011, Oxford University Press

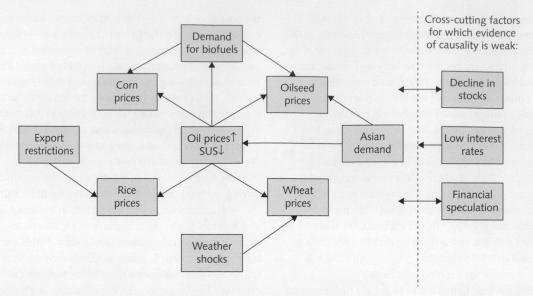

Figure C Summary model of the main factors contributing to the 2006–2008 food price increases.

Source: Headey, D., Fan, S., 2008. Anatomy of a crisis: the causes and consequences of surging food prices. Agricultural Economics 39, 375–391. Copyright © 2008, John Wiley and Sons .

taking the key explanations from Figure B and showing how they interlink.

The debate on the causes of food price volatility, including the 2006–2008 crisis, continues (Headey, 2011; Tadasse et al., 2016). The analytical approaches vary, but the suddenness and size of the 2006–2008 spikes, along with the impacts on food security, poverty, and politics, suggest that it is a wicked problem (a dynamic, complex, intractable problem; see Section 1.5 in *Chapter 1: Introduction*). By exploring the connections between different factors, researchers hope that they may be able to identify at least some of the risks before these problems arise again (Ingram, 2011; Ishii-Eiteman, 2009).

12.2 What are the different types of food systems?

Having defined what a food system is, we now identify different types, based on two main categorizations: (1) a categorization between mainstream food systems and alternative food systems, and (2) a categorization based on the scale of the system.

12.2.1 Mainstream versus alternative food systems

A first important distinction exists between **mainstream** (or conventional) **food systems** and **alternative food systems**. Mainstream food systems are typified by the intensive use of technology, including machinery, chemical fertilizers, and pesticides. These systems produce on a large scale for an international market, using supermarkets as their main outlet for consumer purchase. They are generally oriented towards the most efficient form of food provisioning for a growing global population (Lang et al., 2009). They are considered the basis for the current successes in global food security (Fresco, 2009).

In contrast, alternative food systems prioritize quality over efficiency and usually pay specific attention to certain aspects in the food chain such as organic production processes, ethics of production, and the relationships between producer and consumer (Goodman et al., 2012). Examples include farmers' markets, producer cooperatives, and community gardens, all of which seek to make food locally and involve the community more. This shortens the supply chain and

Figure 12.6 A local 'farmer's market' in Kalimantan, Indonesia. Consumers and food producers meet and transact on river boats.
Source: © David Ehrhardt.

makes it more transparent (Feagan, 2007; Sonnino and Marsden, 2005) (see Figure 12.6 for a market with short supply chains). Alternative food systems explicitly distinguish themselves from conventional food systems. Case study 12.1 and Box 12.2 provide examples of alternative food systems.

In your experience

Do you purposely buy Fairtrade products? If so, why? If not, what is the main reason that prevents you from buying them? Do you think schemes like these are sufficient to make our food systems fairer and more sustainable?

Although the distinction between mainstream and alternative food systems appears clear-cut, the boundary is not always easy to draw. Take the example of organic farming in California. It originally began as an alternative food system, but in recent years, as Californian land prices increased, intensification based on conventional technology became an economic necessity. Organic farmers were pushed to become more and more conventional (Sonnino and Marsden, 2005). Often, farmers now use principles from both mainstream and alternative food systems, resulting in a broad range of heterogenous food systems (Spaargaren et al., 2012).

CASE STUDY 12.1

Fairtrade International

Fair trade is a modern example of a very recognizable alternative food system. Here we refer to two versions of fair trade: (1) 'Fairtrade' as the certification system by Fairtrade International (see Figure A.a), and (2) 'fair trade' as the larger ideal of a fairer global trade system. The aim of both versions is to reform trade to give disadvantaged farmers a fair wage for their work, decent working conditions, and sustainable livelihoods (Raynolds, 2000). By doing so, they aim to reduce labour exploitation and other negative social impacts of food systems (see Figure A.b). There are many organizations pursuing fair trade, with a large number falling under the umbrella organization of Fairtrade International.

The first fair trade initiatives date from the 1960s and 1970s when shops in high-income countries started to sell cane sugar and coffee from Tanzanian and Angolan cooperatives to socially conscious consumers. This distribution channel resulted in relatively high prices, limiting its growth potential. In the late 1980s, alternative European trade organizations started to use labelling to integrate fair-trade products into mainstream retail shops. The first Fairtrade

certificate was introduced in the Netherlands in 1988 (named 'Max Havelaar', after a 1860 novel that criticized farmer exploitation in the Dutch Indies, now Indonesia) and later became one of the three founding organizations of the Fairtrade Labelling Organization, now Fairtrade International.

Certification is a popular way to influence food consumption. In the case of Fairtrade, it is built on the twin assumptions that: (1) producers would like to follow Fairtrade criteria because it is in their interest, and (2) many consumers would like to buy these products because they prefer a fairer food system. Fairtrade International then uses certification to: (1) ensure that their producers, traders, and other workers abide by the Fairtrade criteria, and (2) signal to consumers which products they can buy if they want to support a fair-trade food system. Section 11.3.2 in *Chapter 11: Consumption* has explained the strengths and limitations of food labelling.

Farmers have to be certified by an independent auditor in order to label their products Fairtrade by following a set of regulations. For small farmers, regulations include limits on farm size and requirements for democratic decision-making

a)

b)

Figure A Through Fairtrade certification (a), Fairtrade International aims to reduce labour exploitation, such as child labour, as seen here on a coffee plantation in Nicaragua (b).

Source: (a) © Fairtrade Foundation 2018; (b) Trocaire/Flickr (CC BY 2.0).

Figure B Fairtrade bananas are now available in many supermarkets around the world.

Table A Significant increase in sales of Fairtrade products between 2004 and 2015 (million Euros)

Country	2004	2015
Austria	16	185
Canada	18	273
France	70	442
Germany	58	978
Italy	25	99
Japan	3	74
The Netherlands	35	223
South Africa	–	19
Switzerland	136	475
United Kingdom	206	2193
TOTAL (including other countries)	832	7300

Source: Fairtrade International (2005; 2016)

within the producer organization, alongside a range of product-specific rules (for example, for cane sugar, cereals, or cocoa—see http://www.fairtrade.net/standards.html for more detail). When certified, farmers may sell their products according to the Fairtrade standards and receive a guaranteed price, which is usually above the standard, non-fair-trade price on international markets. The flow of products is closely monitored to ensure that what is sold in the supermarket actually originates from certified farmers.

The growth of Fairtrade products has been substantial; between 2004 and 2015, Fairtrade sugar increased from 1964 to 154,287 metric tons, and Fairtrade bananas (see Figure B) from 80,641 to 553,047 (Fairtrade International, 2016). Moreover, if you look at the sales of Fairtrade

products in high- and middle-income countries around the world (see Table A), you can see marked increases globally. Given this expansion, the Fairtrade certification scheme appears successful as a marketing strategy.

However, despite the growth in Fairtrade sales, the initiative has been criticized in relation to its impact on food systems reform. The high costs in certification (both administrative and financial) make it less attractive for producers to engage in Fairtrade (Getz and Shreck, 2006; Shreck, 2008). Another criticism is that Fairtrade has forgotten its original aim of radically transforming the global trade system by mainstreaming Fairtrade practices and has instead become a niche in the global food system (Geysmans et al., 2017; Jaffee, 2010; Renard, 2005).

BOX
12.2

Alternative food systems

Mainstream food systems are relatively easy to imagine—with their reliance on economies of scale, industrial technologies, specialization, and global trade—but alternative food systems are both more diverse and more difficult to define. Case study 12.1 provided one example of an approach to building an alternative food system—Fairtrade—which emphasizes fair living wages for producers and distributors (as opposed to the cheapest possible products and the highest possible profits). But there are other types, each of which deviates from the mainstream in its own way. Two examples are organic and cooperative food systems.

Organic food systems are centred on the principles of organic agriculture (see Section 3.3 in *Chapter 3: Pollution*) (see Figure A.a), but their ambitions exceed food production in that they aim for a food system that is both socially just and ecologically responsible (Torjusen et al., 2001). They

diverge from the mainstream in their focus on ecological and ethical principles in agriculture and animal husbandry, as well as strong restrictions on the use of GMOs, synthetic pesticides and fertilizers, growth enhancers, antibiotics, and other 'non-organic' elements in the process of food production. Box 11.1 in *Chapter 11: Consumption* provides more information on the perceptions and realities of contemporary organic food systems.

Cooperative food systems (see Figure A.b) centre on the cooperative as an organization, in which communities' own systems collectively make decisions democratically (Sumner et al., 2016). Like organic food systems, cooperative systems are based on principles that go beyond efficiency and financial profit, including community development, democratization, good jobs, and environmental sustainability. They rely on producers, distributors, and consumers collaborating, rather than competing.

a)

b)

Figure A Organic product shelf in (a) a supermarket and (b) a branch of The Co-operative Food supermarket in the UK. All Co-operative Food shops are retail outlets for food produced by members of Britain's Co-operative Food societies such as The Co-operative Group (see https://www.co-operative.coop/).

Source: (a) U.S. Air Force illustration by Airman 1st Class Randahl J. Jenson; (b) theco-operative/Flickr (CC BY 2.0).

12.2.2 Scale: from local to global food systems

A second important way to categorize food systems is **scale** (Ericksen, 2008; FCRN, 2017), ranging from local to global. Of course, food systems are often connected across levels, but identifying which systems operate on which levels is a useful way of making sense of the way food systems work. Local systems keep food production and consumption geographically close together and supply chains short. Examples include farmers' markets, urban farming, and community gardens in high-income countries, as well

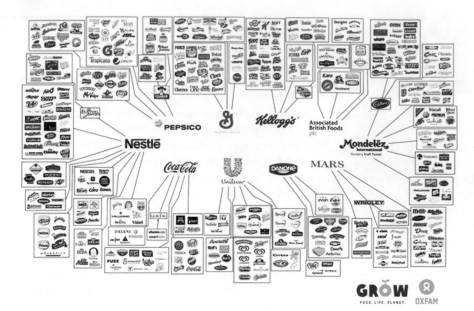

Figure 12.7 Chart showing the ten largest food and beverage companies, according to Oxfam, as well as the brands that these companies control. According to Oxfam, they represent around 10% of the entire world economy and control the majority of the global food system.

Source: Joki Gauthier for Oxfam 2012.

as small-scale food systems for subsistence farming in low- and middle-income countries. Note that many of these local systems are also alternative food systems and, as we will discuss later, localization of food systems is one strategy to make them more sustainable.

Globally, food systems are dominated by large multinational corporations. They work in different sectors, from food processing and sales to agricultural production and biotechnology, but together, they control enormous parts of the global food system. For example,

Oxfam (2013) identified the ten largest food and beverage companies and mapped out the number of brands that they control (see Figure 12.7). Furthermore, four large agribusinesses (ADM, Bunge, Cargill, and Dole) control up to 90% of the global trade in grains (Murphy et al., 2012), and 53% of the global market in crop seeds is controlled by three agribusinesses (Monsanto, Dupont, and Syngenta) (Harris, 2013). Food controversy 12.1 discusses some of the implications of the market power of these agribusinesses.

FOOD
CONTROVERSY
12.1

The power of agribusiness in food systems

The term **agribusiness** refers to businesses engaged in large-scale farming and agricultural production (Davis and Goldberg, 1957). It includes farming, but also other parts of the supply chain: processing, transport, distribution, and sales of agricultural products, as well as the development of new seeds. In recent decades, agribusinesses have grown dramatically. The size of these businesses has given rise to controversies about the dangers of corporate control of the world's food supply. How realistic are these fears? There is

no simple answer, but we can highlight some of the main arguments in the debate.

First, transparency and accountability—given the size of these companies and their multinational reach, national governments struggle to control them. Oxfam (2013) has expressed concern for the lack of ethical oversight of the ten largest food and beverage companies (see https://www.behindthebrands.org for more details). In fact, through lobbying, such companies can often influence government

Figure A Protest against the Monsanto Protection Act in New York City, 2013.

Source: waywuwei/Flickr.

policy using their economic clout as leverage (see Food controversy 13.1 in *Chapter 13: Governance*). For example, in 2013, President Obama signed into law the *Farmer Assurance Provision*, influenced by biotechnology firm Monsanto, which, according to critics, effectively made it impossible for the US government to restrict the production and sale of GM seeds (derisively called the *Monsanto Protection Act* by critics) (see Figure A).

Second, there are concerns over the dominance of agribusinesses in agricultural markets, particularly through their research and development. Monsanto again provides a good illustration. The company is a global powerhouse in the food system. Among its products are different plant breeds (including GM plants), pesticides, and artificial fertilizers. It has an annual turnover revenue of nearly US$15 billion and employs over 20,000 people. It also has a leading role in the development of GM seeds and organisms (see Box 3.7 in *Chapter 3: Pollution*) and in the production of glyphosate, a controversial herbicide.

The sheer size of agribusinesses like Monsanto leads critics to worry about their market power—that is, their ability to influence agricultural markets, for example by raising the price of their products or using intellectual property rights (patents) over seeds and crops to exclude farmers from the benefits of their products. As an example of the latter, by 2013, Monsanto had sued over 400 local farmers for infringing on their patents for crop seeds in the USA alone (Harris, 2013). These lawsuits resulted in at least $23 million in payments to Monsanto but, more importantly, signalled to other farmers that they have to pay the premium on these crop seeds or face the court. This has been controversial in low-income countries where farmers are often poor and unable to afford these seeds (see, for example, Reuters, 2018a, b).

But how can Monsanto, or other technology companies, survive if they are not able to claim ownership over their inventions through patents? Patents are common in all areas of research and development, and for compelling reasons. Monsanto itself argues that: (1) inventors have a moral right to the profits of their inventions, (2) they need the profits because they reinvest up to $3 million per day in research and development and spend generously on philanthropy through the Monsanto Fund, and (3) enforcement is necessary to prevent rewarding those who free-ride on the contributions of others (see also *Chapter 14: Collective Action*) (Monsanto, n.d) (for Monsanto philanthropy, see http://www.monsantofund.org).

In your experience

Do you think it is fair that Monsanto claims profits from their inventions? If so, what do you think are reasonable profits to claim? How would you enforce limits to these profits? Finally, would your opinion be different for technology companies that conduct research and development, for example Apple or Samsung?

A food system scale which has received increasing attention is the city (see also Case study 7.1 in *Chapter 7: Energy*)—not least because the urban population is projected to reach 66% of the global population by 2050. Urban growth presents particular food system challenges, including an increase in meat consumption in cities, as well as higher consumption of (highly) processed food (Dixon et al., 2007; Satterthwaite et al., 2010). In response, there has been an increase in city-level initiatives to manage food systems, for example through **city region food systems** (see, for example, FAO et al., 2016). Often, urban food policies target specific health or environmental challenges (for example, obesity or waste), but in some cases, they aim for more integrated food system reform (IPES-Food, 2017) (see Figure 12.8).

a)

b)

c)

Figure 12.8 City region food systems aim to enhance the sustainability and efficacy of food systems in and around cities, for example by better connecting (a) urban farming, (b) peri-urban greenhouse agriculture, and (c) rural agriculture to urban populations in the Netherlands. Source: (a) Kaz Alting/Wikimedia Commons (CC BY-SA 3.0).

12.3 How do we make food systems more sustainable?

What are sustainable food systems? The UN High Level Panel of Experts on Food Security and Nutrition (2014) defines sustainable food systems as those that deliver '*food security and nutrition for all in such a way that the economic, social and environmental bases to generate food security and nutrition for future generations are not compromised*'. This definition is so broad as to be aspirational, rather than operational (Carlsson et al., 2017), but it does underline that sustainability comprises more than reducing environmental damage; it is also about protecting the economic and social resources in society (Ingram, 2016).

With this definition in mind, we now consider two key questions in the pursuit of food sustainability. How do we measure sustainability in food systems (Section 12.3.1)? And what strategies are available to make food systems more sustainable (Section 12.3.2)? These are highly complex questions, as you have learnt throughout this book, but they are essential to make food systems more sustainable. To make the questions tangible, we provide empirical examples on how people have tried to both quantify and enhance the sustainability of food systems. But keep in mind that there are many other possibilities.

BOX
12.3

The local trap in transitioning to sustainable food systems

The term **local trap** warns against the notion that local food systems would, by default, be more sustainable than non-local ones (Born and Purcell, 2006). Many scholars and activists express a preference for local systems, to the point where there is now a global movement to 'eat locally' (see Figure A). But the problem for food sustainability is that this local preference is not only based on food system impacts on the environment. Much like the debate on GM foods (see Food controversy 11.1 in *Chapter 11: Consumption*), the debate on local food includes many different arguments about community cohesion, accountability, self-determination, and other aspects that, according to some, make local food intrinsically better than food produced elsewhere (even irrespective of its environmental impacts).

However, this conflation of the effects of local food with its possible intrinsic value may be a misconception. First, it is unclear what local food precisely means. To some, it is a precise geographic notion, in which food production and consumption are geographically close (for example, located in the same country). But to others, local food is a political concept that opposes itself to the mainstream, increasingly globalized food system to counteract social marginalization and environmental damage (Peters et al., 2009).

Second, even if we define local in geographic terms, it is not clear that local food systems are more sustainable or socially just than others. For example, as described in Section 6.2.1 in *Chapter 6: Climate Change*, food miles do not generally comprise a large proportion of total food emissions; in some cases, food that has travelled a long distance is actually better than growing locally, because conditions in other countries can be more appropriate for growth and transport can be organized in resource-efficient ways. And while, in some cases, local food has resulted in more inclusive and democratic food governance, in others, it has led to marginalization and exclusion of stakeholders from control over food systems (Born and Purcell, 2006). There are also many products for which local food systems are difficult to imagine, for example in the case of globally traded goods like palm oil (see Case study 12.2). Localization may therefore be a good strategy for promoting sustainability in some food systems, but it is unlikely to be a 'one-size-fits-all' solution.

a)

b)

Figure A Is the local potato seller in Nigeria (a) more sustainable than its competitor the South African supermarket chain Shoprite (b)? And what does the answer to this question tell us about the extent to which 'local food systems' are generally preferable over others?
Source: (a) © David Ehrhardt; (b) Jamie Tubers/Wikimedia Commons (CC BY-SA 4.0).

CASE
STUDY
12.2

Roundtable on sustainable palm oil

In Case study 2.1 in *Chapter 2: Biodiversity* and Food controversy 7.1 in *Chapter 7: Energy*, we discussed the large impacts of oil palm plantations and the development of several initiatives to make the palm oil industry more sustainable. One of the initiatives was the Roundtable on Sustainable Palm Oil (RSPO), an industry-led voluntary membership association (see Figure A). The RSPO has developed a global sustainable palm oil certification scheme that contains a set of principles and criteria addressing environmental and social issues. This certification scheme was decided upon after a series of stakeholder meetings/roundtables (Nikoloyuk, 2009; RSPO, 2013). The agreement of all stakeholders, including large palm oil producers, was a major step, but the RSPO intends to develop its scheme into the main sustainability standard for palm oil production worldwide.

Since its introduction, the RSPO standard has been adopted by many oil palm growers; certified production has grown from 200 million kg in 2008 to 11.5 billion kg in 2017, which is 19% of global palm oil production. The RSPO occupies a central place in the global palm oil supply system, defining the criteria for 'sustainable palm oil' and connecting environmentally relevant information flows between growers, processors, NGOs, and consumers (Cheyns, 2011; McCarthy et al., 2012; Nikoloyuk et al.,

2010; Pesqueira and Glasbergen, 2013; von Geibler, 2012). Given this, it looks from the outside to constitute an effective approach at making the palm oil system more sustainable.

But at the same time, the RSPO is criticized for alleged greenwashing (see Section 11.3.2 in *Chapter 11: Consumption*), because the Roundtable has certified palm oil companies whose sustainability credentials may have been suspect (Greenpeace, 2007). This practice underlines an important lesson for food system reform— on the one hand, the RSPO may effectively bring global stakeholders together, advocate the idea of responsible palm oil in global policy debates, and effectively draw up sustainability guidelines for palm oil businesses. This helps to make some parts of the global palm oil system more sustainable. However, on the other hand, enforcing full implementation of these guidelines in the remote places where palm oil is produced, like Kalimantan (Indonesia; see Figure B), is often beyond the capacity of the RSPO (Silva-Castañeda, 2012). Such gaps in the efficacy of the RSPO (and other voluntary initiatives like it) can likely only be filled through other governance approaches, including legal reform and law enforcement by (local) governments (McCarthy, 2012).

a)

b)

Figure A There has been significant societal concern about palm oil production (a), especially on the loss of biodiversity. The RSPO certification (b) scheme aims to promote the sustainable production of palm oil.

Source: (a) pixelthing/Flickr (CC BY-SA 2.0); (b) © Roundtable on Sustainable Palm Oil.

a)

b)

Figure B Forest in Kalimantan, Indonesia (a) with one of its native inhabitants (b), both of which are threatened by the expansion of oil palm plantations.

Source: © David Ehrhardt.

12.4 **Future outlook**

What can (and should) future food systems look like if they are to provide food security in a sustainable way? One way of addressing such questions is through **scenario analysis,** which uses a highly simplified model of the major drivers of food system sustainability to sketch out different possible futures. Scenario analysis is useful as a tool to think about the range of possible futures for which policymakers and other stakeholders might plan, as well as the paths that lead there. Scenario analysis is based on assumptions and variables. The assumptions are about future trends that you are quite certain will continue, for example increased population growth or climate change impacts. The variables are a limited number of uncertain trends, called drivers.

As an illustration, Figure 12.9 presents the results of a scenario analysis for future food systems, conducted by the World Economic Forum (WEF), an NGO that organizes annual meetings, bringing together over 2000 leaders from businesses, politics, academia, and other sectors to discuss the state of the world (WEF, 2017). The WEF exercise had several key assumptions about future trends important to food systems, including population increases and migration, growing inequality, malnutrition, depletion of natural resources, and climate change.

As drivers, experts selected two variables: the demand shift (x-axis in Figure 12.9) and markets (y-axis). The demand shift refers to consumer demand for food, in particular the extent to which people begin to demand more sustainable food; markets refer to the extent to which goods and technology will be traded freely in future. Together, these two drivers produce four future food systems (corners 1–4 in Figure 12.9):

1. High connectivity and resource-intensive consumption (unchecked consumption, in which high economic growth coincides with environmental destruction)

2. High connectivity and resource-efficient demand (open-source sustainability, with highly globalized and innovative food systems)

3. Low connectivity and resource-intensive consumption (survival of the richest, with low economic growth and high inequality) and

4. Low connectivity and resource-efficient demand (local is the new global, a fragmented world of efficient affluence versus 'hunger hotspots').

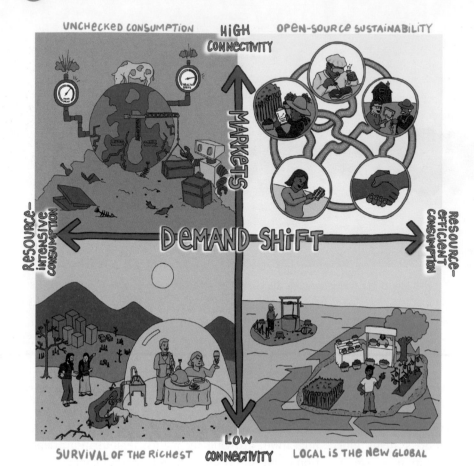

Figure 12.9 Four future food system scenarios developed by the WEF, based on two factors: the efficiency of resource consumption and the level of connectivity of the global food system.

Source: © 2018 World Economic Forum.

Sustainability in this analysis depends strongly on consumer demand—both scenarios in the left half of Figure 12.9 (futures 1 and 3) are unsustainable, while the two on the right are not (futures 2 and 4). But the sustainable scenarios also contain risks—in particular, the risk of inequality, as some will be better placed to benefit from food systems than others (WEF, 2017).

Connect the dots

The 'local is the new global' WEF scenario in Figure 12.9 has clear links to the localization debate discussed earlier in this chapter (see Box 12.3). Which themes from this book do you recognize in the WEF's scenario analysis? Which assumptions would you question, based on this book, and what others could you add? Can you select two other drivers than those chosen by the WEF and sketch out different future worlds for food systems?

Beyond the WEF's analysis, what other possible futures have been identified for food systems? We have already encountered one in fair trade, aiming to change the economic logic of food supply chains. Fair trade's fundamental economic reformism is connected to the idea of **food sovereignty**—a food system model that promotes shifting power from agribusiness and political interests to the producers and consumers of food (Patel, 2009). Sovereign food systems are those that are controlled by the people who produce, distribute, and consume food, rather than by distant multinational corporations and powerful policymakers. This sovereignty should ensure fair and sufficient incomes along the entire supply chain, particularly for producers, but more fundamentally, it is also hoped to result in better decisions about what foods to produce and how to use natural resources sustainably by making better use of local knowledge of agricultural techniques and resource management (Patel, 2009).

All future scenarios for food systems depend on the way in which food will be governed—that is, the ways in which stakeholders work together (or not) to address issues of food sustainability. Conventionally, food governance regulates supply chains through national, government-enforced policies. However, as food systems have become more complex and globally connected, governance approaches have followed suit and transformed (Hospes and Brons, 2016). The future of food systems will now strongly depend on the way in which food governance develops further—the theme that we turn to in *Chapter 13: Governance.*

● QUESTIONS

12.1 What are the core building blocks of food supply chains and of food systems? How do they differ?

12.2 What is the difference between mainstream and alternative food systems?

12.3 What has been the role of Fairtrade in making food systems more sustainable?

12.4 Do you know whether your diet is sustainable? On what indicators do you score best? What would be most difficult for you in maintaining a sustainable diet, and how would you go about it?

12.5 Is local food always more sustainable than food produced elsewhere? Explain.

● FURTHER READING

Eakin, H., Connors, J. P., Wharton, C., Bertmann, F., Xiong, A., Stoltzfus, J. 2017. Identifying attributes of food system sustainability: emerging themes and consensus. *Agriculture and Human Values* 1–17. doi:10.1007/s10460-016-9754-8. **(A study providing clear insights into the diversity of food system sustainability criteria, which identifies six dominant knowledge domains and their discourses on what needs to be sustained in the food system.)**

Ericksen, P. J. 2008. Conceptualizing food systems for global environmental change research. *Global Environmental Change* **18**, 234–245. **(A key work in food systems literature, providing a comprehensive conceptualization of a food system with its interactions, activities, and outcomes, from a socio-ecological systems perspective.)**

High Level Panel of Experts on Food Security and Nutrition (HLPE). 2017. *Nutrition and Food Systems.* Rome: High Level Panel of Experts on Food Security and Nutrition of the Committee on World Food Security. **(A key policy document at the highest institutional level (FAO), which analyses how food systems influence people's dietary patterns and nutritional outcomes and highlights effective policies and programmes.)**

● REFERENCES

Abbott, P., Borot de Battisti, A. 2011. Recent global food price shocks: causes, consequences and lessons for African governments and donors. *Journal of African Economies* **20**, i12–i62.

Alexander, P., Brown, C., Arneth, A., Finnigan, J., Moran, D., Rounsevell, M.D.A. 2017. Losses, inefficiencies and waste in the global food system. *Agricultural Systems* **153**, 190–200.

Barrientos, S., Dolan, C. 2012. *Ethical Sourcing in the Global Food System.* Abingdon: Earthscan.

Berazneva, J., Lee, D.R. 2013. Explaining the African food riots of 2007–2008: an empirical analysis. *Food Policy* **39**, 28–39.

Born, B., Purcell, M. 2006. Avoiding the local trap: scale and food systems in planning research. *Journal of Planning Education and Research* **26**, 195–207.

Bukeviciute, L., Dierx, A., Ilzkovitz, F. 2009. *The Functioning of the Food Supply Chain and Its Effect on Food Prices in the European Union.* Occasional Papers No. 47. Office for Infrastructures and Logistics of the European Communities.

Carlsson, L., Callaghan, E., Morley, A., Broman, G. 2017. Food system sustainability across scales; A proposed local-to-global approach to community planning and assessment. *Sustainability* **9**, 1061–1075.

Cheyns, E. 2011. Multi-stakeholder initiatives for sustainable agriculture: limits of the 'inclusiveness' paradigm. In: Ponte, S., Gibbon, P., Vestergaard, J. (eds). *Governing Through Standards: Origins, Drivers and Limitations.* Houndmills: Palgrave MacMillan, pp. 210–235.

Cohen, M.J., Garrett, J.L. 2010. The food price crisis and urban food (in)security. *Environment and Urbanization* **22**, 467–482.

Davis, J., Goldberg, R. 1957. *A Concept of Agribusiness.* Boston, MA: Division of Research, Graduate School of Business Administration, Harvard University.

Delaney, A., Evans, T., McGreevy, J., Blekking, J., Schlachter, T., Korhonen-Kurki, K., Tamás, P.A., Crane, T.A., Eakin, H., Förch, W., Jones, L., Nelson, D.R., Oberlack, C., Purdon, M. 2016. *Strengthening the Food Systems Governance Evidence Base: Supporting Commensurability of Research Through a Systematic Review of Methods.* Working Paper no. 167. Copenhagen: CGIAR Research Program on Climate Change,

Agriculture and Food Security (CCAFS). Available at: http://www.ccafs.cgiar.org [accessed 9 November 2018].

Dernini, S., Meybeck, A., Burlingame, B., Gitz, V., Lacirignola, C., Debs, P., et al. 2013. Developing a methodological approach for assessing the sustainability of diets: the Mediterranean diet as a case study. *New Medit* **2013**, 28–36.

Deutsche Gesellschaft für Internationale Zusammenarbeit (GIZ), Food and Agriculture Organization of the United Nations (FAO), RUAF Foundation. 2016. *City Region Food Systems and Food Waste Management: Linking Urban and Rural Areas for Sustainable and Resilient Development.* Available at: http://www.fao.org/3/a-i6233e.pdf [accessed 9 November 2018].

Dixon, J., Omwega, A.M., Friel, S., Burns, C., Donati, K., Carlisle, R. 2007. The health equity dimensions of urban food systems. *Journal of Urban Health* **84**, 118–129.

Eakin, H., Connors, J.P., Wharton, C., Bertmann, F., Xiong, A., Stoltzfus, J. 2017. Identifying attributes of food system sustainability: emerging themes and consensus. *Agriculture and Human Values* 1–17. doi: 10.1007/s10460-016-9754-8.

Ericksen, P.J. 2008. Conceptualizing food systems for global environmental change research. *Global Environmental Change* **18**, 234–245.

Fairtrade International. 2005. *Annual Report 2004/2005.* Available at: https://www.fairtrade.net/about-fairtrade/annual-reports.html [accessed 26 April 2018].

Fairtrade International. 2016. *Annual Report 2015/2016.* Available at: https://www.fairtrade.net/about-fairtrade/annual-reports.html [accessed 26 April 2018].

FCRN Foodsource. 2017. Available at: http://foodsource.org.uk/ [accessed 9 November 2018].

Feagan, R. 2007. The place of food: mapping out the 'local' in local food systems. *Progress in Human Geography* **31**, 23–42.

Fresco, L.O. 2009. Challenges for food system adaptation today and tomorrow. *Environmental Science & Policy* **12**, 378–385.

Friedmann, H. 2007. Scaling up: bringing public institutions and food service corporations into the project for a local, sustainable food system in Ontario. *Agriculture and Human Values* **24**, 389–398.

Getz, C., Shreck, A. 2006. What organic and Fairtrade labels do not tell us: towards a place-based understanding of certification. *International Journal of Consumer Studies* **30**, 490–501.

Geysmans, R., de Krom, M.P.M.M., Hustinx, L. 2017. 'Fairtradization': a performative perspective on Fairtrade markets and the role of retail settings in their enactment. *Consumption Markets & Culture* **20**, 539–556.

Godfray, H.C.J., Beddington, J.R., Crute, I.R., Haddad, L., Lawrence, D., Muir, J.F., Pretty, J., Robinson, S., Thomas, S.M., Toulmin, C. 2010. Food security: the challenge of feeding 9 billion people. *Science* **327**, 812–818.

Goodman, D., DuPuis, E.M., Goodman, M.K. 2012. *Alternative Food Networks: Knowledge, Practice, and Politics.* London: Routledge.

Greenpeace. 2007. *How the Palm Oil Industry is Cooking the Climate.* Available at: https://www.greenpeace.org/archive-international/Global/international/planet-2/report/2007/11/cooking-the-climate-full.pdf [accessed 30 May 2018].

Harris, P. 2013. *Monsanto Sued Small Farmers to Protect Seed Patents, Report Says.* The Guardian. Available at: https://www.theguardian.com/environment/2013/feb/12/monsanto-sues-farmers-seed-patents [accessed 27 April 2018].

Headey, D. 2011. Rethinking the global food crisis: the role of trade shocks. *Food Policy* **36**, 136–146.

Headey, D., Fan, S. 2008. Anatomy of a crisis: the causes and consequences of surging food prices. *Agricultural Economics* **39**, 375–391.

Headey, D.D., Martin, W.J. 2016. The impact of food prices on poverty and food security. *Annual Review of Resource Economics* **8**, 329–351.

Hedden, W.P. 1929. *How Great Cities Are Fed.* Boston, MA: D.C. Heath and Co.

Hendrickson, M. K., Heffernan, W.D. 2002. Opening spaces through relocalization: locating potential resistance in the weaknesses of the global food system. *Sociologia Ruralis* **42**, 347–369.

Hospes, O., Brons, A. 2016. Food system governance: a systematic literature review. In: Liljeblad, A.K.J. (ed). *Food Systems Governance: Challenges for Justice, Equality, and Human Rights.* Abingdon and New York: Routledge, pp. 13–41.

Ingram, J. 2011. A food systems approach to researching food security and its interactions with global environmental change. *Food Security* **3**, 417–431.

Ingram, J. 2016. Sustainable food systems for a healthy world. *Sight and Life* **30**, 28–33.

Ingram, J., Ericksen, P., Liverman, D. (eds). 2010. *Food Security and Global Environmental Change.* London and Washington: Earthscan.

IPES Food. 2017. *What Makes Urban Food Policy Happen? Insights From Five Case Studies.*

Ishii-Eiteman, M. 2009. Reorienting local and global food systems: institutional challenges and policy options from the UN Agricultural Assessment. In: Clapp, J., Cohen, M.J. (eds) *The Global Food Crisis: Governance Challenges and Opportunities.* Waterloo, ON: Wilfrid Laurier University Press, pp. 217–236.

Ivanic, M., Martin, W. 2008. *Implications of Higher Global Food Prices for Poverty in Low-Income Countries.* Policy Research Working Paper No. WPS4594. Washington, DC: World Bank.

Jaffee, D. 2010. Fairtrade standards, corporate participation, and social movement responses in the United States. *Journal of Business Ethics* **92**, 267–285.

Johns, T., Powell, B., Maundu, P., Eyzaguirre, P.B. 2013. Agricultural biodiversity as a link between traditional food systems and contemporary development, social integrity and ecological health. *Journal of the Science of Food and Agriculture* **93**, 3433–3442.

Kloppenburg, J., Hendrickson, J., Stevenson, G.W. 1996. Coming in to the foodshed. *Agriculture and Human Values* **13**, 33–42.

Kremer, P., DeLiberty, T.L. 2011. Local food practices and growing potential: mapping the case of Philadelphia. *Applied Geography*, **31**, 1252–1261.

Lang, T., Barling, D., Caraher, M. 2009. *Food Policy: Integrating Health, Environment and Society.* Oxford: Oxford University Press.

McCarthy, J.F. 2012. Certifying in contested spaces: private regulation in Indonesian forestry and palm oil. *Third World Quarterly* **33**, 1871–1888.

McCarthy, J.F., Gillespie, P., Zen, Z. 2012. Swimming upstream: local Indonesian production networks in 'globalized' palm oil production. *World Development* **40**, 555–569.

Milan Urban Food Policy Pact (n.d.). Available at: http://www.milanurbanfoodpolicypact.org/ [accessed 11 October 2017].

Monsanto (n.d.). *Why does Monsanto Enforce Its Patents?* Available at: http://www.monsanto.ca/ourcommitments/Pages/WhydoesMonsantoenforeitsPatents.aspx [accessed 30 April 2017].

Mount, P. 2012. Growing local food: scale and local food systems governance. *Agriculture and Human Values* **29**, 107–121.

Murphy, S., Burch, D., Clapp, J. 2012. *Cereal Secrets: The World's Largest Grain Traders and Global Agriculture*. Oxfam International. Available at: https://www.oxfam.org/sites/www.oxfam.org/files/rr-cereal-secrets-grain-traders-agriculture-30082012-en.pdf [accessed 12 November 2018].

Nikoloyuk, J. 2009. *Sustainability Partnerships in Agro-Commodity Chains. A Model of Partnership Development in the Tea, Palm Oil and Soy Sectors*. Utrecht: Copernicus Institute for Sustainable Development and Innovation.

Nikoloyuk, J., Burns, T.R., de Man, R. 2010. The promise and limitations of partnered governance: the case of sustainable palm oil. *Corporate Governance*, **10**, 59–72.

Oxfam. 2013. *Behind the Brands: Food Justice and the 'Big 10' Food and Beverage Companies*. 166 Oxfam Briefing Paper. Available at: http://www.behindthebrands.org/images/media/Download-files/bp166-behind-brands-260213-en.pdf [accessed 30 May 2018]

Patel, R. 2009. Food sovereignty. *Journal of Peasant Studies* **36**, 663–706.

Pesqueira, L., Glasbergen, P. 2013. Playing the politics of scale: Oxfam's intervention in the Roundtable on Sustainable Palm Oil. [Risky natures, natures of risk]. *Geoforum* **45**, 296–304.

Peters, C.J., Bills, N.L., Wilkins, J.L., Fick, G.W. 2009. Foodshed analysis and its relevance to sustainability. *Renewable Agriculture and Food Systems*, **24**, 1–7.

Pierson, P. 2004. *Politics in Time: History, Institutions, and Social Analysis*, 1st ed. Princeton, NJ: Princeton University Press.

Raynolds, L.T. 2000. Re-embedding global agriculture: the international organic and fair trade movements. *Agriculture and Human Values*, **17**, 297–309.

Renard, M.-C. 2005. Quality certification, regulation and power in Fairtrade. [Certifying Rural Spaces: Quality-Certified Products and Rural Governance]. *Journal of Rural Studies* **21**, 419–431.

Reuters. 2018a. *India Set to Cut Monsanto's GM Cotton Seed Royalties by 20 Percent*. Reuters. Available at: https://www.reuters.com/article/us-india-cotton-monsanto/india-set-to-cut-monsantos-gm-cotton-seed-royalties-by-20-percent-sources-idUSKCN1GJ1SY [accessed 14 April 2018].

Reuters. 2018b. *Monsanto Loses Indian Legal Battle Over GM Cotton Patents*. Available at: https://www.reuters.com/article/us-india-monsanto-nsl/monsanto-loses-indian-legal-battle-over-gm-cotton-patents-idUSKBN1HI2MV [accessed 14 April 2018].

Roundtable on Sustainable Palm Oil (RSPO). 2013. *RSPO Principles and Criteria for the Production of Sustainable Palm Oil, 2013*. Kuala Lumpur: RSPO.

Satterthwaite, D., McGranahan, G., Tacoli, C. 2010. Urbanization and its implications for food and farming. *Philosophical Transactions of the Royal Society B: Biological Sciences* **365**, 2809–2820.

Shreck, A. 2008. Resistance, redistribution, and power in the Fairtrade banana initiative. In: Wright, W., Middendorf, G. (eds). *The Fight over Food. Producers, Consumers, and Activists Challenge the Global Food System*. University Park, PA: Penn State University Press, pp. 121–144.

Silva-Castañeda, L. 2012. A forest of evidence: third-party certification and multiple forms of proof—a case study of oil palm plantations in Indonesia. *Agriculture and Human Values* **29**, 361–370.

Sonnino, R., Marsden, T. 2005. Beyond the divide: rethinking relationships between alternative and conventional food networks in Europe. *Journal of Economic Geography* **6**, 181–199.

Spaargaren, G., Loeber, A., Oosterveer, P. 2012. Food futures in the making. In: Spaargaren, G., Oosterveer, P., Loeber, A. (eds). *Food Practices in Transition. Changing Food Consumption, Retail and Production in the Age of Reflexive Modernity*. New York and London: Routledge, pp. 312–338.

Sumner, J., McMurtry, J.J., Renglich, H. 2016. Leveraging the local: cooperative food systems and the local organic food co-ops network in Ontario, Canada. *Journal of Agriculture, Food Systems, and Community Development* **4**, 47–60.

Tadasse, G., Algieri, B., Kalkuhl, M., von Braun, J. 2016. Drivers and triggers of international food price spikes and volatility. In: Kalkuhi, M., von Braun, J., Torero, M. (eds). *Food Price Volatility and Its Implications for Food Security and Policy*. Cham: Springer, pp. 59–82.

Torjusen, H., Lieblein, G., Wandel, M., Francis, C.A. 2001. Food system orientation and quality perception among consumers and producers of organic food in Hedmark County, Norway. *Food Quality and Preference* **12**, 207–216.

Trienekens, J.H., Wognum, P.M., Beulens, A.J.M., van der Vorst, J.G.A.J. 2012. Transparency in complex dynamic food supply chains. *Advanced Engineering Informatics, Network and Supply Chain System Integration for Mass Customization and Sustainable Behavior* **26**, 55–65.

United Nations Environment Programme (UNEP); Westhoek, H., Ingram, J., Van Berkum, S., Özay, L., Hajer, M. (eds). 2016. *Food Systems and Natural Resources*. A Report of the Working Group on Food Systems of the International Resource Panel.

von Geibler, J. 2012. Market-based governance for sustainability in value chains: conditions for successful standard setting in the palm oil sector. *Journal of Cleaner Production* **56**, 39–53.

World Economic Forum (WEF). 2017. *Shaping the Future of Global Food Systems: A Scenario Analysis*. Available at: http://www3.weforum.org/docs/IP/2016/NVA/WEF_FSA_FutureofGlobalFoodSystems.pdf [accessed 30 May 2018].

13

Governance

How can food systems be governed to promote sustainability?

Gerard Breeman and David Ehrhardt

Chapter Overview

- Groups involved in governance and decision-making, or those influenced by decisions, are often called stakeholders (such as governments, NGOs, trade unions, businesses, and other societal actors). They 'hold' a 'stake' in the outcome.

- Governance consists of activities stakeholders undertake to use their power and authority to solve societal problems.

- Governance is both (1) a problem-solving process and (2) a political process in which stakeholders use power to try to promote their own interests.

- Problem-solving happens through the making and enforcing of rules, from formal laws and policies to guidelines and informal norms (unwritten rules that people follow). The policy cycle is a useful model to describe the process of decision-making.

- Power is defined as the ability to make others do what you want. It has three main aspects: direct and explicit influence during moments of decision-making (for example, 'do what I say'), influence over the agenda setting process (for example, 'this is what we want to investigate and how we will do it'), and influence over people's preferences (for example, 'here is what you should believe').

- Food governance is a wicked problem, complicated by technical complexity, multiple stakeholders, boundary conflicts, and the need for constant adaptation.

Introduction

If humanity produces enough food for everyone, why is 12% of the world population undernourished? Why are sustainable food systems that limit environmental damage and resource depletion not commonplace? And why are consumers faced with scandals about food quality and health hazards (see Figure 13.1)? Many of the answers to these questions lie in the laws, policies, guidelines, and other societal rules that structure our food systems. **Food governance** describes the process in which public and private stakeholders make these rules, with the twin aims to both address problems in food systems and promote public and private interests.

This chapter introduces the basics of governance and explores the main reasons why food governance is difficult. In *Chapter 12: Food Systems*, you learned

Figure 13.1 UK newspaper headlines about the 'horse meat scandal' in 2013, when it was discovered that many foods that advertised as beef contained horse meat instead.

Source: raver_mikey/Flickr (CC BY 2.0).

about the complicated nature of global supply chains, bringing together many different products, processing methods, and distribution channels. You also saw how these supply chains involve many stakeholders and are embedded in wider social, political, economic, and environmental contexts and how these elements all interact with each other.

The key challenge of food governance is to come up with rules and enforcement mechanisms that push individuals and organizations in food systems to behave more sustainably. Rules in this sense are broadly defined as socially constructed constraints on the behaviour of individuals or organizations, which may be written down and explicitly enforced (such as laws and regulations) or implicitly understood and followed (for example, social norms). Throughout this book, you have seen examples of what such rules could look like, from dietary guidelines to stimulate sustainable and healthy diets (see Section 8.4.3 in *Chapter 8: Nutrition*) to the laws and policies that structure how USAID provides tied food aid across the world (see Food controversy 10.1 in *Chapter 10: Food Aid*). But knowing the right rules to fix a problem is only one part of the story. The other parts involve organizational and political questions—who has the authority to make the rules and subsequently enforce them? How do we convince decision-makers of our solution? And how do we deal with stakeholders who disagree with it?

Without examples, questions like these may seem abstract and insignificant, compared to the immense scientific challenge of figuring out the technical solutions to food sustainability. But they are not. To illustrate the importance of food governance, as well as highlight some of its problems and the creative solutions that stakeholders have come up with, Case study 13.1 tells the story of Brazilian soy governance and the tension between economic growth and deforestation.

CASE STUDY 13.1

The hopeful and worrying story of Brazilian soy

We return to the Amazonian forests from where Ms Smith's beef steak originated in *Chapter 6: Climate Change*. Instead of beef, we now look at the crop often used to feed cattle—soy (see Figure A). Soybeans are used for biofuels and foods such as soy milk, soy sauce, miso, tempeh, and tofu. However, the majority of soy production is utilized to feed our livestock (Kroes and Kuepper, 2015). Even if you do not eat tofu, you probably indirectly consume more soy than you might think; the World Wildlife Fund (WWF) estimates that Europeans eat an average

61 kg every year, mostly indirectly, when soy was used to feed livestock (Kroes and Kuepper, 2015).

While soybeans originated in East Asia, the top three producers are now in the Americas (USA, Brazil, and Argentina) and the majority of soy used in Europe now comes from Latin America. The most recent estimate of Brazil's soybean-growing area is around 300,000 km²—over 3.5% of its total arable land (8.51 million km²). This is an area about the size of Italy.

Figure A Soybeans.
Source: CSIRO (CC BY 3.0).

Figure B illustrates the trip that many of these soybeans have to make before they arrive in Europe—produced and harvested in remote rural areas in Brazil, in particular in the state of Mato Grosso in the Amazon rainforest (see Figures B and C); transported by truck for hundreds of kilometres to the closest access points of major rivers such as the Amazon; and then shipped to the nearest seaport and on to European ports such as Rotterdam and other European cities (for example, Vienna). A trip of easily 11,000 km—all to end up in the nearest European pig troth, soy latte, or miso soup.

Aside from the food miles involved in this supply chain, it has resulted in tremendous deforestation throughout South America, particularly in the Amazonian rainforests. This has many impacts, including contributions to climate change and loss of biodiversity, but also destruction of the livelihoods of indigenous communities. As Brazil's soybean production is expected to grow by more than 2.5% per year until 2026, these challenges are also expected to keep growing (OECD/ FAO, 2017). Making Brazilian soy sustainable is therefore a serious governance problem. Moreover, it is complicated by the opposing economic and environmental interests involved, by the fact that the supply chain crosses national boundaries, and by tensions between local and national governments in Brazil.

Yet Brazilian soy governance has had some remarkable successes. Between the mid 1990s and 2006, the expansion of soybean production caused record levels of deforestation; a Greenpeace report estimated that between 2003 and 2006, close to 70,000 km² of Amazonian forest were destroyed— roughly the size of the Netherlands and Belgium together

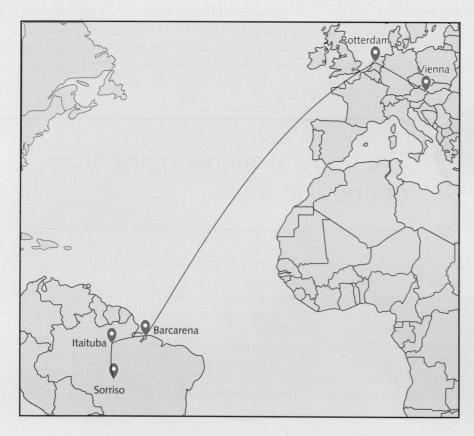

Figure B The transport involved in part of the supply chain of Brazilian soy to European markets.

Source: Map data: Google.

Figure C A soy field in Mato Grosso where there used to be rainforest.

Source: © Bruno Kelly/Greenpeace.

(Greenpeace, 2006). But this report had surprisingly positive consequences; it became a trigger for ABIOVE (Brazilian Vegetable Oil Industry Association) and ANEC (National Association of Cereal Exporters), two business associations representing producers of 92% of Brazilian soybeans, to sign a Soy Moratorium, a voluntary zero-deforestation agreement.

The Moratorium committed soy traders to not buy soy from lands that were deforested after July 2006. In the years following its signing, deforestation rates dropped by 70%—a staggering outcome, given the scale of the problem (Nepstad et al., 2014). There were many contributing factors to this success, including punitive law enforcement by the federal government, other interventions in supply chains, restrictions in credit for soy producers, and expansion of legally protected areas. But the Moratorium has played a significant role and, as such, is a powerful example of sustainable food governance involving governments, businesses, and other stakeholders (Gibbs et al., 2015; Nepstad et al., 2014).

Despite this success, a 70% reduction is not the same as halting deforestation, let alone reforesting. Moreover, it is not clear that the current trends can continue in the face of the projected growth in soybean production and political developments in Brazil. In fact, in 2017 Dutch news reports suggest the opposite, as the Dutch government and engineering companies are collaborating with Brazilian businesses and government to dramatically expand production, build a 1100 km 'soy railway' in the middle of the Amazon, and build over 60 new sea ports. Although these projects are yet to be realized, there are early indications that these developments are again coming at the expense of the Amazonian forests, as well as their indigenous communities (Fearnside, 2017; Kuijpers, 2018).

Pause and think

The story of Brazilian soy is a good illustration of both the importance of sustainable food governance and its challenges. This chapter will explain in more detail what sustainable food governance is and why it is difficult. But based just on reading this case study, why is governing soy supply chains important if we care about sustainability? What are the rules that different stakeholders in Case study 13.1 came up with to make soy more sustainable? Who was able to implement these rules? And what does this case tell you about the challenges of sustainable food governance?

13.1 What is governance?

Governance refers to the act of governing. Intuitively, you might associate it with the actions of the government, in a top–down process where politicians and civil servants design and implement laws and policies (see Box 13.1). However, our definition of governance is much broader, involving all the activities undertaken by societal organizations such as governments, private companies, unions, NGOs, and international organizations to solve societal problems (Hajer et al., 2004). This definition of governance allows for the incorporation of many different ways of governing, including, but not limited to, those where the government is involved.

How does governance work in practice? As a brief introduction, we want to highlight the two key dimensions of the governance process: (1) governance

BOX 13.1

Who is the government?

Although we use terms like government, politicians, or the state frequently in everyday language, it is not always clear what they mean. **Governments** are the organizations that are legally mandated to govern countries, often through a constitution; we use the **state** as a synonym for government. The shape of governments can vary considerably, from China's one-party state or Egypt's military dictatorship to the democratic and multination monarchy of the UK.

Most governments have a **separation of powers**, that is a division between the executive, legislature, and judicial branches of the government. This separation is designed to create accountability within the government and prevent the concentration of power. The specifics of this separation can differ, but the USA is a helpful illustration to outline the basic roles of each of the branches.

In the USA, the executive, such as the President and Cabinet (see Figure A.a), is the organization that holds ultimate responsibility for the governance of a state. It enforces laws and implements policy through ministries and other government bureaucracies (see Figure A.b) and sometimes can make laws by decree. The legislature is a (more or less) representative assembly that makes the laws of a country. The US Congress is an

a)

b)

c)

d)

Figure A The government of a country is often seen to comprise an executive branch (such as the US President and Cabinet) (a) with policy implementation bureaucracies that fall under it (such as the US Department of Agriculture) (b), a legislative branch (for example, the US Congress) (c), and a judiciary (for example, the US Supreme Court and lower courts) (d).

Source: (a) djc/Flickr (CC BY-SA 2.0); (b) Michael Kranewitter/Wikimedia Commons (CC BY-SA 3.0); (c) Koreanet/Flickr (CC BY-SA 3.0); (d) Joe Ravi/Wikimedia Commons (CC BY-SA 3.0).

example of a representative legislature and comprises the Senate and the House of Representatives (see Figure A.c). The judiciary applies and interprets the laws made by the legislature and enforced by the executive. In the USA, this function is occupied by the court system, illustrated by the Supreme Court in Washington, DC (see Figure A.d). Of course, most governments are also multi-level systems (see Section 13.2), in which provincial, municipal, or local governments replicate parts of these three branches at sub-national levels.

as a deliberative process of decision-making, rule-making, and problem-solving, and (2) governance as a political process in which stakeholders use power to advance their interests.

13.1.1 Governance as problem-solving through rule-making

As you have seen in Part 1 of this book, the natural world works according to natural laws—causal patterns that determine, for example, the direction in which water flows or the process through which carbon emissions cause climate change. Human behaviour, in contrast, is only partly determined by such natural laws, for example by the biological functioning of our body or the neural connections in our brains. Our behaviour is also affected by socially constructed rules that we make in families, groups, and societies. These rules are often referred to as **institutions** (North, 1990)—not to be confused with organizations (as the term is often understood to mean in everyday language).

Institutions can be officially codified in government law or informally maintained as norms in social groups. They need to be enforced in order to be effective—that is, someone needs to ensure that people who break a rule are punished. With effective enforcement, rules become constraints on human behaviour; in other words, people start taking the consequences of rule-breaking into account when they are deciding on a course of action. Enforcement can be done by third parties, such as the police and judiciary, but also by group members among themselves. For example, many societies have a strong norm against eating noisily (or the opposite—in some societies, it is a sign of approval of the food). Enforcement of these rules happens socially, through the disapproving stares or comments of those sitting around you at dinner.

In your experience

If you think about your everyday life, what are the rules that affect your behaviour? Are any rules specific to food choices and eating behaviour? How are these rules enforced? Is the enforcement effective?

Some rules are limited to families or small social groups (for example, friends); others apply to large groups or societies, and some are even globally accepted (for example, thou shalt not kill). Governance is the process by which large groups and societies make rules and enforce them in order to solve problems that affect large parts of their population, or even people beyond their group boundaries. A helpful way to look at this process is as policy-making—the design and implementation of rules, regulations, and laws (Howlett et al., 2009; North, 1990). The process of policymaking can be captured in the **policy cycle** that involves an iteration of five stages: agenda setting, policy formulation, decision-making, policy implementation, and policy evaluation (see Figure 13.2).

The policy cycle is not intended as a comprehensive description of the way real policies or institutions are made; rather, it is a heuristic (that is, rule of thumb) that outlines the basic steps of a policy-making process. Its logic is simple. The process begins with the recognition of a problem through, for instance, debates in the media or at a ministry (agenda setting). Subsequently, different solutions are formulated (policy formulation) and a decision is made. Policy implementation follows, after which the results are monitored through a policy evaluation process and then fed back into the agenda setting phase (Howlett et al., 2009).

Policymaking can be a simple process, in which a small group of decision-makers deliberate and go

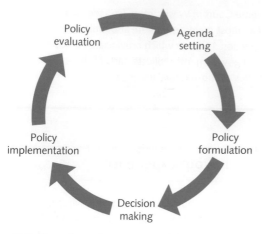

Figure 13.2 The policy cycle.

Source: Derived from Howlett et al., 2009. *Studying Public Policy: Policy Cycles and Policy Subsystems*. Oxford University Press Canada. By permission of Oxford University Press Canada.

through all the steps by themselves. But the large governance problems with which we deal in this book often require extensive involvement of different societal organizations, such as businesses, NGOs, scientific experts, and civil servants. The term **stakeholder** can be used to describe all those with an interest in the governance of a particular issue and the power to exert some influence over the process (Freeman et al., 2010). Including a range of stakeholders in the policy cycle is necessary both for the efficacy and for the legitimacy of the process, but as we will see in Section 13.1.3, stakeholder involvement is also one of the key challenges to effective food governance.

Pause and think

The policy cycle is a helpful tool to better understand the process of government policymaking (see Figure 13.3 for examples of such food policies). However, many rules that structure food systems are not formal policies or laws; rather, they are norms, social codes, or agreements between groups of people or organizations. Think, for example, of cultural dietary restrictions, sharing norms in a religious community, or norms of trust and reciprocity between farmers in a cooperative. Do you think the logic of the policy cycle can also be applied to the way these rules are made? Why (not)?

13.1.2 Governance as politics

Looking at governance as rule-making highlights the deliberative and rational aspects of the process, as well as one of its functions—designing solutions for societal problems, in our case related to food sustainability. But this view of governance ignores its political nature. While there is much disagreement about the best definition of politics (see Figure 13.4), most scholars would associate it with the acquisition, distribution, and use of **power** to advance interests. Power is defined as the ability of one person, group, or organization to influence the behaviour of others, in pursuit of their **interests** (Bachrach and Baratz, 1962). Interests are the things that people or organizations want and are synonymous with preferences (discussed in more detail in

a)

b)

Figure 13.3 Nutrition education for children (a) and checking food temperatures for safety reasons (b), both policies aimed at addressing societal problems related to food.

Source: (a) United States Department of Agriculture; (b) U.S. Air Force photo by Senior Airman Alexis Siekert/Released.

a)

b)

c)

Figure 13.4 Different aspects of politics, including elections, in this case in South Africa (a); parliaments, in this case the Japanese Diet (b); and public protests, in this case against gun violence in US schools (c). What are other important aspects of politics in your experience?

Source: (a) GCIS via governmentza/Flickr (CC BY-ND 2.0); (b) Chris 73/Wikimedia Commons (CC BY-SA 3.0); (c) northlight/Shutterstock.com.

Chapter 14: Collective Action), but remember that such interests are not the same as what people or organizations may objectively need (see, for example, Hsee and Hastie, 2006).

In your experience

If you think about the things you want, for example in the food you consume, are they always the things that are best for you? In other words, are your interests (or preferences) a reflection of your 'best interest'? How would you know? And what does it mean if they are not?

Power is often said to have three 'faces', or ways in which it can operate (Lukes, 2004). The first face of power is present in situations where explicit decisions need to be made and determines who gets to influence the outcome of the decision-making process. The second face precedes the first, in a way, by determining what decisions make it onto the agenda and, conversely, which issues never get discussed in the first place. The third face involves the capacity of one person to influence what others think or want. Think of a television celebrity who has enough charisma and public exposure to convince people that what he says is true—irrespective of what they themselves might think or want. With the third face, powerful people or organizations can shape outcomes without even having to think about agenda setting or explicit decision-making.

As an example, think back to *Chapter 11: Consumption* and the basic decision consumers have to make in the supermarket—do I buy sustainable products or not? The first face of power might involve a parent checking their children's shopping basket and forcing those who have bought unsustainable products to return them to the shelf. The second face of power shows up when the store manager of the supermarket decides to remove all unsustainable products from the shelves, thus taking the decision 'off the agenda' of the shopping consumer. The third face of power, finally, could be the store manager working with a local celebrity to convince consumers that they should buy sustainably.

In your experience

Over whom do you have power? For example, friends or family members, subordinates at work, or children in the sports team you are coaching? Referring to the different 'faces' of power, what kind of power do you have over them? What can you make them do? And under what conditions?

13.1.3 What is food governance?

Food governance is both a deliberative process of making rules to solve food problems and a political game in which different stakeholders in food systems use their power to promote specific interests, often at the expense of others. Governance aimed at sustainable food systems (sustainable food governance) therefore has two goals that overlap with these two dimensions of the governance process:

Goal 1. Finding the right rules that promote more sustainable behaviour throughout food systems (that is, food problem-solving) and

a)

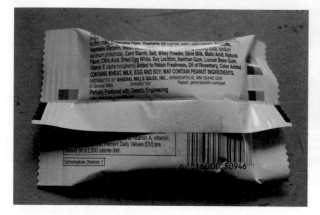

b)

Figure 13.5 Examples of food labels identifying GM contents in food products, mandatory for some products in the EU and many other countries around the world.

Goal 2. Finding ways to organize food governance processes in such a way that they facilitate the design, implementation, and efficacy of these rules (that is, managing food politics).

In other words, some rules structure the behaviour of stakeholders in food systems (Goal 1), and others determine the politics and process of governance itself (Goal 2). Most of the chapters in Parts 1 and 2 of the book have addressed Goal 1. Think, for example, of drafting the right rules for GM crops to reduce the impacts of pollutants on the environment (see Box 3.7 in *Chapter 3: Pollution*), putting limitations on air-freighting of foods to reduce carbon emissions (see Case study 6.1 in *Chapter 6: Climate Change*) or promoting the use of nudging techniques or food labels in supermarkets to promote eco-friendly consumption (see Section 11.3.2 in *Chapter 11: Consumption*) (see Figure 13.5).

A large part of sustainable food governance involves governments and other stakeholders formulating and implementing rules for food systems like those suggested in previous chapters. They can cover a wide range of issues, including agricultural support policies (see Case study 13.2), food safety standards, and policies of food aid and trade regulations (see Section 10.3 in *Chapter 10: Food Aid*). But in addition, stakeholders need to agree on how to organize food governance itself (Goal 2). As *Chapter 12: Food Systems* suggested, reforming governance processes can be an effective, and perhaps even necessary, strategy in pursuing food sustainability. This will therefore be the focus of most of the rest of this chapter.

CASE
STUDY
13.2

The European Union Common Agricultural Policy

The Common Agricultural Policy (CAP) was established in 1962 in the EU and is one of the largest food policies in the world, aiming to '*ensure that agriculture remains sustainable and competitive*' (EC, 2017). The CAP's focus on agricultural production has several reasons, including the interests of many EU member states to keep their agricultural sectors competitive, while also keeping small-scale farmers in business. It also has a geopolitical strategic advantage in ensuring that it produces sufficient food for its own population (that is, to not depend on food imports). The CAP is implemented by the European Commission, part of the executive branch of the EU's governance structure (see Figure A). It was first agreed on in 1962 but has been reformed considerably since then (see also Box 13.2).

In pursuit of the overall objective of sustainable and competitive agriculture, the key function of the CAP is to provide subsidies (financial support) to:

- Increase agricultural productivity by promoting technical progress
- Ensure a fair standard of living for the agricultural community
- Stabilize markets
- Assure the availability of supplies
- Ensure that supplies reach consumers at reasonable prices.

The CAP has many specific policy instruments to achieve these goals, including payments directly to farmers if they meet standardized environmentally friendly practices (see Figure B); buying and subsidizing of certain agricultural products; funding opportunities to foster the development

of rural areas more broadly; and a series of measures that attempt to link the CAP's policies to other EU policies (EC, 2017). Annually, around 39% of the entire EU budget is spent on the CAP (approximately €59 billion annually).

The five CAP goals seem to be broadly supportive of sustainable food systems. However, if we look more closely, they are also a good illustration of the tensions involved in designing and implementing such large-scale and ambitious policy frameworks. For example, if we care about environmental impacts, should technical progress in agriculture come at the expense of the fair standards of living of small-scale, extensive farmers? Should reasonable food prices come at the expense of food quality or animal welfare standards? And is agriculture worth 39% of the EU budget, at the expense of, for example, climate change mitigation or promoting sustainable food systems in other ways, including in urban environments?

The point here is not to say that the CAP is good or bad for food sustainability—that question may be too simplistic for a meaningful answer. Rather, the tensions between the basic aims of the CAP and its stated goal

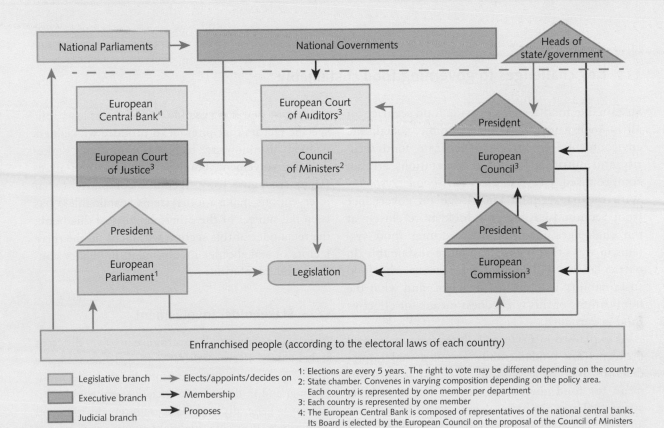

Figure A Simplified flow chart of decision-making in the EU, highlighting the executive (blue), legislative (orange), and judiciary (grey) branches and the relationships between them, the national governments, and the European electorate.

Source: 111Alleskönner/Wikimedia Commons (CC BY-SA 3.0).

a) b) c)

Figure B Agriculture supported by the CAP, from left to right: (a) grapes from Greece; (b) cattle from France; (c) cheese from the Netherlands.

Source: (b) Walpole/Wikimedia Commons (CC BY-SA 3.0).

of promoting food sustainability demonstrate why food sustainability is complex—it is not only technically challenging to design rules that actually achieve the stated aims, but the policy also has to please many stakeholders, operate across national boundaries and societal levels, and change along with the regional and global context in which it operates (see Box 13.2). These complexities are the subject of Section 13.3.

13.2 **Key challenges to governing food sustainably**

Sustainable food systems provide food security for all in such a way that the economic, social, and environmental resources to generate food security and nutrition for future generations are not compromised (see *Chapter 12: Food Systems*). With the background knowledge on governance from Section 13.1, we now look more closely at the challenges involved in governing food systems in ways that make them more sustainable. In particular, we will look at the reasons why food sustainability is a 'wicked problem' and what the implications of this wickedness means for effective food governance.

As explained in *Chapter 1: Introduction*, wicked problems are fundamentally complex and unpredictable, and change continuously. What does this mean for food governance aiming for sustainable systems? There are many definitions of wickedness and complexity, and some go so far as to suggest that they produce fundamentally unsolvable problems (Head and Alford, 2015). While this debilitating complexity may characterize some social systems, we suggest that it is possible to identify at least four specific factors that interact to produce wickedness in food sustainability: technical complexity, stakeholder involvement, boundary conflicts, and adaptation (see Figure 13.6). Technical complexity (what interventions make food systems sustainable?) has been the subject of the other chapters of this book; the remainder of this section will focus on the three factors of stakeholder involvement, boundary conflicts, and adaptation.

13.2.1 **Stakeholder involvement**

As noted previously, all governance involves different stakeholders. Even in dictatorships, leaders engage with experts or other advisors in their decision-making, and in democratic societies, stakeholder inclusion is generally seen as a requirement for good governance. Why? We highlight two reasons related to the efficacy of the decision-making process and its legitimacy, before turning to some of the risks involved in stakeholder inclusion.

Figure 13.6 Technical complexity (a), multiple stakeholders (b), boundary conflicts (c), and adaptation (d) are four main sources of complexity and wickedness in food governance.

Source: (c) Kingroyos/Wikimedia Commons (CC BY-SA 3.0).

Benefits of stakeholder inclusion: efficacy and legitimacy

First, stakeholder inclusion can improve the efficacy of decision-making. Stakeholders often have specific expertise and information that are important to the design and implementation of effective policies. Moreover, increasing the number or adjusting the type of stakeholders may prevent communication **blockages**—moments in the policymaking cycle when progress is hindered by established, yet unproductive, interaction patterns between decision-makers.

When do such blockages arise? Governance issues are usually dominated by a small group of stakeholders (for example, private entrepreneurs, experts, public officials, lobbyists, scientists) who all know each other, meet at the same venues, and are familiar with each other's positions. These groups are known as policy communities or issue networks (Marsh and Rhodes, 2002). Repeated interactions within these groups lead to the same kind of arguments; as a result, the policy community falls prey to groupthink and stagnation and, ultimately, communication blockages (Janis, 1972). New stakeholders may help to deblock these stagnations.

For example, Box 13.2 describes how different stakeholder configurations in the 2006–2008 CAP reform debates led to blocked patterns of thinking and the exclusion of alternative ideas or stakeholders. The European Commission's decision to organize very broad, open public consultations for the subsequent rounds of CAP reform (2010–2012 and 2018–2020) can be understood in this context as an attempt to break open the debate.

In addition to efficacy, the second reason why stakeholder inclusion is important for governance relates to the **legitimacy** of the decision-making process. We define legitimacy as the extent to which stakeholders accept the authority of people who have decision-making power, as well as the process of decision-making itself. It is a requirement for effective decision-making, because if stakeholders feel

BOX
13.2

Blockages in the CAP reform debates

Since its introduction in 1962, the CAP has been the subject of continuous political and policy debate. Main issues have been the CAP's efficacy in terms of its own stated goals, the internal division of costs and benefits between EU member states, and its wider effect on global agricultural markets and food prices, for example by stimulating European overproduction. As a result of these debates, there have been repeated efforts to reform the CAP. In 1992, to reduce overproduction, for example, support for cereals and beef was reduced, along with other reforms; in 2006, similar reductions were applied to sugar beets (see Figure A), of which the EU is the largest global producer.

Every round of CAP reform is an intense process of policymaking, because it is technically difficult and because of the many stakeholders involved—all EU member states, along with a range of different sectoral interests and other lobby groups. There are therefore moments of possible communication blockages, as researchers from Wageningen found out in their study of the CAP reform debates in 2006–2008 (Werkman and Termeer, 2007). Through interviews with stakeholders and observations of the interactions between them, the researchers distinguished six different stakeholder groups, which each have their own interests, values, and inflexible patterns of reasoning:

1. Agricultural entrepreneurs and agribusiness focused on growth
2. Small farmers focused on protecting the status quo
3. Environmentalist stakeholders (for example, environmental NGOs and development organizations)
4. Radical reformers, often promoting structural reforms (for example, fair trade, food sovereignty—see *Chapter 12: Food Systems*)
5. Stakeholders focused on pragmatic problem-solving (for example, public officials and scientists) and
6. Stakeholders aimed at rural development (for example, local government officials).

As the debates progressed, all these subgroups of stakeholders became more and more entrenched in their own circular arguments. As an example, we can take the pattern of reasoning of the second group (small farms focusing on the status quo). From the perspective of the small farmers, the continuation of the CAP subsidies was absolutely necessary, and they did not expect this to be changed during the reform. The fixed pattern of reasoning was: (1) small farmers are known to have a poor income, (2) their income support can be justified because they not only develop

Figure A Sugar beets, a source of refined sugar, produced in the EU and subsidized by the CAP.

Source: Stanzilla/Wikimedia Commons
(CC BY-SA 3.0).

commercial activities, but they also maintain the landscape simply by farming on small farms, and (3) it would be much more expensive if other organizations would take care of the landscape.

The result of this type of reasoning was a blockage with the debates with other stakeholders in the policy domain. The bigger farms, for example, had a different pattern of reasoning and did not think small farmers were poor or that subsidies were necessary. In contrast, they wanted to enhance innovation. All in all, the six different patterns of reasoning were entangled in such a way that they resulted in a constantly repeated pattern of exchanging the same arguments with the same stakeholders over and over again. While these arguments remained valid, they did not allow for the introduction of new or competing ideas, nor for particularly productive discussions between the various stakeholder groups (Werkman and Termeer, 2007). As a result, large parts of the status quo of the CAP were maintained, although a wide variety of small adjustments were made to satisfy all different stakeholders.

the decision-making process is illegitimate, they are unlikely to accept the outcomes and implement the rules.

Stakeholder inclusion impacts legitimacy because it determines whose interests are heard and included in the decision-making process. Stakeholders who are heard and included in governance efforts usually have influence over agenda setting and the rules that are developed. Those who are not included risk not being heard and, more importantly, have no influence over agenda setting or the subsequent phases in the policy cycle (see Figure 13.2). As such, they are less likely to accept the outcomes of the decision-making process as legitimate.

Risks of stakeholder inclusion: efficiency, value conflicts, and framing

Efficacy and legitimacy are strong reasons for including all relevant stakeholders in a governance process. But this also has risks, in particular related to efficiency, value conflicts between stakeholders, and divergent framing of policy problems. Efficiency is a simple concern—the more stakeholders included in a decision-making process, the larger and more time-consuming this process will become. But it also reduces the chances of a consensus decision, given the value conflicts that exist between stakeholders (see Box 13.3).

BOX 13.3

Two examples of value conflicts underlying sustainable food

Value conflicts are common in food governance, because food systems contain many things that people care about deeply—from food security and quality to the sustainability of food systems and the labour standards of producers (Hospes, 2014). We highlight two conflicts as illustrations here. These example conflicts are simplified binary oppositions; real food sustainability dilemmas can relate to many more than two distinct values. But even in their simplified form, the examples illustrate the irreconcilable nature of value conflicts and the resulting challenge for effective food governance.

Production efficiency versus animal welfare. The first conflict comes from the fact that production efficiency concerns are often seen to clash with concerns over animal welfare. For example, battery cages for egg-laying hens can contain 18 hens per m² and are a very efficient way to produce large amounts of eggs on a small area (see Figure A). They use little space, produce cheap eggs, and spread fewer diseases, and the waste can be managed efficiently. However, the battery cages are seriously criticized and, in fact, banned in the EU, because of animal welfare reasons (see Food controversy 15.1 in *Chapter 15: Summary*). The hens are constrained in their natural behaviour, harm each other out of boredom, develop physical deformities, and spread fine particles from their bedding and waste around the barns.

Food security versus overproduction. Many countries support their farmers by either imposing trade barriers or providing direct income support using subsidies. The goals of these policies, such as the CAP in the EU, are to increase agricultural productivity internally, ensure the availability of local food supplies, and establish reasonable consumer prices for citizens of these countries. Subsidizing agriculture and imposing trade barriers may indeed safeguard food security, but they also result in market failures. For example, if farmers receive a fixed price for their produce that exceeds the market price, they will produce more than the market demands, resulting in overproduction such as the European 'milk lakes' and 'butter mountains' of the 1970s through to the 2000s (Howarth, 2000) (see Figure B). This overproduction was a direct result of the CAP, as it subsidized dairy production with billions of euros each year, keeping milk output above EU demand and prices at twice the level of the world market (Elinder, 2005).

Figure A Battery cages for egg-laying hens entail a value conflict between production efficiency and animal welfare.

Source: ק. איתמר/Wikimedia Commons.

Figure B Because of subsidies, farmers continued to produce large quantities of milk, which resulted in overproduction, and large quantities of butter being produced for long-term storage. This overproduction was referred to as the 'butter mountain'.

Source: dpa picture alliance / Alamy Stock Photo.

Besides efficiency concerns and value conflicts, stakeholder inclusion is also complicated by the fact that stakeholders frame governance problems differently. **Framing** is an activity that refers to the attempts by stakeholders to select and highlight specific aspects of an issue and their importance (McAdam et al., 1996). Stakeholders have different interests and goals and, as a result, put forward different problem definitions.

Framing can have serious effects on the ways in which problems are discussed, analysed, and addressed through policy (see also Section 14.2.1 in *Chapter 14: Collective Action*). For example, climate change can be defined as a problem of air pollution caused by the release of GHGs from transportation, industry, and agriculture; as a water problem resulting in flooding and droughts; or as a political problem of insufficient transparency about the implementation of global climate change agreements (see Figure 13.7). Each of these frames gives a different meaning to climate change, presenting it as a different kind of problem with different kinds of solutions, for example reducing the use of polluting technologies, mitigating the effects of flooding through innovative urban development, or improving the monitoring and transparency of government efforts to implement climate change policies.

So stakeholder inclusion is a necessity for effective and legitimate food governance, but it is complicated by efficiency problems, value conflicts, and competing problem frames between stakeholders. Moreover, it is a never-ending process, as stakeholders who have been excluded try to influence the policy cycle from outside. This form of political activity is often referred to advocacy and includes media campaigns, public speaking, conducting research and publishing the results, and lobbying (see Food controversy 13.1).

a)

b)

c)

Figure 13.7 Climate change can be framed as a problem caused by agriculture (a), for example by cattle releasing methane, as a problem causing flooding in countries such as Bangladesh (b), or as a problem made possible by the political failure to monitor the implementation of policies that will help to achieve the goals set in the 2015 Paris Climate Accords (c).

Source: (b) masudananda/Flickr (CC BY-ND 2.0).

FOOD CONTROVERSY 13.1

The tail wagging the dog?

'30,000 lobbyists and counting: is Brussels under corporate sway?'

(Traynor, 2014)

'New US food guidelines show the power of lobbying, not science'

(Duhaime-Ross, 2016)

These two headlines are about a controversial phenomenon in policymaking circles around the world—the power of lobbyists (see also Food controversy 8.1 in *Chapter 8: Nutrition*). **Lobbying** is a form of advocacy, in which interest groups try to influence the policymaking process in their favour by hiring lobbying firms, law firms, former civil servants, or other individuals with the right skills and connections with other stakeholders.

These lobbyists then connect with policymakers in order to promote the interests of their clients. Ultimately, policymakers retain the authority to make the final decision, but there is widespread concern that the tail is wagging the dog (that is, that lobbyists are controlling policymakers, rather than policymakers using lobbyists to make better decisions) (see Figure A).

Why is this controversial? Policymakers and politicians argue that lobbying is necessary, because it brings information into the policy cycle, something policymakers are unable to do by themselves because they have neither the required expertise nor the time to do research. Advocacy thus produces better decisions, proponents argue. Moreover, advocacy in principle allows all kinds of interest groups to influence the policy cycle, including businesses, but also NGOs, trade unions, and religious or ideological organizations. As such, it could also be seen as a way to enhance the legitimacy of the policy process.

Yet many people fear that lobbying prioritizes the interests of businesses that can afford to hire the most effective lobbyists at the expense of scientific evidence or less wealthy interest groups. These worries are compounded by concerns about what is called the revolving door—civil servants or politicians who, after leaving the public sector, use their networks and expertise as lobbyists. This revolving door not only raises ethical questions, but is also thought to enhance the impact of lobbying and thus increase the benefits for the wealthy. And it is big business; a 2007 ranking of the top 50 lobbyists in the USA identified 34 with public experience, and 56% of the revenue of private US lobbying firms between 1998 and 2008 was generated by such revolving-door lobbyists (Blanes i Vidal et al., 2012).

Corporate spending on lobbying suggests that companies think the practice is effective. According to OpenSecrets. org, a website tracking the influence of money on American politics, agribusinesses spent close to $132 million on lobbying in 2017 in the USA alone (see Figure B). This made agribusiness the ninth biggest spending industry in the USA, with health (including pharmaceutical companies) topping the chart at over $561 million. Reliable figures for the EU are more difficult to get, but in 2012, it was estimated that agribusiness lobbyists outnumbered their colleagues working for other sectors by four to one (Chambers, 2016).

Does this prove the power of lobbyists (and through them, money) over policymaking? Interestingly, scientific evidence presents a more nuanced picture. One large study looked at the decisions on 98 policy issues made by the US Congress to explore the influence of lobbying on the American policy cycle (Baumgartner et al., 2009). Its results were complicated. Lobbyists appear to have a strong influence on

a)   b)   c)

Figure A Big Tobacco (a) and Big Pharma (b) are two of the most notorious lobby movements in the USA, representing the interests of tobacco and pharmaceutical companies, respectively; but some warn that Big Food (c) is also on the rise.

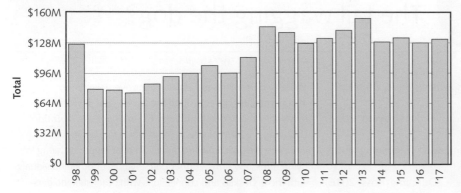

Figure B Annual spending on lobbying by agribusiness firms in the USA.

Source: OpenSecrets.org.

agenda setting. The top lobby priorities in 1999–2002 were health, environment, and transportation, while the public cared most about crime and family policy, employment and taxation, and education (Baumgartner et al., 2014). Congress discussed the lobby priorities at far greater length than the concerns of the public.

Moreover, the study showed that lobbying can influence policymaking in other ways. In particular, it may be important for interest groups to ensure the persistence of the status quo and prevent new policy proposals from reaching the legislature's agenda (Baumgartner et al., 2014). Remember the different faces of power—money might be better spent on maintaining an advantageous policy (second face) than on affecting explicit decisions once they have entered the policy cycle (first face). Lobbyists can be instrumental in this (Baumgartner et al., 2014).

All this seems to confirm the idea that lobbyists and, through them, corporate interests and their money influence policymaking. But the study also showed that it was not

primarily by spending money that lobbyists bought agenda items or policy decisions. Instead, it mattered more that they had high- or mid-level government allies (for example, if they were 'revolving-door' lobbyists). In other words, connections mattered more than money—at least for influencing explicit policy decisions and changes in policy.

This suggests that less wealthy interest groups can also successfully advocate their interests in the policy cycle, as long as they focus on building and exploiting the right connections. Moreover, innovation also plays a role, as the complexity of policymaking in a multi-level polity like the EU allows for NGOs and other advocacy groups to discover new ways of lobbying effectively. For example, in the political struggles around the implementation of Natura 2000, the largest coordinated network of protected areas in the world, a strategic alliance of the WWF and one part of the EU bureaucracy allowed the NGO to become more influential than the opposing coalition of influential land users (Weber and Christophersen, 2002).

 Pause and think

Stakeholder inclusion is necessary, but it also makes governance difficult. A stakeholder or network analysis can help to shed light on the nature of particular policymaking processes, by mapping out the stakeholders and their interests, focus, problem frames, and key strategic issues (Bryson, 1995). Try to conduct a basic stakeholder analysis of the governance of a sustainability issue in your immediate social environment (for example, waste recycling in your student housing). Would including all relevant stakeholders produce effective governance? Why (not)?

13.2.2 Boundary conflicts

In addition to the role of stakeholders, food governance is also complex because it suffers from **boundary conflicts** between different (parts of) organizations. All organizations have boundaries—that is, dividing lines between members and non-members and between issues under its jurisdiction and those outside it (Aldrich and Herker, 1977). While many activities of organizations are within these boundaries, organizations also engage in activities across them.

Boundary conflicts arise when multiple organizations (or different parts of the same organization) engage with the same issue, but without effective coordination. Common problems are either inter-organizational conflict or issues 'falling between the cracks', as all organizations expect others to take care of the problem (see also free-riding in *Chapter 14: Collective Action*). Wicked problems are notorious for producing boundary conflicts between stakeholders, in particular around the following four boundaries: policy domains, time horizons, scales, and societal–public governance relations (Termeer and Dewulf, 2014; Termeer et al., 2015).

Policy domain conflicts

Policy domains are the parts of governance systems that are organized around policy issues such as health, education, or agriculture (Burstein, 1991). Each governance system has its own policy domains and, generally, its organizational structure reflects these domains (for example, through ministries of health, education, or agriculture). Specific policy issues, however, can cross domain boundaries. Water quality and availability, as an example of vital resources for food production and consumption, are important to different policy domains, including

a)

b)

c)

Figure 13.8 Some of the main stakeholders in water quality in California: water sanitation facilities (a), farmers (b), and water-bottling companies (c). They highlight the diversity of policy domains that applies to water—from regulation on water infrastructure to restrictions on agricultural pollution and safety regulations for bottled water.

Sources: (a) U.S. Air Force photo/Ken Wright; (c) Dana Payne/Wikimedia Commons (CC BY-SA 3.0).

infrastructural development, agriculture, and food safety (see Figure 13.8).

Organizations can address policy domain conflicts by integration. For food, this has been the case in Germany since 1949 where the Ministry of Food and Agriculture combines agricultural policy with food safety, nutrition, food sustainability, and other food-related issues. The logic of such integration is that preventing policy domain conflicts will allow for more efficient policymaking and fewer problems 'falling between the cracks' of different parts of government bureaucracy. Of course, these gains should be weighed against the risks involved in reorganization, for example the loss of expertise and the costs involved in the reorganization process. Moreover, the trade-offs are complicated further by the need for

constant adaptation in addressing wicked problems (see Section 13.2.3).

Time horizon conflicts

The second type of boundary conflict arises when an issue has multiple time horizons—that is, when it has different short- and long-term consequences or requires different short- and long-term solutions. Large agribusinesses, for example, contain tension between their short-term marketing incentives for profit and their long-term interests in innovative research and development. Such conflicts are particularly problematic in governance, given the short time horizons of politicians, who generally want to design policies that can be implemented and that produce quick results during their time in office (Pierson, 2004). Food sustainability, as you have seen throughout this book, is a problem that needs long-term solutions that are slow to implement and take effect.

Separation is one effective way of dealing with time horizon conflicts, for example through information barriers within organizations or by setting up a body that works independently from the political–electoral cycle such as an agency with its own structural financing and legal protection. This prevents policy decisions from being captured by the **short-termism** of politicians. For example, the water management programme in the Netherlands, which prepares the Dutch water infrastructure for the consequences of climate change, has its own fund and agency and can therefore be expected to suffer less from timescale conflicts (Van der Steen et al., 2016).

Scale conflicts (governance level and geography)

The third type of boundary conflicts arise when an issue cuts across different geographical areas or governance levels. Both pose problems of authority and responsibility—which public stakeholder in which area, or at which level (local, provincial, or national government), should be held accountable for the implementation and monitoring of sustainability solutions? A crop disease, for instance, does not stop at the border of a town but affects neighboring regions and may become a national or even international problem. In general, two different forms of governance have been developed to respond to these conflicts (Hirst, 2000; Pierre and Peters, 2000; Rhodes, 1997; Van Kersbergen and Van Waarden, 2004).

1. Multi-level governance: a layered governance systems in which public or private stakeholders at different governance levels partake. Case study 13.3 shows an example of multi-level governance of the Baltic Sea fisheries involving both public and private stakeholders.

2. Global governance: collaborative governance between country governments and multilateral organizations. Food security and environmental problems, such as climate change, are often defined at the global level. In addressing these global problems, states are facilitated by international organizations such as the FAO. Membership of multilateral organizations is made up of individual country governments, which pay fees to cover the organizational running costs.

Society–public governance conflicts

The fourth type of boundary conflicts arise when private stakeholders, such as citizens or NGOs, work together with public stakeholders, such as provinces or national governments. Cooperation between different societal partners poses problems of authority and responsibility. For instance, to what extent should a government be held responsible for the reduction of food waste or the loss of biodiversity? A possible way out is to leave all governance to private stakeholders. This results in self-governance, where citizens or private companies organize themselves because a government does not want to, or cannot, provide a service they think is needed. Section 10.3 in *Chapter 10: Food Aid* provided several examples of communities organizing their own food-sharing institutions to enhance food security, such as food banks and informal sharing arrangements.

Another framework that may help to deal with this type of conflict is network governance. A network consists of private and public stakeholders and is explicitly horizontally (or bottom-up) organized. As a result, there is no clear hierarchy in the governance process; coordination is achieved by contracts and other voluntary agreements. The Soy Moratorium from Case study 13.1 is one example of network governance, as stakeholders decided together to commit to collective action to reduce deforestation. Similarly, in global organic food networks, businesses and NGOs use certification standards to facilitate their collaboration and ensure that the principles of certification are implemented along the chain without having to rely on external (for example, government) enforcement (Raynolds, 2004).

 In your experience

Which types of governance outlined in Section 13.2.2, do you experience in your everyday life? Are these types effective? What does effective governance mean to you?

CASE STUDY 13.3

Multi-level decision-making in the Baltic Sea

The Baltic Sea is part of the Atlantic Ocean and borders eight EU countries and Russia (see Figure A). From the development of the Hanseatic League trading network in the thirteenth century onwards, it has been an important trading route; today, it remains one of Russia's export lines for petroleum. Beyond sea trade, the Baltic Sea's economic value is in fishing and tourism. In terms of fishing, 90% of the fish caught commercially are cod, herring, and sprat (source of 'sardines'), but salmon and eel are also significant; the annual amount of fish caught is estimated at over 1 billion kg—just under half of what is caught in the North Sea annually (Burns and Stöhr, 2011). While many of the fish stocks are in a sustainable condition, a recent report warns that the flounder, Atlantic cod, and Atlantic salmon are outside sustainable limits and subject to overfishing (Froese et al., 2016).

Given the many countries bordering the Baltic and the fact that one is outside the EU (Russia), multi-level governance is necessary to manage this natural resource. It consists of two sets of policies: (1) the EU's Common Fisheries Policies (CFP) established in 1983, and (2) a bilateral agreement between the EU and Russia. To avoid overcomplication, we focus on the former. With over 2000 specific rules, the CFP is one of the world's most complex fishery policies; Figure B is a schematic representation of the governance structure that

the CFP has produced in the Baltic, which gives an indication of how complex these governance structures can become. The top of the figure represents the EU level; the middle refers to the national level, and the lower part of the figure to the fishermen at the local level. The left side of the figure represents the multi-level interaction between governments, the middle of the figure the stakeholder interactions, and the right side the expert and science information channels.

In short, the Council of Ministers takes the final and formal decisions on the fishery policy in the Baltic Sea. These decisions are transferred through the national government down to the fishermen (left side of the figure). But this is only part of the process; the Council's decisions are prepared by the European Commission, which uses scientific advice (right side of the figure) and the advice from stakeholder groups (middle of the figure) (see Burns and Stöhr, 2011 for more detail).

Thus, Baltic fishery governance includes stakeholders at local, national, and regional levels, and decisions and information flow up and down through different levels. While there is debate about the extent to which this particular CFP-led governance scheme effectively promotes sustainability in the Baltic Sea, there is little doubt about the appropriateness of the general approach of multi-level governance in cases such as this.

a)

b)

Figure A Map of the Baltic Sea (a), bordering nine countries that all engage in fishing (b) but are bound by different national and EU regulations.

Source: (a) NormanEinstein/Wikimedia Commons (CC BY-SA 3.0).

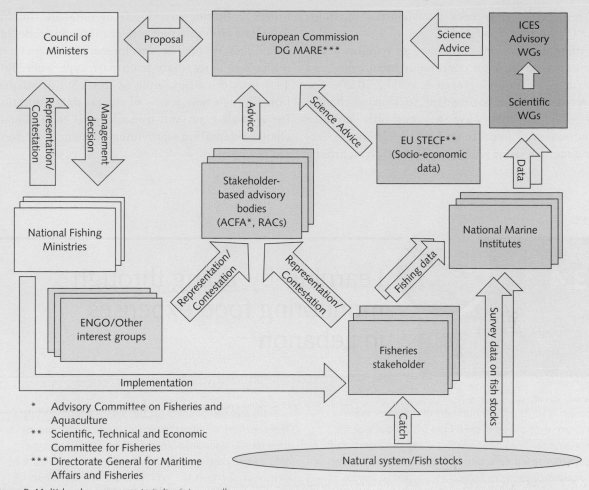

Figure B Multi-level governance in Baltic fishery policy.

Source: Redrawn from Burns, T. R., & Stöhr, C. (2011). Power, knowledge, and conflict in the shaping of commons governance. The case of EU Baltic fisheries. *International Journal of the Commons*, 5(2), 233–258. CC BY 3.0.

13.2.3 Adaptation

The third source of complexity in food system governance is adaptation, which refers to the fact that governance systems need to change along with their society and natural environment, without losing existing expertise. In essence, adaptation is a challenge of continuously collecting information on governance efficacy and adapting to this information. It means having effective monitoring mechanisms in place and being prepared organizationally to cope with both short-term shocks and long-term shifts. This may be called governance resilience (resilience is used across many disciplines; see ecosystem resilience in *Chapter 2: Biodiversity*).

Stakeholders can learn from short-term shocks by setting up emergency response protocols in times when there is no emergency. This can involve writing response manuals, running 'fake' scenario operations (say, an outbreak of food-based diseases), and providing personnel with training in emergency protocols. The WHO, for example, has emergency response systems to fight epidemics; the EU has rapid response systems to fight animal disease outbreaks, and the World Food Programme (WFP) has protocols to respond to emergency humanitarian situations (see *Chapter 10: Food Aid*).

Adaptation to long-term changes requires building expertise through continuous monitoring and 'learning by doing' (Termeer et al., 2015). For governance systems to adapt to climate change, for example, they have to constantly monitor environmental impacts and use modelling to create climate change scenarios. This involves, among others, collecting new kinds of data and linking (big) data with different observation and modelling techniques to predict developments that may be difficult to extrapolate from past experience. Box 13.4 provides an example of this kind of adaptation through new forms of digital data collection for a smaller, yet still difficult, goal than climate change adaptation—providing efficient WFP food aid in Lebanon.

BOX 13.4 — Learning-by-doing through monitoring food expenses in Lebanon

Lebanon is host to large numbers of refugees, many of whom were driven there by the Syrian civil war, continuing tensions between Israel and the Palestinians, and insecurity within Lebanon. The WFP (see *Chapter 10: Food Aid*) provides food aid to displaced people. But rather than providing food in kind (for example, by delivering bags of food to refugee camps), the WFP in Lebanon distributes digital food vouchers to refugees who spend them at local grocery shops. The grocers then receive the cash amounts from the WFP (Flaemig et al., 2017).

WFP's Lebanese programme is considerable in size; every month, between US$18 and US$22 million are distributed to refugees through vouchers. As of early 2017, the programme was assisting 700,000 Syrian, 54,000 Lebanese, and 23,000 Palestinian refugees. The vouchers they receive work like debit cards, which the recipients can use (up to $27 a month) in over 500 shops around the country. This approach has obvious advantages for the local economy, since it stimulates both the grocers and local food production.

But it also has benefits for the WFP's adaptive capacity. The vouchers are digital payment cards, which means their transactions are automatically recorded and the data can be used for monitoring purposes (see Figure A). First of all, it allows the WFP to monitor movements of the refugees. But it also allows researchers to search for anomalous patterns in real time, by mapping out high volatility in store voucher-based sales. Based on prior research, they knew that this was one indicator of stores committing fraud, particularly by giving out cash to the refugees, rather than food items (which is against WFP regulations).

The visualization in Figure A depicts store anomalies across Lebanon in 2014. Mapping out these data helps the WFP to identify stores that potentially are breaking the regulations of the aid programme and thus helps WFP's learning capacities about their food voucher programmes. As such, it is an example of using a new technology—digital food vouchers—to track the uptake of food aid, thus facilitating adaptation and learning for the aid provider.

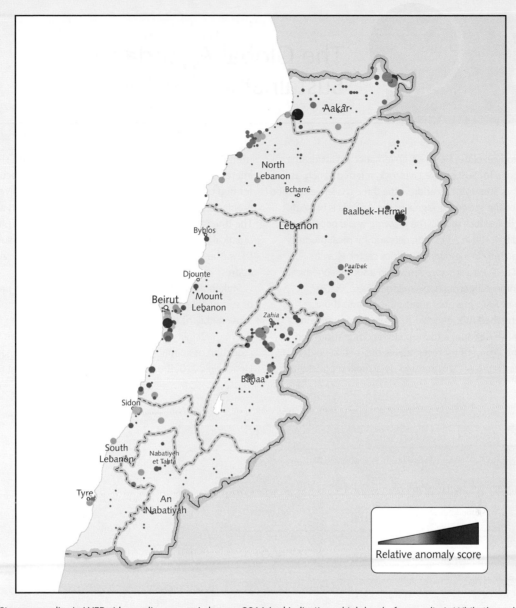

Figure A Store anomalies in WFP aid spending across Lebanon, 2014 (red indicating a high level of anomalies). While these data are not proof of fraud, they are used to direct field checks and thus make learning-by-doing cheaper for the WFP's cash transfer projects.

Source: Flaemig et al., 2017.

CASE STUDY 13.4

The Global Agenda for Sustainable Livestock

This case study explores how one governance initiative—the Global Agenda for Sustainable Livestock, referred to here as the Agenda—has attempted to address the three challenges to food governance. The governance goal of the Agenda is to create a sustainable livestock sector. The growing world population is projected to result in a growing demand for meat products and consequently in a growing livestock sector. How to accommodate livestock sector growth in a socio-economically and environmentally sustainable manner (FAO, 2018)?

A group of FAO member states and organizations, facilitated by the FAO, decided to address this problem and set up the Agenda, a multi-stakeholder platform led by a guiding group of representatives of six stakeholder groups (see Figure A). It is assisted by a small support team

from the FAO. Out of this platform, and based on shared interests, different smaller groups with different focus areas formed to enhance the understanding of specific livestock problems through dialogue, consultation, and joint analyses. The three focus areas developed a wide variety of pilots, practical projects, and local stakeholder dialogues such as the sustainable use of grassland. But how does the Agenda address the governance challenges?

Stakeholders. Different stakeholders define the problem of sustainable livestock in different ways. For example, some highlight the necessity to improve production and the marketability of products, while others focus on the reduction of animal diseases or the processing of manure. Hence, in 2010, the Dialogue Group decided to set up a

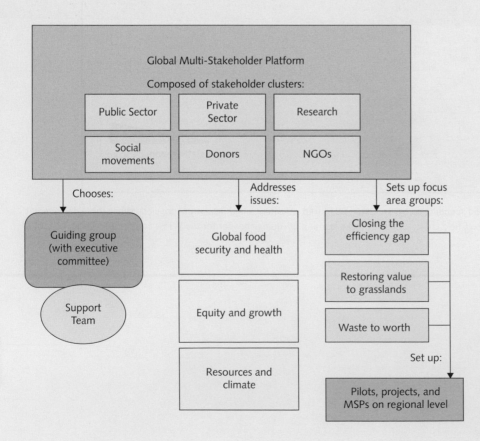

Figure A Multi-stakeholder approach in the Global Agenda for Sustainable Livestock.

Source: Breeman, G., Dijkman, J. & Termeer, C. Food Sec. (2015) 7: 425.

multi-stakeholder platform which would meet every other year. During the meetings, stakeholders are encouraged to discuss their problem frames through organizing brainstorming and break-out sessions. These interactions require trust, along with a tolerance of other people's problem definitions.

The Agenda's strategy to build trust is simple—to engage in extensive and continuous networking with all relevant stakeholders. It tries to be present in many different policymaking centres, international organizations, conferences, and other places where they can build political support, collect information about associated problems, learn what is new in other domains, and try out the feasibility of new ideas. For example, members of the Agenda sponsor discussions at the Global Forum for Food and Agriculture, and they cooperate with the International Planning Committee on Food Sovereignty, the Global Civil Society Mechanism, the World Alliance of Mobile Indigenous People, the International Indian Treaty Council, the International Union of Food Workers, la Via Campesina, and Maela.

The Agenda's approach to communication blockages is to designate a specific part of the organization—the support team—to address or prevent them when debating or negotiating about the future of sustainable livestock. The support team is focused on monitoring discussions and, if a blockage appears, responds with the aim of opening up the discussion. Furthermore, the Agenda also has a large turnover of attendees and participants, thus preventing the formation of groups, which might result in stagnation and reaffirmation of the same ideas.

Boundary conflicts. To deal with boundary conflicts, the Agenda constantly moves between the different scales and levels of government. Through its concrete projects, the Agenda connects global dialogue to local action. A pasture management project in Kyrgyzstan, for example, engages different layers of governance and demands a scale-bridging attitude from the local leadership (see Figure B). Furthermore, the Agenda supports and co-organizes regional livestock sector multi-stakeholder platforms, for example in South and South East Asia (with Dairy Asia) and in Central America.

Adaptation. Finally, the Agenda's governance system aims to cope with short- and longer-term changes through the flexibility in its organizational structure, in particular by continuously building new networks of stakeholders and venues where they can meet and discuss their projects, as well as the larger problems of sustainable livestock and climate change. Initially, the Agenda consisted of only one platform, but over time, different types of organizational elements were added such as a guiding group, focus areas, local multi-stakeholder platforms, and concrete projects. The result is that stakeholders meet each other in different venues on different topics and keep acquiring new information.

Figure B Kyrgyzstan fields with traditional nomadic tents.

Source: Nowak Lukasz/Shutterstock.com.

13.3 Future outlook

This chapter has introduced core concepts of food governance and key challenges that complicate it. Thinking about these challenges helps you understand food governance better, but it might also help to outline ways to make it more effective. To explore this, we present Case study 13.4 on the open multi-stakeholder dialogue of the Global Agenda for Sustainable Livestock. It is an example of food governance addressing the three challenges of stakeholder inclusion, boundary conflicts, and adaptation. With the backdrop of population growth and dietary change with an increasing demand for meat and other animal products, this platform aims to enhance the sustainable development of the livestock sector (Cummins, 2004).

As with many other governance domains, food governance is linked to the major societal trends of globalization and climate change. These trends are likely to determine the future challenges within the domain of food governance. Ongoing globalization will continue to trigger multi-level governance dynamics and produce changing dietary and eating

Pause and think

Think critically about the Agenda's approach to the three governance challenges. What are the strengths of their efforts? And the weaknesses? What do you think the Agenda can contribute to making livestock food systems sustainable? How does it compare to the governance approaches applied in Brazil's soy supply chain or the Baltic fishery sector?

patterns (see also *Chapter 8: Nutrition*). Food governance efforts will need to respond to all these new claims and demands. Furthermore, the increasing interdependencies between different problems will require policy integration across domains and geographical borders. This will require new organizational reforms, such as the examples given in this chapter, the creation of ministries of food, or the enhancement of network governance in food supply chains (de Vries et al., 2016).

● QUESTIONS

13.1 Define governance, and relate it to the concepts of politics and institutions.

13.2 Can you explain, with an example, all three faces of power?

13.3 What challenges cause the wickedness of food sustainability? Does it mean it is unachievable?

13.4 How can the following governance challenges be addressed?

　　a) Value conflicts

　　b) Boundary conflicts

　　c) Communication blockages

　　d) Shifting circumstances

13.5 Search for three party manifestos in your own country, and read the paragraphs concerning food or agricultural governance. What types of governance do you think the various parties have in mind when they were writing these paragraphs? To what extent, and in what ways, are they aiming to address the challenges you mentioned for Question 3?

● FURTHER READING

D'Angelo, P., Kuypers, J. A. (eds). 2010. *Doing News Framing Analysis: Empirical and Theoretical Perspectives*. New York, NY: Routledge. **(Informative book about the background and methods of framing analysis of news media sources.)**

Hoppe, R. 2011. *The Governance of Problems: Puzzling, Powering and Participation*. Bristol: Policy Press. **(Analyses of the often difficult relation between technical solutions, political feasibility, and citizen's involvement in governance schemes.)**

Kersbergen, K. V., Waarden, F. V. 2004. 'Governance'as a bridge between disciplines: cross-disciplinary inspiration regarding shifts in governance and problems of governability, accountability and legitimacy. *European Journal of Political Research* **43**, 143–171. **(Offers a good overview of the background of the concept 'governance'.)**

Rhodes, R. A. 1997. *Understanding Governance: Policy Networks, Governance, Reflexivity and Accountability*. Buckingham: Open University Press. **(Provides an interesting** overview of governance institutions in central, local, and European Union government with methodological innovations and theoretical analysis.)

Termeer, C. J., Dewulf, A., Breeman, G., Stiller, S. J. 2015.

Governance capabilities for dealing wisely with wicked problems. *Administration & Society* **47**, 680–710. (**Offers an in-depth understanding of the governance challenges when dealing with complex (wicked) problems such as food governance.**)

● REFERENCES

Aldrich, H., Herker, D. 1977. Boundary spanning roles and organization structure. *Academy of Management Review* **2**, 217–230.

Bachrach, P., Baratz, M.S. 1962. Two faces of power. *American Political Science Review* **56**, 947–952.

Baumgartner, F.R., Berry, J.M., Hojnacki, M., Kimball, D.C., Leech, B.L. 2014. Money, priorities, and stalemate: how lobbying affects public policy. *Election Law Journal* **13**, 194–209.

Baumgartner, F.R., Berry, J.M., Hojnacki, M., Leech, B.L., Kimball, D.C. 2009. *Lobbying and Policy Change: Who Wins, Who Loses, and Why*. Chicago, IL: University of Chicago Press.

Blanes i Vidal, J., Draca, M., Fons-Rosen, C. 2012. Revolving door lobbyists. *American Economic Review* **102**, 3731–3748.

Breeman, G.E., Dijkman, J., Termeer, C.J.A.M. 2015. Enhancing food security through a multi-stakeholder process: the global agenda for sustainable livestock. *Food Security* **7**, 425–435.

Bryson, J.M. 1995. *Strategic Planning for Public and Nonprofit Organizations: A Guide to Strengthening and Sustaining Organizational Achievement*. San Francisco, CA: Jossey-Bass.

Burns, T.R., Stöhr, C. 2011. Power, knowledge, and conflict in the shaping of commons governance. The case of EU Baltic fisheries. *International Journal of the Commons* **5**, 233–258.

Burstein, P. 1991. Policy domains: organization, culture, and policy outcomes. *Annual Review of Sociology* **17**, 327–350.

Chambers, A. 2016. *The lobbying of the EU: how to achieve greater transparency?* London: Civitas. Available at: http://www.civitas.org.uk/content/files/Anthony-Chambers-EU-lobbying.pdf [accessed 30 May 2018].

Cummins, A. 2004. The Marine Stewardship Council: a multi-stakeholder approach to sustainable fishing. *Corporate Social Responsibility and Environmental Management* **11**, 85–94.

De Vries, H., Bekkers, V., Tummers, L. 2016. Innovation in the public sector: a systematic review and future research agenda. *Public Administration* **94**, 146–166.

Duhaime-Ross, A. 2016. *New US Food Guidelines Show the Power of Lobbying, Not Science*. The Verge. Available at: https://www.theverge.com/2016/1/7/10726606/2015-us-dietary-guidelines-meat-and-soda-lobbying-power [accessed 5 May 2018].

Elinder, L.S. 2005. Obesity, hunger, and agriculture: the damaging role of subsidies. *BMJ* **331**, 1333.

European Commission (EC). 2017. *CAP at a Glance*. Available at: https://ec.europa.eu/agriculture/cap-overview_en [accessed 19 April 2018].

Fearnside, P. 2017. *Business as Usual: A Resurgence of Deforestation in the Brazilian Amazon*. YaleEnvironment360. Available at: https://e360.yale.edu/features/business-as-usual-a-resurgence-of-deforestation-in-the-brazilian-amazon [accessed 30 May 2018].

Flaemig, T., Sandstrom, S., Caccavale, O.M., Bauer, J.M., Husain, A., Halma, A., Poldermans, J. 2017. *Using Big Data to Analyse WFP's Digital Cash Programme in Lebanon, Humanitarian Practice Network*. Available at: http://odihpn.org/blog/using-big-data-to-analyse-wfps-digital-cash-programme-in-lebanon/ [accessed 2 July 2017].

Food and Agriculture Organization of the United Nations (FAO). 2018. *Global Agenda for Sustainable Livestock*. Available at: http://www.livestockdialogue.org/ [accessed 24 March 2018].

Freeman, R.E., Harrison, J.S., Wicks, A.C., Parmar, B.L., De Colle, S. 2010. *Stakeholder Theory: The State of the Art*. Cambridge: Cambridge University Press.

Froese, R., Garilao, C., Winker, H., Coro, G., Demirel, N., Tsikliras, A., Dimarchopoulou, D., Scarcella, G., Sampang-Reyes, A. 2016. *Exploitation and Status of European Stocks*. Available at: http://oceanrep.geomar.de/34476/ [accessed 30 May 2018].

Gibbs, H.K., Rausch, L., Munger, J., Schelly, I., Morton, D.C., Noojipady, P., Soares-Filho, B., Barreto, P., Micol, L., Walker, N.F. 2015. Brazil's soy moratorium. *Science* **347**, 377–378.

Greenpeace. 2006. *Eating Up the Amazon*. Available at: https://www.greenpeace.org/archive-international/Global/international/planet-2/report/2006/7/eating-up-the-amazon.pdf [accessed 30 May 2018].

Hajer, M.A., Van Tatenhove, J.P.M., Laurent, C. 2004. *Nieuwe vormen van governance, een essay over nieuwe vormen van bestuur met een empirische uitwerking naar de domeinen van voedselveiligheid en gebiedsgericht beleid*. RIVM rapport 500013004/2004, RIVM De Bilt.

Head, B.W., Alford, J. 2015. Wicked problems: Implications for public policy and management. *Administration & Society* **47**, 711–739.

Hirst, P. 2000. Democracy and governance. In: Pierre, J. (ed.). *Debating Governance. Authority, Steering and Democracy*. Oxford: Oxford University Press, pp. 13–35.

Hospes, O. 2014. Food sovereignty: the debate, the deadlock, and a suggested detour. *Agriculture and Human Values* **31**, 119–130.

Howarth, R. 2000. The CAP: History and attempts at reform. *Economic Affairs* **20**, 4–10.

Howlett, M., Ramesh, M., Perl, A. 2009. *Studying Public Policy: Policy Cycles and Policy Subsystems*. Oxford: Oxford University Press.

Hsee, C.K., Hastie, R. 2006. Decision and experience: why don't we choose what makes us happy? *Trends in Cognitive Sciences* **10**, 31–37.

Janis, I.L. 1972. *Victims of Groupthink: A Psychological Study of Foreign-Policy Decisions and Fiascoes*. Boston, MA: Houghton Mifflin Company.

Kroes, H., Kuepper, B. 2015. *Mapping the Soy Supply Chain in Europe*. Available at: http://assets.wnf.nl/downloads/mapping_the_soy_supply_chain_in_europe_wnf_12_may_2015_final_1.pdf [accessed 30 May 2018].

Kuijpers, K. 2018. *Nederland bouwt mee aan een sojaroute die de Amazone nog verder vernielt*. Trouw. Available at: https://www.trouw.nl/groen/nederland-bouwt-mee-aan-een-sojaroute-die-de-amazone-nog-verder-vernielt~a5dc16b7/ [accessed 30 May 2018].

Lukes, S. 2004. *Power: A Radical View*. Basingstoke: Palgrave Macmillan.

Marsh, D., Rhodes, R. 2002. Policy communities and issue networks. *Social Networks: Critical Concepts in Sociology* **4**, 89.

McAdam, D., McCarthy, J.D., Zald, M.N. 1996. *Comparative Perspectives on Social Movements: Political Opportunities, Mobilizing Structures, and Cultural Framings*. Cambridge: Cambridge University Press.

Nepstad, D., McGrath, D., Stickler, C., Alencar, A., Azevedo, A., Swette, B., Bezerra, T., DiGiano, M., Shimada, J., Motta, R.S. da, Armijo, E., Castello, L., Brando, P., Hansen, M.C., McGrath-Horn, M., Carvalho, O., Hess, L. 2014. Slowing Amazon deforestation through public policy and interventions in beef and soy supply chains. *Science* **344**, 1118–1123.

North, D.C. 1990. *Institutions, Institutional Change and Economic Performance*. Cambridge: Cambridge University Press.

Organisation for Economic Cooperation and Development (OECD)/ Food and Agricultural Organization (United Nations) (FAO). 2017. *OECD-FAO Agricultural Outlook 2017–2026*. Paris: OECD Publishing. Available at: http://dx.doi.org/10.1787/agr_outlook-2017-en [accessed 9 November 2018].

Pierre, J., Peters, B.G. 2000. *Governance, Politics and the State*. London: MacMillan.

Pierson, P. 2004. *Politics in Time: History, Institutions, and Social Analysis*. Princeton, NJ: Princeton University Press.

Raynolds, L.T. 2004. The globalization of organic agro-food networks. *World Development* **32**, 725–743.

Rhodes, R.A. 1997. *Understanding Governance: Policy Networks, Governance, Reflexivity and Accountability*. Buckingham: Open University Press.

Termeer, C., Dewulf, A. 2014. Scale-sensitivity as a governance capability: observing, acting and enabling. In: Padt, F., Opdam, P., Polman, N., Termeer, C. (eds). *Scale-sensitive Governance of the Environment*. Oxford: Wiley-Blackwell, pp. 38–55.

Termeer, C. J., Dewulf, A., Breeman, G., Stiller, S. J. 2015. Governance capabilities for dealing wisely with wicked problems. *Administration & Society* **47**, 680–710.

Traynor, I. 2014. *30,000 Lobbyists and Counting: Is Brussels Under Corporate Sway?* The Guardian. Available at: https://www.theguardian.com/world/2014/may/08/lobbyists-european-parliament-brussels-corporate [accessed 30 May 2018].

Van der Steen, M., Chin-A-Fat, N., Vink, M., van Twist, M. 2016. Puzzling, powering and perpetuating: Long-term decision-making by the Dutch Delta Committee. *Futures* **76**, 7–17.

Van Kersbergen, K., van Waarden, F. 2004. 'Governance' as a bridge between disciplines: cross-disciplinary inspiration regarding shifts in governance and problems of governability, accountability and legitimacy. *European Journal of Political Research* **43**, 143–171.

Weber, N., Christophersen, T. 2002. The influence of non-governmental organisations on the creation of Natura 2000 during the European Policy process. *Forest Policy and Economics* **4**, 1–12.

Werkman, R.A., Termeer, C.J.A.M. 2007. *Het Nederlandse debat rondom landbouw, landschap en het gemeenschappelijk landbouwbeleid*. Wageningen: Wageningen University, Public Administration and Policy Group.

Collective Action

How do collective action problems hinder the transition to sustainable food systems?

David Ehrhardt, Thijs Bosker, and Caroline Archambault

Chapter Overview

- Many problems that hinder food sustainability, such as resource overexploitation and carbon emissions, are collective action problems: situations in which people fail to cooperate, even though it would be in their (collective) interest to do so.

- Understanding collective action problems requires a model of human decision-making. In this chapter, we use a model of bounded rationality, in which people attempt to act rationally but are constrained (or bounded) by cognitive biases and information problems.

- People act on the basis of bounded rationality in the pursuit of goods, which may be classified as private goods, club goods, common-pool resources, and public goods along the dimensions of excludability and depletability.

- The tragedy of the commons describes a scenario in which non-excludable goods are overexploited or underprovided. This chapter presents three models that help to explain why this scenario can occur: the discrepancy between individual costs and collective benefits, the problem of free-riding, and the model of the prisoner's dilemma.

- These models offer explanations, as well as possible solutions, to the collective action problem, focused either on changing people's preferences or the rules (institutions) under which they live.

Introduction

Why did we hunt the American buffalo to near extinction in the 1800s? Why are we continuing to overexploit crucial resources for food, such as fish stocks (see Figure 14.1a) or fertile topsoil, even though we know the devastating long-term consequences for food security and the environment? Why are we not switching to renewable sources of energy for food production (see Figure 14.1b)? And why are the technologies we have developed to reduce the negative environmental impacts of food production implemented so unequally around the world?

One reason for these failures to govern food systems sustainably is a simple one: it is very, very difficult. As *Chapter 12: Food Systems* described in detail, food systems are highly complex, featuring a multitude of scales and actors, which evolve continuously with improved technologies, developing socioeconomic conditions, and a changing environment. Actors with an interest in making food systems more sustainable may not have sufficient power to elicit change, or they may be ignorant about what to do, given the complexity of the systems.

But complexity and ignorance are only part of the story, as food systems are also subject to the political dynamics of power, interests, and **incentives** (often, the controversies outlined in each chapter are related to these dynamics). Together, these forces create situations in which actors seem unable or unwilling to enhance food sustainability, even when they have the means to do so. For example, you are probably aware of certain food choices that would make your diet more sustainable (see *Chapter 8: Nutrition*), and yet do you take all of them into account when you are shopping for dinner? This chapter explores some of the core explanations for situations like these from the social–scientific literature, as well as potential solutions.

Connect the dots

Thinking back to *Chapter 11: Consumption*, what are the main explanations for food consumption choices? Keep them in mind as you read through this chapter, and consider how they relate to the political–economic explanations provided here.

We first outline how a major challenge in achieving sustainable food production lies not only in ignorance or complexity, but also in our decision-making tendencies and our struggles to act collectively, defined as **collective action problems** (Section 14.1). Subsequently,

a)

b)

Figure 14.1 (a) Why are we still overfishing many of the world's fish stocks, even though we know it is not sustainable? (b) And why are we still using fossil fuels, such as coal, to produce power, even though we know the impacts on climate change? These are examples of collective action problems.

Source: (b) Tony Webster/Flickr (CC BY-SA 2.0).

Section 14.2 provides key definitions and assumptions on which the literature on collective action problems relies, particularly with regard to the nature of human decision-making and the nature of the goals people want to achieve with their actions. Section 14.3 presents three key models that highlight why different situations may give rise to collective action problems: the tragedy of the commons, free-riding, and the prisoner's dilemma. Finally, we use these models to turn to possible solutions to these collective action problems, and we discuss some of the more promising avenues for future research in this field (Section 14.4).

14.1 What challenges do collective action problems pose for food sustainability?

Although technological innovation has kept a Malthusian catastrophe at bay (see *Chapter 1: Introduction*) and allowed us access to unprecedented amounts of food, the dream of a fully sustainable global food system remains elusive. It is challenged, for example, by the loss of biodiversity (see *Chapter 2: Biodiversity*), the input of contaminants into ecosystems (see *Chapter 3: Pollution*), the impact of food production on climate change (*Chapter 6: Climate Change*), and malnutrition (see *Chapter 8: Nutrition*). For many people around the world, these problems feel like a fact of life, leading us to shake our heads in sympathy for the poor and for future generations—and move on. After all, environmental degradation, the loss of species, and malnutrition are not new problems, so what are we going to do? Better to focus

on fighting the battles we can win than dwelling on problems we do not know how to fix.

But is it really a matter of not knowing how to fix these problems? We have managed to produce sufficient food for our growing population, and we know how to prevent malnutrition. For example, global food availability has risen from 2220 kcal per person per day in the early 1960s to 2790 kcal in 2006–2008 (FAO, 2014). In low- and middle-income countries, this increase has been even more pronounced, increasing from 1850 kcal per person per day in the early 1960s to over 2640 kcal in 2006–2008 (FAO, 2014). We already possess the scientific and technical knowledge to reduce GHG emissions to stop climate change (Pacala and Socolow, 2004) (see Figure 14.2). We also have a clear understanding of many of the natural

a)

b)

Figure 14.2 Solutions to reduce carbon emissions include wind turbines as source of renewable energy (a), as well as more natural solutions such as stopping tropical deforestation (for example, in Madagascar) (b).

Source: (b) Cunningchrisw/Wikimedia Commons (CC BY-SA 4.0).

resources that are overexploited, including major fish stocks and fossil fuels, and we know ways of managing them sustainably (Pacala and Socolow, 2004; Pauly et al., 2002; Worm et al., 2009).

So our failure to effectively produce food sustainably is no longer (just) a technological problem or an issue of complexity and ignorance, but rather a problem of human decision-making. We seem to collectively choose not to enhance the sustainability of our food production. Our challenge is therefore no longer just to find technologies or other ways to sustainably produce sufficient healthy food for billions of people, but rather to find ways of working together to make sure existing knowledge is actually utilized effectively.

In the social sciences, this challenge is often referred to as a **collective action problem**, a situation in which people fail to cooperate, even though it would be in their (collective) interest to do so. To better understand why collective action problems occur, we will first discuss some key assumptions we make in this chapter about the nature of human decision-making processes. We will then move on to models for explaining collective action problems and approaches to solving them.

14.2 How do we make decisions?

What drives the way people make decisions? Do they act impulsively on emotions or deliberatively on a cost–benefit analysis of their interests and options, or do they act habitually based on what others are doing? What do people want from their choices and actions? Our answers to each of these questions reflect our assumptions about the nature of human decision-making. Importantly, they all have different implications for the way we might explain human behaviour, including the difficulties of collective action; therefore, we need to state and clarify them. In particular, we want to specify two sets of assumptions about the rational nature of human decision-making and the goals that people want to achieve with their decisions—bounded rationality and goals in the form of different types of goods people strive for.

14.2.1 What is bounded rationality?

The first assumption in this chapter is that people generally act in a rational way—they have goals in mind and strive consciously to act in ways that allow them to achieve these goals. This core assumption has been the subject of decades of deep academic contention (see, for example, Hechter, 1994; Kahneman, 2003; MacDonald, 2003; Tversky and Kahneman, 1986). In some cases, it seems more accurate to describe people's decisions as driven by emotions, considerations of appropriateness, as people act in ways that they think are normal, or simply as a result of authority when someone tells someone else what to do. Moreover, even when people do try to act rationally, they often do not have all the information, and personal biases prevent them from acting in a way that most efficiently achieves their objectives. Together, these limits to rationality have led social scientists to describe rationality in human decision-making as **bounded** (Simon, 1991).

Information is key to understanding how rational thought processes are bounded. For example, consider all the information that goes into your decision to have either a croissant or yogurt for breakfast: the flavours of both products, their price, their impact on your health, your own immediate cravings, the preferences of your friends and partner, and so on (see Figure 14.3). And even if you decide that you prefer yogurt, is organic yogurt better for the environment than non-organic yogurt? To answer this question, you need a whole suite of additional information, for example about the ways

In your experience

Cost–benefit analysis is a key approach to rational decision-making. It systematically estimates and compares the costs and benefits of different options such as different actions to undertake or goods to consume. The outcome of the comparison helps decision-makers to identify the most efficient way to achieve certain benefits, that is, with the fewest costs as a result. In your food choices, do you make cost–benefit analyses of different options available to you? How do you estimate their costs and benefits? And how do you compare them?

a)

b)

c)

d)

Figure 14.3 Even making a seemingly simple decision on what to have for breakfast requires a considerable amount of information to assess which might be the healthiest and most environmentally friendly option. Should it be (a) a croissant, (b) yogurt, (c) breakfast cereals, or (d) a full English breakfast? And if you choose one of these options, for example the yogurt, should it be organic or non organic?

Source: (c) frankieleon/Flickr (CC BY 2.0).

in which the two yogurts impact on important environmental variables such as climate change, biodiversity loss, or water scarcity. And this excludes additional questions which relate to animal welfare, personal health, and so on. The amount of information needed to make this decision suddenly becomes quite overwhelming—and to make the best possible decision, you need all this information to be perfect, because flaws or gaps could result in a less-than-perfect breakfast choice.

Luckily, you probably do not really need to make the 'perfect' choice for breakfast, and therefore, perfect information is an unnecessary luxury for simple decisions such as this. For breakfast, most people have sufficient information to arrive at a 'good enough' outcome that satisfies the majority of their desires. But there are many situations where information is insufficient, even

to arrive at such 'good enough' outcomes, as illustrated by the choice between organic or non-organic yogurt. Assessing the relative impacts of these two types of food on the environment requires detailed, complex information that is unavailable to most people.

Pause and think

What are the different social and environmental costs of each type of breakfast in Figure 14.3? Could you easily identify which is the 'best' (or 'worst') in terms of food sustainability? Thinking back to *Chapters 2 to 7*, what types of environmental impacts would you expect for each of these products?

Even in the age of the Internet, there remain significant **information gaps**, as well as **information asymmetries** (that is, when some parties to a transaction have more information than others; see Akerlof, 1970) between individuals and groups. Particularly concerning are high-level policy issues such as those discussed in this book. Such issues are often only fully understood through extensive education and training. They are also likely to be characterized by complexity; as a result, even well-informed people are often fundamentally uncertain about the possible (unanticipated) consequences of any actions they make take.

For example, there are still individuals who deny human-caused global climate change, even in the face of overwhelming evidence in support of its existence. While we are certain that CO_2 affects our climate, we cannot predict all the consequences or the impact of drastic reductions in carbon emissions. As a result of the academic disagreements and complexity of the problems, even the most well-informed individuals will have to weigh different viewpoints on an issue and make up their own mind as to their relative merits. This requires a range of linguistic, mathematical, and critical thinking skills, and it is also seriously time-consuming.

As a result, people often turn to mental shortcuts and **heuristics** (rules of thumb) that allow them to make quick decisions—which, as it turns out, are riddled with cognitive biases (see, for example, Kahneman, 2003; Tversky and Kahneman, 1986). There are well over 100 documented cognitive biases covering the full range of human activities, such as the **availability bias** where we only take into account recent information we have heard, and **confirmation bias** or the selective gathering and neglecting of (or giving undue weight to) information in order to support a previously held belief (Nickerson, 1998)

A recent World Development Report (World Bank, 2015) highlights just how pervasive these biases are, even in the world of highly educated and self-reflexive professionals. For example, it shows how highly complex problems, such as enhancing the sustainability of food systems, reduce people's ability to think critically—'as the number of options increases, people's ability to accurately evaluate the different options declines' (World Bank, 2015). Instead, people increasingly rely on heuristics, for example by choosing the simplest option, rather than the best one, or by generalizing from a single observation (see Figure 14.4).

In the East, it could be the COLDEST New Year's Eve on record. Perhaps we could use a little bit of that good old Global Warming that our Country, but not other countries, was going to pay TRILLIONS OF DOLLARS to protect against. Bundle up!

7:01 PM - 28 Dec 2017

Figure 14.4 People, including powerful individuals and opinion makers such as US President Trump, rely on heuristics, for example by taking one observation of a cold New Year's Eve on the US East Coast as evidence to suggest that global warming is not real.

Source: Twitter @realDonaldTrump.

They also become increasingly influenced by the way in which a problem is posed (the so-called **framing effect**), as well as susceptible to confirmation bias (World Bank, 2015).

Cognitive biases, such as these, can have serious consequences for decisions on food consumption, food production, and food policymaking. Most research on this subject has focused on food consumption, and it shows the importance of rather simple heuristics and biases in food choices (Scheibehenne et al., 2007). For example, food choices are likely affected by **attention bias**, that is, an individual's tendency to perceive and see things that they already tend to think about a lot (Castellanos et al., 2009; Werthmann et al., 2011). Moreover, food choices also suffer from **optimistic bias**, in which people feel themselves to be less at risk from various hazards than the 'average' person (Sharot, 2011).

In food choices, this optimistic bias has been demonstrated for nutritional risks (Miles and Scaife, 2003; Shepherd and Raats, 2006). Similarly, optimistic bias might play a role in people's estimates of the environmental risks and their associated behaviour. This can even happen at the nation level; a study in 18 countries found a spatial optimistic bias ('things are better here than there') in individuals' assessment of current environmental issues (Gifford et al., 2009). More work needs to be done to map out in more detail the various biases at play in food systems and their impact on food sustainability, but there is no reason to think that people would be less vulnerable to cognitive biases in decisions about food than about other spheres of life.

In sum, there are many reasons to doubt the assumption that human decision-making is always fully rational, and an assumption of bounded rationality is more compelling. This assumption is a necessary building block for many prominent explanatory models for collective decision-making, including those presented in Section 14.3.

14.2.2 Goals: a typology of goods

Our second assumption relates to the nature of the goals that people want to achieve when they make bounded-rational decisions. There has been a long debate about the best way to define the goals that people aim to achieve. Some define these goals narrowly as the maximization of economic **utility**, that is, the individual satisfaction resulting from the consumption of goods and services (Mankiw, 2014). Others argue that this model excludes important human goals such as friendship, teamwork, and altruism. These are valued for other reasons than the mere utility we derive from their 'consumption'. For the purposes of this chapter, though, we will assume that all the goals that people strive to achieve through decision-making can be described as **goods**. These can include a variety of valuable things, such as tangible objects like iPhones or decent food, but also services and intangibles like love, friendship, or political recognition of a local language.

Building on this assumption, we follow classical economics in differentiating four types of goods that are defined using the two variables of **excludability** and **depletability** (or **rivalrousness**, as this dimension is sometimes referred to). Excludability refers to the extent to which the owner(s) of the good can exclude others from its benefits, while depletability indicates whether or not one person's consumption of a good diminishes its benefits for other potential consumers (that is, 'uses up' the good) (see Table 14.1).

Table 14.1 Typology of goods based on excludability and depletability

	Depletable	Non-depletable
Non-excludable	Common-pool resources	Public goods
Excludable	Private goods	Club goods

Private goods, such as food or land, are excludable and depletable; if the owner consumes the good, no one else can. Imagine an apple tree. It grows in your privately owned backyard and the benefit you derive from it is to eat the juicy apples. You can exclude others from eating your apples and when you eat an apple, it is gone. No one else can eat that apple.

Some excludable goods, however, do not run out if one person uses them; imagine that the apple tree stands in a botanical garden and visitors, who pay the entrance, can derive the benefit of walking the paths and seeing the tree. This good is excludable, only available to those who pay the entrance fee, but it is non-depletable, as someone else walking these paths and looking at the tree doesn't prevent you from doing the same. Such goods are called **club goods** (or artificially scarce goods).

Public goods are neither excludable nor depletable, making them accessible for all. If the apple tree stood in a park that was open to all, this would make its viewing a public good. Other examples of public goods are clean air, lighthouses, or the water in our oceans used for shipping goods.

Finally, there are the **common-pool resources** (or common goods), which are non-excludable goods that are depleted when consumed at unsustainable levels—think of the apples that grow on our apple tree in the public park. If we derive benefits from eating the apples and we are not careful about how many we eat, we can strip the tree clean and compromise the growth of future apples. Clean drinking water, fish stocks, or the environment generally are all examples of common-pool resources.

We will use this classification as an important piece in the puzzle of collective action problems in food sustainability. In particular, much of the analysis will revolve around the idea that food sustainability involves the provision and consumption of non-excludable goods—from the common-pool resources of biodiversity, freshwater, and fertile soils to the public good of a fully sustainable food system. Yet it is also important to recognize the limits of the classification. For example, Ronald Coase (1974) famously argued that even lighthouses, often considered as an example of public goods, could, and in fact did, exist effectively as private goods—thus questioning whether public goods existed at all (see Box 14.1).

BOX 14.1

Are lighthouses public goods?

Lighthouses, and the maritime security they provide, appear to be classic examples of a public good and are often used to demonstrate the usefulness of the concept of a public good. Their benefits are non-excludable, since any ship that sails past can see them, and non-depletable, as their light cannot run out through consumption. Given these features, we would expect lighthouses to be underprovided by private actors in the market—since the benefits are non-excludable, companies cannot make a profit from their provision. This is the basic argument that Ronald Coase, Nobel Laureate in Economics in 1991, set out to explore through the case of lighthouses in eighteenth- and nineteenth-century England

(Coase, 1974) (see Figure A). He suggests that, at specific points in English history, there were conditions under which private lighthouses, in fact, did exist, for example by charging ships to contribute to them when they dock in a port.

Coase thus challenges the theoretical classification of lighthouses as public goods and, more broadly, the usefulness of public good theory. In recent years, various authors have taken up the challenge of analysing the historical provision of lighthouses and their implications for the public good theory (Barnett and Block 2007; Bertrand, 2006; van Zandt, 1993), and the debate remains unresolved.

Figure A An English lighthouse near Hunstanton, England. A classic example of a public good or not?

Source: By Photochrom Print Collection [Public domain], via Wikimedia Commons.

Moreover, to say that there are four types of goods does not imply that everything can be neatly categorized as a single type; classification depends on specific rules and context within which a good is consumed, as seen in the shifting classification of our apple tree based on context and the benefits derived. Water is another great example—if it is publicly provided and for free, but scarce, say in the case of a village with a well, it is a common-pool resource, as it is not excludable but depletable (see Figure 14.5a). However, if

water is not scarce and its use is not excludable, water is a public good (see Figure 14.5b)

In some places, water is produced as a club good—there is plenty of water available, but its use is made excludable, for example through fees. In this case, water becomes either a private good (if your water consumption leaves less for others such as with bottled water) or a club good (if it is non-depletable such as the water from your tap if you pay a flat-rate fee for unlimited use) (see Figure 14.5c and d, respectively).

Figure 14.5 Example of water as a different type of good. Water can be a common-pool resource (a) and a public good (b), but also a private good (c) or a club good (d).

Source: (b) Darwin Bell/Flickr (CC BY 2.0); (c) J. Weeks [Public domain], via Wikimedia Commons.

Classifying a good into one of the categories mentioned in Table 14.1 is thus not always straightforward and requires a detailed understanding of the rules and context within which a good is produced and consumed.

Despite the shortcomings of the goods classification, the goods matrix in Table 14.1 illustrates that there are different types of goods and, by extension, goals for which people might aim. As we will see in Section 14.3, the specific characteristics of some of these goods, in particular common-pool resources and public goods, can help to explain why acting collectively is often difficult.

14.3 How can we explain collective action problems?

Why do we find it difficult to act to achieve our collective interests? One obvious hypothesis is that we have many competing interests, referred to as **competing incentives**. An example: while we know that eating less meat reduces carbon emissions and thus reduces impacts on the environment (incentive 1), many individuals have to weigh all this against the pleasure of eating a good steak (incentive 2; see Figure 14.6), and if someone is cooking us dinner, we may even have to weigh all this against the possible social displeasure of annoying the cook by requesting an unplanned vegetarian substitute (incentive 3). These are competing incentives, or competing reasons to act in specific ways, given our beliefs and preferences.

a)

b)

Figure 14.6 Competing incentives can make it hard to make the environmentally right choice, for example deciding between an environmentally more sustainable veggie burger (a) or a steak (b), which, by many, is considered tastier.

Source: (a) https://www.columbusunderground.com/bareburger-introduces-sustainable-vegetarian-impossible-burger-ls1.

Competing incentives are one plausible reason for people's failure to act in ways that preserve the environment and enhance sustainable food production—they have a shared interest in achieving certain collective goals, but these incentives compete with others that, for whatever reason, prevail. In reality, many collective action problems are not quite like this; can we really say, for example, that the future of the environment, and perhaps even the survival of our planet, is less important to us than the enjoyment of a good steak? Probably not, and yet we continue to eat steak.

To explain these kinds of seemingly irrational situations without letting go of our assumption of bounded rationality, we need to go beyond the model of simple competing incentives. In some cases, cognitive biases might provide a sufficient explanation for our decision to choose steak over our planet's future. For example, people have the tendency to 'discount' future gains and losses and value them less than similar gains and losses in the present (Hardisty and Weber, 2009). As a result, people may well undervalue the long-term losses resulting from environmental degradation in their decisions about food. Yet biases such as these are unlikely to be sufficient explanations for the enormity of the collective action problems that hinder food sustainability. To explain this, we need to turn to other explanations, focusing either on power and inequality (see Food controversy 14.1) or on the models explaining collective action problems.

FOOD
CONTROVERSY
14.1

Food politics and global capitalism

The explanations of consumption choices (see *Chapter 11: Consumption*), competing incentives, and cognitive biases that we have seen so far can help us understand the reasons why (isolated) individuals might make decisions that hinder the sustainability of food systems. This micro-level focus, however, may obscure explanations that operate on a larger, more macro scale. Some researchers argue that many of the problems in contemporary food systems (for example unsustainable environmental practices, labour exploitation, unequal global food distribution) are symptoms of the underlying economic system within which they are embedded—capitalism (see, for example, Holt-Gimenez, 2017; Klein, 2015).

Figure A. Anti-capitalist protests in London, 2013.

Source: Photo by George Rex/Flickr (CC BY-SA 2.0).

Although difficult to define, capitalism is seen as the political–economic system of private property, trade, competition, and profit maximization. Like arguments about the incompatibility of capitalism and effective approaches to reverse climate change (Klein, 2015), it might be that healthy and sustainable food systems are impossible in a capitalist context (Holt-Gimenez, 2017). At the heart of such critiques of capitalism is the idea that it prioritizes profits over other valuable things such as justice, sustainability, and equality. In food systems, capitalism is thus thought to ignore, or even worsen, problems like the exploitation of vulnerable people (for example, poor farmers), the unsustainable use of resources, and the unequal distribution of food across the world.

Take the example of property rights, the cornerstone of capitalist economies. In classical economics, private property rights provide owners, for example farmers, incentives to invest in their property and innovate to make it more productive. This benefits these farmers individually, as they receive more profit, but it also benefits society at large because it makes food better and/or cheaper.

For these reasons, many economists argue that private property rights—and capitalism more broadly—have been a source of economic growth and development, including radically improved food security (Acemoglu and Robinson, 2011; De Soto, 2000; North, 1990). But this argument rests on several crucial assumptions that do not always match the reality of food systems. For example, it assumes that food systems are most efficient if property is privatized. In reality, however, many food producers rely on collective property rights (see Case study 14.1 for an example from Kenya). Moreover, there is an implicit assumption that all those who want access to private property, such as land, can have it, and that once they have it, they all have equal opportunities for investment and increased profits. However, in reality, land is scarce and privatization favours those who are already wealthy, deepening inequalities that, over time, become increasingly difficult to redress.

What are the implications of the anti-capitalist critique? For some, it means that making food systems more just, equitable, and sustainable requires no less than a full-scale replacement of the system. But is this desirable? It was the capitalist logic, after all, that drove the Industrial Revolution and the dramatic subsequent improvements in food production that have allowed us to escape the Malthusian catastrophe (see *Chapter 1: Introduction*). Moreover, it was capitalism that has allowed millions of poor people to have been lifted out of poverty in recent decades (see also *Chapter 8: Nutrition*). Might radically replacing capitalism therefore not amount to throwing out the baby with the bath water?

Figure 14.7 Cows grazing on the Selsley Common in the UK.
Source: Sharon Loxton/Wikimedia Commons (CC BY-SA 2.0).

Pause and think

Short of a full-scale, anti-capitalist revolution, one way to reform capitalist economies in pursuit of sustainability could involve ensuring that consumers pay the social and environmental costs of products when they buy them in the supermarket. In more technical terms, this means internalizing more of the external costs (or **externalities**) of unsustainable food systems into the costs of production and prices of food products. Can you explain how this approach would work? And could you think of other reform strategies?

In the remainder of this chapter, we turn to three models of collective action problems to explain why food systems remain unsustainable: the tragedy of the commons, free riding, and the prisoner's dilemma. Each of these models explicitly explains why non-cooperation is a rational outcome, regardless of the immense long-term costs that result from it. In other words, even if actors are interested in sustainability and have the capacity to achieve it, there are powerful forces hindering their ability to cooperate to actually achieve it.

Such collective action problems, in the terminology of the goods matrix (see Table 14.1), exist when the good that is being overconsumed or underprovided is non-excludable. Garrett Hardin (1968) famously described these problems in his article on the **tragedy of the commons**. He describes a 'commons', a field available to everyone (see Figure 14.7), in which:

'[a]s a rational being, each herdsman seeks to maximize his gain. Explicitly or implicitly, more or less consciously, he asks, "What is the utility to me of adding one more animal to my herd?" [. . .] the rational herdsman concludes that the only sensible course for him to pursue is to add another animal to his herd. And another; and another . . . But this is the conclusion reached by each and every rational herdsman sharing a commons. Therein is the tragedy. Each man is locked into a system that compels him to increase his herd without limit—in a world that is limited. Ruin is the destination toward which all men rush, each pursuing his own best interest in a society that believes in the freedom of the commons. Freedom in a commons brings ruin to all.'

Hardin (1968)

Hardin, much like Malthus almost two centuries earlier, focused his article on the risks associated with population growth and our collective inability to constrain it. However, his observation can be applied to virtually all common-pool resources, suggesting that there is something about these types of goods that puts them at serious risk of overexploitation. For example, the management of the global commons, including our atmosphere and the oceans, is an area in which 'the tragedy' has proven particularly relevant. The important question, however, is what causal mechanism is behind this overexploitation. The remainder of this section will present several models that can help to specify the causal mechanisms behind Hardin's tragedy. Subsequently, we will consider the kinds of solutions suggested by each of these models.

14.3.1 What is Hardin's mechanism for the tragedy?

Hardin (1968) himself suggests a mechanism behind the tragedy that revolves around the difference between the benefits to the individual herder and the costs to the entire community of herders. There are two aspects to Hardin's explanation that could account for the tragedy. First, he suggests that individual herders choose to add animals because the benefits of increasing their herd accrue directly and fully to them, while the costs of overgrazing are shared among all herders. This suggests a net benefit to the individual herder in response to adding animals—at least in the short term (Ostrom, 1990).

Second, Hardin suggests that herders face an incentive to add animals to their herd because they will incur the costs of doing so in the long term, while the benefits accrue immediately. Thus, besides the net benefit attached to increasing one's herd, herders might also weigh immediate gains as more important than future costs. This assumption is supported by psychological research focused on the common human tendency to discount the future, as discussed previously. Interestingly, the paper by Hardin did trigger some unwanted side effects, with many governments restricting the use of commons by herdsmen, even if this affected their traditional livelihoods (see Case study 14.1).

CASE STUDY 14.1

The tragedy of the commoners? Collectively managing mobility in the privatized rangelands in Southern Kenya

The Maasai of Southern Kenya have long practised pastoralism, that is, the rearing of livestock, as their primary livelihood (see Figure A). Given the environmental conditions in which they reside, where grazing and water resources are scattered variably and unpredictably across a vast landscape, livestock mobility has been key to pastoral success (see Figure B). To ensure this mobility, lands have long been managed (in)formally communally (as common property). However, recent pressures have pushed the Maasai to privatize rangelands with devastating impacts on their environment and their livelihood. In a twist to Hardin's predictions, privatization brought about a tragedy to the commoners (and their commons).

As Kenya's population rose dramatically over the last century and the pressure on land increased accordingly, the country experimented with several land governance systems in an attempt to sedentarize (that is, to settle down) the mobile Maasai and develop their livelihoods. The fear of Hardin's tragedy of the commons loomed large. The group ranch model, in which a community collectively owns land and manages their livestock, was one such experiment introduced in the early 1960s. Although collective in nature, it was seen by the Kenyan state as a first step in the eventual transition from communal to private ownership of Maasai rangelands. And by the early 1980s, indeed, most of the group ranches in Kenya had voted to privatize

Figure A Maasai herdsmen with their goats near a watering hole. Pastoralism has been the mainstay of the Maasai livelihood and traditionally involves a mobile lifestyle in pursuit of grazing and water.
Source: © Thijs Bosker.

Figure B Fences, such as this fence made of thorny shrubs, hinder the movement of livestock.

Source: © David Ehrhardt.

(Mwangi, 2009). Today, the overwhelming majority of rangelands in Southern Kenya are held under private title.

Why would the Maasai choose to embrace land privatization, a tenure form with clear risks of compromising livestock and wildlife mobility, damaging savannah ecosystems, and threatening the future of pastoralism as a sustainable livelihood pursuit? Why, in other words, did they abandon the commons, which had proved a successful and sustainable land governance strategy for generations?

Decisions to vote in favour of privatization were varied and illustrate competing incentives and information gaps, as well as perhaps cognitive biases. Many voted for privatization because they were unsure what the government would do to this land if it were not protected under a well-recognized form of property rights. A history of dispossession fomented a deep mistrust of the Kenyan state, and privatization provided a sense of security against neighbouring encroaching ethnic groups pushing their agricultural frontiers, as well as elites grabbing land to consolidate wealth and power. This illustrates the impact of competing incentives, between, on the one hand, maintaining communal land ownership and, on the other, protecting one's control over the land from others (that is, the Kenyan state or other Kenyan communities).

The choice to privatize evidently also reflected other competing incentives within Maasai communities. For example, some women, although not permitted officially to vote, desired a more sedentary and permanent lifestyle and pushed their husbands to vote for privatization. Many young men, similarly, imagined a new and less intensive pastoralism. Others simply needed the money, made possible from selling land or resources on their land, to meet immediate needs like hospital bills and school fees.

Finally, for many subdivision supporters, a vote for privatization was not necessarily a vote to dismantle the commons. Many believed they could reap the benefits of *de jure* (legal/official) privatization, while maintaining *de facto* (in practice) common property management. Many hoped, perhaps revealing insufficient information and/or an optimism bias, that the large majority would act collectively and that the few who defected, by excluding others from their private land, would not pose significant threat to mobility and resource access.

Unfortunately, privatization of group ranches in Kenya has resulted in considerable enclosure, widespread land sales, and fencing of property and dramatic impacts on livestock mobility in the area. Livestock must now use heavily eroded public corridors to access major communal water points and seasonal grazing, and there have been severe additional impacts because of climate change (see Figure C). Most

a)

b)

Figure C (a) Maasai livestock using a heavily eroded corridor to get to water. (b) Climate change is another threat to livestock, with massive loss of animals as a result of lack of water and grass.

Source: (a) Caroline Archambault; (b) Geerte Verduijn.

significantly, however, enclosure has demanded a new governance system where permission is required to move animals onto another person's property.

Many efforts, formal and informal, are under way to try and re-create the commons and revive cooperation for collective management of private rangelands (Archambault, 2016; Lesorogol, 2010). In Maasai communities, these efforts have focused on: (1) changing incentives, by making it costly to individuals to enclose and exclude others from their resources, and (2) enhancing information on how enclosure is disruptive both socially and ecologically. NGOs and local leaders are running information campaigns, particularly targeting group ranches that are still under decision, but also advocating in former group ranches for greater cooperation. Through these initiatives, informal norms of solidarity and inter-dependence are being stressed.

14.3.2 How does free-riding explain collective action problems?

Another mechanism behind the tragedy of the commons can relate to **free-riding**. Free-riding occurs when individuals who benefit from resources (for example, clean air, biodiversity conservation, stable climate) do not contribute to their production or protection because they have a reasonable expectation that others will. As a result, the resource is overconsumed or even fully depleted. All cases of cooperation are susceptible to free-riding if the benefits from production or consumption of the good in question are not excludable—that is, if the good is a public good or a common-pool resource (Olson, 1965).

The negative impacts of free-riding depend on the extent to which it affects group behaviour. An extremely negative case occurs when everyone in a group free-rides, because they expect (incorrectly, as it turns out) other group members to take action. As a result, the collective good is completely over-exploited (common-pool resource) or not provided at all (public good). In other cases, however, nearly everyone takes action to conserve a resource, except a few group members who benefit from the actions taken by the others. In this situation, the non-excludable good might be almost optimally used, even though some players are taking advantage of the efforts of others.

A good example of the latter is the collective action taken to protect whales. Due to technological advances in the twentieth century, whales were harvested at an unsustainable rate, putting many of the great whale species (for example, the northern and southern right whales, fin whales, and blue whales) at risk of extinction. To conserve whales, an international moratorium against whaling came into effect in 1985–1986. The moratorium states: '*catch limits for the killing for commercial purposes of whales from all stocks for the 1986 coastal and the 1985/86 pelagic seasons and thereafter shall be zero.*' This moratorium was accepted by nearly all whaling nations, nearly reducing whale landings to zero. Only a few countries, including Japan, Norway, and Iceland, continue whaling, thereby free-riding on the other countries that stopped whaling (see Figure 14.8).

In your experience

Are you a free-rider? Can you think of situations where you did less than you should have, because you expected others would do the work for you? Under what conditions are you more or less likely to show this kind of behaviour?

Figure 14.8 Most nations have stopped whaling to conserve whales, while nations including Japan free-ride and continue to catch whales, as seen in this figure, under the claim of scientific research.

Source: Australian Customs and Border Protection Service (CC BY-SA 3.0).

The free-riding logic can also be applied to private goods with negative **externalities**. These are costs that accrue to a third party but are not included in the good's price (see *Chapter 11: Consumption* for a detailed discussion of externalities). In food, negative externalities often relate to the environmental costs of the production or consumption process that are not included in the price the consumer pays. Examples of externalities in food production are discussed in *Chapters 2 to 7*, such as loss of biodiversity, pollution, emission of GHGs, water shortages due to overextraction, and soil degradation (Godfray et al., 2010).

In food consumption, the benefits of a 'good' food choice in terms of its environmental impacts are non-excludable; the costs are private. As such, unless environmentally friendly food choices are also the ones that people themselves prefer, free-riding is likely to be a problem (and more so in large groups than small ones; see Box 14.2).

BOX 14.2 Group size and free-riding

There is a famous argument that small groups are twice blessed, because their members face both social and economic incentives to cooperate in the provision of public goods and/or in the protection of common-pool resources (Olson,1965). It is a simple rule of thumb—the larger the group, the greater the risk of free-riding (Olson 1965) (see Figure A).

This is partly due to the costs of monitoring each other and peer pressure (social incentives); in small groups, it is relatively simple for group members to monitor who is contributing what and to enforce norms on acceptable contributions. In large groups, with less contact between members and a more diffuse sense of group identity, it may be very hard to monitor each other's actions and rather

ineffective to pressure each other to conform to loosely defined 'group' expectations. Imagine a small village where everyone knows each other. They all rely on the same main water source. It is much more difficult to free-ride, by, for example, overusing the water or polluting it, when you will be recognized and very likely shamed. In contrast, in a densely populated urban slum, where you are a lot more anonymous and your (in)actions are hidden by all the people who use the same water source, you will likely not be noticed nor punished for depleting it.

But group size also matters because it affects the impact of individual efforts and the relative returns group members get on their contributions (economic incentives). In small groups, individual efforts are likely worth more (and often more

a)

b)

Figure A Group size matters. In large urban settlements, such as this slum in Mumbai (a), it is impossible to know everyone. This will likely result in a higher chance of free-riding, compared to a small rural Indian village (b) where most people know each other.

Source: (a) Sthitaprajna Jena/Flickr (CC BY-SA 2.0); (b) Deeptrivia/Wikimedia Commons (CC BY-SA 3.0).

necessary) in order to produce a 'good enough' public good; moreover, smaller groups, by definition, involve larger relative returns to effort than larger ones. Group size also affects the perceived impact, for example, in the depletability of goods—if large groups are involved in the consumption of a common-pool resource, individual group members may feel that their own consumption is negligible and, in consequence, the good is a non-depletable public good. If they treat it as such, they will free ride on the protection of others. In the urban slum, your extraction may really feel like a mere drop in the bucket.

Olson pessimistically concludes, 'unless the number of individuals in a group is quite small, or unless there is coercion or some other special device to make individuals act in their common interest, rational, self-interested individuals will not act to achieve their common or group interests' (Olson, 1965). This has dramatic implications for the global environmental issues we are facing as a result of unsustainable food systems; they are especially likely to suffer from free-riding, since they involve extremely large groups—the whole of humanity.

14.3.3 How does the prisoner's dilemma explain collective action problems?

A third famous model for teasing out the mechanisms that underlie our collective action problems is the prisoner's dilemma, which has a similar structure to Hardin's tragedy of the commons but points to a different mechanism. Imagine two individuals who have been brought in by the police. Both of them can be imprisoned for a minor felony but are suspected of a major felony. The prisoners are questioned by the police and are given the option to testify against the other prisoner (that is, defect) or keep quiet (that is, cooperate). If they both cooperate with each other,

they go to prison for the minor felony. If they both defect they are both sentenced for the major felony. However, if one cooperates but the other defects, the defector will be released quickly, while the cooperator will be convicted of the major felony (see Box 14.3)

This puts the prisoners in a dilemma: to defect or to cooperate. Given the way that the payoffs are structured, they each face strong incentives to defect—with the result that they both end up being prosecuted and jailed for the major felony. (see Box 14.3). Together, they would have been better off keeping quiet, but without prior trust or communication between the prisoners, they are unlikely to blindly trust the other's cooperation.

BOX 14.3 Formalizing the prisoner's dilemma through game theory

Social–scientific models can be expressed in different ways, for example in words, stories, graphs, or mathematical equations. **Game theory** provides powerful tools for formalizing models from words into graphics and mathematics. This can have several advantages; most importantly, it facilitates checking its rigour and internal logic. The prisoners' dilemma is a good example to illustrate the basics of the approach. The consequences of one prisoner's choice depends on the choice by the other, as shown in the payoff matrix below

(Ostrom, 1990: 217). The rules of the game are as follows:

- Players are rational and wish the shortest sentence possible for themselves.
- The game is non-cooperative, that is, players cannot communicate with each other and do not know the moves of other players.
- Players have complete information, that is, players know the rules of the game and all the outcomes in the payoff matrix.

What is the most likely outcome of this game? The central insight from the prisoner's dilemma resembles closely what Hardin told us in the tragedy of the commons—that people's individual incentives may not lead them to act in ways that produce the best outcome for their group. If players are individually rational, therefore, their **dominant strategy** (that is, the strategy that brings them the best result, irrespective of the other player's strategies) will always be to confess and aim for the 3-month sentence. But together, this means the players end up in the third-best scenario—confess–confess, with an 8-year sentence for each player.

	Prisoner 2	
Prisoner 1	*Not confess*	*Confess*
Not confess	1 year each	10 years for prisoner 1 3 months for prisoner 2
Confess	3 months for prisoner 1 10 years for prisoner 2	8 years each

Figure A Payoff matrix of a prisoner's dilemma game, showing all the benefits and costs of the options in the game.

Source: *Governing the Commons: The Evolution of Institutions for Collective Action* by Elinor Ostrom. © Cambridge University Press 1990.

Pause and think

What might be the advantages of formalizing models in the way suggested in Box 14.3? In what situations would you try to do this? When might it not be necessary or helpful? Are there any weaknesses in the assumptions?

While the story applies this dilemma to confessing to crimes, we could just as easily consider two herders in the commons, like in Hardin's article, or two multinational fishing companies deciding whether to overexploit certain fish stocks (see Case study 14.2). Their respective payoffs might look very similar to those in the prisoner's dilemma, in which cooperation results in a sustainable exploitation of the resource, but defection looks more attractive to each one in isolation. The result? Defection at the expense of public goods or common-pool resources that are essential to a sustainable food system.

Studies in the 1980s provide a powerful example of a prisoner's dilemma in India, related to the sustainable use of water for irrigation of crops (Wade, 1988). In this system, farmers did not believe that refraining from taking water out of the irrigation canals (see Figure 14.9) would result in sufficient water later on. As a result, they used the water resources unsustainably.

Figure 14.9 Irrigation canal in a peanut field in India.

Source: Seratobikiba/Wikimedia Commons (CC BY-SA 4.0).

Of course, it is important to remember that the prisoner's dilemma model has very specific conditions that may not be easily met in real world situations (Ostrom, 1990). For example, the traditional prisoner's dilemma is only played once; if, on the contrary, the game is repeated, experimental evidence shows that players will develop a reciprocal 'tit-for-tat' strategy, in which they mirror each other's choices, resulting in more cooperation (Axelrod, 1987; Bo, 2005; Bo and Frechette, 2011; Press and Dyson, 2014).

In a way, the model of the prisoner's dilemma therefore contains hints about ways to fix the irrational outcomes it produces. For example, by slightly altering the payoff matrix or changing some of the rules of the game to promote repeat interactions. This is true for all models and mechanisms we have identified as drivers of collective action problems—as they explain problems, they also suggest their solutions. In Section 14.4, we illustrate this with reference to the theories presented above.

Pause and think

What are the differences between Hardin's mechanisms, free-riding, and the prisoner's dilemma? Can you think of examples for each of them? Are there situations that could be analysed using all three models?

CASE STUDY 14.2

Two examples of how to deal with overfishing

The current exploitation of fish stocks in the world's oceans is a prime example of a tragedy of the commons that has led to the unsustainable use of resources. Overexploitation has resulted in a remarkable change in the ownership of open water over the last decades. Here we will discuss two of these changes: the development of exclusive economic zones (EEZs) and marine protection areas (MPAs).

From a common-pool resource to a private good: exclusive economic zones

An **exclusive economic zone** (EEZ) is a sea zone over which a state or country has special rights when it comes to the exploitation of resources, including fish stocks. In 1950, the EEZ was first implemented as a 4-mile zone near the coast of all countries of the world—the territorial waters. For Iceland, a country which heavily relied on cod fishing to support their economy, this meant that the majority of the fish stock on which they depended was outside their territorial waters. During this period, fisheries worldwide were industrialized, resulting in factory ships which can catch high volumes of fish. This resulted in the risk of overfishing. In fact, Iceland noticed a sharp decline in overall fish stocks and decided to unilaterally increase their territorial waters to 12 miles (see Figure A for the progression of Iceland's EEZ).

The majority of countries respected this decision, with the exception of the British fishing fleet. The result was a serious conflict between Iceland and the UK—the First Cod War (1958). Heavily supported by their navy (30 navy ships and thousands of sailors), the UK decided to continue fishing within the 12-mile zone of Iceland. This resulted in a response by the Icelandic government, who mobilized the Icelandic coastguard's seven boats and approximately 100 sailors.

To protect themselves from the Icelandic coastguard, the British started to fish in 30 × 30-mile zones, flanked by naval ships. However, this method of fishing was much less effective, resulting in a dramatic reduction in fish landings, eventually resulting in the UK agreeing on the 12-mile zone.

According to Iceland, however, this was not enough to protect the fish stock from overexploitation. In 1971, Iceland again increased the size of its EEZ, now to 50 miles (see Figure A). This time, both the UK and West Germany objected to the decision and refused to respect the territorial claim by Iceland. The result was the Second Cod War (1972). Again, naval ships were sent to Iceland to protect the British and West German fishing fleets from the Icelandic coastguard.

As a response to the continued fishing, the coastguard started cutting the lines of fishing nets, resulting in massive financial damage to the fishing fleets. Iceland again came out on top, and the 50-mile zone was accepted by the UK and West Germany.

In 1975, Iceland increased its territorial waters again, this time to a 200-mile zone (see Figure A), which resulted in the Third Cod War. More nets were cut, with more confrontations between the UK and Iceland (see Figure B), but Iceland came out on top again. Now the entire 200-mile zone is an EEZ, which is prescribed in the United Nations Convention on the Law of the Sea.

With the development of the EEZ, the status of resources changed. Before the EEZ was in place, the fish around Iceland were a common-pool resource, accessible to anyone and susceptible to unsustainable exploitation. With the implementation of the EEZ, these rights have defaulted to a specific country, making these resources private goods, which can be governed by just one nation (see Figure C).

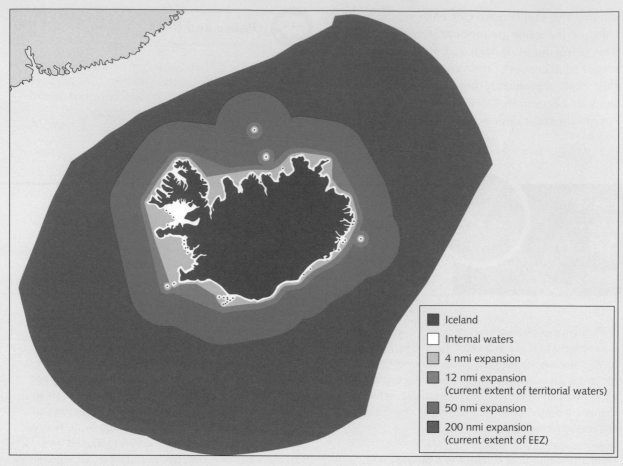

Figure A Expansion of the Icelandic EEZ.

Source: Kjallakr at English Wikipedia/CC BY-SA 3.0.

Figure B A collision between a vessel from the Icelandic coastguard and a British navy ship during the Third Cod War.

Source: Kjallakr at English Wikipedia/ CC BY-SA 3.0.

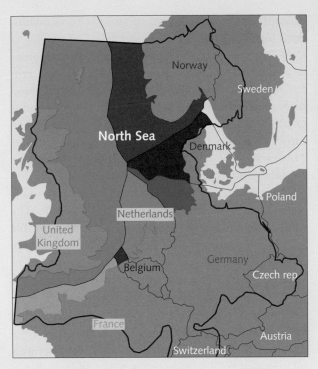

Figure C The complex distribution of EEZs within the North Sea.

From a public good to a club good: marine protection areas
Another example is the development of **marine protected areas** (MPAs) (see Figure D). MPAs are protected areas in our seas and oceans, and even large freshwater bodies, in which human activity is restricted. This protects the natural, and sometimes

Figure D The Great Barrier Reef, a hotspot of biodiversity and an important example of an MPA.

cultural, resources of a region. It can result in limits on fisheries, ecotourism (for example, whale watching), and ocean prospecting (for example, for oil).

Before the development of MPAs, marine resources were often non-excludable and non-depletable (although some resources within them were depletable, for example fish stocks or oil fields), making them a public good. Effectively, MPAs are no longer public goods but are turned into club goods, similar to wildlife reserves and national parks in terrestrial ecosystems. The resources are still not depletable but are excludable, making protection much more manageable. In 2006, the UN ambitiously set a target of conservation of 10% of the world's ecological regions as MPAs by 2012, a goal which still has not been attained (in 2017, only 6.35% of the global oceans were covered as MPAs).

14.4 Solving collective action problems?

Ideally, models such as the ones discussed previously simultaneously provide effective explanations for, as well as potential solutions to, complex problems in society. We have seen examples of how models can be used to explain problems related to food sustainability; in this section, we will demonstrate how they can be used to derive solutions for these problems.

For example, if the cause of a collective action problem lies in externalities or information asymmetries, the solution is to change the incentives or enhance access to information. In the case of consumption of the endangered Atlantic bluefin tuna,

for example, we can argue that overfishing of tuna is driven by a combination of the following:

1. The relatively high price tuna can fetch on the global market (US$3.1 million was paid in 2019 for a 278 kg bluefin tuna), which underestimates the environmental costs (reflecting externalities), and

2. The lack of public awareness of the environmental consequences of tuna fishing (showing an information gap).

A solution would be initiating public information campaigns to show people the true cost of their behaviour (internalizing the environmental costs and

Figure 14.10 As a result of public awareness campaigns by NGOs on sustainable fisheries, many retailers in Europe now (also) sell products which are deemed more sustainable, for example products with the ecolabel Marine Stewardship Council (MSC), which sets a standard for sustainable fishing, and traceable supply chains to these fisheries.

Source: Courtesy of Marine Stewardship Council (MSC).

improving information). A variety of public information campaigns to promote sustainable fisheries are undertaken by NGOs, including the WWF and Greenpeace (Leadbitter and Benguerel, 2014). As a result of the increased public awareness, as well as other reasons, retailers have taken action to replace products with more sustainable alternatives (see Figure 14.10).

Table 14.2 summarizes the three collective action models presented in Section 14.3, their main causes, and some possible solutions. The solutions listed here are by no means exhaustive, but even so, they indicate that there are often multiple ways to address the same causes. Table 14.2 also suggests that solutions often relate to one of two things—either changing the things that people want (what economists call their preferences) or changing the 'rules of the game', a

concept that social scientists often refer to as **institutions** (see *Chapter 13: Governance*).

An example of changing preferences are campaigns to promote the reduction of food waste or other pro-environmental food choices. While the logic behind such an approach is sound, changing people's minds in practice is often much harder than it appears in theory (see also *Chapter 8: Nutrition* and *Chapter 11: Consumption*). Changing institutions also has an appealing intuitive logic; if rules affect behaviour and we want to alter behaviour (for example, by promoting cooperation), changing rules might work. The rules we should change depend on the situation; Table 14.2 suggests some general directions. For example, privatization (that is, promoting excludability of goods through private property rights; see Case study 14.2) or third-party coercion may help address Hardin's tragedy and free-riding, while facilitating repeat interactions may reduce the prisoner's dilemma.

In theory, institutional change may be a powerful means of solving collective action problems. But is there any reason to think that changing institutions is easier than changing preferences? Work by the important economist Elinor Ostrom and colleagues (1999) suggests that institutional change can be effective if it focuses on '*restricting access and creating incentives (usually by assigning individual rights to, or shares of, the resource) for users to invest in the resource instead of overexploit it.*' Fishermen in Turkey, for example, have long had an elaborate system in dividing access to the most productive areas in the sea, ensuring that the commons are managed in a sustainable way (Ostrom, 1990). Similarly, community-based institutional arrangements

Table 14.2 Models, causes, and solutions of collective action problems

Model	Main causes	Possible solutions
Hardin's mechanisms	• Benefits are individual, costs are shared collectively • Benefits are short-term, costs are long-term	• Privatization: makes costs individual • Third-party coercion: create short-term individual costs to non-cooperation • Information campaigns: emphasize the impact of long-term costs
Free-riding	• Benefits are non-excludable • Costs are shared • Group size	• Privatization: makes benefits excludable and costs individual • Monitoring and third-party coercion: increase costs of non-cooperation • Reduced group size (decentralize)
Prisoner's dilemma	• Payoffs incentivize defecting • Rules of the interaction prevent coordination (for example, build-up of trust)	• Alter payoff matrix to incentivize cooperation over defection • Change critical interaction patterns, for example by facilitating repeat interactions

are effectively managing water resources in various contexts around the world (Tang, 1991). And qanats have been used for millennia by communities to irrigate arid and semi-arid lands (see Case study 14.3). Local, context-specific institutions can thus prove to be effective solutions to the problems of collective action.

Yet throughout the literature on common-pool resource management, it is also evident that there are strict and necessary conditions for the development and implementation of effective and sustainable rules. Ostrom (1990) formulated eight design principles for community-based natural resource management, which may serve as a starting point for thinking concretely about the rules that can help to prevent Hardin's tragedy (see Box 14.4). Furthermore, it remains an open question whether these bottom–up, local solutions can be 'scaled up' to the global problems of environmental sustainability.

CASE STUDY 14.3

The qanat: how communities have successfully produced food in the desert for millennia

Imagine living in an agricultural community in Persia (now Iran) several thousand years ago. Water in this region is a scarce commodity, but of vital importance to irrigate crops. How do you, as a community, ensure sufficient access to water for farmers? And how do you ensure that everyone gets their fair share?

In many arid regions, you will find strange wormhole-like structures (see Figure A)—qanats. These structures were first developed in Persia around 3000 years ago, from where they spread all over the world. They are found in Asia, the Arabian Peninsula, Northern Africa, Southern Europe, and even the Americas (under a different name, but based on the

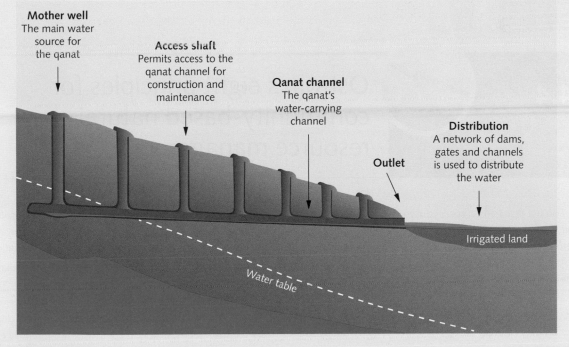

Mother well
The main water source for the qanat

Access shaft
Permits access to the qanat channel for construction and maintenance

Qanat channel
The qanat's water-carrying channel

Distribution
A network of dams, gates and channels is used to distribute the water

Outlet

Irrigated land

Water table

Figure A A cross-section of a qanat.

Source: Samuel Bailey/Wikimedia Commons (CC BY 3.0).

a)

b)

Figure B An underground tunnel (a) brings water from the Mother well to the agricultural fields where it is used for irrigation (b).

Source: (a) NAEINSUN/Wikimedia Commons (CC BY-SA 3.0); (b) Taguelmoust/Wikimedia Commons (CC BY-SA 3.0).

same approach). Qanats are built to irrigate crops in these dry regions, and some of them have been in use for more than 2500 years. They are community-run and provide an excellent example of how to efficiently govern a common-pool resource.

In brief, qanats are underground tunnel systems which are used to transport water from underground aquifers (see *Chapter 4: Water*) to agricultural areas (see Figure A). The origins of qanats are wells (referred to as the Mother well), which are transformed into artificial springs (see Figure B). A long, slightly sloping canal is dug from the original well,

towards the agricultural area (see Figure B). This system also allows water to move to agriculture fields, without the need for pumping, and has proved a very efficient way to grow crops in dry regions. As the water is transported underground, there is minimal loss due to evaporation. Moreover, the wormholes are actually airshafts. These are needed to provide air to workers who enter the qanat during construction and maintenance of the system. In this way, the qanat did not only provide water to irrigate the crops, but also drinking water to the community.

BOX 14.4

Ostrom's eight principles for community-based natural resource management

Based on her research on successful examples of community-based natural resource management, such as the qanat system in Case study 14.3, Ostrom (1990) formulated eight principles for the design of effective ways of managing common-pool resources that have been tested and further developed by Cox et al. (2010):

1. Make sure to have well-defined boundaries around a community of users and the resource system used by the community.

2. Make sure any agreements regarding the use of a resource (appropriation rules) align with the rules for providing and maintaining it, as well as with any relevant local conditions.

3. Ensure that most of the individuals affected by the management of the resources can participate in making and modifying the management rules.

4. Ensure that monitoring of the management rules is done by the community, or that there is clear accountability towards the community.

5. Have a sanctioning system in place where sanctions progress incrementally, based on either the severity or the repetition of violations (graduated sanctions).

6. Develop low-cost and accessible conflict resolution mechanisms for disputes about management of the resource.

7. Ensure that external governments do not challenge the rights of local users to create their own institutions.

8. Organize resource governance activities in nested levels, with smaller resource systems embedded within larger and even larger ones.

14.5 Future outlook

'Some of the most difficult future problems, however, will involve resources that are difficult to manage at the scale of a village, a large watershed, or even a single country. [. . .] Management of these resources depends on the cooperation of appropriate international institutions and national, regional, and local institutions.'

Elinor Ostrom et al. (1999)

The tragedy of the commons and the mechanisms that explain it are, in a way, timeless problems that are likely to continue to plague large-scale efforts to promote the sustainability of our food systems. Ostrom's pioneering work, as well as the large literature that this has spawned, helps to outline means of improving the ways in which common-pool resources in food systems are managed. Perhaps the various city-level attempts at promoting sustainable alternative food systems are opportunities to experiment with the implementation of these lessons, given that many cities have effective forms of bottom-up decision-making (see *Chapter 12: Food Systems* for more details).

Yet there is a fundamental tension between the local, bottom-up nature of these systems (as well as the long time frames in which they are typically developed) on the one hand and the rapidly globalizing nature of our food systems on the other. Where small-scale, local alternative food systems might be developed that are effectively sustainable, it is unclear whether Ostrom's principles are of much use for thinking about 'sustainabilizing' food on a global scale.

Moreover, several global trends should make us cautious against an overly optimistic attitude towards the potential of bottom-up institutional design as a silver bullet to promote sustainable food systems. First, in many of the poorer parts of the world, rapid population increases put pressure on existing systems of resource management, as more and more people want to benefit from them. Second, the simultaneous spread of global capitalism and the increasing reach of multinational agricultural corporations, as well as governments, have pushed for privatization of natural resources—often to the detriment of the locally sustainable forms of resource management. Finally, as food systems become increasingly global, it becomes more and more difficult to apply lessons from bottom–up, locally sensitive institutional solutions. Ostrom et al. (1999) recognize the severity of these challenges, but solutions are not so readily identifiable.

Connect the dots

Looking back to the discussion of alternative and sustainable food systems in *Chapter 12: Food Systems*, how does that discussion connect to the problems of collective action and their possible solutions analysed here? What lessons does the collective action approach presented here teach you about the best ways to 'sustainabilize' food systems?

● QUESTIONS

14.1 Beyond (bounded) rationality, what alternative models for human decision-making can you think of?

14.2 Explain how roads can be all four different types of goods.

14.3 Can you think of examples of the tragedy of the commons that are not like the prisoner's dilemma?

14.4 What does the case study of Maasai rangeland

privatization teach you about the strengths and limitations of collective action models?

14.5 If you analyse the case of overfishing as a collective action problem as a prisoner's dilemma, what would you suggest as solutions, based on the premises of the model? Would these solutions change if you think the situation reflects free-riding or Hardin's mechanisms?

● FURTHER READING

Hardin, G. 1968. The tragedy of the commons. *Science* **162**, 1243–1248. (Classic, succinct explanation of the tragedy of the commons.)

Ostrom, E., Burger, J., Field, C.B., Norgaard, R.B., Policansky, D. 1999. Revisiting the commons: local lessons, global challenges.

Science **284**, 278–282. (Discussion of the ways in which the lessons from bottom-up resource management might be applied to global challenges of managing large-scale resources that depend on international collaboration.)

● REFERENCES

Acemoglu, D., Robinson, J.A. 2011. *Why Nations Fail: The Origins of Power, Prosperity, and Poverty*. New York, NY: Broadway Business.

Akerlof, G.A. 1970. The market for 'lemons': quality uncertainty and the market mechanism. *Quarterly Journal of Economics* **84**, 488–500.

Archambault, C. 2016. Re-creating the commons and re-configuring Maasai women's roles on the rangelands in the face of fragmentation. *International Journal of the Commons* **10**, 728–746.

Axelrod, R. 1987. The evolution of strategies in the iterated prisoner's dilemma. In: Bicchieri, C., Jeffrey, R., Skyrms, B. (eds). *The Dynamics of Norms*. Cambridge: Cambridge University Press, pp. 1–16.

Barnett, W., Block, W. 2007. Coase and Van Zandt on lighthouses. *Public Finance Review* **35**, 710–733.

Bertrand, E. 2006. The Coasean analysis of lighthouse financing: myths and realities. *Cambridge Journal of Economics* **30**, 389–402.

Bó, P.D. 2005. Cooperation under the shadow of the future: experimental evidence from infinitely repeated games. *American Economic Review* **95**, 1591–1604.

Bó, P.D., Fréchette, G.R. 2011. The evolution of cooperation in infinitely repeated games: experimental evidence. *American Economic Review* **101**, 411–429.

Castellanos, E.H., Charboneau, E., Dietrich, M.S., Park, S., Bradley, B.P., Mogg, K., Cowan, R.L. 2009. Obese adults have visual attention bias for food cue images: evidence for altered reward system function. *International Journal of Obesity* **33**, 1063–1073.

Coase, R. H. 1974. The lighthouse in economics. *Journal of Law & Economics* **17**, 357–376.

Cox, M., Arnold, G., Villamayor Tomás, S. 2010. A review of design principles for community-based natural resource management. *Ecology and Society* **15**, 38.

Food and Agriculture Organization. 2014. *FAO Statistical Yearbook 2012*. Available at: http://www.fao.org/docrep/015/i2490e/i2490e03a.pdf [accessed 12 November 2017].

Gifford, R., Scannell, L., Kormos, C., Smolova, L., Biel, A., Boncu, S., Corral, V., Güntherf, H., Hanyu, K., Hine, D., Kaiser, F.G., Korpela, K., Lima, L.M., Mertig, A.G., Mira, R.G., Moser, G., Passafaro, P., Pinheiro, J.Q., Saini, S., Sako, T., Sautkina, E., Savina, Y., Schmuck, P., Schultz, W., Sobeck, K., Sundblad, E.-L., Uzzell, D. 2009. Temporal pessimism and spatial optimism in environmental assessments: an 18-nation study. *Journal of Environmental Psychology* **29**, 1–14.

Godfray, H.C.J., Beddington, J.R., Crute, I.R., Haddad, L., Lawrence, D., Muir, J.F., Pretty, J., Robinson, S., Thomas, S.M., Toulmin, C. 2010. Food security: the challenge of feeding 9 billion people. *Science* **327**, 812–818.

Hardin, G. 1968. The tragedy of the commons. *Science* **162**, 1243–1248.

Hardisty, D.J., Weber, E.U. 2009. Discounting future green: money versus the environment. *Journal of Experimental Psychology: General* **138**, 329–340.

Hechter, M. 1994. The role of values in rational choice theory. *Rationality and Society* **6**, 318–333.

Holt-Giménez, E. 2017. *A Foodie's Guide to Capitalism: Understanding the Political Economy of What We Eat*. New York, NY: NYU Press.

Kahneman, D. 2003. Maps of bounded rationality: psychology for behavioral economics. *American Economic Review* **93**, 1449–1475.

Klein, N. 2015. *This Changes Everything: Capitalism vs. the Climate*. New York, NY: Simon & Schuster.

Leadbitter, D., Benguerel, R. 2014. Sustainable tuna: can the marketplace improve fishery management? *Business Strategy and the Environment* **23**, 417–432.

Lesorogol, C.K. 2010. Creating common grazing rights on private parcels: how new rules produce incentives for cooperative land management. *Cooperation in Economy and Society* **2010**, 239–258.

Lusk, J. L., Nilsson, T., Foster, K. 2007. Public preferences and private choices: effect of altruism and free riding on demand for environmentally certified pork. *Environmental and Resource Economics* **36**, 499–521.

MacDonald, P.K. 2003. Useful fiction or miracle maker: the competing epistemological foundations of rational choice theory. *American Political Science Review* **97**, 551–565.

Mankiw, N.G. 2014. *Principles of Macroeconomics*, 7th ed. Stamford: Cengage Learning.

Miles, S., Scaife, V. 2003. Optimistic bias and food. *Nutrition Research Reviews* **16**, 3–19.

Mwangi, E. 2009. Property rights and governance of Africa's rangelands: a policy overview. *Natural Resource Forum* **33**, 160–170.

Nickerson, R.S. 1998. Confirmation bias: a ubiquitous phenomenon in many guises. *Review of General Psychology* **2**, 175–220.

North, D.C. 1990. *Institutions, Institutional Change, and Economic Performance. The Political Economy of Institutions and Decisions*. Cambridge: Cambridge University Press.

Olson, M. 1965. *The Logic of Collective Action*. Cambridge: Harvard University Press.

Ostrom, E. 1990. *Governing the Commons: The Evolution of Institutions for Collective Action*. Cambridge: Cambridge University Press.

Ostrom, E., Burger, J., Field, C.B., Norgaard, R.B., Policansky, D. 1999. Revisiting the commons: local lessons, global challenges. *Science* **284**, 278–282.

Pacala, S., Socolow, R. 2004. Stabilization wedges: solving the climate problem for the next 50 years with current technologies. *Science* **305**, 968–972.

Pauly, D., Christensen, V., Guenette, S., Pitcher, T.J., Sumaila, U.R., Walters, C.J., Watson, R., Zeller, D. 2002. Towards sustainability in world fisheries. *Nature* **418**, 689–695.

Press, W.H., Dyson, F.J. 2014. Iterated prisoner's dilemma contains strategies that dominate any evolutionary opponent. *Proceedings of the National Academy of Sciences of the United States* **109**, 10409–10413.

Scheibehenne, B., Miesler, L., Todd, P.M. 2007. Fast and frugal food choices: uncovering individual decision heuristics. *Appetite* **49**, 578–589.

Sharot, T. 2011. The optimism bias. *Current Biology* **21**, R941–R945.

Shepherd, R., Raats, M. (eds). 2006. *The Psychology of Food Choice*. Wallingford: CABI.

Simon, H.A. 1991. Bounded rationality and organizational learning. *Organization Science* **2**, 125–134.

Soto, H.D. 2000. *The Mystery of Capital: Why Capitalism Triumphs in the West and Fails Everywhere Else*. New York, NY: Basic Books.

Tang, S.Y. 1991. Institutional arrangements and the management of common-pool resources. *Public Administration Review* **51**, 42–51.

Tversky, A., Kahneman, D. 1986. Rational choice and the framing of decisions. *Journal of Business* **59**, S251–S278.

van Zandt, D.E. 1993. The lessons of the lighthouse: 'government' or 'private' provision of goods. *Journal of Legal Studies* **22**, 47–72.

Wade, R. 1988. The management of irrigation systems: How to evoke trust and avoid prisoner's dilemma. *World Development* **16**, 489–500.

Werthmann, J., Roefs, A., Nederkoorn, C., Mogg, K., Bradley, B.P., Jansen, A. 2011. Can(not) take my eyes off it: Attention bias for food in overweight participants. *Health Psychology* **30**, 561–569.

World Bank. 2015. *World Development Report 2015: Mind, Society, and Behavior*. Available at: http://www.worldbank.org/en/publication/wdr2015 [accessed 9 November 2018].

Worm, B., Hilborn, R., Baum, J.K., Branch, T.A., Collie, J.S., Costello, C., Fogarty, M.J., Fulton, E.A., Hutchings, J.A., Jennings, S., Jensen, O.P., Lotze, H.K., Mace, P.M., McClanahan, T.R., Minto, C., Palumbi, S.R., Parma, A.M., Ricard, D., Rosenberg, A.A., Watson, R., Zeller, D. 2009. Rebuilding global fisheries. *Science* **325**, 578–585.

15

Summary
A view towards the future

Paul Behrens, Thijs Bosker,
and David Ehrhardt

'Prediction is hard, especially about the future.'

Niels Bohr

Chapter Overview

- Considerable evidence is building to suggest that humanity is 'overshooting' Earth's ecological boundaries, part of which can be attributed to food production and consumption. On top of this, collectively, we are failing to meet global food needs.

- There are tensions between the social needs of equitable (and maximal) food security and sustainable food systems. Any policy on food sustainability will therefore produce winners and losers. It is crucial to map out, as accurately as possible, the effects of various food sustainability policies.

- Transdisciplinary research is necessary in order to better understand the complex causal connections that explain food security and food sustainability on a global level. Advances are being made in this respect—including the Doughnut model, which visualizes social and ecological boundaries—but much more work remains to be done.

- We will need normative frameworks to help navigate the trade-offs inherent to these policy choices. The Golden Rule provides a starting point but also raises new questions to be answered.

Introduction

In 1972, a group of scientists, commissioned by the Club of Rome (an organization of politicians, diplomats, scientists, economists, and business leaders), published an influential book called *The Limits to Growth* (Meadows, 1972). As the name suggests, the scientists explored the effects of exponential economic and population growth on a planet with finite resources and land (see Figure 15.1). To do this, they constructed a model that projected the trends established in the middle twentieth century onwards. Shockingly, two of the

a)

b)

Figure 15.1 The Club of Rome rang alarm bells about our ever-expanding human population, highlighting limits to growth. Cities have seen tremendous growth, resulting in mega cities, such as Tokyo (a), and inequality, such as seen in a slum in Nairobi (b), both of which have their own specific challenges to feed their populations sustainably.

Source: LuxTonnerre/Flickr (CC BY 2.0); (b) St. Aloysius Gonzaga High School Journalism Club/Flickr (CC BY 2.0).

three scenarios they looked at ended in the overshoot and collapse of human civilization and the environment by 2025 and 2040, respectively.

If this sounds familiar, that is because the argument is very similar to that of the Malthusian catastrophe—a scenario described in the introduction of this book. In Malthus' logic, exponential population growth outstrips the ability of food systems to provide sufficient food. Since the Malthusian catastrophe was avoided, detractors might say that the limits to growth is a similarly flawed concept, as it does not take into account transformational technological fixes in the efficiency and recycling capability of the economy and food systems. But one of the main questions posed by this book is whether it is better to live with self-imposed restrictions and within limits, or is it better to carry on growing in the hope that technology can overcome the limits we face?

In 1972, these were still open questions, as the predicted ecological overshoot was yet to transpire. Now, it has been over 40 years since the publication of *The Limits to Growth*, so we can see whether it has stood the test of time (so far at least—as even Malthus' predictions were right for a few decades). In an update to their analysis in 2005, the team behind *The Limits to Growth* argued that not only had their model stood the test of time, but also humanity is now in overshoot mode, unsustainably using up Earth's resources for economic growth (see Figure 15.2) (Meadows et al., 2004).

Along with several overall scenarios, *The Limits to Growth* also projected the growth of the population, consumption, and pollutants and the depletion of some

Figure 15.2 A landfill in Northern India, which highlights our unsustainable use of resources and the impacts on the local environment and biodiversity.

Source: A. J. T. Johnsingh, WWF-India and NCF/Wikimedia Commons (CC BY-SA 3.0).

resources. Several other follow-up studies have found that we are following many of the projections in *The Limits to Growth* closely (Bardi, 2015). Projections for food, services, industrial output per capita, along with resource use, and pollutants are all in agreement. However, other projections were off. For example, where *The Limits to Growth* projected the end of oil by hitting resource limits, technological developments have improved extraction techniques, resulting in a new boom for oil and gas producers in so-called 'unconventional' resources (this includes the shale gas revolution in the USA). Perhaps most importantly, there was an environmental factor that was only hinted at in *The Limits to Growth*—climate change, as it was before we fully understood what was happening.

Many people take a dim view of these prognostications, suggesting that human innovation and ingenuity will always succeed against such limits by increasing yields, improving technologies, substituting products, and reaching better efficiencies (examples of this line of argument can be read in Steven Pinker's book *Enlightenment Now*). Recently, Charles C. Mann described the people on either side of the debate as Wizards and Prophets, with Wizards citing the successes of humankind, and Prophets the excesses. In general, Wizards highlight technological fixes, while Prophets emphasize the need to change our lifestyles and reduce our consumption.

Whichever side of this debate you are on, the questions raised by Malthus and *The Limits to Growth* continue to be of great importance to society. As highlighted throughout this textbook, active research is conducted to quantify the limits of our natural systems, which, in turn, feeds into heated debates on how to move forward.

In your experience

Based on the description given in the introduction, do you consider yourself a Wizard or a Prophet? For example, when it comes to diets, perhaps you are a Prophet, as you think a lifestyle change is necessary (see *Chapter 8: Nutrition*). Or you can be a Wizard, because you think we can come up with new technological innovations to produce healthy and sustainable diets. Perhaps you are a mix, depending on the issue at hand?

The goal of this chapter is to visit and re-emphasize the need for a transdisciplinary approach to sustainable food production. We then discuss how the information in this book can be used to make decisions and how these decisions often result in winners and losers (that is, some will gain and others will lose in the transition). Finally, we will return to exploring projections and look at the future challenges we face.

15.1 Can a transdisciplinary approach to sustainability be useful?

In this book, we have presented aspects of food sustainability across the disciplinary spectrum from natural, social, and governance perspectives. We have linked concepts across these disciplines to highlight the transdisciplinary nature of studying food sustainability. For example, we have seen how current food systems are contributing significantly to climate change. In turn, climate change has a strong influence on biodiversity, environmental pollution, agricultural productivity, and access to healthy and nutritious food. We have seen that the extent to which food systems are sustainable depends not only on available food production and distribution technologies, but also on consumers' health requirements, their personal and cultural preferences, and the ways in which governance is able to engage in collective action. As such, the book has illustrated the 'wickedly' complex nature of food systems and has presented many of the disciplinary and interdisciplinary connections necessary in developing sustainable solutions (see Figure 15.3).

Figure 15.3 Slash-and-burn agriculture, as seen here in Thailand, is an example of an environmental issue requiring a transdisciplinary approach. We have to understand the an environmental impacts when a tropical rainforest is removed, but also the importance of slash-and-burn on the livelihoods of the local community, as well as the globalized market, driving the development of mono-culture plantations to provide food to Western countries.

Source: mattmangum/Flickr (CC BY 2.0).

It is not enough to be aware of the connections, however, as this provides only a cursory understanding of the issues. As previous examples illustrate, it is important to actually build knowledge across a range of different research fields. Many stakeholders, including scientists and policymakers, realize the importance of integrating different disciplines to address challenges related to sustainable food production. For this reason, science is becoming more interdisciplinary. However, this transition first happens by taking small steps, by combining neighbouring fields in the natural sciences or social sciences, with few links between the natural and social sciences (Porter and Rafols, 2009).

New interdisciplinary fields between disciplinary neighbours are appearing more frequently, unifying previously disparate areas of research. For example, in the natural sciences, a new field called hydro-biogeochemistry combines four disciplines—hydrology, biology, geology, and chemistry—to give insights into how soil, freshwater, and marine environments respond to different stressors across the four disciplines. Findings from this work have resulted in a shift on how we think of food production. For example, by understanding variations in zinc (a plant micronutrient) in different hydrological and soil conditions, scientists are able to investigate the impact of climate change on future micronutrient availability (Washington, 2017).

In the social sciences, interdisciplinary fields often arise within specific geographical areas (area studies), shared theoretical or epistemological frameworks (behavioural sciences, institutionalism), or as a result of policy challenges such as water management. To take the field of water management as an example, it connects civil engineering and hydrology with the study of governance in order to analyse technical water management solutions (such as surge barriers or water sanitation facilities). A canonical example is integrated watershed management (IWM), as discussed in Section 4.5 of *Chapter 4: Water*. Watersheds are often not confined to local, regional, or even national borders. The governance of the resource requires policymakers, community members, businesses, and natural scientists from many stakeholder groups to work together. Fields like these have changed the way in which we look at resource management for sustainable food. For example, interdisciplinary studies of water

management have suggested new solutions for the sustainable management of river basins (Hoekstra, 2010) or the pro-poor management of drinking water in low-income country cities such as Jakarta in Indonesia (Bakker, 2007).

Fortunately, increasing attention is being given to the type of interdisciplinary work described in the previous paragraphs, with major scientific journals publishing broader work encapsulating more fields. In addition, as these neighbouring fields increasingly collaborate, opportunities arise for connections across more distant disciplines.

Researchers are beginning to formulate theoretical models that connect even broader disciplines. One example is the Doughnut model for economic growth, shown in Figure 15.4 (Raworth, 2012). On the outer ring of the doughnut, we have the environmental pressures from biodiversity loss to ocean acidification (this is the same as the concept of planetary boundaries by Steffen et al., 2015). As we go further beyond the boundary, the red area extends further 'above' the ecological ceiling. In the Doughnut model, we can see that biodiversity loss (as discussed in *Chapter 2: Biodiversity*), climate change (see *Chapter 6: Climate Change*), land conversion (see *Chapter 2: Biodiversity*, *Chapter 5: Soils*, *Chapter 6: Climate Change*, and *Chapter 7: Energy*), and nitrogen and phosphorus flows (see *Chapter 3: Pollution*) all exceed ecological boundaries. On the inner circle of the doughnut, the social foundations are shown, including peace and justice (related to *Chapter 9: Food Security*), political voice (see *Chapter 13: Governance* and *Chapter 14: Collective Action*), and social equality (see *Chapter 9: Food Security* and *Chapter 10: Food Aid*). Shortfalls in the foundations appear as red sections towards the centre of the doughnut.

Connect the dots

See Figure 15.4, and pick a social foundation from inside of the Doughnut model. Now, imagine improvements in this foundation (for example, better access to health care). What impacts might this have on the outside of the Doughnut model? For example, research shows that although life span increases, parents have fewer children. The net result might be a reduction in the ecological impacts on the outside of the Doughnut model.

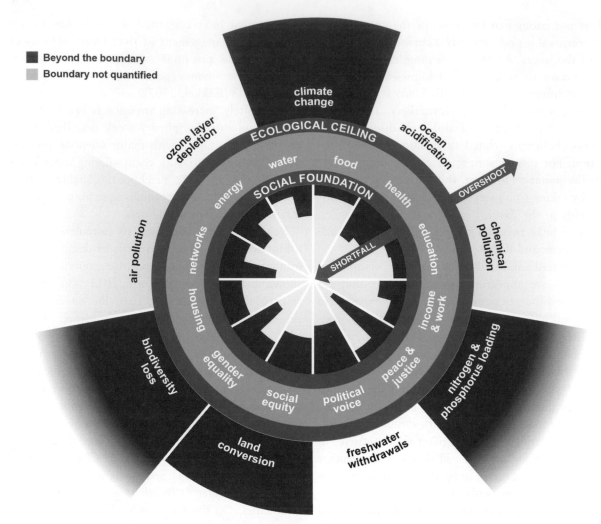

Figure 15.4 The Doughnut model for combining social and planetary boundaries, depicting in red social shortfalls and ecological overshoots, and in green the ecologically safe and socially just space.

Source: Raworth, Kate. (2017). A Doughnut for the Anthropocene: humanity's compass in the 21st century. *The Lancet Planetary Health*. 1. e48-e49. © 2017 Kate Raworth. Published by Elsevier Ltd. CC BY.

The Doughnut model assumes that there is a green, ecologically safe and socially just space in between overshoot and shortfall, a space that would also include sustainable food systems. It integrates concepts in a very transdisciplinary way, but it still requires a lot of analytical and empirical work—for example, in building indicators that can be monitored.

15.2 Making decisions in a complex world

These sorts of transdisciplinary models may allow decisions to be made with input across relevant academic disciplines. In some cases, input from many researchers can lead to decisions that unequivocally deliver positive social and environmental changes (win-wins). For example, research on Nationally Recommended Diets (NRDs) has highlighted that healthier diets resulted in lower environmental impacts in high-income countries (Behrens et al., 2017). However, in some cases, such obvious win-win scenarios do not exist.

For example, in that same study, switching to NRDs in low- and middle-income countries would be beneficial for health but would result in increased environmental impacts. In such scenarios, how do we balance the needs of people with that of the planet? How do we make trade-offs when we are faced with mutually exclusive, complex decisions? And how do we stay inside the ecologically safe and socially just space, as illustrated by the Doughnut model?

The United Nations Sustainable Development Goals (SDGs) are arguably the most complete and widely supported collection of societal ambitions and are a good illustration of the synergies and trade-offs that interdisciplinary approaches highlight. These goals provide a set of quantitative and qualitative targets for the achievement of sustainable global development, from gender equality to access to food (UN, 2015). They are full of trade-offs; take, for example, the tension between boosting growth (Goal 8), which increases consumption but can make it harder to meet environmental targets (for example, Goal 13) (think about trying to stay inside the 'safe space' of the Doughnut model). Or consider how increasing access to electricity may improve the ability to pump water for irrigation and provide food for communities (Goals 2 and 7) but might deplete water resources (Goal 6).

Figure 15.5 shows the interactions between SDGs, from goals that are indivisible (success in one goal is contingent on another) to goals that cancel each other out (achievement of one goal makes another impossible), and provides examples of the strength of each interaction (Nilsson et al., 2016).

Transdisciplinary research can help inform decisions in cases where trade-offs are possible or apparent. For example, if we achieve increased equality worldwide (SDG 10), the trade-off will be increased environmental pressures (in SDGs 6, 15, and 13; see Figure 15.6) and could result in around 4% higher carbon emissions (Scherer et al., 2018). This type of transdisciplinary knowledge can be used by policymakers to drive additional carbon mitigation needed to compensate for these increases.

Pause and think

In the 'Example' column on Figure 15.5, the interactions of pairs of SDGs are mapped and classified (for example, indivisible, consistent, and cancelling). Find several more examples of interactions within the goals. The full set of SDGs are available in an easy-access format at: http://www.undp.org/content/undp/en/home/sustainable-development-goals.html

While there is a limit to what (even perfect) scientific knowledge can do to help elucidate decisions when choices interact negatively, what it can do is clarify who stands to gain from certain choices and who the losers will be. In a rapidly globalizing world, these are never straightforward scenarios, but increasingly accurate estimations of the different knock-on effects of policy choices are a necessary foundation for all trade-off choices.

15.3 Winners and losers in the global transition to sustainability

In any complex policy decision, there will be people who win from the changes and people who lose. This is because, in all of these possible futures, there are specific local and sectoral impacts, which will necessarily create global and local winners and losers. For example, climate changes resulting from GHG emissions disproportionally impact regions in Africa over regions in Northern Europe and America (IPCC, 2013). In many cases, populations that have already lost out in global development stand to lose even more, while wealthy

regions like northern Norway and Canada may actually stand to gain through increased agricultural yields from increasing temperatures due to climate change (highlighted in *Chapter 6: Climate Change* and in Uleberg et al., 2014) (see Figure 15.7).

To take a sectoral example, in the transition to renewable energy, individuals will inevitably lose jobs in the fossil fuel sector, while other individuals will find jobs in the solar industry (often in different locations requiring the movement of families and support networks). The response to this changing landscape

Interaction	Name	Explanation	Example
+3	Indivisible	Inextricably linked to the achievement of another goal.	Ending all forms of discrimination against women and girls is indivisible from ensuring women's full and effective participation and equal opportunities for leadership.
+2	Reinforcing	Aids the achievement of another goal.	Providing access to electricity reinforces water-pumping and irrigation systems. Strengthening the capacity to adapt to climate-related hazards reduces losses caused by disasters.
+1	Enabling	Creates conditions that further another goal.	Providing electricity access in rural homes enables education, because it makes it possible to do homework at night with electric lighting.
0	Consistent	No significant positive or negative interactions.	Ensuring education for all does not interact significantly with infrastructure development or conservation of ocean ecosystems.
−1	Constraining	Limits options on another goal.	Improved water efficiency can constrain agricultural irrigaion. Reducing climate change can constrain the options for energy access.
−2	Counteracting	Clashes with another goal.	Boosting consumption for growth can counteract waste reduction and climate mitigation.
−3	Cancelling	Makes it impossible to reach another goal.	Fully ensuring public transparency and democratic accountability cannot be combined with national-security goals, Full protection of natural reserves excludes public access for recreation.

Figure 15.5 The interactions of various synergistic or antagonistic Sustainable Development Goals.

Source: Nilsson, Måns & Griggs, Dave & Visbeck, Martin. (2016). Policy: Map the interactions between Sustainable Development Goals. Nature. 534. 320-322. Copyright © 2016, Springer Nature.

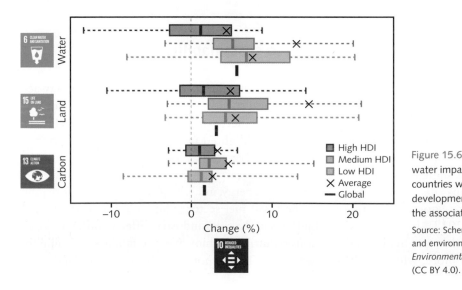

Figure 15.6 The change in carbon, land, and water impacts as a result of increasing equality in countries with a low, medium, and high human development index. The square pictures show the associated SDGs for each of the trade-offs.

Source: Scherer et al. 2018. Trade-offs between social and environmental sustainable development goals. *Environmental Science and Policy* 90, 65–72. (CC BY 4.0).

a)

b)

Figure 15.7 Opposing trends in the impact of climate change; in some regions, climate change opens up areas for agriculture; for example, in wealthy countries such as Canada and Norway (a). In contrast, in many other regions, climate change will have devastating impacts on people who are already struggling to make ends meet—for example, in the Horn of Africa (b).

Source: (b) Oxfam East Africa/Flickr (CC BY 2.0).

can have important political implications. Just think back to the 2016 US election race where the concerns of 'coal country' appeared to be an important factor in the campaign of President Trump (NPR, 2016).

If we are to take radical decisions to improve food sustainability, who will be the losers, both in the short and longer term? In 'wickedly' complex questions (see *Chapter 14: Collective Action*) like how to enhance food sustainability, it is practically impossible to predict winners and losers at all societal levels, across all sectors, and in all places. But even as the effects of policy choices become observable, they are often only the starting point of endless scholarly and political debates about their real nature and impacts. With potentially dramatic political and economic consequences, and hence powerful engagement from special interest groups, debates can end up in confusion, with endless blame displacement, and, ultimately, damaging inaction. With so many different organizations and people involved, Jamieson (2010) suggests that '*today we face the possibility that the global environment may be destroyed yet no one will be responsible.*'

This is not good enough. Alongside studying the best approaches to making food sustainable and identifying the winners and losers of such approaches, we also need to develop arguments to guide trade-offs and to describe the obligations by which the winners have to compensate the losers. In the next and final section, we will highlight the starting point of such arguments—the Golden Rule—and some of the complexities of applying it to food sustainability.

15.4 The wrongs and the rights: ethics for food sustainability

What are the ethical ramifications of both the status quo and plausible future developments of food sustainability? There are many different ethical frameworks available to pose this question, and we have touched upon some in this textbook (for example, utilitarianism, as described in *Chapter 14: Collective Action*). Some of the most established and fundamental frameworks are incompatible in both theory and practice, but there is one axiom that is a very common starting point: '*Do unto others as they would do to you*'.

This is often termed the Golden Rule and is a concept fundamental to many religious and other ideologies of fairness and reciprocity, likely dating back to at least as far as the Ancient Egyptians, around 1500 BC. While the principle has strong intuitive support, it leaves open the question: 'To whom do we have this moral obligation?' This is a particularly crucial issue in thinking about long-term ecological sustainability (see Food controversy 15.1 for a discussion on intergenerational and inter-species obligations).

The Golden Rule and its application to future generations and other species

The Golden Rule is a form of reciprocal altruism, which appears in many religious texts and ethical traditions. It is often taught to children through didactic literature (see Figure A). The Golden Rule is applicable not just at the individual level, but also at the group level. Some examples include:

- Socrates: *Treat your inferior as you would wish your superior to treat you*
- Judaism: *You shall not take vengeance or bear a grudge against your kinsfolk. Love your neighbour as yourself: I am the lord* (Leviticus 19:18)
- Christianity: *Do to others what you want them to do to you* (Matthew 7:12)
- Islam: *As you would have people do to you, do to them; and what you dislike to be done to you, don't do to them* (Kitab al-Kafi)

- Hinduism: *One should never do that to another which one regards as injurious to one's own self*—Brihaspati, Mahabharata (Anusasana Parva, Section CXIII, Verse 8)
- Confucianism: *What you do not wish for yourself, do not do to others*.

An important question for food sustainability is whether the Golden Rule also applies to future generations, or even to other species. Research on intergenerational reciprocity has suggested that how a previous generation has acted will influence the behaviour of the present generation towards a future generation, both when it comes to benefits and burdens (Wade-Benzoni, 2002). Given that sustainability is a long-term concern that has important ramifications on future generations, it would stand to reason that our Golden Rule obligations extend to them. At the same time, accepting

Figure A The cover of a popular book teaching children about the Golden Rule.

Source: Taken from The Berenstain Bears and the Golden Rule by Stan Berenstain Copyright © 2008 by Berenstain Bears, Inc. Used by permission of Zondervan. www.zondervan.com.

these obligations seriously has enormous implications for our current behaviour that are likely to run into strong public resistance (see Barry, 1997 and Meyer, 2017 for an overview of arguments about intergenerational justice).

The question of the moral status of future people is one of the most important controversies you may never have heard of. In efforts to model the future impacts of policy decisions, researchers and economists have to make a decision on how to weigh future generations' benefits against the present costs; this is called a 'discount rate'. This is mostly a moral statement, and the choice of discount rate is a decision made by the researcher or policymaker. Currently, many researchers discount quickly (at around 4–5% per year), assuming that we should wait and let future people fix our problems (on the basis that they will be richer and smarter). However, remember that exponentials are powerful—the result of compounding this rate over time suggests that the entire output of the 2016 global economy in 200 years' time would be worth less than $1 per person per year now. This has resulted in some prominent researchers suggesting that current economic-climate models are 'grossly misleading' (Stern, 2016).

Along with the controversy surrounding future generations, there is another controversial question regarding the Golden Rule—does it also apply to other species? Should we consider the orangutans displaced from their natural habitat when we create an oil palm plantation using the Golden Rule? Or the insects currently experiencing sharp declines in many regions? The inclusion of other species is often used by animal rights activists as an argument to highlight animal suffering (for example, in intensive animal farming such as the pig industry) (see Figure B.a). Many people are horrified at the thought of eating a dog, and yet we eat pigs, which are generally considered to have similar, if not greater, intelligence to dogs. There is a very polarized discussion between different groups, encouraging people to get out and march for animal rights (see Figure B.b).

Recent animal cognition research is showing more intelligence across many species than previously thought. Ultimately though, whether you feel this is relevant still depends on your initial perspective on the status of animals and the significance of their suffering (see Gruen, 2017 for an overview of different positions in this debate).

a)

b)

Figure B Does the Golden Rule also apply to other species? For example, to animals used for food production? And does this mean that intensive pig farming (a) is unethical? Going further, would animals that are kept to provide any food (think milk or eggs) be viewed as animal oppression (b)?

Source: (a) United States Environmental Protection Agency; (b) https://www.freethevoiceless.com/welcome/.

So, in this very simplified view of being ethical by treating others as we would like to be treated ourselves, how are we currently doing? Perhaps the clearest example is the prevalence of poverty in the world. Approximately 767 million people live under the extreme poverty line (defined as an income of less than $1.90 per person per day) and are struggling to meet homeostatic needs (such as regulating the body's heat, water, and energy requirements). This is in a period in which over 33% of all food is wasted in food production, processing, and consumption (FAO, 2011) (see Figure 15.8).

Having said this, there have been enormous developments in the reduction of poverty in recent decades.

a)

b)

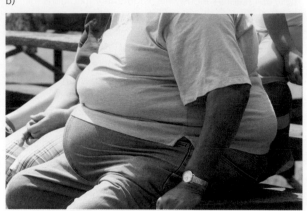

Figure 15.8 Contrasting issues: (a) food insecurity in thes Horn of Africa where individuals sometimes struggle to meet homeostatic needs, and (b) obesity as a growing global issue.

Source: (a) Oxfam East Africa/Flickr (CC BY 2.0); (b) www.welovecostarica.com/Flickr (CC BY 2.0).

Between 1980 and 2015, the number of people surviving on below $1.90 per day reduced from 44% to 9.6%, driven predominantly by astonishing growth in the incomes of people in China. However, there is still a long way to go from meeting basic homeostatic needs (at $1.90 per day) to earning an equitable income (the global average today is $27 per person per day). Moreover, such equity is unlikely to be sustainable, as growth in affluence (and food consumption) is one of the largest contributors to environmental pressures.

15.5 Looking to the future

This tension between the social need to reduce poverty and the ecological need to reduce carbon emissions brings us back to the Doughnut model from Figure 15.4. We may feel, intuitively, that we have an ethical obligation to help today's poor people to rise to contemporary Western standards of living, yet this is likely to result in a serious (further) ecological overshoot. Given the population of countries like China (1.4 billion) and India (1.5 billion), along with the billion people in Africa, the future survival of human societies as we know them is a battle that will be waged outside high-income countries. It is not a battle that can be won without high-income countries diverting significant resources and technology to lower-income countries, countries which will also suffer much of the future environmental damage (see Case study 15.1).

What all this growth means is the most important unknown of modernity. Will it mean overshoot and collapse, along the lines of Malthus or the Club of Rome, or a sustainable transition which will move to zero, or even negative, environmental degradation and equitable food security throughout the world? Will we fail to meet climate targets, resulting in huge changes in global and local climates? Or will we a transition from fossil fuels to renewable energies and low-carbon alternatives? Will some early ideas be able to scale up and provide a significant amount of food sustainably (see Figure 15.9)? And how would these improvements compare to baseline conditions before globalization (see Box 15.1)?

Global trends in population growth and increasing affluence are essential starting points in addressing questions about the future of food sustainability. In the introductory chapter of this textbook, we reflected on the unprecedented global population growth over the last 100 years. Although the human population is growing at a slower and slower rate, the population will likely continue to grow for the next century or more, likely reaching somewhere between 9.5 and

Figure 15.9 Urban container farming in Brooklyn, New York. Plants are grown under specific lighting in shipping containers.
Source: © Square Roots.

CASE STUDY 15.1

Life on the delta: a story of two countries

The Netherlands and Bangladesh both sit on large, low-lying deltas. Both countries have high population densities and have to fight erosion, salinization, and rising sea levels. But this is where the similarities between the two countries end (see Figure A).

The Netherlands is a rich, developed country, having benefited from early industrialization and colonialism.

Bangladesh is a much poorer country, whose industrialization was delayed and which was the subject of British colonization. Since the Netherlands went through the Industrial Revolution earlier, it has contributed much more to global warming—the same warming that threatens both low-lying countries due to sea level rise. In the Netherlands, sea level rise will

a)

b)

Figure A Living in a delta in Bangladesh (a) and the Netherlands (b) results in different challenges, due to different levels of affluence.
Source: (a) Arne Hückelheim/Wikimedia Commons (CC BY-SA 3.0); (b) Wulfson/Wikimedia Commons (CC BY-SA 3.0).

be managed for many decades to come by upgrades to sea defences and the massive Delta-works project which protects the country with dams, sluices, locks, dykes, levees, and storm surge barriers, at a cost of billions of Euros (see Figure B.a). In Bangladesh, urban centres experience regular flooding due to inadequate defences (see Figure B.b).

This is a clear example of inequality in the outcomes of environmental damage. The very people who have done the least to contribute to the problem are also the ones with the least resources to deal with the consequences. It is likely that Bangladesh will be unable to manage the inundation by the sea without significant help from overseas. This would

include investment, technology, and skilled workers. More pessimistically, if things are left to continue, there will be a massive retreat from the delta as the sea levels rise, with huge numbers of people migrating to India and beyond. This may present grave challenges to those countries, which are already struggling to supply the local populations' needs (see Figure C). It will also have serious impacts on existing geopolitical tensions such as those between India and Pakistan.

This may seem speculative, but it is, in fact, just a direct continuation of current trends without any intervention. Connections here can easily be made with ethical concerns highlighted in this chapter.

a)

b)

Figure B (a) The Dutch Delta-works, a vast set of water management defences. (b) Flooding in Bangladesh in 1991. The insufficient dykes can be seen at the edge of the river. The areas where the defences were breached can be detected easily.

Source: (a) Vladimír Šiman/Wikimedia Commons (CC BY 3.0).

Figure C There is massive erosion in the Ganges River Delta, which is linked to rising sea levels due to climate change. This results in the loss of fertile land for growing crops, significantly threatening the livelihoods of the local inhabitants.

Source: http://www.ipsnews.net/2011/11/bangladesh-demands-climate-justice/.

BOX 15.1

Shifting baseline syndrome

If we want to describe the impacts humans have had on the environment, we need some sort of reference point, or baseline, from which to measure change. But what should that baseline be? Should this be the pre-industrial age? Or before the start of the Industrial Revolution? Or when the Club of Rome published their report on *The Limits to Growth*?

Setting the correct baseline can be tricky and can have large consequences, as demonstrated by Daniel Pauly, a fisheries scientist. He introduced the term 'shifting baseline syndrome' (Pauly et al., 1998). Pauly puts forward the argument that '*each generation of fisheries scientists accepts as a baseline the stock size and species composition that occurred at the beginning of their careers, and uses this to evaluate changes. When the next generation starts its career, the stocks have further declined, but it is the stocks at that time that serve as a new baseline.*' As a result, there is a gradual shift in the baseline, and thus acceptable number of fish, over generations.

Many follow-up studies have confirmed this syndrome in the fisheries sciences. For example, a study in the Gulf of California recorded a rapid shift in baselines among different generations of fishermen (Saenz-Arroyo et al., 2005). The study found that, compared to younger fishermen, older fishermen identified five times as many species and four times as many fishing sites that were at some point abundant or reproductive but were now depleted (see Figure A).

Most work to date has linked the shifting baseline syndrome to fisheries; however, the concepts extend beyond this research field. For example, the shifting baseline has also been confirmed in biodiversity conservation (Papworth et al., 2009).

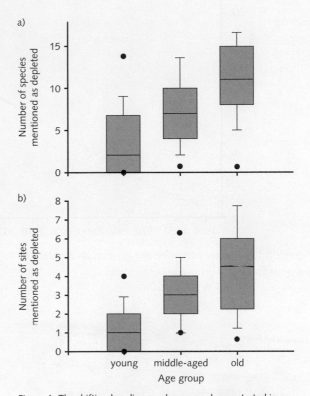

Figure A The shifting baseline syndrome, as demonstrated in different age groups of fishermen in the Gulf of Mexico. The figure demonstrates the perception of fishermen in different age groups, and shows the older the fisherman, the more they perceive fish stocks (a) and sites (b) as depleted compared to younger age groups.

Source: Saenz-Arroyo, A., Roberts, C., Torre, J., Carino-Olvera, M., Enriquez-Andrade, R., 2005. Rapidly shifting environmental baselines among fishers of the Gulf of California. *Proc. R. Soc. B Biol. Sci.* 272, 1957–1962.

11 billion people in 2100 (see Figure 15.10). It is important to note that the best ways for population to reduce does not include 'top–down' government population control (such as China's one-child policy), but the education of women, increasing wealth, and improving health services.

These population increases mean that our food system has to adapt to feed an additional 2.2–3.7 billion people over the coming decades. In addition to population growth, the increase in affluence is already changing dietary patterns. For example, meat consumption is projected to increase significantly in many developing countries (see Figure 15.11), causing increased environmental impacts (see *Chapter 2: Biodiversity*, *Chapter 3: Pollution*, *Chapter 6: Climate Change*, and *Chapter 8: Nutrition*).

But what does all this mean? We are already reaching all sorts of limits in the use of global

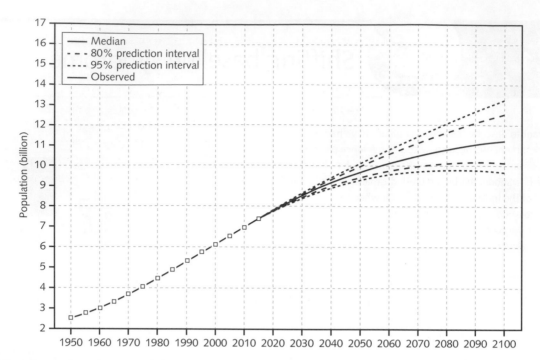

Figure. 15.10 Projected population growth from 2017 to 2100 by the UN.

Source: UN Department of Economic and Social Affairs.

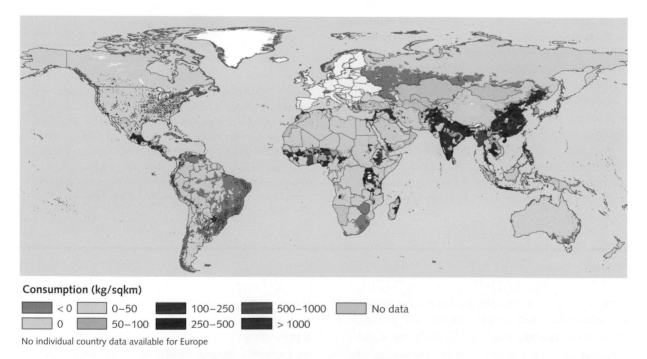

Consumption (kg/sqkm)

■ < 0	□ 0–50
□ 0	▨ 50–100

■ 100–250	▨ 500–1000	□ No data
■ 250–500	■ > 1000	

No individual country data available for Europe

Figure 15.11 Map showing projected global increases of demand for beef from 2000 to 2030. Legend indicates kg per km² demand increase. Developing countries of Latin America, Africa, and Asia exhibit the highest levels of demand increase. Data for Europe were not available.

Source: Food and Agriculture Organization of the United Nations, 2011. Mapping supply and demand for animal-source foods to 2030, by T.P. Robinson & F. Pozzi. Animal Production and Health Working Paper. No. 2. Rome. http://www.fao.org/docrep/014/i2425e/i2425e00.pdf. Reproduced with permission.

food resources with our present population size and affluence—with 70% of freshwater used in agriculture (Koehler, 2008), around 25% of terrestrial NPP (Krausmann et al., 2013), and at least 33% of ice-free land now used in food production (Ramankutty et al., 2008). This is notwithstanding other critical issues such as loss of biodiversity and climate change.

At the start of this chapter, we discussed the findings of the Club of Rome's *Limits to Growth*. Unfortunately, we have little reason to be optimistic, since—as we have outlined throughout this book—the problems are large-scale (global) and intractable (hard to govern and to fix). We have to simultaneously raise the wealth and prosperity of the majority of the world's population, while drastically reducing our environmental impacts. Given the short-term thinking prevalent in politics, businesses, and even choices made by individuals, long-term thinking will not come easily and may take a long time to develop.

In textbooks such as this, or indeed in any text discussing the environment and sustainability, there is always a temptation to put an optimistic 'spin' at the end of the book, leaving the reader with a sense of optimism about the future (and also, presumably, about the book). In fact, we have already discussed this optimism bias in this textbook, as a pernicious source of large-scale collective action problems (see *Chapter 14: Collective Action*). While there are, of course, trends about which to be optimistic, we will resist the temptation to highlight them here. Rather, we want to underline that we are an exponential species—we create problems in an exponential way, but we also (sometimes) fix them in an exponential way. And, while the complexity of food sustainability can be seen as daunting (to the point of being incapacitating), the complexity also suggests that there are solutions that have, as yet, not been identified. We can only realize a better world through a transition to sustainable systems, avoiding as much damage as possible. Whether or not we do this is crucial, urgent, and entirely in our hands.

● QUESTIONS

15.1 What is meant by 'overshoot' in the 1972 report *The Limits to Growth*?

15.2 What does current scientific evidence suggest about the extent to which humanity is 'overshooting'?

15.3 Give three examples of interdisciplinary fields across the sciences and social sciences.

15.4 What is the Golden Rule? Does it help to make decisions regarding the transition to food sustainability? Why (not)?

15.5 Is there much basis to be optimistic about imminent transitions to sustainable and equitable food systems?

● FURTHER READING

Christian, D. 2011. *Maps of Time: An Introduction To Big History*, 3rd ed. Berkeley, CA: University of California Press. **(One of the first books to attempt 'big history', a large-scale overview of human civilizations over the entire history of the planet.)**

Grinspoon, D. 2016. *Earth in Human Hands: Shaping Our Planet's Future*. New York, NY: Grand Central Publishing, Hachette. **(An excellent perspective on the power of human civilization's ability to alter the planet and a reflection on the responsibility of humanity too.)**

Mann, C. 2017. *The Wizard and the Prophet*. New York, NY: Picador. **(A very readable history of environmental movements over time. Pits the Wizards (innovation optimists) against the Prophets (environmental scientists) to investigate the issues in water, energy, food, and climate change.)**

Meadows, D.H.M. 1972. *The Limits to Growth*. The Club of Rome. doi:10.1111/j.1752-1688.1972.tb05230.x. **(The famous book discussed in this chaper outlining a new—at the time—global model for investigating the growth in human civilization and its impact on nature.)**

● REFERENCES

Bakker, K. 2007. Trickle down? Private sector participation and the pro-poor water supply debate in Jakarta, Indonesia. *Geoforum* **38**, 855–868.

Bardi, U. 2015. Limits to growth. In: *International Encyclopedia of the Social & Behavioral Sciences*. pp. 138–143.

Barry, B. 1997. Sustainability and intergenerational justice. *Theoria* **45**, 43–65.

Behrens, P., Kiefte-de Jong, J., Bosker, T., De Koning, A., Rodrigues, J.F.D., Tukker, A. 2017. Evaluating the environmental impacts of dietary recommendations. *Proceedings of the National Academy of Sciences of the United States* **114**, 13412–13417.

Food and Agricultural Organization (FAO). 2011. *Global Food Losses and Food Waste—Extent, Causes and Prevention*. Rome: FAO.

Gruen, L. 2017. The Moral Status of Animals. In: *Stanford Encyclopedia Philosophy*. Available at: https://plato.stanford.edu/entries/moral-animal/ [accessed 9 November 2018].

Hoekstra, A.Y. 2010. The global dimension of water governance: why the river basin approach is no longer sufficient and why cooperative action at global level is needed. *Water* **3**, 21–46.

Intergovernmental Panel on Climate Change (IPCC), Stocker, T.F., Qin, D., Plattner, G.-K., Tignor, M., Allen, S.K., Boschung, J., Nauels, A., Xia, Y., Bex, V., Midgley, P.M. (eds). 2013. *Climate Change 2013: The Physical Science Basis*. Contribution of Working Group I to the Fifth Assessment Report of the Intergovernmental Panel on Climate Change. Cambridge and New York, NY: Cambridge University Press.

Jamieson, D. 2010. Climate, nature and ethics: an editorial essay. *Wiley Interdisciplinary Reviews: Climate Change* **1**, 621–623.

Koehler, A. 2008. Water use in LCA: managing the planet's freshwater resources. *International Journal of Life Cycle Assessment* **13**, 451–455.

Krausmann, F., Erb, K.-H., Gingrich, S., Haberl, H., Bondeau, A., Gaube, V., Lauk, C., Plutzar, C., Searchinger, T.D. 2013. Global human appropriation of net primary production doubled in the 20th century. *Proceedings of the National Academy of Sciences of the United States* **110**, 10324–10329.

Meadows, D., Randers, J., Meadows, D. 2004. *A Synopsis: Limits to Growth—30 Year Update*. Abingdon: Earthscan.

Meadows, D.H.M., 1972. *The Limits to Growth*. The Club of Rome. doi:10.1111/j.1752-1688.1972.tb05230.x

Meyer, L. 2017. *Intergenerational Justice*. In: *Stanford Encyclopedia of Philosophy*. Available at: https://plato.stanford.edu/entries/justice-intergenerational/ [accessed 9 November 2018].

Nilsson, M., Griggs, D., Visback, M. 2016. Map the interactions between Sustainable Development Goa. *Nature* **534**, 320–322.

NPR. 2016. *Where Coal was King, Pa. Voters Hope Trump Rejuvenates Their Economy*. Available at: https://www.npr.org/2016/11/18/502539469/where-coal-was-king-pa-voters-hope-trump-rejuvenates-their-economy [accessed 9 November 2018].

Papworth, S.K., Rist, J., Coad, L., Milner-Gulland, E.J. 2009. Evidence for shifting baseline syndrome in conservation. *Conservaion Letters* **2**, 93–100.

Pauly, D., Christensen, V., Dalsgaard, J., Froese, R., Torres, F. 1998. Fishing down marine food webs. *Science* **279**, 860–863.

Porter, A.L., Rafols, I. 2009. Is science becoming more interdisciplinary? Measuring and mapping six research fields over time. *Scientometrics* **81**, 719–745.

Ramankutty, N., Evan, A.T., Monfreda, C., Foley, J.A. 2008. Farming the planet: 1. Geographic distribution of global agricultural lands in the year 2000. *Global Biogeochemical Cycles* **22**, GB1003.

Raworth, K. 2012. Comment A Doughnut for the Anthropocene: humanity' s compass in the 21st century. *The Lancet Planetary Health* **1**, e48–e49.

Saenz-Arroyo, A., Roberts, C., Torre, J., Carino-Olvera, M., Enriquez-Andrade, R. 2005. Rapidly shifting environmental baselines among fishers of the Gulf of California. *Proceedings of the Royal Society of London B: Biological Sciences* **272**, 1957–1962.

Scherer, L., Behrens, P., de Koning, A., Heijungs, R., Sprecher, B., Tukker, A. 2018. Trade-offs between social and environmental sustainable development goals. *Environmental Science and Policy* **90**, 65–72.

Scherer, L., Heijungs, R., Koning, A. De, Tukker, A. n.d. Extended Data Environmental Impact Inequality among the Rich and the Poor, 1–6.

Steffen, W., Richardson, K., Rockström, J., Cornell, S., Fetzer, I., Bennett, E., Biggs, R., Carpenter, S. 2015. Planetary boundaries: guiding human development on a changing planet. *Science* **348**, 1217.

Stern, N. 2016. Economics: current climate models are grossly misleading. *Nature* **530**, 407–409.

Uleberg, E., Hanssen-Bauer, I., van Oort, B., Dalmannsdottir, S. 2014. Impact of climate change on agriculture in Northern Norway and potential strategies for adaptation. *Climate Change* **122**, 27–39.

United Nations. 2015. *Transforming Our World: the 2030 Agenda for Sustainable Development*. United Nations General Assembly. 47098, 1–35.

University of Washington. 2017. *Rice Grain Quality and Climate Change*. Available at: https://www.uwhydrobiogeochem.com/rice-grain-quality-climate-change/ [accessed 12 November 2018].

Wade-Benzoni, K.A. 2002. A golden rule over time: reciprocity in intergenerational allocation decisions. *Academy of Management Journal* **45**, 1011–1028.

GLOSSARY

'4 N' framework Explanation for why people continue to choose to eat meat, in that to many meat-eaters, it is **Natural, Normal, Necessary**, and **Nice**.

Abiotic components All non-living chemical and physical parts of our environment, including water, light, temperature, and nutrients.

Absorption The percentage of sunlight absorbed at the Earth.

Acute toxicity Toxic effect which occurs almost immediately after exposure to a pollutant (often at a high dose).

Adaptation What we do when something has resulted in a bad outcome and we need to compensate for it.

Aflatoxins Toxins produced by fungi in certain crops when they are not properly dried after harvest. Consumption of aflatoxins can result in impacts ranging from long-term liver and immune diseases to even poisoning in high doses.

Agribusiness Business engaged in large-scale farming and agricultural production.

Agrobiodiversity The variety and variability of animals, plants, and microorganisms that are used directly (crops, livestock) or indirectly (non-harvested species that support agro-ecosystems and biocontrol) to support agricultural ecosystems.

Aid dependency A situation in which aid recipients come to rely on aid flows for some of their basic needs and activities.

Albedo The amount of light reflected away from the surface of the Earth.

Alternative food system Food system developed as alternatives to mainstream global food systems. Alternative food systems are a broad category but are generally characterized by a concern with quality and a sense of paying specific attention to aspects such as organic, fair trade, and local production, particularly valuing direct relationships between producer and consumer.

Anthropocene Newly proposed geological epoch indicating the significant impacts of humans on the environment. Global climate and environmental processes are being driven by human activities.

Anthropogenic climate change The change brought about by the addition of more greenhouse gases from human activities.

Aquiclude A dense sedimentary layer of clay or rock that is relatively impermeable.

Aquifer Porous and permeable bodies of rock or sedimentary deposits with an ability to both store and transmit water.

Asymmetric information See *Principal agent problem*.

Attention bias An individual's tendency to perceive and see things that they already tend to think about a lot.

Attitude–behaviour gap A disconnect between the consumer's knowledge and their behavioural choices.

Autotrophs (also called producers) Organisms that capture energy and combine this with inorganic components in their environment to produce complex organic compounds.

Availability bias A situation in which we put undue weight on recent information we have received.

Bankfull discharge The amount of discharge (streamflow) that fills up the channel but does not flow overbank (not a true flood event).

Baseflow Groundwater that seeps into a river channel.

Bioaccumulation Bioaccumulation occurs when an organism absorbs a xenobiotic compound at a rate faster than that at which the compound is lost by elimination.

Biocapacity The total amount of resources available in a specific area.

Biodiversity 'The variety of life on Earth, it includes all organisms, species, and populations; the genetic variation among these; and their complex assemblages of communities and ecosystems' (IUCN, 2016)

Biodiversity hotspots Areas or ecosystems with exceptionally high biodiversity.

Bioenergy Any energy that is produced from the growth of biological matter over relatively short periods (does not include the biological matter that formed fossil fuels over many millions of years).

Bioenergy carbon capture and sequestration (BECCS) The combination of bioenergy and carbon capture, resulting in net negative emissions (or a reduction of carbon dioxide in the atmosphere).

Biomagnification A process in which a xenobiotic compound increases its concentrations in the tissues of organisms as it moves up in a food chain.

Biomimicry The imitation of natural systems within food production systems.

Biotic components All living, or once living, parts of an ecosystem, including plants, animals, bacteria, and decomposing matter.

Boomerang paradigm This paradigm posited that a pollutant does not disappear when we emit or discharge it into the environment, but, at some point and at some place, it will resurface, just like a boomerang.

Boundary conflicts Tensions between multiple organizations (or different parts of the same organization) that engage with the same issue, but without effective coordination.

Bounded rationality An assumption about the nature of human decision-making that highlights that, while individuals strive to rationally weigh the costs and benefits of the decisions they make, they are constrained by information problems, cognitive biases, and limited time.

Capillary flow The ability of water to move upwards through the soil in small, narrow spaces, even against gravitational forces.

Carbon capture The removal of carbon dioxide, predominantly from the atmosphere, which may include bioenergy carbon capture or direct carbon capture, among other techniques.

Carbon neutral The idea that the whole cycle of a product does not release a net emission of carbon and that any emissions are made up of sequestrations.

Cardiovascular disease A range of diseases which affect the heart or blood vessels (for example, heart attack, angina, etc.). May include issues with blood vessels outside of the heart, such as a stroke which is an issue of poor blood flow to the brain.

Carrying capacity The maximum number of organisms a habitat can support sustainably over the long term. The unit is the number of organisms per area of land (for example, acre or hectare).

Cash crops Crops that are grown to be sold by farmers, and not intended for the farmer to eat.

Centre pivot irrigation This consists of a rotating sprinkler, and the timing and volume of water delivery are much more precise than irrigation by flooding.

Chronic aid Long-term aid that responds to long-lasting crises such as poverty or food insecurity.

Chronic Persistent and long-lasting.

Chronic toxicity Toxic effect that manifests after a longer period of exposure to a pollutant (often at lower doses).

Circular economy An economy in which economic productivity does not come at the expense of resource depletion and both ecosystem functioning and human well-being are maximized.

City region food system Linking a city to its peri-urban and rural hinterland—a regional landscape across which flows of people, goods, and ecosystem services are managed.

Civil society A term used to describe organizations that are neither the state nor private businesses and often represent specific groups or interests.

Club goods Excludable and non-depletable goods.

Cold-chain The ability to keep food cold throughout its lifespan. This is critically important for highly perishable foods to prevent spoilage and ensure food safety.

Collective action problem A situation in which people fail to cooperate, even though it would be in their (collective) interest to do so.

Common-pool resources Non-excludable and depletable goods.

Communicable diseases Diseases that can be spread from animal to animal (where animal includes humans).

Communication blockages Moments in the policymaking cycle when progress is hindered by established, yet unproductive, interaction patterns between decision-makers.

Competing incentives A situation in which an individual, or a group, faced with a decision has multiple incentives that each motivates the decision-maker to perform a different action.

Confined aquifers Also called artesian aquifers, they are separated from the surface by an aquiclude.

Confirmation bias The selective gathering and neglecting of (or giving undue weight to) information in order to support previously held beliefs.

Consumption The use of products and services. Sustainable consumption would be the use of products and services which do not degrade the environment.

Consumption-based accounting The accounting of environmental impacts from an individual, no matter where the impacts took place. Includes the impacts in other countries due to imported goods (for example, Brazilian beef consumed in Europe).

Contaminant A compound in the environment which is found at elevated concentrations above the natural background level for the area and for the organism.

Cost–benefit analysis Approach to decision-making that estimates and compares all the costs and benefits of different options to decide between them.

Cradle-to-gate Calculation of the impacts of animal rearing from the point of growth to the point at which the animal leaves the farm (does not include impacts from post-production).

Culturally appropriate food Food that is consistent with one's individual cultures.

Dead zones Large areas in estuaries with a reduction in oxygen levels (hypoxia or anoxia), resulting in minimal animal diversity because of the lack of oxygen.

Deflation Fine-grained humus-rich prairie topsoil which is selectively eroded by wind, a process known as deflation.

Demand or demand-side Indicates a consumption-based perspective, whereby people's choices or use of technology influence the demand of a good or service. For example, demand-side energy technologies, such as smart fridges, can reduce the need for increasing electricity supply by shifting time of use.

Demand The requirements of users of energy or any other resource or product.

Depletability (of goods) The extent to which consumption depletes a good for other potential consumers. Synonym: *Rivalrousness*.

Determinants (of human decisions) Factors that affect human decisions.

Dilution paradigm This paradigm posited that the solution to pollution can be found in dilution. In other words, contamination is washed away and comes in larger waterbodies where it is diluted to low concentrations that do not affect biota and might even be purified due to the self-regulatory principles of our planet.

Discharge Volumetric rate of streamflow, usually expressed in cubic metres per second (m^3/s).

Domestic food assistance transfer of resources by governments to enhance food security of their own citizens (or residents).

Dominant strategy (in game theory) A strategy that brings a player in a game the best outcome, regardless of other players' strategies.

Double burden of malnutrition When undernutrition and overnutrition occur in the same population, and even the same family.

Drainage divide A topographic border defined by the elevation (often a line of hills or mountains) that encloses a watershed.

Eco-chic A consumer lifestyle that emerged as a result of our modern, fast-paced lifestyles, in combination with the politics of environmentalism and socio-spiritual trends that aim to reconnect human beings with nature.

Eco-consumerism Environmentally friendly, 'green' consumer behaviour.

Ecological footprint The amount of biologically productive land and sea used by an individual. The unit is the area of land per individual.

Ecological niche The position and role of a species within its environment, including how it reproduces, how it meets its need for water, food, and shelter, its predators, and what kind of environmental conditions it can tolerate (for example, temperature and humidity).

Ecosystem An ecosystem consists of a community of organisms that interact with one another (also see *Biotic components*) and with their environment of non-living matter and energy (also see *Abiotic components*). Most commonly, we define ecosystems as distinct areas such as a forest, a lake, or the tundra.

Ecosystem approach Method of conservation which targets conservation at the level of communities, habitats, or entire ecosystems, rather than individual species.

Ecosystem functioning Ecological processes that control the fluxes of energy, nutrients, and organic matter in the environment.

Ecosystem resilience The ability of an ecosystem to withstand environmental pressures and to recover quickly from any disturbances.

Ecosystem services Services and benefits that people obtain from ecosystems and that are provided in the natural environment.

Efficiency The amount of energy retained in an energy conversion, usually in percentage of the initial energy available.

Emergency aid Short-term aid in response to urgent crises.

Empty calories Products with little to no nutritional value, and their excess consumption is closely related to the obesity epidemic.

End-of-pipe solutions Solutions aimed to clean up pollutants that have already been formed and released into the environment.

Energy The potential for one system to do work on another system. Work can include anything which can be used, in theory, to make motion.

Energy poverty Lack of access to energy services, limiting the ability of people to engage in activities other than time-intensive manual labour and threatening human survival (for example, if energy for heating or cooling is unavailable). Lack of access to electricity is one key indicator of energy poverty (either because it is too expensive or not available).

Energy transition Refers to the necessary changes in energy systems to avoid climate change and other environmental problems.

Entomophagy The human consumption of insects as food.

Entropy The concept that, in any energy conversion, we will always lose some useful energy which is lost to the environment.

Environmental flows A management concept to maintain sufficient streamflow to sustain various stakeholder, geomorphic, and ecological functions.

Environmental risk assessment (or ecological risk assessment) A structured approach to quantitatively or qualitatively estimating the risk related to a specific threat to species or ecological structures.

Environmental risk management A process in which information from an environmental risk assessment is used by risk managers to determine the course of action, also taking other factors into account such as economic, social, or legal constraints.

Epidemiology The discipline which investigates diseases, including where they are, how they spread and, how they are controlled.

Essential amino acids Amino acids that cannot be built from other building blocks by an organism. They have to be ingested directly, rather than converted from other products.

Eutrophication The process in which increased availability of one or more limiting growth factors for photosynthesis (for example, nutrient fertilizers or carbon dioxide) results in excessive plant and algal growth.

Evapotranspiration A collective process referring to evaporation directly from surface water bodies, soil, and the surface of plant leaves.

Excludability (of goods) The extent to which the owner of a good can exclude others from its consumption.

Exclusive economic zone (EEZ) A sea zone over which a state or country has special rights when it comes to the exploitation of resources, including fish stocks.

Externalities Benefits or costs of the production or consumption of a good that accrue to a third party but are not included in a good's price.

First-generation biofuels Fuels which have been produced from the edible, easy-to-access parts of plants, which can compete with food production.

Flood pulses Refer to the seasonal floodings that allow for an exchange of water, sediments, nutrients, and organisms between rivers and different floodplains environments.

Flows Refer to resources which flow through the environment and cannot be stored directly.

Food access One of the four pillars of food security. The ability to physically acquire food in a given place. The most common form of food insecurity is a lack of food access for economic reasons.

Food affordability The primary cause of food insecurity globally. Affordability depends on the amount of disposable income an individual or family has, relative to the cost of food in a given place.

Food aid The voluntary transfer of resources for the purpose of enhancing food security in a specific population.

Food availability One of the four pillars of food security. The physical availability or supply of food in a given place. A famine is the ultimate display of food insecurity as a result of a lack of food availability.

Food banks Civil society organizations that collect, sort, and redistribute food donations to food-insecure people.

Food chain A linear network of organisms that feed on each other.

Food cultures Refer to practices, attitudes, and beliefs, as well as networks and institutions, surrounding the production, distribution, and consumption of food.

Food deficit country A country that is unable to grow enough food to feed its population.

Food governance Process in which stakeholders in food systems aim to address problems and promote their own interests by making rules.

Food labels Provide product information to consumers, while also shaping consumer perceptions of products.

Food miles The assessment of impacts resulting from transportation of food from one place to another.

Food scares Moments of escalating public anxiety (and media attention) over the safety of food products.

Food security Defined by the United Nations as, 'When all people, at all times, have physical, social, and economic access to sufficient, safe, and nutritious food to meet dietary needs for a productive and healthy life.'

Food sovereignty The idea that the people who produce, distribute, and consume food should control the mechanisms and policies of food production and distribution, rather than the corporations and market institutions they believe have come to dominate the global food system.

Food stability One of the pillars of food security. Refers to the consistent capacity for food to be available, accessible, and utilized amid changing conditions.

Food stamps Vouchers given to eligible individuals, which can be used in exchange for food items in stores.

Food supply chain The sequence of processes and organizations involved in the production, processing, and distribution of food items.

Food surplus countries Countries that are able to grow enough food to feed their population and may trade or ship surplus food to other places.

Food system The complete set of people, institutions, activities, processes, and infrastructures involved in producing and consuming food for a given population.

Food utilization One of the pillars of food security. Involves how the body digests food, how food is prepared, and whether food is safe to eat, which ultimately influences the health of people.

Food web A network of food chains indicating the interactions between different organisms.

Food-for-work (FFW) In-kind food aid that is conditional on recipients providing labour (in other words, temporary jobs, often in construction, in which salaries are paid in food).

Foodshed The potential of a specific geographic region for producing food for its population.

Fortification The practice of deliberately increasing the content of a specific micronutrient in a food product.

Framing An activity that refers to the attempt of actors to select and highlight specific aspects of an issue and their importance. It is used to emphasize certain aspects of a situation or a problem that serves the interests of stakeholders.

Framing effect A situation in which people's decisions are influenced by the way in which a problem is posed or framed.

Free-riding A situation in which a group of people have a shared interest in maintaining or providing a non-excludable good. Given that the costs are private but the benefits shared, group members face an incentive to overexploit, or not to contribute to, the shared good if they have a reasonable expectation that others will bear the costs for them.

Front-of-pipe solutions (or front-end solutions) Solutions aimed to prevent pollution by removing the pollutant before it is released into the environment. Most commonly applied to point-source pollution.

Game theory Analytical approach that uses mathematical models to understand cooperation and conflict between rational decision-makers.

Global governance Collaboration between nation states and international organizations which seek to develop collaborative activities that enhance global goals, especially in the domains of food security, climate change, and environmental problems.

Global warming potentials (GWPs) A way of comparing different greenhouse gases with different warming powers.

Goods Tangible or intangible consumables, including many services, that provide utility to the consumer—colloquially, things that people want.

Governance All the activities undertaken by societal organizations (for example, governments, private companies, unions, non-governmental organizations, and international organizations) to solve societal problems.

Governance The use of power and authority to set rules and influence people's behaviour. Often associated with the state and the formal government of a country but, in fact, can involve a wide range of actors, from the decentralized self-governance of village democracies to the strict hierarchies of personalized dictatorships.

Governments Organizations that are legally mandated to govern countries, often through a constitution. Synonym: *State*.

Green advertising An advertisement that meets one or more of the following criteria: (1) explicitly or implicitly addresses the relationship between a product/service and the biophysical environment, (2) promotes a green lifestyle with or without highlighting a product/service, and (3) presents a corporate image of environmental responsibility.

Green Revolution The period between 1961 and 2011 during which agricultural land use has increased from 40 million to 181 million hectares. In addition, scientific innovations resulted in a significant increase in yields per hectare.

Greenhouse gases (GHGs) An atmospheric gas which absorbs and radiates light in the infrared. Gases which contribute to the greenhouse effect.

Greenwashing Occurs where companies use their PR and marketing strategies to deviate the consumer's attention, promoting a superficial perception of environmentally friendly products, when, in reality, just one or several components of the production/distribution processes are sustainable.

Gross primary production (GPP) The amount of chemical energy (that is, organic carbon structures such as glucose) captured by photosynthesizing organisms. Expressed in mass of carbon produced per unit area per year (for example, grams of organic carbon per m^2 per year).

Gully Semi-permanent erosional features that range approximately from 1 to 10 m in width and depth.

Heterotrophs (also called consumers) Organisms which are not able to produce their own organic compounds and therefore depend on the energy captured by autotrophs.

Heuristics Simple 'rules of thumb' that people construct to facilitate their capacity to quickly make decisions in complex situations.

Holism An approach in which systems and their properties are studied as a whole, with the assumption that the whole is greater than the sum of its parts.

Holocene Geological epoch starting 10,000 years ago, following the Pleistocene. The Holocene is an interglacial period, with a very stable climate and temperature.

Human Appropriated Net Primary Production (HANPP) The proportion of total net primary production in a transformed natural ecosystem which is used by humans.

Humus The chemically stable remains of organic matter after decomposition of plant organic matter through microbial activities.

Hydraulic civilization Ancient agrarian societies oriented around the management of water resources by government institutions.

Hydraulic infrastructure Usually considered 'hard' engineering structures that modify landscapes and regulate water resources, including dams, dykes (levees), canals, pumps, and irrigation systems.

Hydrograph A graph used by hydrologists to examine the impacts of different agricultural practices, plotting time (*x*-axis) and discharge (*y*-axis).

Hydrologic cycle The global distribution and transfer of water across different states and places, including terrestrial, oceanic, and atmospheric.

Hydroponics The growing of crops outside the soil.

Incentive Something that motivates an individual or group to take certain action or make a particular decision.

Infiltration The downward movement of water through soil and rock.

Infiltration The downward percolation of water through the soil or rock matrix.

Information asymmetry A situation in which some parties to a transaction have more information than others.

Information gaps Situations in which people or organizations lack information that is important to decisions they have to make.

Insolation The amount of sunlight in power per square metre hitting the surface of the Earth.

Institutions Rules that constrain people's decisions.

Institutions Societal rules that constrain behaviour.

Integrated watershed management An adaptive river basin management that integrates different levels of stakeholders and strives for sustainable management of water resources and associated ecological and economic activities while reducing flood risk.

International food aid Voluntary transfer of resources by governments to enhance food security of recipients across national borders.

Invasive species (alien or non-native species) 'Organisms (usually transported by humans) which successfully establish themselves in, and then overcome otherwise intact, pre-existing native ecosystems' (UN, 2016).

IPAT equation An equation to quantify impacts of humans on the environment. This formula states that environmental impacts by humans (*I*) are the product of the human population size (*P*), the affluence of individuals (*A*), and the (harmful and beneficial) impacts of technology (*T*).

Keystone species Species which have a large impact on their environment relative to their abundance, thereby playing a key role in maintaining the structure and functioning of ecosystems.

Land degradation The decline in the quality and quantity of landscape by physical and/or chemical processes that decrease the ability of the land to sustain natural biodiversity and support human uses such as food production. Land degradation is especially driven by human activities such as improper agricultural practices and irrigation. Two key forms of land degradation include accelerated soil erosion and salinization. Reducing land degradation is a key element of the United Nations' Sustainable Development Goals.

Land grabbing The buying or leasing of large pieces of land, often by companies and governments.

Land sharing The objectives of conserving biodiversity and wildlife-friendly food production are integrated in the same landscape.

Land sparing The objectives of conserving biodiversity and wildlife-friendly food production are spatially segregated within the landscape, for example by combining high-yield agriculture with the protection of natural habitats elsewhere.

Land use change Changing the purpose and use of land from one activity to another. Often used to describe alterations in nature as a response to the change (that is, in biodiversity, climate, pollution, etc.).

Law of energy conservation Energy is never created or destroyed, just converted to different forms. When we say 'use' in reference to energy, that is 'we use 520 EJ of energy a year'; this is short hand for 'we convert 520 EJ of energy from original sources to useful outputs for us'.

Legitimacy The extent to which stakeholders accept the authority of people who have decision-making power, as well as the process of decision-making itself.

Lethal effect Effect which occurs after acute or chronic exposure of organisms to a contaminant, resulting in mortality events.

Life cycle assessment (LCA) A method for assessing the impacts of products by adding up different steps in production (and possibly use and disposal, depending on the boundaries).

Litter Dead organic matter at the soil surface that has not decomposed (sometimes called duff) and may include twigs, grasses, and leaves.

Lobbying. A form of advocacy in which privately contracted individuals or organizations engage with the policy-making process to advance the interests of their clients.

Local trap The notion that local by default means more sustainable, conflating the scale of a food system with its desired outcome.

Localization Promoting local food systems.

Loess A wind-blown deposit of mineral matter, primarily silt (0.0625–0.004 mm), which accumulated over thousands of years during the Pleistocene and comprises about 10% of Earth's surface.

Macronutrients Provide most of the energy that the body needs, including proteins, carbohydrates, and fats.

Mainstream food system Food system that relies on intensive use of technology, including machinery, chemical fertilizers, and pesticides, producing on a large scale for a global market, using supermarkets as their main outlet towards consumers.

Malnutrition A deficiency or imbalance of nutrient intake, which can be related to either under- or overnutrition.

Malthusian catastrophe Theory introduced by Thomas Malthus (1766–1834), highlighting the risk of a future where the population would exceed the amount of food available and would then collapse in famine and war.

Marine protected areas (MPAs) Protected areas in our seas and oceans in which human activity is restricted.

Mediterranean diet A diet eaten predominantly in southern Europe with a high intake of vegetables and olive oil and a moderate consumption of protein.

Micronutrients Provide important vitamins, minerals, and trace elements which are important for different systems in the body such as the immune system, the sensory system, the skeleton, blood, etc.

Mitigation any effort which is made to avoid a bad outcome.

Multi-level governance A wide variety of private and public organizations interacting with each other on different layers of governments and society, within the same societal sector or policy domain. They are not necessary hierarchically positioned.

Multilateral food aid Food aid donated by governments through multilateral organizations (such as the World Food Programme).

Natural greenhouse effect The warming of the atmosphere due to natural greenhouse gases, making the planet habitable.

Nature-inclusive agriculture A type of agriculture which is based on a resilient food ecosystem, in which biodiversity provides key ecosystem services.

Net primary production (NPP) Gross primary production minus the rate at which autotrophs use some of the acquired chemical energy for respiration. This can be thought of as the 'free' energy for new biomass or reproduction.

Network governance A wide variety of private and public organizations interacting with each other within the same societal sector or policy domain. They do not have one central organization nor are they hierarchically positioned.

Nitrogen deposition The process which describes the deposition (or input) of reactive nitrogen (N_2 which is available to plants) from the atmosphere to aquatic and terrestrial ecosystems.

No-till agriculture An agricultural practice that does not break up the soil by tillage and is associated with increased infiltration and soil organic matter.

Non-communicable diseases Diseases that cannot be spread from animal to animal (where animal includes humans).

Non-point source pollution Diffuse form of pollution which comes from many different sources simultaneously, for example metal run-off from urban environments and sediment run-off from logged forests.

Non-renewable resource A resource which does not renew itself in a time frame which is meaningful to humans (for example, centuries to millions of years). The usage of non-renewable resources will reduce the available stock until the stock is depleted.

Nudging The use of people's cognitive boundaries, biases, and other weaknesses in small interventions that incentivize particular forms of behaviour.

Nutrition transition The development of diets as nations economically develop.

Nutritional status Refers to the balance between food intake and the processes of growth and health, which can be measured by food intake or measures of body composition and biomarker.

Official development assistance (ODA) All government aid designed to promote the economic development and welfare of low- and middle-income countries.

Optimistic bias The feeling that people view themselves to be less at risk from various hazards than the 'average' person.

Pedogenesis The process of soil formation, which takes centuries to millennia, making soil a non-renewable resource.

Permeability The 'connectedness' of individual pore spaces, which is an important physical characteristic (porosity is the other) of an aquifer which controls the ability to store and transmit water.

Persistent organic pollutants (POPs) Organic compounds that persist in the environment (as they are resistant to environmental degradation) that can bioaccumulate in organisms and that pose a risk of adverse health effect to organisms.

Pescetarian diet A diet excluding meat, dairy, and eggs, but allowing fish.

Physiological Anything related to the body.

Planetary boundary A level that indicates a safe operating space for humanity for an environmental process, including climate change, freshwater use, biochemical cycles.

Pleistocene Geological epoch covering 2.6 million to 10,000 years ago. During this period, the Earth's climate experienced large fluctuations in temperature—ice ages were followed by short, warm periods.

Point-source pollution A source of pollution with a clearly defined discharge point to the environment, for example a smoke stack or a sewage discharge.

Policy cycle A model for policymaking that involves an iteration of five stages agenda setting, policy formulation, decision-making, policy implementation, and policy evaluation.

Policy domains The parts of governance systems that are organized around policy issues.

Pollutant A compound in the environment whose nature, location, or quantity produce undesirable environmental effects.

Pollution The presence of substances (for example, chemicals or plastics) or energy (for example, light or heat) in the environment resulting in undesirable environmental effects.

Population density The total number of individuals (human or otherwise) in a population living in a defined unit of space (for example, Earth, a lake, or a nature reserve).

Porosity The space between rock or sedimentary particles, which is an important physical characteristic (permeability is the other) of an aquifer which controls the ability to store and transmit water.

Porosity The space between rock or sedimentary particles, which is an important physical characteristic (permeability is the other) of an aquifer which controls the ability to store and transmit water.

Post-production Processes after production such as food processing, distribution, and cooking.

Power The ability of one person, group, or organization to influence the behaviour of others, in pursuit of their interests.

Power The rate of conversion of energy.

Power density There are many definitions of power density, depending on the issue looked at; the definition used here refers to the amount of power produced per square metre of land required for an energy system.

Pre-production Processes that take place before the growing of plants or rearing of animals such as manufacture of fertilizers, pesticides, etc.

Price sensitivity The weight attached to price in the consumer valuation of a product's overall attractiveness or utility.

Principal agent problem A situation in which a principal hires an agent for a specific task, and the agent cheats the principal. If there is some tension in the interests between the principal and agent (*split incentives*) and the principal does not have enough information about the agent's capabilities or actions (*asymmetric information*), the agent is likely to cheat the principal.

Private goods Excludable and depletable goods.

Production Processes during the growth of crops, farming of fish, and rearing of animals.

Production-based accounting The accounting of environmental impacts at the national border. Does not include impacts in other countries due to trade in goods.

Programme aid Untargeted government-to-government aid.

Progressive The opposite of a regressive policy (often tax), which has a proportionally beneficial impact on higher-income groups than lower-income groups.

Public goods Non-excludable and non-depletable goods.

Public governance Governance by public organizations that belong to the state such as governments, ministries, law enforcement agencies, and other public bureaucracies.

Recharge zone The land area where precipitation can infiltrate into the soil and rock to replenish the aquifer with water.

Reductionism An approach in which systems are explained in terms of their individual constituent parts and their interactions.

Reflection The percentage of sunlight reflected back from the surface of the Earth.

Regressive A policy (often tax) which has a proportionally worse impact on lower-income groups than higher-income groups.

Relative sea level rise The summation of absolute sea level rise plus local ground subsidence.

Renewable resource A resource which can renew in a relative short time frame and therefore, when used sustainably, does not run out.

Resource flows Renewable resources that, when used sustainably, do not run out. Sunlight is an example of a resource flow.

Resource stocks Resources that require time to accumulate. An important example are fossil fuels, requiring millions of years at high temperatures and pressures to accumulate.

Rills Concentrated paths of run-off usually approximately 10–50 cm wide and approximately 1–10 cm deep.

Risk factors Variables which have some association with an increased risk of disease and/or infections. Used in epidemiology.

Rivalrousness (of goods) See *Depletability*.

Salinization The process when denser saline groundwater displaces freshwater and degrades water quality. This occurs when groundwater withdrawal is not balanced with recharge.

Salinization The transfer of dissolved salts with upward moving capillary water, from lower horizons to upper soil horizons. Salinization reduces soil quality and is a serious form of land degradation.

Scale The spatial, temporal, quantitative, or analytical dimensions used to measure and study any phenomenon. In food systems, scales run from local to regional to national to global, and tend to be highly interconnected.

Scenario analysis Analytical approach that uses a highly simplified model of the major drivers of a phenomenon to sketch out different possible futures.

Second-generation biofuels Fuels which are produced from the non-edible parts of plants and which do not directly compete with food production.

Self-governance Self-organization of societies and communities, beyond the market and short of the state.

Separation of powers The division between the three main branches of government executive, legislature, and judiciary.

Sequestration The capture and long-term storage of chemical compounds, including nitrogen, phosphorus, and carbon.

Sequestration The storage of a substance, used here specifically to mean the storage of carbon dioxide which has been captured by any method.

Sheet wash The incremental removal of fine layers (approximately 1 mm thick) of topsoil, especially soil particles that were already dislodged by raindrop impact.

Short-termism Short-term thinking, even when problems might require a longer-term view.

Sinks These are the locations or activities which draw emissions away from another system. For example, in climate science, a major carbon dioxide sink is growing forested areas.

Smallholders Small farmers who typically have less than 2 ha (slightly less than 5 acres) of land to produce their food and food for income. Globally, it is estimated that 2.5 billion people are smallholders, and many people in low-income countries are smallholders.

Social networks The connections you have with other people, for example through extended family (kinship) ties, friendship, religious affiliations, schooling, or work.

Soil degradation The loss of beneficial soil functions, primarily occurring because of a reduction in soil quality and a physical loss of soil by erosion.

Soil erosion the physical loss of soil, triggered by wind- and water-driven processes.

Soil horizons Distinctive 'layers' within the soil defined by physical and chemical properties, specifically soil colour, texture, structure, and chemistry, which differ from soil horizons above or below. The thickness of soil horizons generally ranges from about 10 cm to 30 cm, and each soil commonly has from three to five horizons.

Soil organic matter (SOM) The amount of plant and animal residue in the soil within various stages of decomposition.

Soil parent material Mineral matter, which comprises about half of soil mass by weight. Derived from weathered rocks or sedimentary deposits.

Soil quality The degree to which soil functions in its natural capacity, which includes its ability to sustain plant and animal productivity while maintaining its hydrologic and biogeochemical roles within the broader landscape.

Soil texture The proportion of sand, silt, and clay mineral matter in the soil.

Soil water budget A visualization tool to determine when a soil moisture deficit is likely to occur, by linking evapotranspiration to rainfall.

Sources These are the originating locations or activity for an emission. For example, in climate science, a major carbon dioxide source is the energy system, and in ecotoxicology, a major source of plastic pollution is the fishing industry.

Species approach Method of conservation which focuses on areas which hold particular species, for instance rare species or keystone species.

Split incentives See *Principal agent problem*.

Spoilers Stakeholders or third parties working in their own interests and against aid efficacy.

Stakeholder All individuals and organizations with an interest in the governance of a particular issue and the power to exert some influence over the process.

State See *Governments*.

Stocks Refer to resources which can be stored over reasonable periods (oil, coal, etc.).

Stunting A form of malnutrition characterized by low height for age.

Sublethal effect Effect which occurs after acute or chronic exposure of organisms to a contaminant and that lowers the fitness of organisms, for example by lowering the reproductive output, changing behaviour, or reducing immune functioning.

Subsidence The sinking (lowering) of the land surface, often caused by compaction of the underlying soil or sedimentary layers or by oxidation and decomposition (loss) of organic surficial layers.

Supplementary feeding In-kind food aid to populations that are food-insecure.

Supply or supply-side Indicates a perspective from the production side of goods and services. Often concerned with production efficiencies, yields, etc.

Supply The ability to supply the requirements of demand.

Surface roughness The roots, twigs, and litter which cover the soil and which buffer the shear stress generated by run-off, as these physical structures interrupt run-off paths.

Surplus dumping Situation in which food producers in donor countries overproduce and donors purchase these surpluses to use as food aid.

Sustainability A situation in which both human and natural systems are able to survive and flourish in the very long-term future.

Sustainable consumption Promotes consumer use of goods and services that allows us to meet our needs, while minimizing the use of natural resource stocks, pollutants, and contaminants, and reducing waste and emissions throughout the entire product life cycle.

Swidden (slash-and-burn) A type of traditional or indigenous agriculture that involves a rotating crop system that includes burning of felled vegetation to temporarily enrich soils for agriculture. Also referred to as shifting cultivation or slash-and-burn agriculture.

Targeting Variable that describes whether food aid reaches the people who actually need it (without flowing to people who do not), at the moment they need it, in the appropriate form, and through efficient supply chains.

Terra pretta A very fertile dark brown or black soil in the Amazon basin that has been enriched with organic carbon by humans.

Tied aid Aid that is offered on the condition that it be used to procure goods or services from the donor of the aid.

Time horizon The time over which we calculate an impact.

Tragedy of the commons Common human tendency to overexploit non-excludable goods.

Trophic levels The positions organisms occupy in a food chain or food web.

UN's Sustainable Development Goals Seventeen interrelated goals for organizing international efforts around the three pillars of sustainable development environmental, economic, and social well-being.

Unconfined aquifers Not separated from the surface by an overhead aquiclude and therefore are not under pressure, and usually at shallower depths (than confined aquifers).

Utility The satisfaction an individual derives from the consumption of a particular bundle of goods.

Vegan diet A diet which excludes meat, poultry, fish, eggs, and dairy products.

Vegetarian diet A non-meat diet which can vary, depending on the approach. Lacto-ovo vegetarian diets exclude meat, fish, and poultry but allow dairy products and eggs. Lacto-vegetarian diets exclude meat, fish, poultry, and eggs but allow milk cheese, and butter.

Waste and disposal What happens to products after consumption. This can include burning, burying, and recycling (among many other options).

Wasting Form of malnutrition characterized by low weight for height.

Water footprint An approach to quantify the total amount (volume) of freshwater used to produce a good or service for a specific user and further classified into blue water (aquifer or surface water body), green water (rain water), or grey water (effluent).

Water table The upper level of the aquifer or zone of saturation.

Watershed (or drainage basin) The combined area for which all water from headwater streams, medium streams, and rivers drains to one single point.

Watersheds (or drainage basins) An area of land defined by a topographic barrier that directs all run-offs to a single outlet via a hierarchical network of rivers.

Well-to-wheel Calculation of the impacts of oil over the life cycle of a product, from extraction (from a well) to use in cars (to the wheel).

Wicked problem A problem that is dynamic, fundamentally complex, and, as a result, incredibly difficult to solve.

Wilting point The minimum amount of soil moisture that sustains plants, with further reductions resulting in wilting.

Xenobiotic compound A compound, typically a synthetic chemical, which is foreign to the body or an ecological system.

INDEX

Boxes, figures, and tables are indicated by italic *b*, *f*, and *t* following the page numbers.

D